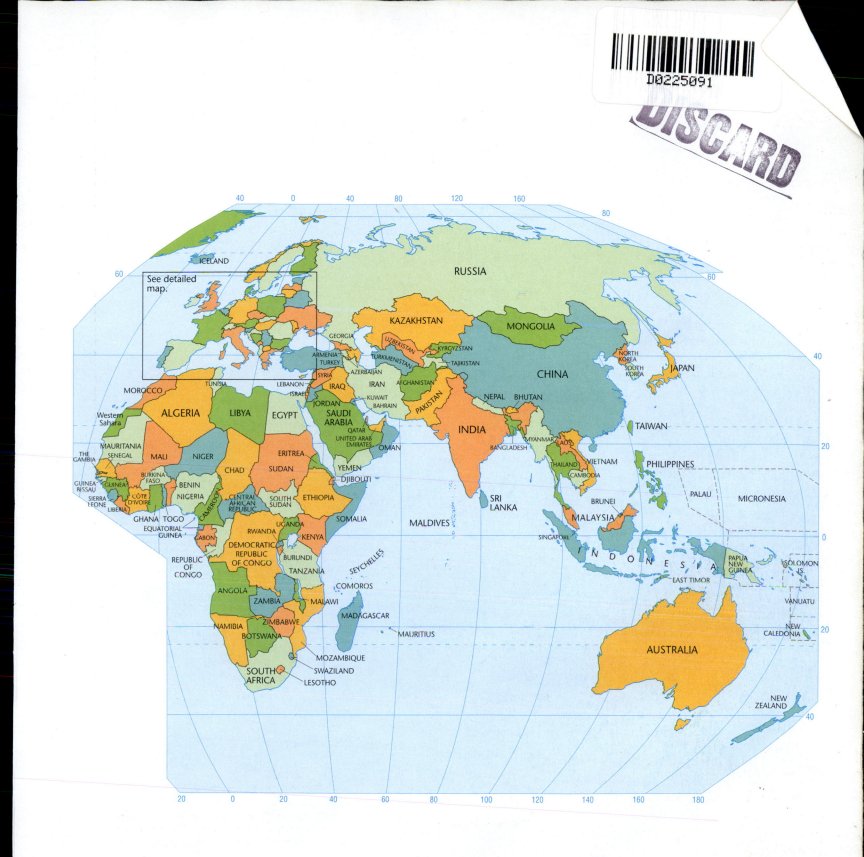

Jordan's
Fundamentals of The Human Mosaic
A Thematic Approach to Cultural Geography

Second Edition

Terry G. Jordan-Bychkov
University of Texas at Austin (late)

Mona Domosh
Dartmouth College

Roderick P. Neumann
Florida International University

Patricia L. Price
Florida International University

Nicole C. James
Research Associate

W. H. Freeman and Company
A Macmillan Higher Education Company

Publisher: Steven Rigolosi

Developmental Editor: Keith E. Harrington

Marketing Manager: Thomas Digiano

Media and Supplements Editor: Stephanie Ellis

Media Producer: Jenny Chiu

Art Director: Diana Blume

Text and Cover Designer: Vicki Tomaselli

Illustration Coordinator: Janice Donnola

Illustrations: Maps.com

Photo Editor: Bianca Moscatelli

Photo Researcher: Christina Micek

Director of Production: Ellen Cash

Project Editor: Robert Errera

Production Coordinator: Susan Wein

Composition: cMPreparé

Printing and Binding: RR Donnelley

Cultural Icons (cover and interior): World map, © Embe2006/Dreamstime.com; Global network,
© Roy Mattappallil Thomas/Dreamstime.com; Hibiscus leaf, Fotofermer/Shutterstock;
Heads, iStockphoto/Thinkstock/Getty Images; Cityscapes, hugolacasse/Shutterstock

Note: Photographs not otherwise credited are the property of Scott, Foresman and Company.

Library of Congress Control Number: 2013930872
ISBN-13: 978-1-4641-1068-9
ISBN-10: 1-4641-1068-9

Printed in the United States of America

First printing

W. H. Freeman and Company
41 Madison Ave.
New York, NY 10010
Houndmills, Basingstoke RG21 6XS, England
www.whfreeman.com/geography

CONTENTS

PREFACE

eography is a diverse academic discipline. It concerns place and region and employs methodologies from the social sciences, humanities, and earth sciences. Geographers deal with a wide range of subjects, from spatial patterns of human occupancy to the interaction between people and their environments. In doing so, they strive for a holistic view of Earth as the home of humankind.

Because the world is in constant flux, geography is an ever-changing discipline. Geographers necessarily consider a wide range of topics and view them from several perspectives. They continually seek new ways of looking at the inhabited Earth. Every revision of an introductory text such as Jordan's *Fundamentals of The Human Mosaic* requires a careful balancing of classic research and attention to recent innovations in the dynamic field that is cultural geography.

FUNDAMENTALS OF THE HUMAN MOSAIC AND
JORDAN'S FIVE CLASSIC THEMES OF CULTURAL GEOGRAPHY

Since its first publication in the mid 1970s, *The Human Mosaic* has been built around five themes that allow students to relate to the most important aspects of cultural geography at every point in the text.

W. H. Freeman & Company, the publisher of the text since its eighth edition, is proud to offer this alternative version of this classic text. Jordan's *Fundamentals of The Human Mosaic* is a concise introduction to the study of cultural geography for those instructors who wish to adopt a streamlined text. At just 11 chapters and 366 pages, Jordan's *Fundamentals of The Human Mosaic* is approximately 25% shorter than the full-length text.

This version of the text uses the "classic" Jordan-Bychkov thematic framework, with its emphasis on cultural (rather than human) geography. These five themes are:

Culture Region **Cultural Diffusion**

Cultural Ecology **Cultural Interaction** **Cultural Landscape**

The five themes are introduced and explained in the first chapter and serve as the framework for the 10 topical chapters that follow. Each theme is applied to a variety of geographical topics: demography, language, race and ethnicity, politics, religion, agriculture, industry, and the city. A small icon accompanies each theme as a visual reminder to students when these themes recur throughout the book.

These five themes of cultural geography allow several key topics to be addressed throughout the textbook, rather than isolated in just one chapter. For example, the important trend of **globalization** is treated under the cultural diffusion theme. The many aspects of **migration** are explored within the cultural diffusion theme as well.

page 16

page 34

Instructors have long told us that the five themes are a highly successful framework for instruction. The *culture region* theme appeals to students' natural curiosity about the differences among places. The dynamic aspect of culture—particularly relevant in this age of globalization and rapid change—is conveyed through the theme of *cultural diffusion*. Students acquire an appreciation for how cultural traits spread (or do not spread) from place to place. *Cultural ecology*, also highly relevant to our age, addresses the complicated relationship between culture and physical environments. *Cultural interaction* permits students to view culture as an interrelated whole, in which one facet acts on and is acted on by other facets. Many of the classic models developed by geographers are covered in the cultural interaction theme. Last, the theme of *cultural landscape* heightens students' awareness of the visible character of places and regions.

FEATURES OF JORDAN'S *FUNDAMENTALS OF THE HUMAN MOSAIC*

To remain fully up to date with the progress of cultural geography, Jordan's *Fundamentals of The Human Mosaic* is adapted from the twelfth edition of *The Human Mosaic*. This brief version offers:

- Streamlined instruction within each chapter, with no additional boxes or sidebars to distract students' attention.

- Shortened coverage of the city in Chapters 10 and 11, allowing instructors to cover urban geography in a manageable number of pages.

Jordan's *Fundamentals of The Human Mosaic* also offers three key features to assist students in learning the concepts:

page 168

page 175

- A **marginal glossary** reviews the key terms on each page.

- **Thinking Geographically** questions—designed to get students thinking critically about geography—accompany each photo and figure, with accompanying answers at the end of the chapter.

- Each chapter concludes with a **key terms list** with page references. Sources are placed at the end of the book rather than at the end of each chapter.

TOPICAL COVERAGE: COMBINING THE TIMELESS AND THE TIMELY

Jordan's *Fundamentals of The Human Mosaic* offers a mixture of the timeless and the timely, combining Terry Jordan's classic and popular examples with up-to-date coverage, data, statistics, new examples, and current research in each chapter. For example, all world maps have been updated to include South Sudan.

Among the new topics included in this edition are:

- An extended discussion of vernacular culture regions (Chapter 1)

- The rise of Facebook through hierarchical and contagious diffusion (Chapter 1)

- Updated coverage of natural hazards and disasters, such as the 2011 Japanese earthquake and tsunami (Chapters 1, 5)

page 193

- New coverage of foodways (world and local food and drink preferences) (Chapter 2)

- Hip-hop music as an example of cultural interaction (Chapter 2)

- A discussion of ethnomedicine and ecology (Chapter 2)

- David Lowenthal's work regarding the "cult of bigness" and the U.S. landscape (Chapter 2)

- Additional coverage of key concepts in population geography: natural decrease, absolute population density, physiological population density, dependency ratio, population growth rate, and population doubling time (Chapter 3)

- New discussion of medical geography and disease diffusion (Chapter 3)

- Urban landscape changes in South America and the rise of shantytowns/favelas (Chapter 3)

- A redrawn, easier-to-understand linguistic family tree (Chapter 4)

- A discussion of texting and global language modification (Chapter 4)

- Additional coverage of toponyms (Chapter 4)

- Expanded discussion of refugees and internally displaced persons (Chapter 5)

- A new discussion of environmental racism in the Seattle area (Chapter 5)

- Clarification of the difference between international political organizations and supranational political organizations (Chapter 6)

- Role of the Internet in political movements, such as the Arab Spring and the Occupy Wall Street movement (Chapter 6)

- New examples of national iconography on the landscape: Queen's Day in the Netherlands and the memorial to victims of the Nanjing Massacre in China (Chapter 6)

- Cultural interaction of religion and the cow population in India (Chapter 7)

- Updated coverage of religious adherence in the United States (Chapter 7)

- New map of agricultural regions in the United States, Figure 8.8 (Chapter 8)

- Extended discussion of aquaculture and the U.S. exclusive economic zone (Chapter 8)

- New map showing ancient sites of domestication for selected crops, Figure 8.17 (Chapter 8)

- Shrinkage of the Aral Sea (Chapter 8)

- New global map of biotech crop countries, Figure 8.28 (Chapter 8)

- Components of the Human Development Index, Figure 9.2 (Chapter 9)

- Updated discussion of the debates and evidence regarding global climate change and the environmental sustainability index (Chapter 9)

- New map of the world's urbanized population, Figure 10.1 (Chapter 10)

- Discussion of Latin American urban landscapes (Chapter 10)

- Modern critiques of the concentric zone, sector, and multiple-nuclei models of the city (Chapter 11)

page 311

MEDIA AND SUPPLEMENTS

Jordan's *Fundamentals of The Human Mosaic* is accompanied by a media and supplements package that facilitates learning and enhances the teaching experience.

Student Supplements

Atlas Options

Rand McNally's Atlas of World Geography, 176 pages

This atlas contains 52 physical, political, and thematic maps of the world and continents; 49 regional, physical, and thematic maps; and dozens of metropolitan-area insert maps. It also contains a section on common geographic questions, a glossary of terms, and a comprehensive 25-page index.

W. H. Freeman Quick Reference Guide

The briefest possible atlas, this double-sided, trifold laminated grid features a complete world map on one side and political mapping exercises on the other. It is the ideal portable world reference guide.

Google Earth Workbook

Google Earth Exercises for Human/Cultural Geography, by Diana Ter-Ghazaryan, University of Miami, and Bradley A. Shellito, Youngstown State University, ISBN: 1-4641-2199-0, 62 pages

This compact workbook provides an introduction to GoogleEarth, along with an activity to accompany each chapter of Jordan's *Fundamentals of The Human Mosaic*. In each activity, students explore Google Earth and answer a set of questions to assess their geographic understanding.

Mapping Workbook

Exploring Human Geography with Maps, second edition, by Margaret Pearce, University of Kansas, and Owen Dwyer, Indiana University–Purdue University, ISBN: 1-4292-2981-0

This full-color workbook introduces students to the diverse world of maps as fundamental tools for exploring and presenting ideas in human and cultural geography. It directly addresses the concepts of Jordan's *Fundamentals of The Human Mosaic,* chapter by chapter, and it includes activities accessible through *The Human Mosaic Online* at www.whfreeman.com/jordanfundamentals2e.

This second edition of *Exploring Human Geography with Maps* provides:

- Nine new activities featuring current topics suggested by geography instructors.
- Web and text updates for all other activities.

- An instructor's guide to help integrate *Exploring Human Geography with Maps* into the curriculum.
- Assessment questions for instructors that draw from both Jordan's *Fundamentals of The Human Mosaic* and *Exploring Human Geography with Maps,* further incorporating map-reading skills into the classroom.
- PowerPoint slides with maps from the text and information from the instructor's guide to help in lectures.

Self-Study on the Web

Jordan's *Fundamentals of The Human Mosaic Online:* www.whfreeman.com/jordanfundamentals2e

The companion Web site serves as an online study guide. The core of the site includes a range of features that encourage critical thinking and assist in study and review:

- A multiple-choice quiz for each chapter
- MapBuilder thematic map-creation software and exercises
- Flashcards with key vocabulary
- Blank outline maps
- Web links and activities

e-Books and Other Value-Priced Options

An e-Book version of Jordan's *Fundamentals of The Human Mosaic* is available in two formats: as an online e-Book from W. H. Freeman and Company or via CourseSmart. For information about the online e-Book, please contact your W. H. Freeman sales consultant. For access to the CourseSmart e-Book, please visit www.coursesmart.com.

An unbound looseleaf version of the text is also available at a discount. To order the looseleaf version, please use ISBN 1-4641-2102-8.

LaunchPad for Jordan, *Fundamentals of the Human Mosaic,* 2e with LearningCurve

www.whfreeman.com/launchpad/jordanfundamentals2e

This comprehensive resource is designed to offer a complete solution for today's classroom. LaunchPad for Jordan, *Fundamentals of the Human Mosaic,* 2e with LearningCurve offers all the instructor and student resources available on the book companion web site, as well as premium resources available only on the Portal:

- An e-Book of Jordan's *Fundamentals of The Human Mosaic* 2E, complete and customizable. Students can quickly search the text and personalize it just as they would the printed version, with highlighting, bookmarking, and note-taking features.
- **A Guide to Using Google Earth** for the novice, plus step-by-step **Google Earth** exercises for each chapter.

- Selected articles from *Focus on Geography* magazine (one for each chapter in the textbook) and accompanying quizzes for each article.
- A wide variety of **mapping activities,** including drag-and-drop matching and MapBuilder (thematic map-building). New **map-learning activities** can be assigned, with grades reporting directly to an online instructor grade book.
- Online **news feeds** for highly respected magazines such as *Scientific American* and *The Economist*.
- **Gallup WorldWind** polling data links (across world regions).
- Full grade book functionality and integration.
- Professional links and listservs.

For more information or to schedule a demo, please contact your W. H. Freeman sales representative.

LEARNING Curve

Learning Curve is a major asset within LaunchPad for Jordan, *Fundamentals of the Human Mosaic*, 2e. It is an intuitive, fun, and highly effective formative assessment tool.

Based on extensive educational research, Learning Curve is a great asset for helping students prepare for lectures and exams. Game-like quizzing motivates students to engage with their course, and reporting tools help instructors identify students' areas of strength and weakness. LearningCurve activities are interactive, personal, and layered. As students work through the Learning Curve activity for each chapter, they move further up Bloom's taxonomy; once they have mastered basic definitions and concepts, they move into analyzing, synthesizing, and integrating content, often with the aid of maps and other visual information. A personalized study plan for each chapter points students to the areas in which they need further study and provides direct links to relevant information in the e-Book.

Instructor Supplements

Videos

W. H. Freeman Human/Cultural Geography 2-DVD set.
ISBN: 1-4292-7392-5

A two-DVD set containing approximately 25 projection-quality video clips is available free to adopters of Jordan's *Fundamentals of The Human Mosaic*. The DVD set also contains an Instructor's Video Guide. All of the videos, along with accompanying quizzes, are also available online in the instructor's resource section on the book's companion web site: **www.whfreeman.com/jordanfundamentals2e**. To view the complete video listing, please go to **www.whfreeman.com/geography**, and to request the DVD set please contact your W. H. Freeman sales consultant.

Test Bank and Presentation Assets

Instructor's Resource Web Site
www.whfreeman.com/jordanfundamentals2e

The instructor's side of the companion Web site contains all the text images available in JPEG format and as Microsoft PowerPoint slides for use in classroom presentations. Images have been optimized for high-quality projection in large classrooms. A separate set of lecture-ready PowerPoint slides is also provided. The instructor resources also contain chapter-by-chapter Microsoft Word test bank files that can be easily modified.

Course Management

All instructor and student resources are also available via Blackboard, Sakai, Moodle, Desire2Learn, Canvas, and Angel to enhance your course. W. H. Freeman and Company offers a course cartridge that populates your course Web site with content tied directly to the book.

Computer Test Bank

Computerized Test Bank, expanded and adapted from the work of Ray Sumner, Long Beach City College; Jose Javier Lopez, Minnesota State University; Roxane Fridirici, California State University–Sacramento; and Douglas Munski, University of North Dakota.
CD-ROM ISBN: 1-4641-2097-8.

Microsoft Word files of the newly revised *Test Bank* are available on the companion Web site. The Test Bank is carefully designed to match the pedagogical intent of the textbook and to include questions from basic memorization to comprehension of concepts. It contains more than 1250 test questions (multiple choice and true/false) and includes more than 150 map-analysis questions. The computerized Diploma test bank is also available on CD-ROM.

ACKNOWLEDGMENTS/REVIEWERS

No textbook is ever written single-handedly or even double-handedly. Many geographers have contributed advice, comments, ideas, and assistance to *The Human Mosaic* over the past 30 years.

Reviewers of the 12th edition of *The Human Mosaic*

Toni Alexander, *Auburn University*
Shaunna Barnhart, *Allegheny College*
Mathias Le Bossé, *Kutztown University of Pennsylvania*
Larry E. Brown, *University of Missouri–Columbia*
Thomas Chapman, *Old Dominion University*
Anthony J. Dzik, *Shawnee State University*
Caroline Faria, *Florida International University*
Tracie Hallock, *Keystone College*

Ellen Hansen, *Emporia State University*
Miriam Helen Hill, *Jacksonville State University*
Megan Holroyd, *University of Kansas*
Richard Hunter, *SUNY Cortland*
Kari B. Jensen, *Hofstra University*
Ronald Kalafsky, *University of Tennessee*
Olaf Kuhlke, *University of Minnesota*, Duluth
Vincent Mazeika, *Ohio University*
Leigh Anne L. Opitz, *University of Nebraska*, Omaha
Debra J. Taylor Matthews, *Boise State University*
Blake L. Mayberry, *University of Kansas*
Neusa McWilliams, *University of Toledo*
Norman Moline, *Augustana College*
Edris J. Montalvo, *Cameron University*
Thomas Orf, *Las Positas College*
Paul E. Phillips, *Fort Hays (Kansas) State University*
Mary Kimsey Tacy, *James Madison University*
Jean Vincent, *Santa Fe College*
Montgomery Walker, *Yakima Valley Community College*

Reviewers of the Earlier Editions

Jennifer Adams, *Pennsylvania State University;* Paul C. Adams, *University of Texas,* Austin; W. Frank Ainsley, *University of North Carolina,* Wilmington; Christopher Airriess, *Ball State University;* Nigel Allan, *University of California,* Davis; Thomas D. Anderson, *Bowling Green State University;* Timothy G. Anderson, *Ohio Wesleyan University;* Patrick Ashwood, *Hawkeye Community College;* Nancy Bain, *Ohio University;* Timothy Bawden, *University of Wisconsin,* Eau Claire; Brad Bays, *Oklahoma State University;* A. Steele Becker, *University of Nebraska,* Kearney; Sarah Bednarz, *Texas A&M University;* Gigi Berardi, *Western Washington University;* Daniel Borough, *California State University,* Los Angeles; Patricia Boudinot, *George Mason University;* Wayne Brew, *Montgomery County Community College;* Michael J. Broadway, *Northern Michigan University;* Sarah Osgood Brooks, *Central Michigan University;* Scott S. Brown, *Francis Marion University;* Craig S. Campbell, *Youngstown State University;* Kimberlee J. Chambers, *Sonoma State University;* Wing H. Cheung, *Palomar Community College;* Carolyn A. Coulter, *Atlantic Cape Community College;* Merel J. Cox, *Pennsylvania State University,* Altoona; Marcelo Cruz, *University of Wisconsin,* Green Bay; Christina Dando, *University of Nebraska,* Omaha; Robin E. Datel, *California State University,* Sacramento; James A. Davis, *Brigham Young University;* Richard Deal, *Western Kentucky University;* Jeff R. DeGrave, *University of Wisconsin,* Eau Claire; Lorraine Dowler, *Pennsylvania State University;* Christine Drake, *Old Dominion University;* Owen Dwyer, IUPUI; Matthew Ebiner, *El Camino College;* James D. Ewing, *Florida Community College;*

Maria Grace Fadiman, *Florida Atlantic University;* Jacksonville; David Albert Farmer, *Wilmington College;* Kim Feigenbaum, *Santa Fe Community College;* D. J. P. Forth, *West Hills Community College;* Carolyn Gallaher, *American University;* Charles R. Gildersleeve, *University of Nebraska,* Omaha; M. A. Goodman, *Grossmont College;* Jeffrey J. Gordon, *Bowling Green State University;* Richard J. Grant, *University of Miami;* Charles F. Gritzner, *South Dakota State University;* Sally Gros, *University of Oklahoma,* Norman; Qian Guo, *San Francisco State University;* Joshua Hagen, *Marshall University;* Daniel J. Hammel, *University of Toledo;* Ellen R. Hansen, *Emporia State University;* Jennifer Helzer, *California State University,* Stanislaus; Andy Herod, *University of Georgia;* Elliot P. Hertzenberg, *Wilmington College;* Deryck Holdsworth, *Pennsylvania State University;* Cecelia Hudleson, *Foothill College;* Ronald Isaac, *Ohio University;* Gregory Jeane, *Samford University;* Brad Jokisch, *Ohio University;* James R. Keese, *California Polytechnic State University;* Artimus Keiffer, *Wittenberg University;* Edward L. Kinman, *Longwood University;* Marti L. Klein, *Saddleback College;* Vandara Kohli, *California State University,* Bakersfield; Jennifer Kopf, *West Texas A&M;* John C. Kostelnick, *Illinois State University;* Debra Kreitzer, *Western Kentucky University;* Olaf Kuhlke, *University of Minnesota,* Duluth; Michael Kukral, *Ohio Wesleyan University;* Hsiang-te Kung, *Memphis University;* William Laatsch, *University of Wisconsin,* Green Bay; Paul R. Larson, *Southern Utah University;* Ann Legreid, *Central Missouri State University;* Peter Li, *Tennessee Technological University;* Ronald Lockmann, *California State University,* Dominiquez Hills; Jose Javier Lopez, *Minnesota State University;* Michael Madsen, *Brigham Young University,* Idaho; Jesse O. McKee, *University of Southern Mississippi;* Wayne McKim, *Towson State University;* Douglas Meyer, *Eastern Illinois University;* Klaus Meyer-Arendt, *Mississippi State University;* John Milbauer, *Northeastern State University;* Cynthia A. Miller, *Syracuse University;* Glenn R. Miller, *Bridgewater State College;* Don Mitchell, *Syracuse University;* Edris J. Montalvo, *Texas State University,* San Marcos; Karen Morin, *Bucknell University;* James Mulvihill, *California State University,* San Bernardino; Douglas Munski, *University of North Dakota;* Gareth A. Myers, *University of Kansas;* David J. Nemeth, *University of Toledo;* James W. Newton, *University of North Carolina,* Chapel Hill; Michael G. Noll, *Valdosta State University;* Ann M. Oberhauser, *West Virginia University;* Stephen M. O'Connell, *Oklahoma State University;* Thomas Orf, *Prestonburg Community College;* Brian Osborne, *Queen's University;* Kenji Oshiro, *Wright State University;* Bimal K. Paul, *Kansas State University;* Eric Prout, *Texas A&M University;* Darren Purcell, *University of Oklahoma;* Virginia M. Ragan, *Maple Woods Community*

College; **Jeffrey P. Richetto,** *University of Alabama;* **Henry O. Robertson,** *Louisiana State University,* Alexandria; **Robert Rundstrom,** *University of Oklahoma;* **Norman H. Runge,** *University of Delaware;* **Stephen Sandlin,** *California State University,* Pomona; **Lydia Savage,** *University of Southern Maine;* **Steven M. Schnell,** *Kutztown University of Pennsylvania;* **Andrew Schoolmaster, III,** *University of North Texas;* **Steven Schnell,** *Kutztown University of Pennsylvania;* **Cynthia S. Simmons,** *Michigan State University;* **Emily Skop,** *University of Texas,* Austin; **Christa Smith,** *Clemson University;* **Anne K. Soper,** *Indiana University;* **Roger W. Stump,** *State University of New York,* Albany; **Ray Sumner,** *Long Beach City College;* **David A. Tait,** *Rogers State University;* **Jonathan Taylor,** *California State University,* Fullerton; **Thomas Terich,** *Western Washington University;* **Thomas M. Tharp,** *Purdue University;* **Benjamin F. Timms,** *California Polytechnic State University;* **Ralph Triplette,** *Western Carolina University;* **Daniel E. Turbeville, III,** *Eastern Washington University;* **Elisabeth S. Vidon,** *Indiana University;* **Ingolf Vogeler,** *University of Wisconsin;* **Timothy M. Vowles,** *Colorado State University;* **Philip Wagner,** *Simon Fraser State University;* **Barney Warf,** *Florida State University;* **Henry Way,** *University of Kansas;* **Barbara Weightman,** *California State University,* Fullerton; **John Western,** *Syracuse University;* **W. Michael Wheeler,** *Southwestern Oklahoma State University;* **David Wilkins,** *University of Utah;* **Douglas Wilms,** *East Carolina State University; and* **Donald Zeigler,** *Old Dominion University.*

Reviewers/Advisors on the Preparation of Jordan's *Fundamentals of The Human Mosaic*, 1st and 2nd Editions

Brad Bays, *Oklahoma State University*
Meera Chatterjee, *University of Akron*
Linc De Bunce, *Blue Mountain Community College*
Roger E. Dendinger, *South Dakota School of Mines and Technology*
Anthony J. Dzik, *Shawnee State University*
Matthew Ebiner, *El Camino College*
Kim Feigenbaum, *Santa Fe College*
Steven Graves, *California State University–Northridge*
Charles Gritzner, *South Dakota State University*
Jeffrey Gordon, *Bowling Green State University*
Thomas J. Karwoski, *Anne Arundel Community College*
Heidi Lannon, *Santa Fe College*
David Nemeth, *University of Toledo*
Jeffrey P. Richetto, University of Alabama
Albert L. Rydant, *Keene State College*
Daniel E. Turbeville III, *Eastern Washington University*
Brad Watkins, *University of Central Oklahoma*
Robert C. Ziegenfus, *Kutztown University of Pennsylvania*

ABOUT THE AUTHORS

Terry G. Jordan-Bychkov was the Walter Prescott Webb Professor in the Department of Geography at the University of Texas, Austin. He earned his Ph.D. at the University of Wisconsin, Madison. A specialist in the cultural and historical geography of the United States, Jordan-Bychkov was particularly interested in the diffusion of Old World culture in North America that helped produce the vivid geographical mosaic evident today. He served as president of the Association of American Geographers in 1987 and 1988 and received an honors award from that organization. He wrote on a wide range of American cultural topics, including forest colonization, cattle ranching, folk architecture, and ethnicity. His scholarly books include *The European Culture Area: A Systematic Geography,* fourth edition (with Bella Bychkova Jordan, 2002), *Anglo-Celtic Australia: Colonial Immigration and Cultural Regionalism* (with Alyson L. Grenier, 2002), *Siberian Village: Land and Life in the Sakha Republic* (with Bella Bychkova Jordan, 2001), *The Mountain West: Interpreting the Folk Landscape* (with Jon Kilpinen and Charles Gritzner, 1997), *North American Cattle Ranching Frontiers* (1993), *The American Backwoods Frontier* (with Matt Kaups, 1989), *American Log Building (1985), Texas Graveyards* (1982), *Trails to Texas: Southern Roots of Western Cattle Ranching* (1981), and *German Seed in Texas Soil* (1966).

Mona Domosh is professor of geography at Dartmouth College. She earned her Ph.D. at Clark University. Her research has examined the links between gender ideologies and the cultural formation of large American cities in the nineteenth century, particularly in regard to such critical but vexing distinctions as consumption/production, public/private, masculine/feminine. She is currently engaged in research that takes the ideological association of women, femininity, and space in a more postcolonial direction by asking what roles nineteenth-century ideas of femininity, masculinity, consumption, and "whiteness" played in the crucial shift from American nation building to empire building. Domosh is the author of *American Commodities in an Age of Empire* (2006); *Invented Cities: The Creation of Landscape in 19th-*Century New York and Boston (1996); the coauthor, with Joni Seager, of *Putting Women in Place: Feminist Geographers Make Sense of the World* (2001); and the coeditor of *Handbook of Cultural Geography* (2002).

Roderick P. Neumann is professor of geography and chair of the Department of Global and Sociocultural Studies at Florida International University. He earned his Ph.D. at the University of California, Berkeley. He studies the complex interactions of culture and nature through a specific focus on national parks and natural resources. In his research, he combines the analytical tools of cultural and political ecology with landscape studies. He has pursued these investigations through historical and ethnographic research mostly in East Africa, with some comparative work in North America and Central America. His current research explores interwoven narratives of nature, landscape, and identity in the European Union, with a particular emphasis on Spain. His scholarly books include *Imposing Wilderness: Struggles over Livelihoods and Nature Preservation in Africa* (1998), *Making Political Ecology* (2005), and *The Commercialization of Non-Timber Forest Products* (2000), the latter coauthored with Eric Hirsch.

Patricia L. Price is associate professor of geography at Florida International University. She earned her Ph.D. at the University of Washington. Connecting the long-standing theme of humanistic scholarship in geography to more recent critical approaches best describes her ongoing intellectual project. From her initial field research in urban Mexico, she has extended her focus to the border between Mexico and the United States and, most recently, to south Florida as a borderland of sorts. Her most recent field research is on comparative ethnic neighborhoods, conducted with colleagues and graduate students in Phoenix, Chicago, and Miami, and funded by the National Science Foundation. Price is the author of *Dry Place: Landscapes of Belonging and Exclusion* (2004) and coeditor (with Tim Oakes) of *The Cultural Geography Reader* (2008).

Painting by Helicopter Tjungurrayi depicting the Great Sandy Desert south of Balgo, Western Australia. (Wangkartu Helicopter Tjungurrayi, copyright Warlayirti Artists.)

Cultural Geography: An Introduction

1

Most of us are born geographers. We are curious about the distinctive character of places and peoples. We think in terms of territory and space. Take a look outside your window right now. The houses and commercial buildings, streets and highways, gardens and lawns all tell us something interesting and profound about who we are as a culture. If you travel down the road or on a jet to another region or country, the view outside your window will change, sometimes subtly, sometimes drastically. Our geographical imagination will push us to look and think and begin to make sense of what is going on in these different places, environments, and landscapes. It is this curiosity about the world—about how and why it is structured the way it is, what it means, and how we have changed it and continue to change it—that is at the heart of human geography. You are already geographers; we hope that our book will make you better ones.

If every place on Earth were identical, we would not need geography, but each is unique. Every place, however, does share characteristics with other places. Geographers define the concept of *region* as a grouping of similar places or of places with similar characteristics. The existence of different regions endows the Earth's surface with a mosaiclike quality. Geography as an academic discipline is an outgrowth of both our curiosity about lands and peoples other than our own and our need to come to grips with the place-centered element within our soul. When professional academic geographers consider the differences and similarities among places, they want to understand what they see. They first find out exactly what variations exist among regions and places by describing them as precisely as possible. Then they try to decide what forces made these areas different or alike. Geographers ask *what? where? why?* and *how?*

Our natural geographical curiosity and intrinsic need for identity were long ago reinforced by pragmatism, the practical motives of traders and empire builders who wanted information about the world for the purposes of commerce and conquest. This concern for the practical aspects of geography first arose thousands of years ago among the ancient Greeks, Romans, Mesopotamians, and Phoenicians, the greatest traders and empire builders of their times. They cataloged factual information about locations, places, and products. Indeed, **geography** is a Greek word meaning literally "to describe the Earth." Not content merely to chart and describe the world, these ancient geographers soon began to ask questions about why cultures and environments differ from place to place, initiating the study of what today we call geography.

> **geography** The study of spatial patterns and of differences and similarities from one place to another in environment and culture.

WHAT IS **CULTURAL** GEOGRAPHY?

As its name implies, *cultural geography* forms one part of the discipline of geography, complementing physical geography, the part of the discipline that deals with the natural environment. To understand the scope of cultural geography, we must first agree on the meaning of *culture*.

Many definitions of culture exist, some broad and some narrow. We might best define **culture** as learned collective human behavior, as opposed to instinctive, or inborn, behavior. These learned traits form a way of life held in common by a group of people. Learned similarities in speech, behavior, ideology, livelihood, technology, value system, and society bind people together. Culture involves a communication system of acquired beliefs, memories, perceptions, traditions, and attitudes that serve to supplement and channel instinctive behavior. In short, as geographer Yi-Fu Tuan has said, culture is the "local, customary way of doing things; geographers write about ways of life."

> **culture** A total way of life held in common by a group of people, including such learned features as speech, ideology, behavior, livelihood, technology, and government; or the local, customary way of doing things—a way of life; an ever-changing process in which a group is actively engaged; a dynamic mix of symbols, beliefs, speech, and practices.

A particular culture is not a static, fixed phenomenon, and it does not always govern its members. Rather, as geographers Kay Anderson and Fay Gale put it, "culture is a *process* in which people are actively engaged." Individual members can and do change a culture, which means that ways of life constantly change and that tensions between opposing views are usually present. Cultures are never internally homogeneous because individual humans never think or behave in exactly the same manner.

(a)

(b)

FIGURE 1.1 Two traditional houses of worship. Geographers seek to learn how and why cultures differ, or are similar, from one place to another. Often the differences and similarities have a visual expression. (a: Rob Crandall/Stock Connection/Alamy; b: Frans Lemmens/Alamy.)

THINKING GEOGRAPHICALLY In what ways are these two structures—one a Catholic church in Honduras (a) and the other a Buddhist temple in Laos (b)—alike and different?

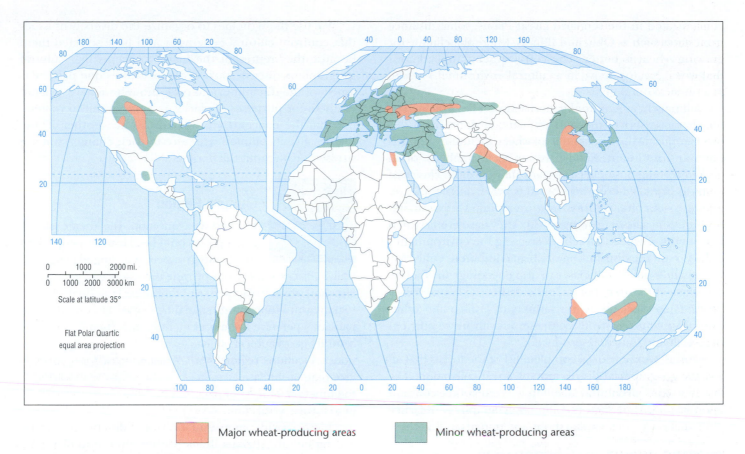

| | Major wheat-producing areas | | Minor wheat-producing areas |

FIGURE 1.2 Areas of wheat production in the world today. These regions are based on a single trait: the importance of wheat in the agricultural system. This map tells us what and where. It raises the question of why.

THINKING GEOGRAPHICALLY What causal forces might be at work to produce this geographical distribution of wheat farming?

Cultural geography, then, is the study of spatial variations among cultural groups and the spatial functioning of society.

cultural geography The study of spatial variations among cultural groups and the spatial functioning of society.

It focuses on describing and analyzing the ways language, religion, economy, government, and other cultural phenomena vary or remain constant from one place to another and on explaining how humans function spatially (**Figure 1.1**). Cultural geography is, at heart, a celebration of human diversity, or as the Russian ethnographer Leo Gumilev wrote, "the study of differences among peoples."

In seeking explanations for cultural diversity and place identity, geographers consider a wide array of factors that cause this diversity. Some of these involve the **physical environment:** terrain, climate, natural vegetation, wildlife, variations in soil, and the pattern of land and water. Because we cannot understand a culture removed from its physical setting, human geography offers not only a spatial perspective but also an ecological one.

physical environment All aspects of the natural physical surroundings, such as climate, terrain, soils, vegetation, and wildlife.

Many complex forces are at work on cultural phenomena, and all of them are interconnected in very complicated ways. The complexity of the forces that affect culture can be illustrated by an example drawn from agricultural geo-graphy: the distribution of wheat cultivation in the world. If you look at **Figure 1.2**, you can see major areas of wheat cultivation in Australia but not in Africa, in the United States but not in Chile, in China but not in Southeast Asia. Why does this spatial pattern exist? Partly it results from environmental factors such as climate, terrain, and soils. Some regions have always been too dry for wheat cultivation. The land in others is too steep or infertile. Indeed, there is a strong correlation between wheat cultivation and midlatitude climates, level terrain, and good soil.

Still, we should not place exclusive importance on such physical factors. People can modify the effects of climate through irrigation; the use of hothouses; or the development of new, specialized strains of wheat. They can conquer slopes by terracing, and they can make poor soils productive with fertilization. For example, farmers in mountainous parts of Greece traditionally wrested an annual harvest of wheat from tiny terraced plots where soil had been trapped behind hand-built stone retaining walls. Even in the United States, environmental factors alone cannot explain the fact that major wheat cultivation is concentrated in the semiarid Great Plains, some distance from states such as Ohio and Illinois, where the climate for growing wheat is better. The cultural geographer knows that wheat has to survive in a cultural environment as well as a physical one.

Ultimately, agricultural patterns cannot be explained by the characteristics of the land and climate alone. Many factors complicate the distribution of wheat, including people's tastes and traditions. Food preferences and taboos, often backed by religious beliefs, strongly influence the choice of crops to plant. Where wheat bread is preferred, people are willing to put great efforts into overcoming physical surroundings hostile to growing wheat. They have even created new strains of wheat, thereby decreasing the environment's influence on the distribution of wheat cultivation. Other factors, such as public policies, can also encourage or discourage wheat cultivation. For example, tariffs (taxes on foreign products) protect the wheat farmers of France and other European countries from competition with more efficient American and Canadian producers.

This is by no means a complete list of the forces that affect the geographical distribution of wheat cultivation. The distribution of all cultural elements is a result of the constant interplay of diverse factors. Cultural geography is the discipline that seeks such explanations and understandings.

THEMES IN CULTURAL GEOGRAPHY

Our study of cultures is organized around five geographical concepts or themes:

- Culture region
- Cultural diffusion
- Cultural ecology
- Cultural interaction
- Cultural landscape

FIGURE 1.3 An Inuit hunter with his dogsled team. Various facets of a multitrait formal region can be seen here, including the clothing, the use of dogsleds as transportation, and hunting as a livelihood system. (Bryan & Cherry Alexander/Arctic Photo.)

THINKING GEOGRAPHICALLY What evidence suggests changes in this region?

We use these themes to organize the diversity of issues that confront cultural geography and have selected them because they represent the major concepts that cultural geographers discuss. Each of them stresses one particular aspect of the discipline, and even though we have separated them for purposes of clarity, it is important to remember that the concepts are related to one another. These themes give a common structure to each chapter and are stressed throughout the book.

Culture **Region**

Phrased as a question, the theme of culture region could be "How are people and their traits grouped or arranged geographically?" Places and regions provide the essence of geography. How and why are places alike or different? How do they mesh together into functioning spatial networks? How do their inhabitants perceive them and identify with them? These are central geographical questions. A **culture region** is a grouping of similar places or the functional union of places to form a spatial unit.

> **culture region** A geographical unit based on characteristics and functions of culture.

Maps provide an essential tool for describing and revealing culture regions. If, as is often said, one picture is worth a thousand words, then a well-prepared map is worth at least 10,000 words to the geographer. No description in words can rival a map's revealing force. Maps are valuable tools particularly because they so concisely portray spatial patterns in culture.

Geographers recognize three types of regions: formal, functional, and vernacular.

Formal Culture Regions A **formal culture region** is an area inhabited by people who have one or more traits in common, such as language, religion, or a system of livelihood. It is an area, therefore, that is relatively homogeneous with regard to one or more cultural traits. Geographers use this concept to map spatial differences throughout the world. For example, an Arabic-language formal region can be drawn on a map of languages and would include the areas where Arabic is spoken, rather than, say, English or Hindi or Mandarin. Similarly, a wheat-farming formal region would include the parts of the world where wheat is a major crop (look again at Figure 1.2).

> **formal culture region** A cultural region inhabited by people who have one or more cultural traits in common.

The examples of Arabic speech and of wheat cultivation represent the concept of formal culture region at its simplest level. Each is based on a single cultural trait. More commonly, formal culture regions depend on multiple related traits (**Figure 1.3**). Thus, an Inuit (Eskimo) culture region might be based on language, religion, economy, social organization, and type of dwellings. The region would reflect the spatial distribution of these five Inuit cultural traits. Districts in which all five of these traits are present would be part of the culture region. Similarly, Europe can be subdivided into several multitrait regions (**Figure 1.4**).

Formal culture regions are the geographer's somewhat arbitrary creations. No two cultural traits have the same

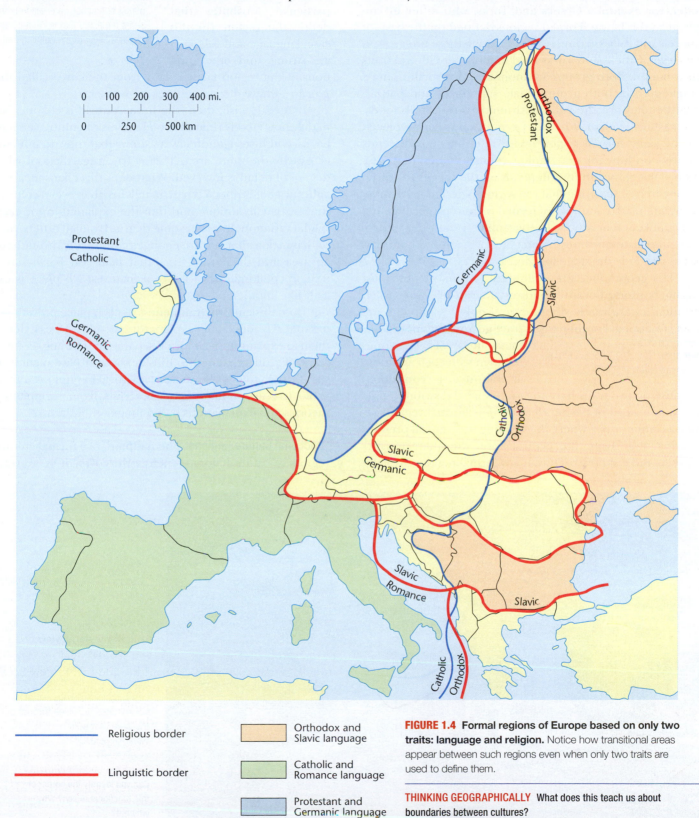

	Religious border
	Linguistic border

Orthodox and Slavic language

Catholic and Romance language

Protestant and Germanic language

FIGURE 1.4 Formal regions of Europe based on only two traits: language and religion. Notice how transitional areas appear between such regions even when only two traits are used to define them.

THINKING GEOGRAPHICALLY What does this teach us about boundaries between cultures?

distribution, and the territorial extent of a culture region depends on what and how many defining traits are used (see Figure 1.4). Why *five* Inuit traits, not four or six? Why not *foods* instead of (or in addition to) dwelling types? Consider, for example, Greeks and Turks, who differ in language and religion. Formal regions defined on the basis of speech and religious faith would separate these two groups. However, Greeks and Turks hold many other cultural traits in common. Both groups are monotheistic, worshipping a single god. In both groups, male supremacy and patriarchal families are the rule. Both enjoy certain folk foods, such as shish kebab. Whether Greeks and Turks are placed in the same formal region or in different ones depends entirely on how the geographer chooses to define the region. That choice in turn depends on the specific purpose of research or teaching that the region is designed to serve. Thus, an infinite number of formal regions can be created. It is unlikely that any two geographers would use exactly the same distinguishing criteria or place cultural boundaries in precisely the same location.

The geographer who identifies a formal region must locate borders. Because cultures overlap and mix, such boundaries are rarely sharp, even if only a single cultural trait is mapped. For this reason, we find **border zones** rather than lines. These zones broaden with each additional trait that is considered

border zone The area where different regions meet and sometimes overlap.

because no two traits have the same spatial distribution. As a result, instead of having clear borders, formal regions reveal a center or *core* where the defining traits are all present. Moving away from the central core, the characteristics weaken and disappear. Thus, many formal regions display a

core-periphery pattern in which the region can be divided into two sections, one near the center, where the particular attributes that define the region (in this case, language and religion) are strong, and other portions of the region farther away from the core, called the *periphery*, where those attributes are weaker.

core-periphery A concept based on the tendency of both formal and functional culture regions to consist of a core, in which defining traits are purest or functions are headquartered, and a periphery that is tributary and displays fewer of the defining traits.

In a real sense, then, the human world is chaotic. No matter how closely related two elements of culture seem to be, careful investigation always shows that they do not cover exactly the same area. This is true regardless of what degree of detail is involved. What does this chaos mean to cultural geographers? First, it tells us that every cultural trait is spatially unique and that the explanation for each spatial variation differs in some degree from all others. Second, it means that culture changes continually throughout an area and that every inhabited place on Earth has a unique combination of cultural features. No place is exactly like another.

Does this cultural uniqueness of each place prevent geographers from seeking explanatory theories? Does it doom them to explaining each locale separately? The answer must be no. The fact that no two hills or rocks, no two planets or stars, no two trees or flowers are identical has not prevented geologists, astronomers, and botanists from formulating theories and explanations based on generalizations.

Functional Culture Regions The hallmark of a formal culture region is cultural homogeneity. Moreover, it is abstract

FIGURE 1.5 Aerial view of Denver. This image clearly illustrates the node of a functional region—here, the dense cluster of commercial buildings—that coordinates activities throughout the area that surrounds it. (Jim Wark/Airphoto.)

THINKING GEOGRAPHICALLY
Can you identify the border of this functional region? Why or why not?

functional culture region A cultural area that functions as a unit politically, socially, or economically.

rather than concrete. By contrast, **a functional culture region** need not be culturally homogeneous; instead, it is an area that has been organized to function politically, socially, or economically as one unit. A city, an independent state, a precinct, a church diocese or parish, a trade area, a farm, and a Federal Reserve Bank district are all examples of functional culture regions. Functional culture regions have **nodes,** or central points where the functions are coordinated and directed. Examples of such

node A central point in a functional culture region where functions are coordinated and directed.

nodes are city halls, national capitals, precinct voting places, parish churches, factories, and banks. In this sense, functional regions also possess a core-periphery configuration in common with formal culture regions.

Many functional regions have clearly defined borders. A metropolitan area is a functional region that includes all of the land under the jurisdiction of a particular urban government (**Figure 1.5**). The borders of this functional region may not be so apparent from a car window, but they will be clearly delineated on a regional map by a line distinguishing one jurisdiction from another. Similarly, each state in the United States and each Canadian province is a functional region, coordinated and directed from a capital, with government control extended over a fixed area with clearly defined borders.

Not all functional regions have fixed, precise borders, however. A good example is a daily newspaper's circulation area. The node for the paper is the plant where it is produced. Every morning, trucks move out of the plant to distribute the paper throughout the city. The newspaper may have a sales area extending into the city's suburbs, local bedroom communities, nearby towns, and rural areas. There its sales area overlaps with the sales territories of competing newspapers published in other cities. It would be futile to try to define exclusive borders for such an area. How would you draw a sales area boundary for *The New York Times?* Its Sunday edition is sold in some quantity even in California, thousands of miles from its node, and it is published simultaneously in different cities.

Functional culture regions generally do not coincide spatially with formal regions, and this disjuncture often creates problems for the functional region. Germany provides an example (**Figure 1.6**). As an independent state, Germany

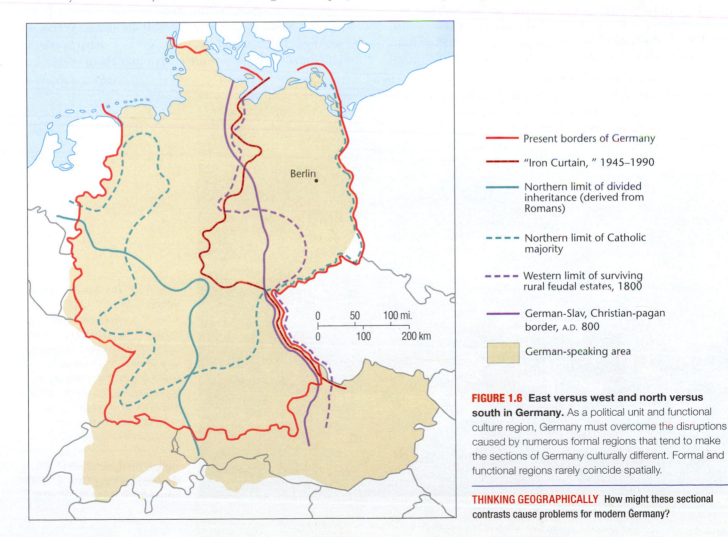

— Present borders of Germany

— "Iron Curtain," 1945–1990

— Northern limit of divided inheritance (derived from Romans)

- - - Northern limit of Catholic majority

- - - Western limit of surviving rural feudal estates, 1800

— German-Slav, Christian-pagan border, A.D. 800

German-speaking area

FIGURE 1.6 East versus west and north versus south in Germany. As a political unit and functional culture region, Germany must overcome the disruptions caused by numerous formal regions that tend to make the sections of Germany culturally different. Formal and functional regions rarely coincide spatially.

THINKING GEOGRAPHICALLY How might these sectional contrasts cause problems for modern Germany?

forms a functional region. Language provides a substantial basis for political unity. However, the formal region of the German language extends beyond the political borders of Germany and includes part or all of eight other independent states. More important, numerous formal regions have borders cutting through German territory. Some of these have endured for millennia, causing differences among northern, southern, eastern, and western Germans. These contrasts make the functioning of the German state more difficult and help explain why Germany has been politically fragmented more often than unified.

Vernacular Culture Regions A **vernacular culture region** is one that is *perceived* to exist by its inhabitants, as evidenced by the widespread acceptance and use of a special regional name. Some vernacular culture regions are based on physical environmental features. For example, there are many regions called simply "the valley." Wikipedia lists more than 30 different regions within the United States and Canada that are referred to as such, places as varied as the Sudbury Basin in Ontario and the Lehigh Valley in Pennsylvania. In the 1980s, "the valley" became synonymous with the San Fernando Valley in Southern California (**Figure 1.7**) and

> **vernacular culture region** A cultural region perceived to exist by its inhabitants; based in the collective spatial perception of the population at large; bearing a generally accepted name or nickname (such as Dixie).

became associated with a type of landscape (suburban), a person (a white, teenage girl, called the "valley girl"), and a way of speaking ("Valspeak").

Examples of vernacular regions based on physical environmental features can be found outside of the United States and Canada as well. One such example is the Outback in Australia. For most people, "the Outback" conjures images of a wild and rugged place dominated by rocky red soil, exotic wildlife, and the majestic *Uluru* (or Ayers Rock) (**Figure 1.8**). This vernacular region loosely corresponds to certain climate zones, soil types, and terrain features. However, like other vernacular regions, the Outback is based more on popular perception than on these physical characteristics. Images of the Outback as a frontier, created in fiction and films like *Crocodile Dundee*, have generated a host of pop culture phenomena including foods (for example, "shrimp on the barbie"), clothing styles, and even well-known phrases like "G'day, mate!" Another global example of a vernacular region is the Riviera on the sunny northwest coast of the Mediterranean Sea (**Figure 1.9**). Although this region corresponds loosely to a particular climatic zone, it is *imagined* by most people as a region of affluence featuring yachts, casinos, and royalty.

Other vernacular culture regions find their basis in economic, political, or historical characteristics. Vernacular regions, like most regions, generally lack sharp borders, and the inhabitants of any given area may claim residence

FIGURE 1.7 The San Fernando Valley. Referred to locally as "the valley," the San Fernando Valley is located in the northwestern area of Los Angeles. Despite its reputation as a white, suburban area, portions of the valley are densely populated, and the area is home to a wide range of peoples from many different backgrounds. (Robert Landau.)

THINKING GEOGRAPHICALLY Can you think of other examples of regions that are locally known as "the valley"?

FIGURE 1.8 **Uluru, or Ayers Rock, in Australia's Northern Territory.** This monolith was first named Ayers Rock in 1873 in honor of Sir Henry Ayers, the (then) Chief Secretary of South Australia. In 1993, it was renamed Uluru/Ayers Rock, incorporating a word from a local Aboriginal language, under a new dual-language naming policy adopted in Australia. (Neale Cousland/Dreamstime.com.)

THINKING GEOGRAPHICALLY Can you think of other vernacular regions based on physical environmental characteristics that are similar to the Outback in terms of their reputation for being a "rugged frontier"?

in more than one such region. They vary in scale from city neighborhoods to sizable parts of continents.

At a basic level, the vernacular culture region grows out of people's sense of belonging and identification with a particular region. By contrast, many formal or functional culture regions lack this attribute and, as a result, are often far less meaningful for people. You're more likely to hear people say, "we're fighting to preserve 'the valley' from further urban development," than to see people rally under the banner of "wheat-growing areas of the world!" Self-conscious regional identity can have major political and social ramifications.

Vernacular culture regions often lack the organization necessary for functional regions, although they may be centered on a single urban node, and they frequently do not display the cultural homogeneity that characterizes formal regions.

FIGURE 1.9 **The principality of Monaco on the Mediterranean Riviera.** Monaco is perceived by most people as a playground for Europe's elite. Luxury yachts and opulent casinos dot the coastline, providing evidence of the "rich and famous" lifestyle defining this vernacular region. (Vladimir Melnik/Shutterstock.)

THINKING GEOGRAPHICALLY How do you think perceptions of the Riviera might differ between people who live or vacation there and people in the rest of the world?

Cultural **Diffusion**

Regardless of type, the culture regions of the world evolved through communication and contact among people. In other words, they are the product of **cultural diffusion,** the spatial spread of learned ideas, innovations, and attitudes throughout an area. The study of cultural diffusion is a very important theme in cultural geography. Through the study of diffusion, cultural geographers can begin to understand how spatial patterns in culture evolved.

> **cultural diffusion** The spread of elements of culture from the point of origin over an area.
>
> **independent invention** A cultural innovation that is developed in two or more locations by individuals or groups working independently.

After all, any culture is the product of almost countless innovations that spread from their points of origin to cover a wider area. Some innovations occur only once, and geographers can sometimes trace a cultural element back to a single place of origin. In other cases, **independent invention** occurs: the same or a very similar innovation is separately developed at different places by different people.

Types of Diffusion Geographers, drawing heavily on the research of Torsten Hägerstrand, recognize several different kinds of diffusion (**Figure 1.10**). **Relocation diffusion** occurs when individuals or groups with a particular idea or practice migrate from one location to another, thereby bringing the idea or practice to their new homeland. Religions frequently spread this way. An example is the migration of Christianity with European settlers who came to America. In **expansion diffusion,** ideas or practices spread throughout a population, from area to area, in a snowballing process, so that the total number of knowers or users and the areas of occurrence increase.

Expansion diffusion can be further divided into three subtypes. In **hierarchical diffusion,** ideas leapfrog from one important person to another or from one urban center to another, temporarily bypassing other persons or rural territories. We can see hierarchical diffusion at work in everyday life

> **relocation diffusion** The spread of an innovation or other element of culture that occurs with the bodily relocation (migration) of the individual or group responsible for the innovation.
>
> **expansion diffusion** The spread of innovations within an area in a snowballing process so that the total number of knowers or users becomes greater and the area of occurrence grows.

> **hierarchical diffusion** A type of expansion diffusion in which innovations spread from one important person to another or from one urban center to another, temporarily bypassing other persons or rural areas.

RELOCATION DIFFUSION

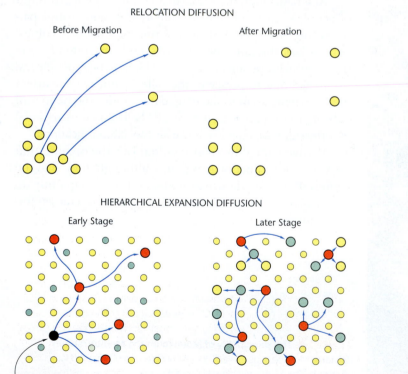

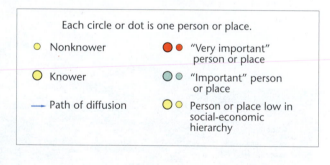

Each circle or dot is one person or place.

- ○ Nonknower
- ○ Knower
- → Path of diffusion
- ●● "Very important" person or place
- ●● "Important" person or place
- ○○ Person or place low in social-economic hierarchy

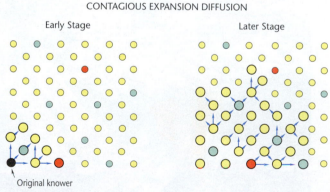

FIGURE 1.10 Types of cultural diffusion. These diagrams are merely suggestive; in reality, spatial diffusion is far more complex. In hierarchical diffusion, different scales can be used so that, for example, the category "very important person" could be replaced by "large city."

THINKING GEOGRAPHICALLY Why would the starting point for hierarchical diffusion frequently be a very important person and a large city at the same time?

by observing the acceptance of new modes of dress or foods. For example, sushi restaurants originally diffused from Japan in the 1970s very slowly because many people were reluctant to eat raw fish. In the United States, the first sushi restaurants appeared in the major cities Los Angeles and New York. Only gradually throughout the 1980s and 1990s did sushi eating become more common in the less urbanized parts of the country. By contrast, **contagious diffusion** is the wavelike spread of ideas in the manner of a contagious disease, moving through-

> **contagious diffusion** A type of expansion diffusion in which cultural innovation spreads by person-to-person contact, moving wavelike through an area and population without regard to social status.

out space without regard to hierarchies. Hierarchical and contagious diffusion often work together. The global spread of social networking provides an example of how these two types of diffusion can reinforce each other. As you can see in **Figure 1.11**, the use of Facebook had spread throughout much of the world by 2009. This diffusion took place first through a hierarchical process wherein the student inventors of Facebook granted access only to fellow students at Harvard University and to other universities in the Boston area. Access was then granted to other Ivy League universities that were hierarchically linked to Harvard. Soon, however, open access was granted and contagious diffusion processes took over. Interest in the site then spread through word of mouth to the rest of the United States and throughout the world.

Sometimes a specific trait is rejected but the underlying idea is accepted, resulting in **stimulus diffusion.**

> **stimulus diffusion** A type of expansion diffusion in which a specific trait fails to spread but the underlying idea or concept is accepted.

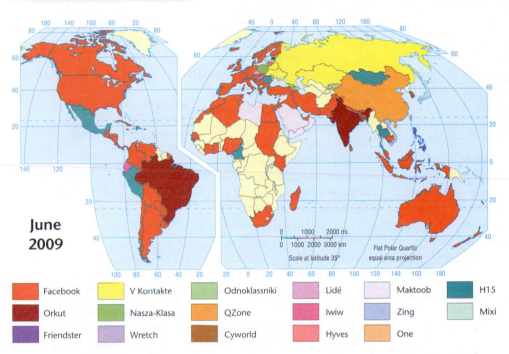

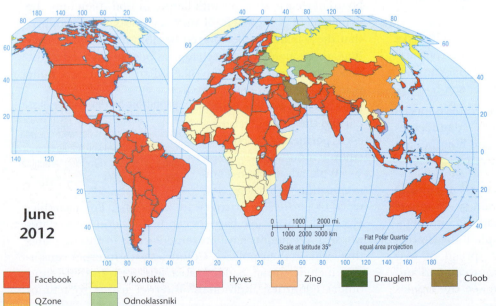

FIGURE 1.11 Dominant social networking site in countries with significant Internet penetration, June 2009 and 2011. By 2009, Facebook was the world's predominant social networking site. However, there were some populous countries whose Internet users still preferred other social networks. Between 2009 and 2011, Facebook use continued to diffuse, coming to dominate social networking in several countries that had previously been loyal to other social networking sites (including India, Mexico, Poland, Libya, and Portugal). (Source: Vincenzo Cosenza.)

THINKING GEOGRAPHICALLY Do you think that Facebook use will continue to diffuse and dominate globally in the future? Why or why not?

For example, early Siberian peoples domesticated reindeer only after exposure to the domesticated cattle raised by cultures to their south. The Siberians had no use for cattle, but the idea of domesticated herds appealed to them, and they began domesticating reindeer, an animal they had long hunted.

In recent decades, all types of cultural diffusion have been affected by advances in communication technology, especially the Internet. The Internet has dramatically changed the rate at which cultural information and ideas can be exchanged between people all over the world without in-person contact. For example, in an instant, a musician or band can share a video on YouTube; clothing and other trendy items can be purchased from far-off locations where new styles are emerging; and political ideas can be exchanged and debated on blogs read by people living under different forms of government. In addition to the Internet, an increase in cellular phone usage has expanded communication, resulting in the development of "text speak," wherein people converse in a new language of acronyms that have even begun to be published in texting dictionaries (LOL, laughing out loud; TTYL, talk to you later).

These new electronic exchanges have inspired geographers to consider how cultural diffusion occurs today as compared to in the past. In a new branch of geography called **cybergeography,** scholars study the structure of the Internet and how people interact on it. They examine how technology is changing cultural diffusion and map online spaces. Here are some examples of recent cybergeographic research:

cybergeography A branch of geography that studies the Internet as a virtual place. Cybergeographers examine locations in cyberspace as sites of human interaction with structures that can be mapped.

- Steve Jones has examined how music is increasingly reaching people via the Internet.

- Sarah Holloway and Gill Valentine have conducted a study of British children's use of the Internet and how these young people adopt "Americanized" trends and ideas as a result of this interaction.

- Eric Gordon in the United States has edited a special edition of the journal *Space and Culture* that focuses on virtual worlds, like Second Life, and how they aid cultural diffusion.

While cybergeographic research focuses on cultural diffusion since the advent of the Internet, geographers have long been concerned with the spread of cultural innovation and have devised theories to explain these processes. Geographers have described the spread of new ideas as being similar to throwing a rock into a pond and watching the spread of ripples. You can see the ripples become gradually weaker as they move away from the point of impact. In the same way, diffusion was thought to become weaker as a cultural innovation moves away from its point of origin. Therefore, diffusion was thought to decrease with distance, with innovations usually being accepted most thoroughly in areas closest to their origin. Time was also considered a factor, because in the past innovations often took more time to diffuse the farther they spread from the origin. This general decrease of acceptance of cultural innovation with both increasing distance and time produced what geographers call **time-distance decay.** However, the Internet and other contemporary forms of mass communication have greatly accelerated diffusion, diminishing the impact of time-distance decay.

time-distance decay The decrease in acceptance of a cultural innovation with increasing time and distance from its origin.

In addition to the gradual weakening or decay of an innovation through time and distance, barriers can retard its spread. **Absorbing barriers** completely halt diffusion, allowing no further progress. For example, in 1998, the fundamentalist Islamic Taliban government of Afghanistan decided to abolish television, videocassette recorders, and videotapes, viewing them as causes of corruption in society. As a result, the cultural diffusion of television sets was reversed, and the important role of television as a communication device to facilitate the spread of ideas was eliminated.

absorbing barrier A barrier that completely halts diffusion of innovations and blocks the spread of cultural elements.

Extreme examples aside, few absorbing barriers exist in the world. More commonly, barriers are permeable, allowing part of the innovation wave to diffuse through but acting to weaken or retard the continued spread. When a school board objects to students with tattoos or body piercings, the principal of a high school may set limits by mandating that these markings be covered by clothing. However, over time, those mandates may change as people get used to the idea of body markings. More likely than not, though, some mandates will remain in place. In this way, the principal and the school board act as a **permeable barrier** to cultural innovations.

permeable barrier A barrier that permits some aspects of an innovation to diffuse through it but weakens and retards continued spread. An innovation can be modified in passing through a permeable barrier.

With advances in communication technology, new types of permeable barriers are emerging. For example, electronic permeable barriers can be found in some states whose governments have prevented access to Internet sites that criticize their policies. Countries with Internet censorship include China, Saudi Arabia, and Iran. In these (and other) countries, individuals' Internet use is sometimes monitored to prevent communications that the government believes may lead to social unrest or organized protest movements. Censorship and monitoring can be considered permeable barriers that restrict the spread of certain ideas while allowing other types of information to diffuse.

REFLECTING ON **GEOGRAPHY**
The Internet has made the diffusion of many forms of cultural change much more rapid and has diminished the effect of time-distance decay. Can you think of examples of time-distance decay continuing to operate in the Internet age?

Three Stages of Innovation Acceptance Acceptance of innovations at any given point in space passes through three distinct stages. In the first stage, acceptance takes place at a steady yet slow rate, perhaps because the innovation has not yet caught on, the benefits have not been adequately demonstrated, or a product is not readily available. During the second stage, rapid growth in acceptance occurs, and the trait spreads widely, as with a fashion style or dance fad. Often diffusion on a microscale exhibits what is called the **neighborhood effect,** which means simply that acceptance is usually most rapid in small clusters around an initial adopter. Due to increasing Internet use, the neighborhood effect is now also being observed in cyberspace. Online neighborhoods function, in many ways, like those in physical space. In other words, people who participate in certain online groups or sites may also demonstrate a neighborhood effect wherein new ideas and cultural innovations are accepted first by a person administering a site, writing a blog, or acting in some other prominent role. From that point, the new ideas or practices are accepted by other people in these online communities. During the third stage of innovation, acceptance, growth shows a slower rate than in the second, perhaps because the fad is passing or because an area (physical or virtual) is already saturated with the innovation.

Although all places and communities hypothetically have equal potential to adopt a new idea or practice, diffusion typically produces a core-periphery spatial arrangement, the same pattern observed earlier in our discussion of regions (see Figure 1.4). Hägerstrand offered an explanation of how diffusion produces such a regional configuration. The distribution of innovations can be random, but the overlap of new ideas and traits as they diffuse through space and time is greatest toward the center of the region and least at the peripheries. As a result of this

neighborhood effect Microscale diffusion in which acceptance of an innovation is most rapid in small clusters around an initial adopter.

overlap, more innovations are adopted in the center, or core, of the region.

In addition to the almost instant communication of ideas through the Internet and other digital media, we see many other examples of diffusion in today's world. These include the rapid movement of goods from the place of production to the place of consumption and the seemingly nonstop movement of money around the globe through digital financial networks. These types of movements through space do not necessarily follow the pattern of core-periphery but instead create new and different types of patterns that are not yet fully studied or defined. Other types of diffusion, such as large-scale movements of people between different regions, can be best thought of as **migration** from one region or country to another through particular routes. In today's globalizing world, with better and faster communication and transportation technologies, many migrants more easily maintain ties to their homelands even after they have migrated, and some may move back and forth between their home countries and those to which they have migrated. These movements and maintenance of ties lead to a phenomenon known as **transnationalism.**

migration The large-scale movement of people between different regions of the world.

transnationalism A phenomenon in which immigrants maintain social and/or economic ties to their place of origin. Often, these relationships include visits "home" and the circular exchange of money and goods.

A Special Case of Diffusion: Globalization How and why are different cultures, economies, and societies linked around the world? Given all these new linkages, why are there so many differences between different groups of people in the world? The modern technological age, in which improved worldwide transportation and communications allow the instantaneous diffusion of ideas and innovations, has accelerated the phenomenon called **globalization.** This term refers to a world increasingly linked, in which international borders are diminished in importance and a worldwide marketplace is created. This interconnected world has been created from a set of factors: faster, cheaper, and more reliable transportation; the almost instantaneous communication that the Internet, cellular phones, and other media have allowed; and the creation of digital sources of information and media.

globalization The binding together of all the lands and peoples of the world into an integrated system driven by capitalistic free markets, in which cultural diffusion is rapid, independent states are weakened, and cultural homogenization is encouraged.

Thus, globalization in this sense is a rather recent phenomenon, dating from the late twentieth century. Yet we know that long before modern times, different countries and different parts of the world were linked. For example, in early medieval times overland trade routes connected China with other parts of Asia; the British East India Company maintained maritime trading routes between England and large portions of South Asia as early as the seventeenth century; and religious and political wars in Europe and the Middle East brought different peoples into direct contact with one another. Some geographers refer to such moments as *early global encounters* and suggest they set the background for contemporary globalization.

The increasingly linked economic, political, and cultural networks around the world might lead many to believe that different groups of people around the globe are becoming more and more alike. In some ways this is true, but actually these new global encounters have enabled an increasing recognition of the differences between groups of people, and some of those differences have been caused by globalization itself. Some groups of people have access to advanced technologies, more thorough health care, and education, whereas others do not. If we mapped certain indicators of human well-being on a global scale, such as life expectancy, literacy, and standard of living, we would find quite an uneven distribution. **Figure 1.12** shows us that different cultures around the world have different access to these types of resources.

HUMAN DEVELOPMENT INDEX

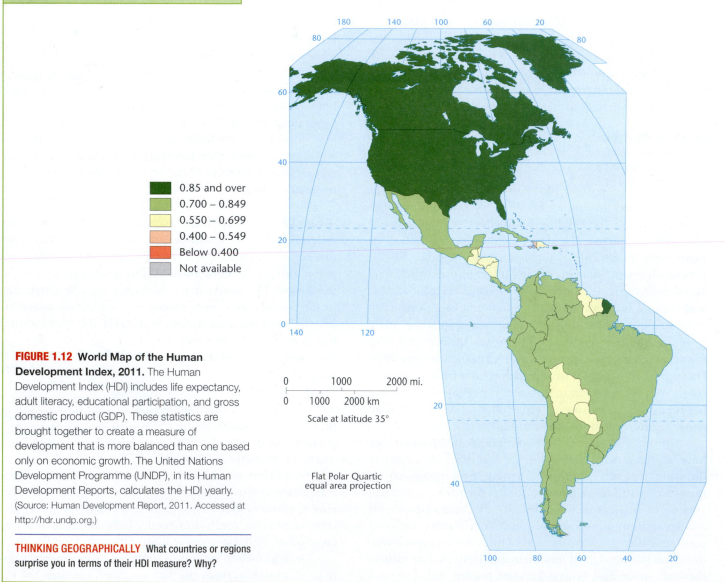

0.85 and over
0.700 − 0.849
0.550 − 0.699
0.400 − 0.549
Below 0.400
Not available

0   1000   2000 mi.

0   1000   2000 km

Scale at latitude 35°

Flat Polar Quartic
equal area projection

FIGURE 1.12 **World Map of the Human Development Index, 2011.** The Human Development Index (HDI) includes life expectancy, adult literacy, educational participation, and gross domestic product (GDP). These statistics are brought together to create a measure of development that is more balanced than one based only on economic growth. The United Nations Development Programme (UNDP), in its Human Development Reports, calculates the HDI yearly. (Source: Human Development Report, 2011. Accessed at http://hdr.undp.org.)

THINKING GEOGRAPHICALLY What countries or regions surprise you in terms of their HDI measure? Why?

uneven development The tendency for industry to develop in a core-periphery pattern, enriching the industrialized countries of the core and impoverishing the less-industrialized periphery. The term is also used to describe urban patterns in which suburban areas are enriched while the inner city is impoverished.

These differences are what geographers mean when they refer to *stages of development*. In Figure 1.12, you can see that some regions of the world have a fairly high Human Development Index (HDI) and some have a relatively low HDI. Scholars often refer to these two types of regions as *developed* (relatively high HDI) and *developing* (relatively low HDI). This inequitable distribution of resources is referred to as **uneven develop-**

ment. We will discuss development in much more detail in Chapters 3 and 9.

Culture, of course, is a key variable in global interactions and interconnections. In fact, as we have suggested, globalization is occurring through cultural media such as film, television, and the Internet (and particularly through social networking sites like Facebook and Twitter). These media provide channels that allow people to exchange information regarding trends in pop culture, product information, and changing lifestyles. In the case of the Internet, this exchange is virtually instantaneous. As a result, people's choices to follow trends or to adopt technology and

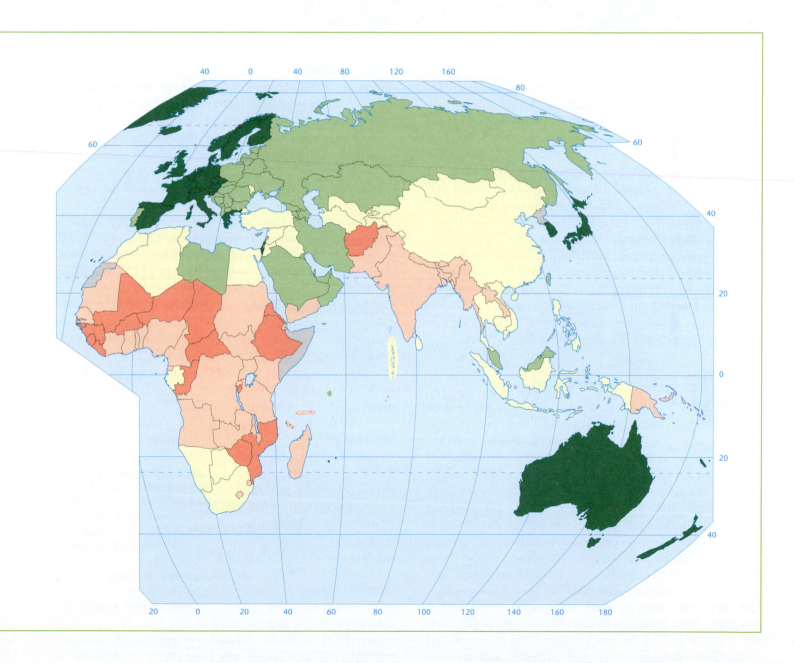

FIGURE 1.13 **A globalized urban street scene in Ghana, West Africa.** A woman uses a traditional mode of transporting goods, while others use gasoline-powered vehicles. (Garoline Penn/Imapact/HIP/The Image Works Ghana.)

THINKING GEOGRAPHICALLY Do you think that an urban street scene in the United States or Europe would display a similar number of influences from Africa? Why or why not?

products from other cultures are more easily and quickly influenced by global changes than ever before. **Figure 1.13** illustrates a diverse blend of global cultural influences on the once traditional cultural landscape of Ghana in West Africa. If we consider culture as a way of life, then globalization is a key shaper of culture and in turn is shaped by it. Some geographers have suggested that globalizing processes and an increase in personal mobility will work to homogenize different peoples, breaking down culture regions and eventually producing a single global culture. Other geographers see a different picture in which new forms of media and communication allow local cultures to maintain their distinct identities, reinforcing the diversity of cultures around the world. Throughout *Fundamentals of The Human Mosaic*, we will return to these issues, considering the complex role of culture and cultures in an increasingly global world.

Cultural **Ecology**

Cultural geographers look on people and nature as interacting. Cultures do not exist in an environmental vacuum; each human group occupies a piece of the physical earth and develops in a specific natural habitat. The cultural geographer must study the interaction between culture and environment in order to understand spatial variations in culture. This is the study of *cultural ecology*.

The term *ecology* was coined in the nineteenth century to refer to a new biological science concerned with studying the complex relationships among living organisms and their physical environments. Two influential geographers who applied the study of these relationships to the field of geography were Friedrich Ratzel (1844–1904) of Germany and Paul Vidal de La Blache (1845–1918) of France. Ratzel trained in the natural sciences and applied ideas from those disciplines to his geographic theories. In his most notable works, *Anthropogeographie* (1882) and *Politische Geographie* (1897), Ratzel advanced the idea that states are like living organisms that expand over the earth with the objective of acquiring "living space" or *lebensraum*. Ratzel went on to theorize that because states are organic and growing, any state's current borders are only temporary, and the expansion of these borders reflects the health of that state.

At the time of Ratzel's writings in Germany, Vidal was already a major force in French geographic thought. Vidal believed that Ratzel's view of states as living organisms was overstated and believed, instead, that they only "resemble living things." He also warned that Ratzel's ideas concerning the expansion of states could lead to dangerous political pressures being placed on the field of geography from outside the discipline. Therefore, he recommended that geographers not adopt a preexisting model of states and territory (as suggested by Ratzel) but rather approach the study of human interactions with the earth with an open mind to include other types of political entities.

Because of the writings of Ratzel and Vidal (as well as other geographers of the time), the idea that geography should concern itself with the study of relationships between humans and their physical environment continued to grow in popularity into the early twentieth century. As a result, by the mid-twentieth century, geographers began to join the term *ecology* with the term *culture* in order to delineate a new field of study, **cultural ecology.** Later, another concept, the **ecosystem,** was introduced to describe a

> **cultural ecology** Broadly defined, the study of the relationships between the physical environment and culture; narrowly (and more commonly) defined, the study of culture as an adaptive system that facilitates human adaptation to nature and environmental change.
>
> **ecosystem** A territorially bounded system consisting of interaction between organic and inorganic components.

territorially bounded system consisting of interacting organic and inorganic components. Plant and animal species were said to be adapted to specific conditions in the ecosystem and functioned to help keep the system stable over time. Thus cultural ecology today is the study of (1) environmental influences on culture and (2) the impact of people, acting through their culture, on an ecosystem.

Cultural ecology implies a two-way street, with people and the environment/ecosystem exerting influence on one another. Put differently, cultural ecology is based on the premise that culture is the human method of meeting physical environmental challenges—that culture is an adaptive system. The term **cultural adaptation** is used in this context. This outlook borrows heavily from the biological sciences, with the assumption that plant and animal adaptations are relevant to the study of humans. Culture serves to facilitate long-term successful nongenetic human adaptation to nature and environmental change. **Adaptive strategy** involves those aspects of culture that serve to provide the necessities of life—food, clothing, shelter, and defense. No two cultures employ the same strategy, even within the same physical environment. Such strategies involve culturally transmitted, or learned, behavior that permits a population to survive in its natural environment.

> **cultural adaptation** The adaptation of humans and cultures to the challenges posed by the physical environment.
>
> **adaptive strategy** The unique way in which each culture uses its particular physical environment; those aspects of culture that serve to provide the necessities of life—food, clothing, shelter, and defense.

Through the years, human geographers have developed various perspectives on the interaction between humans and the land. Four schools of thought have developed: environmental determinism, possibilism, environmental perception, and humans as modifiers of the Earth.

Environmental Determinism

During the late nineteenth and early twentieth centuries, many geographers accepted **environmental determinism,** the belief that the physical environment is the dominant force shaping cultures and that humankind is essentially a passive product of its physical surroundings. Humans are clay to be molded by nature. Similar physical environments produce similar cultures.

> **environmental determinism** The belief that cultures are directly or indirectly shaped by the physical environment.

For example, environmental determinists believed that peoples of the mountains were predestined by the rugged terrain to be simple, backward, conservative, unimaginative, and freedom loving. Dwellers in the desert were likely to believe in one God but to live under the rule of tyrants. Temperate climates produced inventiveness, industriousness, and democracy. Coastlands pitted with fjords produced great navigators and fishers. Environmental determinism had serious consequences, particularly during the time of European colonization in the late nineteenth century. For example, many Europeans saw Latin American native inhabitants as lazy, childlike, and prone to vices such as alcoholism because of the tropical climates that cover much of this region. Living in a tropical climate supposedly ensured that people didn't have to work very hard for their food. Europeans were able to rationalize their colonization of large portions of the world in part along climatic lines. Because the natives were "naturally" lazy and slow, the European reasoning went, the natives would benefit from the presence of the "naturally" stronger, smarter, and more industrious Europeans who came from more temperate lands.

Determinists overemphasize the role of environment in human affairs. This does not mean that environmental influences are inconsequential or that geographers should not study such influences. Rather, the physical environment is only one of many forces affecting human culture and is never the sole determinant of behavior and beliefs.

Possibilism

Since the 1920s, **possibilism** has been the favored view among geographers. Possibilists claim that any physical environment offers a number of possible ways for a culture to develop. In this way, the local environment helps shape its resident culture. However, a culture's way of life ultimately depends on the choices people make among the possibilities that are offered by the environment. These choices are guided by cultural heritage and are shaped by a particular political and economic system. Possibilists see the physical environment as offering opportunities and limitations; people make choices among these to satisfy their needs. **Figure 1.14** on page 18 provides an interesting example: the cities of San Francisco and Chongqing both were built on similar physical terrains that dictated an overall form, but differing cultures led to very different street patterns, architecture, and land use. In short, local traits of culture and economy are the products of culturally based decisions made within the limits of possibilities offered by the environment.

> **possibilism** A school of thought based on the belief that humans, rather than the physical environment, are the primary active force; that any environment offers a number of possible ways for a culture to develop; and that the choices among these possibilities are guided by cultural heritage.

Most possibilists think that the higher the technological level of a culture, the greater the number of possibilities and the weaker the influences of the physical environment. Technologically advanced cultures, in this view, have achieved some mastery over their physical surroundings. Geographers Jim Norwine and Thomas Anderson, however, warn that even in these advanced societies "the quantity and quality of human life are still strongly influenced by the natural environment," especially climate. They argue that humankind's control of nature is anything

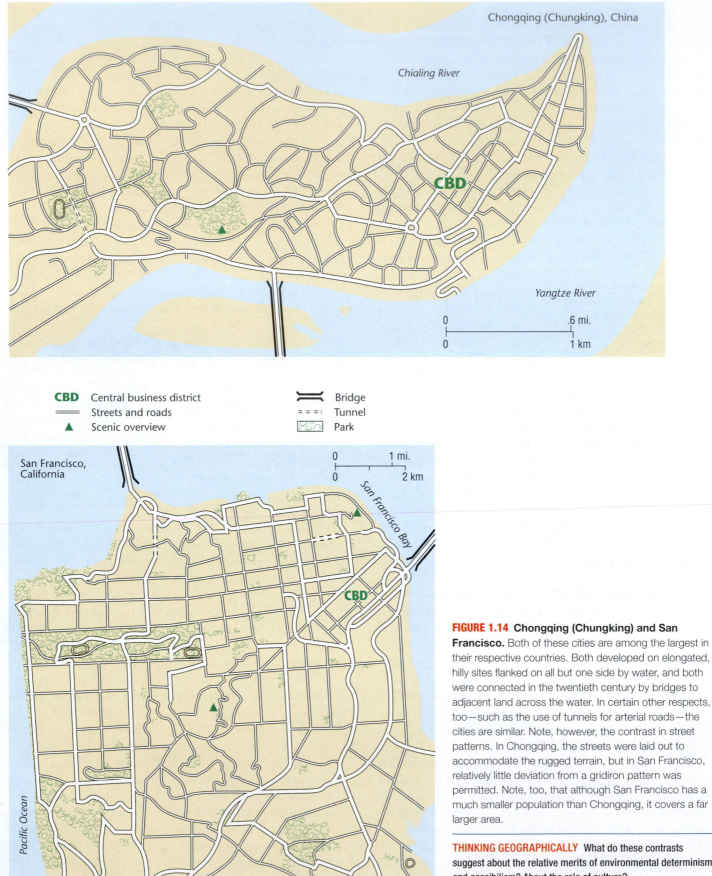

CBD Central business district   Bridge
    Streets and roads   Tunnel
▲ Scenic overview   Park

FIGURE 1.14 Chongqing (Chungking) and San Francisco. Both of these cities are among the largest in their respective countries. Both developed on elongated, hilly sites flanked on all but one side by water, and both were connected in the twentieth century by bridges to adjacent land across the water. In certain other respects, too—such as the use of tunnels for arterial roads—the cities are similar. Note, however, the contrast in street patterns. In Chongqing, the streets were laid out to accommodate the rugged terrain, but in San Francisco, relatively little deviation from a gridiron pattern was permitted. Note, too, that although San Francisco has a much smaller population than Chongqing, it covers a far larger area.

THINKING GEOGRAPHICALLY What do these contrasts suggest about the relative merits of environmental determinism and possibilism? About the role of culture?

but supreme and perhaps even illusory. One only has to think of the devastation caused by the 2010 earthquake in Haiti and the 2011 Japanese earthquake and tsunami to underscore the often-illusory character of the control humans think they have over their physical surroundings.

Environmental Perception Another approach to cultural ecology focuses on how humans perceive nature. Each person and cultural group has mental images of the physical environment, shaped by knowledge, ignorance, experience, values, and emotions. To describe such mental images, cultural geographers use the term **environmental perception.** Whereas the possibilist sees humankind as having a choice of possibilities in a given physical setting, the environmental perceptionist declares that the choices people make will depend more on what they perceive the environment to be than on the actual character of the land. Perception, in turn, is colored by the teachings of culture.

> **environmental perception** The belief that culture depends more on what people perceive the environment to be than on the actual character of the environment; perception, in turn, is colored by the teachings of culture.

Some of the most productive research done by environmental perceptionists has been on the topic of **natural hazards,** such as flooding, hurricanes, volcanic eruptions, earthquakes, insect infestations, and droughts. All cultures react to such hazards and catastrophes, but the reactions vary greatly from one group to another. Some people reason that natural disasters and risks are unavoidable acts of the gods, perhaps even divine retribution. Often they seek to cope with the hazards by placating their gods. Others hold government responsible for taking care of them when hazards yield disasters. Many cultural groups living in industrialized states view natural hazards as problems that can be solved by technological means. A common manifestation of this belief has been the

> **natural hazard** Inherent danger present in a given habitat, such as flooding, hurricanes, volcanic eruptions, or earthquakes; often perceived differently by different peoples.

widespread construction of dams, levees, and sea walls to prevent flooding. However, history shows that these measures do not always ensure protection. For example, during Hurricane Katrina in 2005, numerous dams and levees in the New Orleans area failed and more than 1800 people lost their lives. Tens of thousands of people were left homeless, and the city has still not fully recovered.

In 2011, another natural disaster demonstrated that human attempts to provide technological safeguards against environmental forces do not always succeed. A magnitude 9.0 earthquake hit Japan, causing a devastating tsunami with up to 49-foot waves. The tsunami crushed sea walls and other built defenses, flooding thousands of roadways and destroying countless homes, schools, and commercial buildings. **Figure 1.15** shows an example of this devastation in Otsuchi, a small coastal town. Immediately following the tsunami, 12,000 of the town's 15,000 inhabitants were either dead or reported missing. Despite early warning systems in place, these (and other) Japanese residents had only 8 to 10 minutes to attempt escape before the wave hit. Region-wide, over 20,000 people died or were reported missing and presumed dead after the disaster. The earthquake and tsunami also did unprecedented damage to infrastructure, including airports, rail lines, and shipping ports. At the Fukushima nuclear facility, two reactors exploded and radiation leaked out, necessitating the evacuation of thousands of residents living near the facility. The massive loss of life, staggering damage to the built environment, and serious nuclear accident resulting from the Japanese earthquake and tsunami caused many other states to question their technological capabilities in the face of natural disasters.

FIGURE 1.15 The remains of Otsuchi after the 2011 Japanese earthquake and tsunami. The total destruction of most of the homes, businesses, and roads in this small town left survivors without shelter, food, water, heat, electricity, or medicine. Roads into the area became impassable, complicating rescue and relief efforts. (Athit Perawongmetha/Getty Images.)

THINKING GEOGRAPHICALLY How do you think the experiences of people surviving the 2011 Japanese disaster will affect their perceptions of the physical environment?

In virtually all cultures, people knowingly inhabit hazard zones, especially floodplains, exposed coastal sites, drought-prone regions, and the environs of active volcanoes. More Americans than ever now live in areas likely to be devastated by hurricanes along the coast of the Gulf of Mexico and atop earthquake faults in California. How accurately do they perceive the hazard involved? Why have they chosen to live there? How might we minimize the eventual disasters? The cultural geographer seeks the answers to such questions and aspires, with other geographers, to mitigate the inevitable disasters through such devices as land-use planning.

Perhaps the most fundamental expression of environmental perception lies in the way different cultures see nature itself. Nature is a culturally derived concept that has different meanings to different peoples. In the **organic view,** held by many traditional groups, people are part of nature. The habitat possesses a soul, is filled with nature-spirits, and must not be offended. By contrast, most Western peoples believe in the **mechanistic view** of nature. Humans are separate from and hold dominion over nature. They see the habitat as an integrated system of mechanisms governed by external forces that can be rendered into natural laws and understood by the human mind.

> **organic view of nature** The view that humans are part of, not separate from, nature and that the habitat possesses a soul and is filled with nature-spirits.

> **mechanistic view of nature** The view that humans are separate from nature and hold dominion over it and that the habitat is an integrated mechanism governed by external forces that the human mind can understand and manipulate.

Humans as Modifiers of the Earth Many cultural geographers, observing the environmental changes people have wrought, emphasize humans as modifiers of the habitat. This presents yet another facet of cultural ecology. In a sense, the humans-as-modifier school of thought is the opposite of environmental determinism. Whereas the determinists proclaim that nature molds humankind and possibilists believe that nature presents possibilities for people, geographers who study the human impact on the land assert that humans mold nature (**Figure 1.16**).

In addition to the deliberate modifications of the Earth through such activities as mining, logging, and irrigation, we now know that even seemingly innocuous behavior, repeated for millennia, for centuries, or in some cases for mere decades, can have catastrophic effects on the environment. Plowing fields and grazing livestock can eventually denude regions. The increasing release of fossil-fuel emissions from vehicles and factories—what are known as *greenhouse gases*—is arguably leading to global warming, with potentially devastating effects on Earth's environment. Clearly, access to energy and technology is the key variable that controls the magnitude and speed of

FIGURE 1.16 Human modification of Earth includes severe soil erosion. This asphalt road in Koh Chang, Thailand, has collapsed due to soil erosion. (Oleg D./Shutterstock.)

THINKING GEOGRAPHICALLY How can we adopt less destructive ways of modifying the land?

environmental alteration. Geographers seek to understand and explain the processes of environmental alteration as they vary from one culture to another and, through applied geography, to propose alternative, less destructive modes of behavior.

Cultural geographers began to concentrate on the human role in changing the face of Earth long before the present level of ecological consciousness developed. They learned early on that different cultural groups have widely different outlooks on humankind's role in changing Earth. Some, such as those rooted in the mechanistic tradition, tend to regard environmental modification as divinely approved, viewing humans as God's helpers in completing the task of creation. Other groups, organic in their view of nature, are much more cautious, taking care not to offend the forces of nature. They see humans as part of nature, meant to be in harmony with their environment.

Cultural **Interaction**

The relationship between people and the land, the theme of cultural ecology, lies at the heart of traditional geography. However, the explanation of human spatial variations also requires consideration of a whole range of cultural and economic factors. The cultural geographer recognizes that all facets of culture are systematically and spatially intertwined or integrated. In short, cultures are complex wholes rather than series of unrelated traits. They form integrated systems in which the parts fit together causally. All aspects of culture are functionally interdependent on one another. The theme of **cultural interaction** addresses this complexity. Cultural interaction recognizes that (1) the immediate causes of some cultural phenomena are other cultural phenomena and (2) a change

> **cultural interaction** The relationship of various elements within a culture.

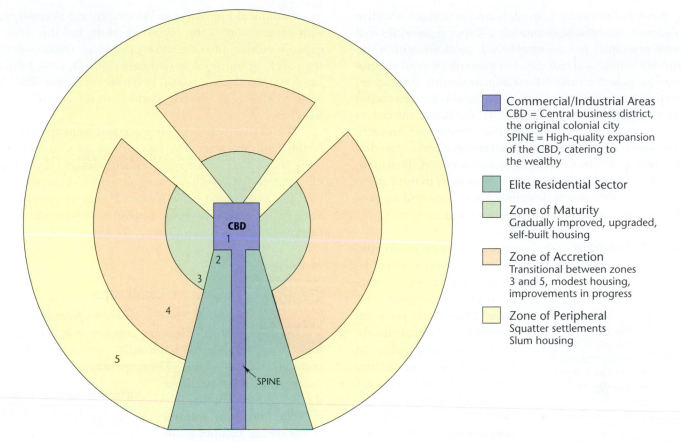

Commercial/Industrial Areas
CBD = Central business district,
the original colonial city
SPINE = High-quality expansion
of the CBD, catering to
the wealthy

Elite Residential Sector

Zone of Maturity
Gradually improved, upgraded,
self-built housing

Zone of Accretion
Transitional between zones
3 and 5, modest housing,
improvements in progress

Zone of Peripheral
Squatter settlements
Slum housing

FIGURE 1.17 A generalized model of the Latin American city.
Urban structure differs from one culture to another, and in many ways
the cities of Latin America are distinctive, having much in common
with one another. Geographers Ernst Griffin and Larry Ford developed
the model diagrammed here to help describe and explain the
processes at work shaping the cities of Latin America. (After Griffin and
Ford, 1980: 406.)

THINKING GEOGRAPHICALLY In what ways would this model not be applicable
to cities in the United States and Canada?

in one element of culture requires an accommodating
change in others.

For example, religious belief has the potential to influence a group's voting behavior, diet, shopping patterns,
type of employment, and social standing. Traditional Hinduism, the majority religion of India, segregates people
into social classes called *castes* and specifies what forms of
livelihood are appropriate for each. The Mormon faith forbids the consumption of alcoholic beverages, tobacco, and
caffeine, thereby influencing both the diet and shopping
patterns of its members.

Geographers have employed two fundamentally different approaches to studying cultural interaction: the scientific and the humanistic.

Social Science Those who view cultural geography as a social
science, also called "modernists," believe that we should apply
the scientific method to the study of people. Emulating physicists and chemists, they devise theories and seek regularities
or universal spatial principles that cut across cultural lines to
govern all of humankind. These principles ideally become

the basis for laws of human
spatial behavior. **Space** is the
word that perhaps best connotes this scientific approach
to cultural geography.

Social scientists face several difficulties. First, they
tend to overemphasize economic factors at the expense of culture. Second, they have no
laboratories in which to test their theories. They cannot run
controlled experiments in which certain forces can be neutralized so that others can be studied. Their response to this deficiency is a technique known
as **model** building. Aware that
in the real world many causal
forces are involved, they set
up artificial model situations
in order to focus on one or
several potential factors. **Figure 1.17** is an example of a model.
Throughout *Fundamentals of The Human Mosaic*, we will see
many of the classic models built by cultural geographers.

space A term used to connote the
objective, quantitative, theoretical,
model-based, economics-oriented type
of geography that seeks to understand
spatial systems and networks through
application of the principles of social
science.

model An abstraction, an imaginary
situation, proposed by geographers to
simulate laboratory conditions so that
they may isolate certain causal forces
for detailed study.

Some culture-specific models are particularly sensitive to cultural variables. Such models still seek regularities and spatial principles but do so relatively modestly within the bounds of individual cultures. For example, several geographers proposed a model for Latin American cities in an effort to stress similarities among them and to understand underlying causal forces (Figure 1.17). No actual city in Latin America conforms precisely to this model. Instead, the model is deliberately simplified and generalized so that Latin American cities can be better understood. This model will look strange to a person living in a city in the United States or Canada, for it describes a very different kind of urban environment based in another culture.

Humanistic Geography If the social scientist seeks to level the differences among people in the quest for universal explanations, the humanistic geographer celebrates the unique-

> **place** A term used to connote the subjective, humanistic, culturally oriented type of geography that seeks to understand the unique character of individual regions and places, rejecting the principles of science as flawed and unknowingly biased.
>
> **topophilia** Love of place; used to describe people who exhibit a strong sense of place.

ness of each region and place. **Place** is the key word connoting a humanistic approach to geography. In fact, the humanistic geographer Yi-Fu Tuan even coined the word **topophilia,** literally "love of place," to describe people who exhibit a strong sense of place and the geographers who are attracted to the study of such places and peoples. Geographer Edward Relph tells us that "to be human is to have and know your place" in the geographical sense. To the humanist, cultural geography is an art rather than a science.

Traditional humanists in geography are as concerned with explanation as the social scientists, but they seek to explain unique phenomena—place and region—rather than seeking universal spatial laws. Indeed, most humanists doubt that laws of spatial behavior even exist. They believe in a far more chaotic world than the scientist could tolerate.

The debate between scientists and humanists is both healthy and necessary. The two groups ask different questions about place and space; not surprisingly, they obtain different answers. Both lines of inquiry yield valuable findings. In *Fundamentals of The Human Mosaic,* we present both sides of the scientist versus humanist debate within the theme of cultural interaction.

Cultural **Landscape**

What are the visible expressions of culture? How are peoples' interactions with nature materially expressed? What do regions actually look like? These questions provide the basis of our fifth and final theme, **cultural landscape.**

> **cultural landscape** The artificial landscape; the visible human imprint on the land.

The human or cultural landscape is composed of all the built forms that cultural groups create in inhabiting Earth—roads, agricultural fields, cities, houses, parks, gardens, commercial buildings, and so on. Every inhabited area has a cultural landscape, fashioned from the natural landscape, and each uniquely reflects the culture or cultures that created it (**Figure 1.18**). Landscape mirrors a

FIGURE 1.18 Terraced cultural landscape of an irrigated rice district in Yunnan Province, China. In such areas, the artificial landscape made by people overwhelms nature and forms a human mosaic on the land. (Yann Layma/The Image Bank/ Getty Images.)

THINKING GEOGRAPHICALLY Why is rice cultivated in such hilly areas in Asia, while in the United States rice farming is confined to flat plains?

culture's needs, values, and attitudes toward Earth, and the cultural geographer can learn much about a group of people by carefully observing and studying the landscape. Indeed, so important is this visual record of cultures that some geographers regard landscape study as geography's central interest.

Why is such importance attached to the human landscape? Perhaps part of the answer is that it visually reflects the most basic strivings of humankind: shelter, food, and clothing. In addition, the cultural landscape reveals people's different attitudes toward the modifications of Earth. It also contains valuable evidence about the origin, spread, and development of cultures because it usually preserves various types of archaic forms. Dominant and alternative cultures use, alter, and manipulate landscapes to express their diverse identities.

Aside from containing archaic forms, landscapes also convey revealing messages about the present-day inhabitants and cultures. According to geographer Pierce Lewis, "The cultural landscape is our collective and revealing autobiography, reflecting our tastes, values, aspirations, and fears in tangible forms." Cultural landscapes offer "texts" that geographers read to discover dominant ideas and prevailing practices within a culture, as well as less dominant and alternative forms within that culture. This reading, however, is often a very difficult task, given the complexity of cultures, cultural change, and recent globalizing trends that can obscure local histories.

Lewis attempted to simplify the reading of cultural landscapes by devising a set of five axioms (or rules) with a number of associated corollaries (something that naturally follows a stated rule). One such axiom in Lewis's work is the axiom of landscape as a clue to culture. This axiom states that cultural landscapes provide strong evidence of the people who live there and their history. In the regional corollary to this rule, Lewis states the following: If one region *looks* substantially different from another region, it is likely that the cultures of the two places are also different. **Figure 1.19a** shows French street names on signs in Montréal, Québec, reflecting the region's early settlement by the French as well as its contemporary use of the French language. **Figure 1.19b**, in contrast, reflects a settlement history dominated by the British and the use of the English language in Canada's British Columbia region. These clues to reading the cultural landscapes in these two regions of Canada lead us to investigate other cultural differences that may exist between them.

REFLECTING ON GEOGRAPHY

As we will learn throughout this book, landscapes are often created from more than one set of cultural values and beliefs, which are often in conflict with one another. How then can we read conflict into the landscape, when it appears so natural and unified? What sort of information would we need?

(a)

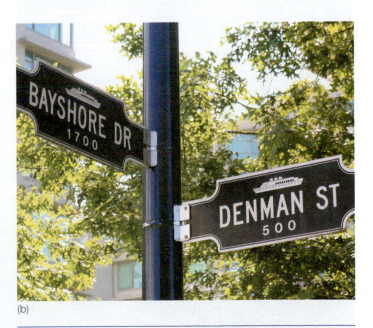

(b)

FIGURE 1.19 Canadian street signs in (a) Montréal, Québec and (b) Vancouver, British Columbia. The street names in these two cities reflect different cultural histories and influences. Montréal is home to many French-speaking people and incorporates many aspects of French culture. Vancouver is a cosmopolitan city with English as its official language. (a: Athit Perawongmetha/Getty Images; b: Hisham Ibrahim/Alamy.)

THINKING GEOGRAPHICALLY What other examples of regional differences in the cultural landscape may provide clues to differences in cultural influence on those regions?

Since the time of Lewis's groundbreaking work in the 1970s, geographers have pushed the idea of reading the landscape further in order to focus on the symbolic and ideological qualities of landscape. In fact, as geographer Denis Cosgrove has suggested, the very idea of landscape itself was ideological, in that its development in the Renaissance served the interests of the new elite class for whom agricultural land was valued not for its productivity but for its use as a visual subject. Land, in other words, was important to look at as a scene, and the actual workings that were necessary for agriculture were thus hidden from these views. If you go to an art museum, for example, it will be difficult, if not impossible, to find in the Italian Renaissance room any landscape paintings that depict agricultural laborers.

Closer to home, we need only look outside our windows to see other symbolic and ideological landscapes.

symbolic landscape Landscape that express the values, beliefs, and meanings of a particular culture.

One of the most familiar and obvious **symbolic landscapes** is the urban skyline. Composed of tall buildings,

many of which house financial service industries, it represents the power and dominance of finance and economics within that culture (**Figure 1.20**). However, other cities are dominated by tall structures that have little to do with economics but more with religion. In medieval Europe, for example, cathedrals and churches rose high above other buildings, symbolizing the centrality and dominance of Roman Catholicism in this culture.

Even the most mundane landscape element can be interpreted as symbolic. The typical middle-class suburban American or Canadian home, for example, can be interpreted as an expression of a dominant set of ideas about culture and family structure (**Figure 1.21**). These homes are often composed of a living room, dining room, kitchen, and bedrooms, all separated by walls. The cultural assumptions built into this division of space include the assumed value of individual privacy (everyone has his or her own bedroom); the idea that certain functions should be spatially separate from others (cooking, eating, socializing, sleeping); and the notion that a family is composed of a mother, father, and children (indicated by the "master"

FIGURE 1.20 Yokohama at dusk. This skyline is a powerful symbol of the economic importance of the world's largest city, Tokyo–Yokohama. (Jose Fuste Raga/CORBIS.)

THINKING GEOGRAPHICALLY What landscape form best symbolizes your town or city?

FIGURE 1.21 **American ranch house.** The horizontally expansive ranch is a common house form in the United States and Canada. (iStockphoto/Thinkstock.)

THINKING GEOGRAPHICALLY Can you think of other countries where ranch houses are common?

bedroom and smaller children's bedrooms). Thus, even the most common of landscapes can be seen as symbolic of a particular culture.

Aspects of the Cultural Landscape As we have seen, the physical content of the cultural landscape is both varied and complex. To better study and understand these complexities, geographical studies focus on three principal aspects of landscape: settlement forms, land-division patterns, and architectural styles.

In the study of **settlement forms,** geographers describe and explain the spatial arrangement of buildings, roads, and other features that people construct while inhabiting an area. One of the basic ways in which geographers categorize settlement forms is to examine their degree of **nucleation,** or the relative density of landscape elements. Urban centers are of course very nucleated, whereas rural farming areas tend to be

> **settlement form** The spatial arrangement of buildings, roads, towns, and other features that people construct while inhabiting an area.
>
> **nucleation** A settlement form characterized by density.
>
> **dispersed** A settlement form in which people live relatively distant from one another.

much less nucleated, what geographers call **dispersed.** Another common way to think about settlement forms is the degree to which they appear standardized and planned, such as the grid form of much of the American West (**Figure 1.22**), versus the degree to which the forms appear to be organic (that is, to have been built without any apparent geometric plan), such as the central areas of most European cities. Thinking about settlement forms in terms of these two basic categories helps geographers begin their analysis of the relationships between cultures and the landscapes they produce.

Land-division patterns indicate the uses of particular parcels of land and reveal the way people have divided the land for economic, social, and political uses. Within a particular nucleated settlement form—a city, for example—you can see different patterns of land use. Some areas are devoted to economic uses, others to residential, political (city hall, for example), social, and cultural uses. Each of these areas can be further divided. Economic uses can include offices for financial services, retail stores, warehouses, and factories. Residential areas are often divided into middle-class, upper-class, and working-class districts and/or are grouped by ethnicity and race (see Chapter 11). Such patterns, of course, vary a great deal from place to place and culture to

> **land-division pattern** The spatial pattern of various types of land use.

FIGURE 1.22 **The town of Westmoreland in the Imperial Valley of California.** It's difficult to find a more regularized, geometric land pattern than this. (Jim Wark/Airphoto.)

THINKING GEOGRAPHICALLY Why do you think geometric patterns dominate the western United States?

(a)

(b)

FIGURE 1.23 **Architecture as a reflection of culture.** (a) This log house near Ottawa, Canada, is a folk dwelling that stands in sharp contrast to the professional architecture of (b) the Toronto skyline. (a: Courtesy of Terry G. Jordan-Bychov; b: Photodisc/Getty Images.)

THINKING GEOGRAPHICALLY What conclusion might a perceptive person from another culture reach (considering the "virtues" of height, durability, and centrality) about the ideology of the culture that produced the Toronto landscape?

culture, as we will see throughout this book. One of the best ways to glimpse settlement and land-division patterns is through an airplane window. Looking down, you can see the multicolored abstract patterns of planted fields, as vivid as any modern painting, and the regular checkerboard or chaotic tangle of urban streets.

Perhaps no other aspect of the human landscape is as readily visible from ground level as the **architectural style** of a culture. Geographers look at the exterior materials and decoration, as well as the layout and design of the interiors. Styles tend to vary both through time, as cultures change, and across space, in the sense that different cultures adopt and invent their own stylistic detailing according to their own particular needs, aesthetics, and desires. Thus, examining architectural style is often useful when trying to date a particular landscape element or when trying to understand the particular values and beliefs that cultures may hold. In North American culture, different building styles catch the eye: modest white New England churches and giant urban cathedrals, hand-hewn barns and geodesic domes, wooden one-room schoolhouses and the new windowless school buildings of urban areas, shopping malls and glass office buildings. Each tells us something about the people who designed, built, or inhabit these spaces. Thus, architecture provides a vivid record of the resident culture (**Figure 1.23**). For this reason, cultural geographers have traditionally devoted considerable attention to examining architecture and style in the cultural landscape.

architectural style The exterior and interior designs and layouts of the cultural and physical landscape.

Key Terms

absorbing barrier, 12
adaptive strategy, 17
architectural style, 26
border zone, 6
contagious diffusion, 11
core-periphery, 6
cultural adaptation, 17
cultural diffusion, 10
cultural ecology, 16
cultural geography, 3
cultural interaction, 20
cultural landscape, 22
culture, 2
culture region, 4
cybergeography, 12
dispersed, 25
ecosystem, 16
environmental
    determinism, 17
environmental
    perception, 19
expansion diffusion, 10
formal culture region, 4
functional culture region, 7
geography, 2
globalization, 13

hierarchical diffusion, 10
independent invention, 10
land-division pattern, 25
mechanistic view of
    nature, 20
migration, 13
model, 21
natural hazard, 19
neighborhood effect, 13
node, 7
nucleation, 25
organic view of nature, 20
permeable barrier, 12
physical environment, 3
place, 22
possibilism, 17
relocation diffusion, 10
settlement form, 25
space, 21
stimulus diffusion, 11
symbolic landscape, 24
time-distance decay, 12
topophilia, 22
transnationalism, 13
uneven development, 15
vernacular culture region, 8

Geography on the Internet

You can learn more about the discipline of geography and the subdiscipline of human geography on the Internet at the following Web sites:

American Geographical Society

http://www.amergeog.org/
America's oldest geographical organization, with a long and distinguished record; publisher of the *Geographical Review*.

Association of American Geographers

http://www.aag.org/
The leading organization of professional geographers in the United States. This site contains information about the discipline, the association, and its activities, including annual and regional meetings.

National Geographic Society

http://www.nationalgeographic.com/
An organization that for more than a century has served to popularize geography with active programs of publishing and television presentations prepared for the public.

Royal Geographic Society/Institute of British Geographers

http://www.rgs.org/
Explore the activities of these allied British organizations, whose collective history goes back to the Age of Exploration and Discovery in the 1800s.

Recommended Books on Cultural Geography

Anderson, Kay, Mona Domosh, Steve Pile, and Nigel Thrift (eds.). 2003. *Handbook of Cultural Geography*. London: Sage Publications. An edited collection of essays that push the boundaries of cultural geography into such subdisciplines as economic, social, and political geography.

Conzen, Michael P. (ed.). 2010. *The Making of the American Landscape*, 2nd ed. New York: Routledge. A collection of essays examining environmental and cultural forces, from early American Indian landscapes to contemporary landscapes of mass culture, that have shaped the visual character and geographical diversity of settled America.

Cosgrove, Denis. 1998. *Social Formation and Symbolic Landscape*. Madison: University of Wisconsin Press. The landmark study that outlines the relationships between the idea of landscape and social and class formation in such places as Italy, England, and the United States.

Foote, Kenneth E., Peter J. Hugill, Kent Mathewson, and Jonathan M. Smith (eds.). 1994. *Re-Reading Cultural Geography*. Austin: University of Texas Press. A beautifully compiled representative collection of some of the best works in American cultural geography at the end of the twentieth century and a useful companion to the book edited by Wagner and Mikesell.

Goudie, Andrew. 2000. *The Earth Transformed: An Introduction to Human Impacts on the Environment*, 5th ed. Cambridge, MA.: MIT Press. A fine introduction to the theme of cultural ecology—in particular, habitat modification—as viewed by a British geographer.

Mitchell, Don. 1999. *Cultural Geography: A Critical Introduction*. New York: Blackwell. An introductory text on cultural geography that emphasizes the material and political elements of the discipline.

Oakes, Timothy S., and Patricia L. Price (eds.). 2008. *The Cultural Geography Reader*. New York: Routledge. A collection of 52 classic and contemporary readings in cultural geography that discusses key geographic concepts and their connections to other disciplines and academic traditions from both the United States and the United Kingdom.

Radcliffe, Sarah (ed.). 2006. *Culture and Development in a Globalizing World: Geographies, Actors, and Paradigms*. London: Routledge. A series of essays that provide case studies from around the world showing the various ways that culture and economic development are integrally related.

Tuan, Yi-Fu. 1974. *Topophilia: A Study of Environmental Perception, Attitudes, and Values*. Englewood Cliffs, NJ: Prentice-Hall. A Chinese-born geographer's innovative and imaginative look at people's attachment to place, a central concern of the cultural approach to human geography.

Journals in Human and Cultural Geography

Annals of the Association of American Geographers. Volume 1 was published in 1911. The leading scholarly journal of American geographers.

Cultural Geographies (formerly known as *Ecumene*). Volume I was published in 1994.

Journal of Cultural Geography. Published semiannually by the Department of Geography, Oklahoma State University, Stillwater, OK. Volume 1 was published in 1980.

Progress in Human Geography. A quarterly journal providing critical appraisal of developments and trends in the discipline. Volume 1 was published in 1977.

Social and Cultural Geography. Volume 1 was published in 2000 by Routledge, Taylor, & Francis Ltd. in Great Britain.

Answers to Thinking Geographically Questions

Figure 1.1: Both mark off sacred space from its surroundings; both are built in traditional styles; both are carefully maintained. The external symbols are different, as are the architectural styles.

Figure 1.2: A major force would be climate: sufficient water (but not too much) and warm temperatures for wheat growing, plus flat land for fields. Another important force is dietary preference.

Figure 1.3: The gun is obviously a piece of introduced technology. The dogs' harnesses also appear to be made from materials brought from elsewhere.

Figure 1.4: Most cultures are separated by a transition zone where characteristics of one culture gradually replace those of the neighboring one, rather than a sharp dividing line.

Figure 1.5: Identification of the border is difficult at best. In the background there appears to be rural farmland, but even that is dependent on the city core—the node—for important services like banking, medical care, legal services, stores, and (because Denver is a state capital) government.

Figure 1.6: A major divide is still the former East Germany, which is poorer, has major pollution problems, and even has slightly different speech patterns from the western part of the country. Although the religious configuration is more complex, after movements of population in the late twentieth century, religion still instills different worldviews and concepts of right and wrong behavior. Differences in landholding laws can be extremely persistent and require special accommodation.

Figure 1.7: Silicon Valley and the Lower Rio Grande in Texas are examples. You may know of other examples from your local area.

Figure 1.8: The American West, Siberia, and the African bush are vernacular regions that are imagined by most people to be wild, remote, hostile environments. They loosely correspond to climate zones but have formally undefined boundaries.

Figure 1.9: Answers will vary, but people who live outside the Riviera define the region by its extreme affluence and "society life," while people who live or vacation there may characterize it more by its desirable climate and amenities than by its affluence.

Figure 1.10: Centers of power, such as government, financial institutions, and cultural institutions, are frequently located in large cities, where individual leaders often live.

Figure 1.11: Answers will vary, but it is likely that Facebook usage will continue to spread and may eventually decline as new social networking forms (and sites) develop and the novelty of Facebook diminishes.

Figure 1.12: Your answers will vary depending on your acquaintance with world regions. The Middle East, especially Turkey; Middle and South America; and China and Southeast Asia may seem surprisingly high.

Figure 1.13: American and European cities probably do not display as many African influences because Americans and Europeans are generally less interested in adopting cultural trends and ideas from Africa than Africans are in copying trends from the United States or Europe. The United States and Europe have longstanding worldwide reputations for being on the vanguard of new ideas and offering affluent lifestyles.

Figure 1.14: Both cities, as they grew, had to face the challenge of connections to a hinterland across considerable bodies of water. However, internally, Chongqing allowed the street pattern to follow the lay of the land because the leaders felt no compulsion to do otherwise. (In flatter areas, Chinese cities were carefully laid out in a square.) San Francisco shows the imposition of the modern grid pattern from the initial Spanish settlement and the public land survey. In this way, leaders faced certain needs but had the possibility of solving the problems in different ways.

Figure 1.15: Answers will vary, but it is likely that people surviving the 2011 Japanese disaster will feel more fearful and wary of the physical environmental forces that could affect their region in the future. They may also become less trusting of technology's ability to protect them against such events.

Figure 1.16: Erosion can be controlled by allowing trees and grass to continue to grow, because they hold the soil in place. Also, clear-cutting promotes erosion, so forests that are more selectively cut can be a controlling factor. Roads can be engineered to keep soil in place and follow contours.

Figure 1.17: In the United States and Canada, the outskirts (the light yellow, labeled 5) tend to house the wealthy and middle-class residents in suburbs. Slums tend to be in the center of town, just outside the central business district. North American cities have less self-built housing.

Figure 1.18: In Asia the pressure of population has led to turning steep slopes into farmland by terracing them (which also helps to control the flow of water). In the United States, there is sufficient farmland (and a smaller population dependent upon it), so there is less pressure for additional farmland created this way. The mainstream American diet is also based on wheat, rather than rice, as the staple grain, so that the market demands less rice production.

Figure 1.19: Answers will vary, but you may find such clues in (1) the architectural styles of houses in a particular region (e.g., wood-frame cabin-style houses in the American Midwest or ranch-style adobe houses in the American Southwest) or (2) the types of ethnic restaurants in a particular region (e.g., Mexican restaurants in the American Southwest, Cuban restaurants in Miami).

Figure 1.20: Answers will vary, depending on landmarks in your home city. Among the possibilities are the hills and major bridges of San Francisco, the skyscrapers of New York, the historic buildings of Boston, and the adobe (and its replication) of Santa Fe.

Figure 1.21: Such houses, often called bungalows or villas, are common in Australia and New Zealand, and they have also been widely built in Great Britain.

Figure 1.22: The Public Land Survey (Township and Range Survey) was devised to make the legal description, and thus the subdivision and sale of, land in the West easy to attract settlers. Americans regarded "taming the wilderness" as a manifest destiny and responsibility. Because cadastral (land-ownership) patterns are so durable, the result has been a landscape of rectangular parcels and fields.

Figure 1.23: Toronto's highest buildings are centrally located, and they are built of durable materials (reinforced concrete and structural steel). The highest buildings are commercial structures, not buildings for religious purposes, indicating that the most important activity is economic. Even residences (high-rise apartment buildings) are tall.

Two American father-daughter couples. (Left: Richard T. Nowitz/Corbis; Right: Rob Gage/Getty Images.)

Folk and Popular Cultures

2

No matter where you live, if you look carefully, you will be reminded constantly of how important the expression of cultural identity is to people's daily lives. The geography of cultural difference is evident everywhere—not only in the geographic distribution of different cultures but also in the way that difference is created or reinforced by geography. For example, in the United States, the history of legally enforced spatial segregation of "whites" and "blacks" has been important in establishing and maintaining cultural differences between these groups.

In Chapter 1 we noted that cultural geographers are interested in studying the geographic expression of difference both among and within cultures. For example, using the concept of formal region, we can identify and map differences among cultures. This sort of analysis is usually done on a very large geographic scale—a continent or even the entire world. Geographers are also interested in analyses at smaller scales. When we look more closely at a formal culture region, we begin to see that differences appear along racial, ethnic, gender, and other lines of distinction. Sometimes

subculture A group of people with norms, values, and material practices that differentiate them from the dominant culture to which they belong.

groups within a dominant culture become distinctive enough that we label them **subcultures.** These can be the result of resistance to the dominant culture or they can be the result of a distinct religious, ethnic, or national group forming an enclave community within a larger culture.

In this chapter, we explore the geographies of cultural difference using two broad categories of classification: popular culture and folk culture. Popular culture, as we will see, is almost, but not quite, synonymous with mass culture and is the dominant form of cultural expression. Folk culture is, to a large degree, distinguished in relation to popular culture.

MANY CULTURES: MATERIAL, NONMATERIAL, FOLK, POPULAR, AND MASS

Cultures are classified using many different criteria. The concept of culture includes both material and nonmaterial

material culture All physical, tangible objects made and used by members of a cultural group. Examples are clothing, buildings, tools and utensils, musical instruments, furniture, and artwork; the visible aspect of culture.

elements. **Material culture** includes all objects or things made and used by members of a cultural group: buildings, furniture, clothing, artwork, musical instruments, and other physical

objects. The elements of material culture are visible. **Nonmaterial culture** includes the wide range of beliefs, values, myths, and symbolic meanings that are transmitted across generations of a given society. Cultures may be categorized and geographically located using criteria based on either or both of these features.

nonmaterial culture The wide variety of tales, songs, lore, beliefs, superstitions, and customs that pass from generation to generation as part of an oral or written tradition.

Let's explore how these criteria are used to identify, categorize, and graphically delineate cultures. According to literary critic and cultural theorist Raymond Williams, *culture* is one of the two or three most complicated words in the English language. In the late eighteenth and early nineteenth centuries, people began to speak of "cultures" in the plural form. Specifically, they began thinking about "European culture" in relation to other cultures around the world. As Europe industrialized and urbanized in the nineteenth century, a new term, *folk culture*, was invented to distinguish traditional ways of life in rural spaces from new urban, industrial ones. Thus, folk culture was defined and made sense only in relation to an urban, industrialized culture. Urban dwellers began to think—in increasingly romantic and nostalgic terms—of rural spaces as inhabited by distinct folk cultures.

The word **folk** describes a rural people who live in an old-fashioned way—a people

folk Traditional, rural; the opposite of "popular."

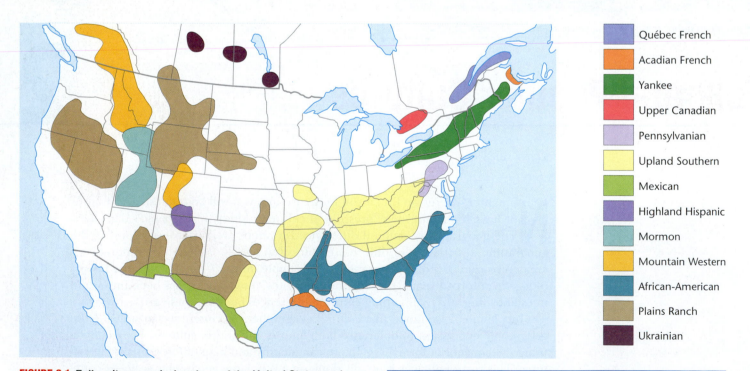

FIGURE 2.1 Folk culture survival regions of the United States and southern Canada. Almost all are now in decay and retreat. The closest modern example of a folk culture, not included on this map, is the Amish culture.

Legend:
- Québec French
- Acadian French
- Yankee
- Upper Canadian
- Pennsylvanian
- Upland Southern
- Mexican
- Highland Hispanic
- Mormon
- Mountain Western
- African-American
- Plains Ranch
- Ukrainian

THINKING GEOGRAPHICALLY What elements of folk cultures remain in such regions? How do they become part of popular culture?

holding onto a lifestyle relatively little influenced by modern technology. **Folk cultures** are small, rural, cohesive, conservative, isolated, largely self-sufficient groups that are homogeneous in custom and ethnicity. In terms of nonmaterial culture, folk cultures typically have strong family or clan structures and highly localized rituals. There is little division of labor other than that between the sexes. Order is maintained through sanctions based in religion or the family, and interpersonal relationships are strong. In material cultural terms, most goods are handmade, and a subsistence economy prevails. Individualism is generally weakly developed in folk cultures, as are social classes. **Folk geography,** a term coined by Eugene Wilhelm, is the study of the spatial patterns and ecology of these traditional groups.

In the poorer countries of the underdeveloped world, some aspects of folk culture still exist, though few if any peoples have been left untouched by the forces of globalization. In industrialized countries, such as the United States and Canada, unaltered folk cultures no longer exist, though many remnants can be found (**Figure 2.1**). Perhaps the closest modern example in North America is the Amish, a German-American farming sect that largely renounces the products and labor-saving devices of the industrial age, though they do practice commercial agriculture. In Amish areas, horse-drawn buggies still serve as a local transportation device, and the faithful own no automobiles or appliances. The Amish central religious concept of *demut*, "humility," clearly reflects the weakness of individualism and social class so typical of folk cultures, and a corresponding strength of Amish group identity is evident. The denomination, a variety of the Mennonite faith, provides the principal mechanism for maintaining order.

Popular culture is generated from and concentrated mainly in heterogeneous urban areas (**Figure 2.2**). Popular material goods, mass-produced by machines

folk culture Small, cohesive, stable, isolated, nearly self-sufficient group that are homogeneous in custom and ethnicity; characterized by a strong family or clan structure, order maintained through sanctions based in the religion or family, little division of labor other than that between the sexes, frequent and strong interpersonal relationships, and material cultures consisting mainly of handmade goods.

folk geography The study of the spatial patterns and ecology of traditional groups.

popular culture A dynamic culture based in large, heterogeneous societies permitting considerable individualism, innovation, and change; features include a money-based economy, division of labor into professions, secular institutions of control, and weak interpersonal ties; used to describe a common set of cultural ideas, values, and practices that are consumed by a population.

FIGURE 2.2 Manifestation of popular culture. Popular culture is reflected in every aspect of life, from the clothes we wear to the recreational activities that occupy our leisure time. (Left: Scott Olson/Getty Images; Right: Mikhael Subotsky.)

THINKING GEOGRAPHICALLY How does modern technology, itself a trait of popular culture, make widespread popular culture possible?

in factories, are consumed by the culture. A cash economy, rather than barter or subsistence, dominates. Relationships among individuals are more numerous but less personal than in folk cultures, and the family structure is weaker. Mass media such as film, print, television, radio, the Internet, and other digital media are more influential in shaping popular culture. People are more mobile, less attached to place and environment. Secular institutions of authority—such as the police, army, and courts—take the place of family and church in maintaining order. Individualism is strongly developed and innovation and change permitted or welcomed. Labor is divided into professions.

These characterizations of popular culture are also sometimes used to describe **mass culture.** However, experts point to slight differences between these two terms. Popular culture is generally used to describe a common set of cultural ideas, values, and practices that are *consumed* by a population. Meanwhile, mass culture usually refers more specifically to a form of culture that is *produced, distributed, and marketed* through mass media, art, and other forms of communication. Despite this difference in focus, for practical purposes, we can imagine that popular culture and mass culture, hand in hand, constitute the form of culture that most people in contemporary, developed societies experience daily.

> **mass culture** A form of culture that is produced, distributed, and marketed through mass media, art, and other forms of communication.

If all of these characteristics seem rather commonplace and "normal," it should not be surprising. You are, after all, firmly enmeshed in the popular culture, or you would not be attending college to seek higher education. The majority of people in the United States, Canada, and other developed nations now belong to the popular rather than the folk culture. Industrialization, urbanization, the rise of formal education, and the resultant increase in leisure time all contributed to the spread of popular culture and the consequent retreat of folklife.

In reality, all of culture presents a continuum, on which "folk" and "popular" represent two extremes. Many gradations between the two are possible.

We can use our five themes of cultural ecology—region, diffusion, ecology, interaction, and landscape—to study both folk and popular culture. Let's begin with culture regions.

FOLK AND POPULAR CULTURE REGIONS

How do cultures vary geographically? Some cultures exhibit major material and nonmaterial variations from place to place with minor variations over time. Others display less difference from region to region but change rapidly over time. For this reason, the theme of culture region is particularly well suited to the study of cultural difference. Formal culture regions can be delineated on the basis of both material and nonmaterial elements.

Material Folk Culture Regions

Although folk culture has largely vanished from the United States and Canada, vestiges remain in various areas of both countries. Figure 2.1 shows culture regions in which the material artifacts of 13 different North American folk cultures survive in some abundance, but even these artifacts are disappearing. Each region possesses many distinctive relics of material culture.

For example, the strongly Germanic Pennsylvanian folk culture region features an unusual Swiss-German type of barn, distinguished by an overhanging upper-level "forebay" on one side (**Figure 2.3**). In contrast, barns are usually attached to the rear of houses in the Yankee folk region. This region is also distinguished by an elaborate form of traditional gravestone art featuring "winged death

FIGURE 2.3 **A multilevel barn with projecting forebay in central Pennsylvania.** Every folk culture region possesses distinctive forms of traditional architecture. Of Swiss origin, the forebay barn is one of the main identifying material traits of the Pennsylvanian culture region. (Courtesy of Terry G. Jordan-Bychkov.)

THINKING GEOGRAPHICALLY How did the idea of this barn come from Switzerland to central Pennsylvania? Why would the extended-forebay design be useful to farmers who were fattening cattle?

FIGURE 2.4 A "scraped-earth" folk graveyard in east Texas. The laborious removal of all grass from such cemeteries is an African-American custom. Long ago, this practice diffused from the African-American folk culture region to Euro-Americans in the southern coastal plain of the United States to become simply a "southern" custom. (Courtesy of Terry G. Jordan-Bychkov.)

THINKING GEOGRAPHICALLY Would removing the vegetation in a cemetery have originally had a practical purpose? Why or why not?

FIGURE 2.5 Beef wheel in the ranching country of the Harney Basin in central Oregon. This windlass device hoists the carcass of a slaughtered animal to facilitate butchering. Derived, as was much of the local ranching culture, from Hispanic Californians, the beef wheel represents the material folk culture of ranching. (Courtesy of Terry G. Jordan-Bychkov.)

THINKING GEOGRAPHICALLY How does this apparatus show the characteristic of folk culture of developing tools to meet an immediate need? Would this tool have been advertised? How do you know?

heads." This motif is most commonly found on graves dating from the seventeenth and eighteenth centuries. In their earliest forms these carvings depicted Death as a gruesome skull and bones in various formations. However, by the early nineteenth century, they were modified to resemble cherub faces with elaborate wings on either side.

The Upland South is noted in part for the abundance of a variety of distinctive house types built using notched-log construction. The African-American folk region displays such features as the "scraped-earth" cemetery, from which all grass is laboriously removed to expose the bare ground (**Figure 2.4**); the banjo, an instrument of African origin; and head kerchiefs worn by women. Stone grist mills with large sails to generate wind power for the grinding of grain help to characterize the Québec French folk region, as does the popularity of pétanque, a bowling game played with small metal balls. The Mormon folk culture region is identifiable by its farm villages clustered in checkerboard patterns and by the presence of distinctive derricks (lifting devices) used to loose-stack hay in agricultural fields. This gathering technique contrasts with baling methods more commonly used outside the Mormon folk culture region. The Western Plains ranch folk culture produced such material items as the beef wheel, a windlass (winch) used during butchering (**Figure 2.5**). These examples of material artifacts are only a few of the many that survive from various folk regions.

Folk Food Regions

Perhaps no other aspect of folk culture endures as abundantly as traditional foods. In Latin America, for example,

folk cultures remain vivid and vital in many regions, with diverse culinary traditions. Each displays distinctive choices of food and methods of preparation. For example, Mexican cuisine features abundant use of chile peppers and maize tortillas in cooking; the Caribbean areas offer combined rice-bean dishes and various rum drinks; in the Amazonian region, monkey and caiman are favored foods; Brazilian fare is distinguished by *cuzcuz* (cooked grain) and sugarcane brandy. These and other Latin American foods derive from American Indians, Africans, Spaniards, and Portuguese.

Is Popular Culture Placeless?

Superficially at least, popular culture varies less from place to place than does folk culture. In fact, Canadian geographer Edward Relph goes so far as to propose that popular culture produces a profound **placelessness,** a spatial standardization that diminishes regional variety and demeans the human

placelessness A spatial standardization that diminishes regional variety; may result from the spread of popular culture, which can diminish or destroy the uniqueness of place through cultural standardization on a local, national, or worldwide scale.

(a)

(b)

FIGURE 2.6 Placelessness exemplified: scenes from almost anywhere in the developed world. (See also Curtis, 1982. Photos, left: Alex Segre/Alamy; right: H. Mark Weidman Photography/Alamy.)

THINKING GEOGRAPHICALLY Guess where these two photos were taken.

spirit. Others observe that one place seems pretty much like another, each robbed of its unique character by the pervasive influence of a continental or even worldwide popular culture (**Figure 2.6**). When compared with regions and places produced by folk culture, rich in their uniqueness (**Figure 2.7**), the geographical face of popular culture often seems expressionless. The greater mobility of people in popular culture weakens attachments to place and compounds the problem of placelessness. Moreover, the spread of McDonald's, Starbucks, CNN, shopping malls, and much else further adds to the sense of placelessness.

Is popular culture truly regionless and placeless? Many cultural geographers are cautious about making such a sweeping generalization. Geographer Michael Weiss, for example, argues in his book *The Clustering of America* that "American society has become increasingly fragmented" and identifies 40 "lifestyle clusters" based on postal zip codes. "Those five digits can indicate the kinds of magazines you read, the meals you serve at dinner," and what political party you support. "Tell me someone's zip code and I can predict what they eat, drink, drive—even think." The lifestyle clusters, each of which is a formal culture region, bear Weiss's colorful names—such as "Gray Power" (upper-middle-class retirement areas), "Old Yankee Rows" (older

FIGURE 2.7 Retaining a sense of place: a hill town in Cappadocia Province, Turkey. This town, produced by a folk culture, exhibits striking individuality. (Courtesy of Terry G. Jordan-Bychkov.)

THINKING GEOGRAPHICALLY How can you tell that this is not a popular culture landscape?

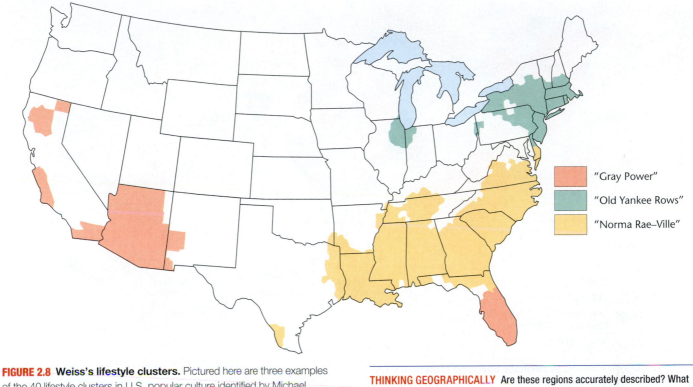

FIGURE 2.8 **Weiss's lifestyle clusters.** Pictured here are three examples of the 40 lifestyle clusters in U.S. popular culture identified by Michael Weiss. Patterns of consumption within popular cultures shift regionally, resulting in "lifestyle clusters." For a description of each lifestyle, see the text.

Map legend:
- "Gray Power"
- "Old Yankee Rows"
- "Norma Rae–Ville"

THINKING GEOGRAPHICALLY Are these regions accurately described? What would you change?

ethnic neighborhoods of the Northeast), and "Norma Rae–Ville" (lower- and middle-class southern mill towns, named for the Sally Field movie about the tribulations of a union organizer in a textile manufacturing town) (**Figure 2.8**). Old Yankee Rowers, for example, typically have a high school education, enjoy bowling and ice hockey, and are three times as likely as the average American to live in a row house or duplex. Residents of Norma Rae–Ville are mostly nonunion factory workers, have trouble earning a living, and consume twice as much canned stew as the national average. In short, a whole panoply of popular subcultures exist in America and the world at large, each possessing its own belief system, spokespeople, dress code, and lifestyle.

REFLECTING ON GEOGRAPHY

Do you live in a "placeless" place in "nowhere USA"? If not, how is a distinctive regional form of popular culture reflected in your region?

Popular Food and Drink

A persistent formal regionalization of popular culture is vividly revealed by what foods and beverages are consumed, which varies markedly from one part of a country to another and throughout different parts of the world. **Figure 2.9** on page 36 illustrates this variation in regard to alcoholic beverage preference. Differences in beverage choice can be attributed to differing cultural influences, including religious taboos regarding the consumption of alcohol. Patterns in beverage choice are also linked to patterns of economic development in many parts of the world. People in less developed regions may choose to consume "other" forms of alcohol (see map key) that can be made cheaply using local ingredients.

Variations in beverage consumption are notable between regions in the United States as well. The highest per capita levels of U.S. beer consumption occur in the West, with the notable exception of Mormon Utah. Whiskey made from corn, manufactured both legally and illegally, has been a traditional southern alcoholic beverage, whereas wine is more common in California.

Foods consumed by members of the North American popular culture also vary from place to place. In the South, grits, barbecued pork and beef, fried chicken, and hamburgers are far more popular than elsewhere in the United States, whereas more pizza and submarine sandwiches are consumed in the North, the destination for many Italian immigrants. Contrasting food styles have developed as part of regionalized pop culture but can also often be traced to

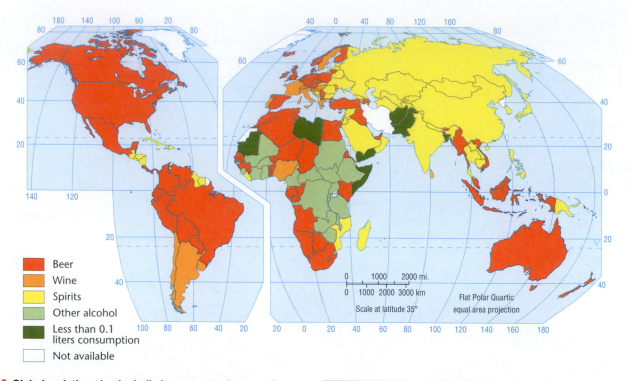

Beer
Wine
Spirits
Other alcohol
Less than 0.1 liters consumption
Not available

FIGURE 2.9 Global variations in alcoholic beverage preference. The consumption of beer and distilled spirits dominates in many countries, while wine drinking is less popular on a global scale. (Source: http://chartsbin.com.)

THINKING GEOGRAPHICALLY Why do you think that a preference for wine drinking is concentrated mainly in Western and Southern Europe and in the southern portion of South America?

immigrant populations who, like the Italians in the Northeast, settled in these areas in times past. **Figure 2.10** illustrates examples of foods that have become well known in various popular culture regions across the United States. One of these regional specialties, fried green tomatoes, became nationally famous following the release of a 1991 movie by the same name that depicts a cinematic view of Southern life.

The spread of global brands such as Coca-Cola and Kentucky Fried Chicken would seem to indicate increasing homogenization of food and beverage consumption. Yet many studies show that such brands have different meanings in different places around the world. Coca-Cola may represent modernization and progress in one place and foreign domination in another. Sometimes multinational corporations have to change their foods and beverages to suit local cultural preferences. For example, in Mumbai, India, McDonald's has had to add local-style sauces to its menus. Rather than a simple one-way process of "Americanizing" Indian food preferences, local consumers are "Indianizing" McDonald's.

Fast food might seem to epitomize popular culture, yet its importance varies greatly even within the United States.

Figure 2.11 on page 38 shows that fast-food consumption dominates in the Southwest and West, while most of the country's middle section partakes of (comparatively) less fast food. Such differences undermine the geographical uniformity or placelessness supposedly created by popular culture. Music provides another example.

Popular Music

Popular culture has spawned many styles of music, all of which reveal geographical patterns in levels of acceptance. Country music provides a good example of this phenomenon. Country music began to develop in the southern United States in the 1920s, incorporating aspects of several folk music traditions as well as cowboy music. Since that time, the popularity of country music has grown immensely, spreading to all 50 states and throughout the world. However, a distinct regional quality in the popularity of country music within the United States can still be observed. Geographer Ben Marsh illustrated this geographic pattern by creating a map based on state-name mentions in country music lyrics as a measure of the musical style's popularity in those states (**Figure 2.12**, page 38).

(a)

(b)

(c)

(d)

FIGURE 2.10 American regional dishes. (a) Chicago-style hot dogs from the Midwest, (b) fried green tomatoes from the South, (c) traditional Chinese foods from the West Coast, and (d) clam chowder from New England represent the wide array of regional popular foodways in the United States. (a: ZUMA Wire Service/Alamy; b: Cartela/Shutterstock; c: DesmondLWF/Shutterstock; d: Og-vision/Dreamstime.com.)

THINKING GEOGRAPHICALLY Can you think of other popular regional foods that you have encountered in places you have lived or traveled?

The cartogram suggests (by relative state size) that country music's popularity is highest in and around its original southern source region. Tennessee, with its capital, Nashville, is the long-time heart of the country-music industry. The popularity of country music extends from this Tennessee core outward through the American South. Notice that the states in the northern tier of the country receive many fewer state-name mentions, indicating the corresponding lower level of country music's popularity in that region.

Vernacular Culture Regions

A **vernacular culture region** is the product of the spatial perception of the population at large—a composite of the mental maps of the people. Such regions vary greatly in size, from small districts covering only part of a city or town to huge, multistate areas. Like most other geographical regions, they often overlap and usually have poorly defined borders.

Almost every part of the industrialized Western world offers examples of vernacular regions based in the popular culture. **Figure 2.13** on page 39 shows some sizable vernacular regions in North America. Geographer Wilbur Zelinsky compiled these regions by determining the most common name for businesses appearing in the white pages of urban telephone directories. One curious feature of the map is the sizable populous district—in New York, Ontario, eastern Ohio, and western Pennsylvania—where no regional affiliation is perceived. Using a different source of information, geographer Joseph

> **vernacular culture region** A culture region perceived to exist by its inhabitants; based in the collective spatial perception of the population at large; bearing a generally accepted name or nickname, such as Dixie.

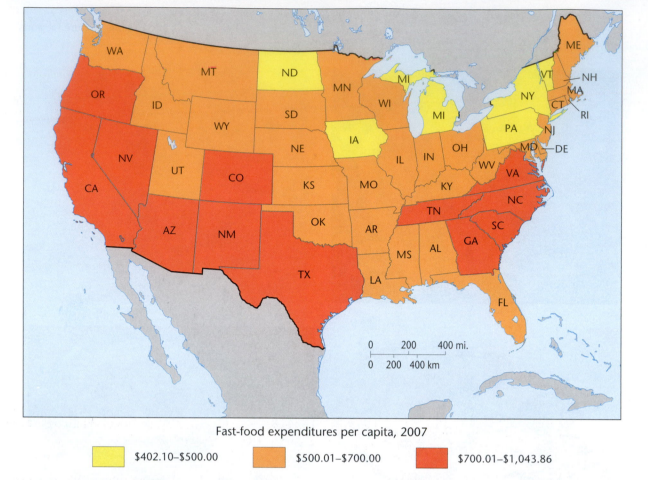

Fast-food expenditures per capita, 2007

$402.10–$500.00	$500.01–$700.00	$700.01–$1,043.86

FIGURE 2.11 Fast-food expenditures per capita, by state, 2007.

(Source: U.S. Department of Agriculture, 2007.)

THINKING GEOGRAPHICALLY What factors do you think might influence the fast-food consumption patterns revealed by this map?

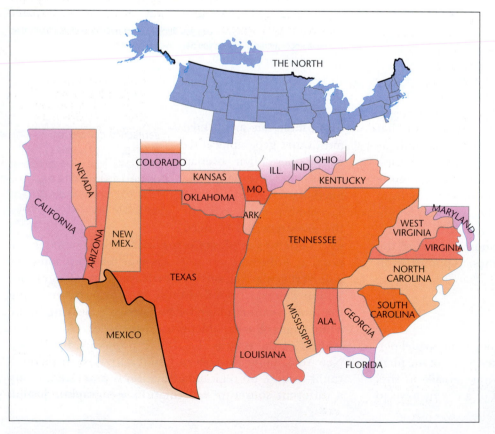

FIGURE 2.12 **States mentioned in country music lyrics.** The core of country music popularity is in Tennessee, Texas, and Louisiana. Country music is much less popular in northern states. (Source: http://bigthink.com; Marsh, Ben. "A rose colored map," *Harper's Magazine*, July 1977.)

THINKING GEOGRAPHICALLY How do you think a cartogram showing frequency of state-name mentions in rap/hip-hop lyrics might differ from the cartogram of state names in country music lyrics?

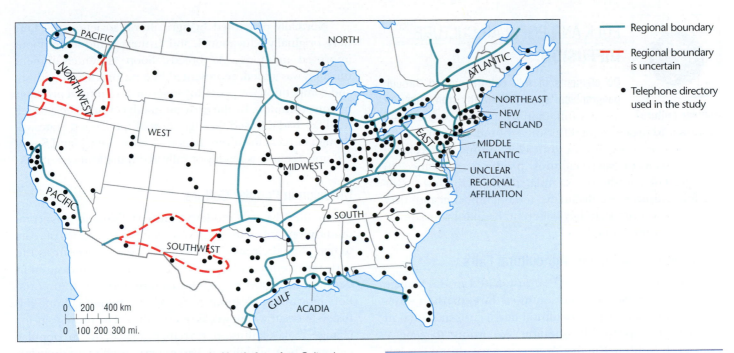

FIGURE 2.13 Some vernacular regions in North America. Cultural geographer Wilbur Zelinsky mapped these regions on the basis of business names in the white pages of metropolitan telephone directories. (Adapted from Zelinsky, 1980a: 14.)

THINKING GEOGRAPHICALLY Why are names containing "West" more widespread than those containing "East"? What might account for the areas where no region name is perceived?

Brownell sought to delimit the popular "Midwest" in 1960 (**Figure 2.14**). He sent out questionnaires to postal employees in the midsection of the United States, from the Appalachians to the Rockies, asking each whether, in his or her opinion, the community lay in the "Midwest." The results identified a vernacular region in which the

residents considered themselves midwesterners. A similar survey done almost 30 years later, using student respondents, revealed a core-periphery pattern for the Midwest (see Figure 2.14). As befits an element of popular culture, the vernacular region is often perpetuated by the mass media, especially radio and television.

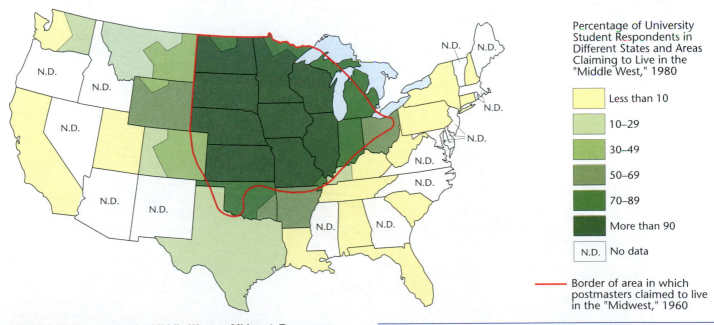

FIGURE 2.14 The vernacular Middle West or Midwest. Two surveys, taken a generation apart and using two different groups of respondents, yielded similar results. (Sources: Brownell, 1960: 83; Shortridge, 1989.)

THINKING GEOGRAPHICALLY Why would such a vernacular region remain fairly stable?

FOLK AND POPULAR CULTURE DIFFUSION

Do elements of folk culture spread through geographical space differently from those of popular culture? Whereas folk culture spreads by the same models and processes of diffusion as popular culture, diffusion operates more slowly within a folk setting. The relative conservatism of such cultures produces resistance to change. The Amish, for example, as one of the few surviving folk cultures, are distinctive today simply because they reject innovations that they believe to be inappropriate for their way of life and values.

Diffusion in Folk Culture: Agricultural Fairs

An example of folk culture that originated in the Yankee region and spread west and southwest by expansion diffusion was the American agricultural fair, a custom rooted in medieval European folk tradition. According to geographer Fred Kniffen, the first American agricultural fair was held in Pittsfield, Massachusetts, in 1810, and the idea then gained favor throughout western New England and the adjacent Hudson Valley (**Figure 2.15**). From that source region it diffused westward into the American heartland, the Midwest, where it gained its widest acceptance.

Normally promoted by agricultural societies, the fairs were originally educational, and farmers could learn about improved methods and breeds. Soon an entertainment function was added, represented by a racetrack and midway, and competition for prizes for superior agricultural products became common. By the early twentieth century, the agricultural fair had diffused through most of the United States, although farmers in culture regions such as the Upland South did not accept it as readily or fully as did the Midwesterners.

Diffusion in Popular Culture

Before the advent of modern transportation and mass communications, innovations usually required thousands of years to complete their areal spread, and even as recently as the early nineteenth century the time span was still measured in decades. In regard to popular culture, modern transportation and communications networks now permit cultural diffusion to occur within weeks or even days. The propensity for change makes diffusion extremely important in popular culture. The availability of devices permitting rapid diffusion enhances the chance for change in popular culture.

An example of the change from slower means of diffusion in folk culture to the rapid global spread of cultural ideas and trends in popular culture can be found in the growth of "Bollywood," the Indian film industry. Though

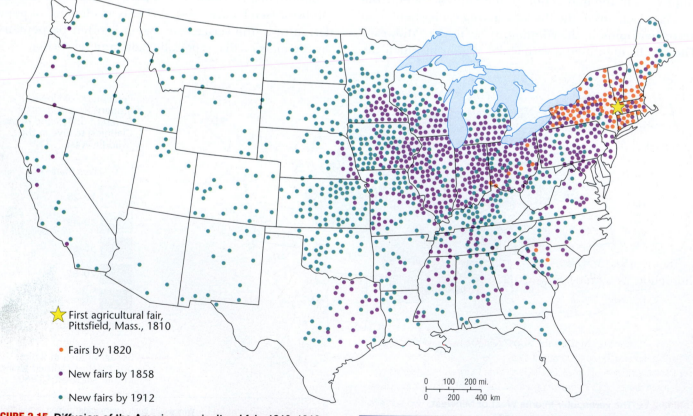

First agricultural fair, Pittsfield, Mass., 1810

• Fairs by 1820

• New fairs by 1858

• New fairs by 1912

FIGURE 2.15 Diffusion of the American agricultural fair, 1810–1910. (After Kniffen, 1951: 45, 47, 51.)

THINKING GEOGRAPHICALLY What type or types of cultural diffusion might have been at work here?

FIGURE 2.16 **A Bollywood film dance scene.** The global popularity of Bollywood illustrates the transformation of a folk culture element into a highly diffused form of popular culture. (AF archive/Alamy.)

THINKING GEOGRAPHICALLY Can you think of other trends in popular culture that have diffused to the United States from folk or popular cultures in other parts of the world?

Bollywood is very familiar to many people around the world in twenty-first-century popular culture, its roots lie in Indian folk dance traditions and Hindi-language music that date back centuries. In the twentieth century, these traditions were transformed into popular entertainment via the development of a massive Indian film industry. **Figure 2.16** illustrates the elaborate costumes and intricate dance steps that have made the Bollywood phenomenon so popular with movie audiences.

Due to the migration of Indian people to many other countries, and due to the recent international success of Indian films like *Slumdog Millionaire*, Bollywood music and dance styles have now diffused through many parts of the world. For example, in the United States, Bollywood dance has not only been featured in mainstream films, it has also become popular as a new fitness activity at many gyms and dance studios. Bollywood dance has grown in popularity on college campuses across the United States as well, with some schools sponsoring Bollywood dance teams that compete nationally against other dance teams.

In the diffusion of all types of popular culture phenomena, hierarchical processes often play a greater role than in folk culture because popular society, unlike folk culture, is highly stratified by socioeconomic class. For example, the spread of McDonald's restaurants—beginning in 1955 in the United States and later internationally—occurred hierarchically for the most part, revealing a bias in favor of large urban markets (**Figure 2.17**). Further facilitating the diffusion of popular culture is the fact that time-distance decay is weaker in such regions, largely because of the reach of mass media.

Sometimes, however, diffusion in popular culture works differently, as a study of Walmart revealed. Geographers

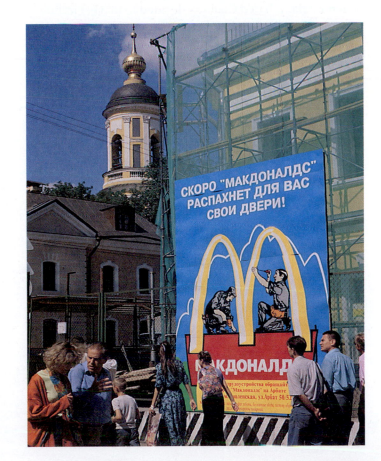

FIGURE 2.17 **Another McDonald's opens in Moscow.** McDonald's, which first spread to Moscow about 1987, has always preferred hierarchical diffusion. Of all McDonald's outlets worldwide today, about 45 percent are located in foreign countries, almost always in large cities. (Courtesy of Terry G. Jordan-Bychkov.)

THINKING GEOGRAPHICALLY Why might McDonald's be considered a prestigious restaurant in countries like Russia and China?

Thomas Graff and Dub Ashton concluded that Walmart initially diffused from its Arkansas base in a largely contagious pattern, reaching first into other parts of Arkansas and neighboring states. Simultaneously, as often happens in the spatial spread of culture, another pattern of diffusion was at work, one Graff and Ashton called *reverse hierarchical diffusion*. Walmart initially located its stores in smaller towns and markets, only later spreading into cities—the precise reverse of the way hierarchical diffusion normally works. This combination of contagious and reverse hierarchical diffusion led Walmart to become the nation's largest retailer in only 30 years.

Advertising

The most effective device for diffusion in popular culture, as Zelinsky suggests, confronts us almost every day of our lives. Commercial advertising of retail products and services bombards our eyes and ears with great effect. Using the techniques of social science, especially psychology, advertisers have learned how to sell us products we do not need. The skill with which advertising firms prepare commercials often determines the success or failure of a product. In short, popular culture is equipped with the most potent devices and techniques of diffusion ever devised.

Commercial advertising is limited in its capacity to overcome all spatial and cultural barriers to homogenization. Cases from international advertising are illustrative. When England-based Cadbury decided to market its line of chocolates and other candies in China, the company was forced to change its advertising strategy. Unlike much of Europe and North America, China had no culture of impulse buying and no tradition of self-service. Cadbury had to change the names of its products, avoid using mass marketing, focus on a small group of high-end consumers, and even change product content.

Place of product origin is also extremely important in trying to advertise and market internationally. Sometimes place helps sell a product—think of New Zealand wool or Italian olive oil—and sometimes it is a hindrance to sales. There are many examples of products having negative associations among consumers because the country of origin has a tarnished international reputation. A good example is South Africa's advertising efforts during the era of apartheid when consumer boycotts of the country's products were common. South African industries had to suppress references to country of origin in their advertising in order to market their products internationally.

Communications Barriers

Although the communications media, including the Internet, create the potential for almost instant diffusion over very large areas, this diffusion can be greatly retarded if access to the media is denied or limited. *Billboard*, a magazine devoted largely to popular music, described one such

barrier. A record company executive complained that radio stations and disk jockeys refused to play punk rock records, thereby denying the style an equal opportunity for exposure. He claimed that punk devotees were concentrated in New York City, Los Angeles, Boston, and London, where many young people had found the style reflective of their feelings and frustrations. Without access to radio stations, punk rock could diffuse from these centers only through live concerts and the record sales they generated. The publishers of *Billboard* noted that "punk rock is but one of a number of musical forms which initially had problems breaking through nationally out of regional footholds." Pachanga, ska, pop gospel, women's music, reggae, and gangsta rap had similar difficulties. Similarly, Time Warner, a major distributor of gangsta rap music, endured scathing criticism from the U.S. Congress in 1995 because of potentially offensive or deleterious aspects of this genre. This criticism eventually led the company to sell the subsidiary label that recorded this form of rap. To control the programming of radio and television, or media distribution generally, is to control much of the diffusionary apparatus in popular culture. The diffusion of innovations ultimately depends on the flow of information.

Government censorship, as opposed to mere criticism, also creates barriers to diffusion, with varying degrees of effectiveness. In 1995, the Islamic fundamentalist regime in Iran, opposed to what it perceived as the corrupting influences of Western popular culture, outlawed television satellite dishes in an attempt to prevent citizens from watching programs broadcast in foreign countries. The Taliban government of Afghanistan went even further, banning all television sets. Control of the media can greatly control people's tastes in, preferences in, and ideas about popular culture. Even so, repressive regimes must cope with a proliferation of communication methods, including the Internet, YouTube, and social media. So pervasive has cultural diffusion become that the insular, isolated status of nations is probably no longer attainable for very long, even under totalitarian conditions.

> ### REFLECTING ON **GEOGRAPHY**
> Because Canadian newspapers devote so much more coverage to international stories than U.S. newspapers, are Americans more provincial than Canadians?

Diffusion of the Rodeo

Barriers of one kind or another usually weaken the diffusion of elements of popular culture before they become ubiquitous. The rodeo provides an example. Rooted in the ranching folk culture of the American West, it has never completely escaped that setting (**Figure 2.18**).

Like so many elements of popular culture, the modern rodeo had its origins in folk tradition. Taking their name from the Spanish *rodear*, "to round up," rodeos began

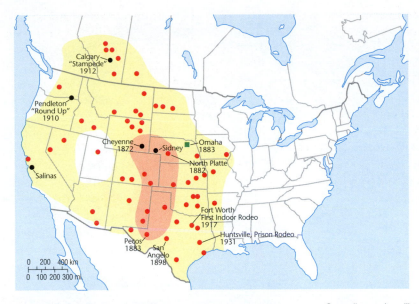

FIGURE 2.18 **Origin and diffusion of the American commercial rodeo.** Derived originally from folk culture, rodeos evolved through formal cowboy contests and Wild West shows to emerge, in the late 1880s and 1890s, in their present popular culture form. The border between the United States and Canada proved no barrier to the diffusion, although Canadian rodeo, like Canadian football, differs in some respects from the U.S. type. (Sources: Frederickson, 1984; Pillsbury, 1990b.)

THINKING GEOGRAPHICALLY   What barriers might the diffusion have encountered?

simply as roundups of cattle in the Spanish livestock ranching system in northern Mexico and the American Southwest. Anglo-Americans adopted Mexican cowboy skills in the nineteenth century, and cowboys from adjacent ranches began to hold contests at roundup time. Eventually, some cowboy contests on the Great Plains became formalized, with prizes awarded.

The transition to commercial rodeo, with admission tickets and grandstands, came quickly as an outgrowth of the formal cowboy contests. One such affair, at North Platte, Nebraska, in 1882, led to the inclusion of some rodeo events in a Wild West show in Omaha in 1883. These shows, which moved by railroad from town to town in the manner of circuses, were probably the most potent agent of early rodeo diffusion. Within a decade of the Omaha event, commercial rodeos were being held independently of Wild West shows in several towns, such as Prescott, Arizona. Spreading rapidly, commercial rodeos had appeared throughout much of the West and parts of Canada by the early 1900s. At Cheyenne, Wyoming, the famous Frontier Days rodeo was first held in 1897. By World War I, the rodeo had also become an institution in the provinces of western Canada, where the Calgary Stampede began in 1912.

Today, rodeos are held in 36 states and three Canadian provinces. Oklahoma's annual calendar of events lists no fewer than 98 scheduled rodeos. Rodeos have received the greatest acceptance in the popular culture found west of the Mississippi and Missouri rivers (see Figure 2.18). Absorbing and permeable barriers to the diffusion of commercial rodeo were encountered at the border of Mexico, south of which bullfighting occupies a dominant position, and in the Mormon culture region centered in Utah.

Blowguns: Diffusion or Independent Invention?

Often the path of past diffusion of an item of material culture is not clearly known or understood, presenting geographers with a problem of interpretation. The blowgun is a good example. A hunting tool of many indigenous peoples, a blowgun is a long hollow tube through which a projectile is blown by the force of one's breath. Geographer Stephen Jett mapped the distribution of blowguns, which he discovered were used in societies in both the Eastern and Western hemispheres, all the way from the island of Madagascar off the east coast of Africa to the Amazon rain forests of South America (**Figure 2.19**, page 44).

Indonesian peoples, probably on the island of Borneo, appear to have first invented the blowgun. It became their principal hunting weapon and diffused through much of the equatorial island belt of the Eastern Hemisphere. How, then, do we account for its presence among Native American groups in the Western Hemisphere? Was it independently invented by Native Americans? Was it brought to the Americas by relocation diffusion in pre-Columbian times? Or did it spread to the New World only after the European discovery of America? We do not know the answers to these questions, but the problem is common to cultural geography, especially in the study of the traditions of nonliterate cultures, which have no written records that might reveal such diffusion. Certain rules of thumb can be employed in any given situation to help resolve the issue.

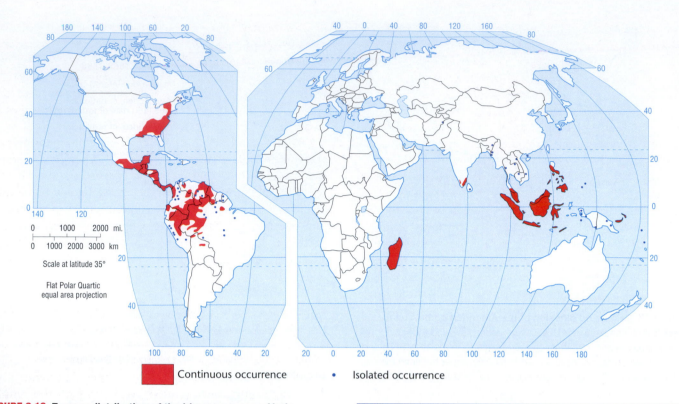

Continuous occurrence   • Isolated occurrence

FIGURE 2.19 **Former distribution of the blowgun among Native Americans, South Asians, Africans, and Pacific Islanders.** The blowgun occurred among folk cultures in two widely separated areas of the world. (Source: Jett, 1991: 92–93.)

THINKING GEOGRAPHICALLY Was this the result of independent invention or cultural diffusion? What kinds of data might one seek to answer this question? Compare and contrast the occurrence in the Indian and Pacific ocean lands to the distribution of the Austronesian languages (see Chapter 4).

For example, if one or more nonfunctional features of blowguns, such as a decorative motif or specific terminology, occurred in both South America and Indonesia, then the logical conclusion would be that cultural diffusion explained the distribution of blowguns.

THE **ECOLOGY** OF FOLK AND POPULAR CULTURES

How is nature related to cultural difference? Do different cultures and subcultures differ in their interactions with the physical environment? Are some cultures closer to nature than others? People who depend on the land for their livelihood—farmers, hunters, and ranchers, for example—tend to have a different view of nature from those who work in commerce and manufacturing in the city. Indeed, one of the main distinctions between folk culture and popular culture is their differing relationships with nature.

Ethnomedicine and Ecology

Folk medicine is often closely linked to the environment. People in folk societies commonly treat diseases and

disorders with drugs and medicines derived from the root, bark, blossom, or fruit of plants. Medical practice using these ingredients is generally passed down orally over many generations. In recent years, these medical traditions have received increased interest in many disciplines, including anthropology, botany, and geography. In addition to studying the natural ingredients that folk cultures use to treat illness and disease, **ethnomedicine** examines the cultural interpretation of health and disease. Most folk cultures interpret these conditions as the result of two basic causes: (1) natural forces, like heat or cold, and (2) unnatural forces, like evil spirits.

ethnomedicine An interdisciplinary area of study that focuses on the natural ingredients and traditional practices used in folk cultures to treat illness and disease, as well as on cultural differences in perceptions of health and disease.

As a result of increasing interest in ethnomedicine and the proven effectiveness of many of its cures and practices, mainstream medical practitioners in many countries (including the United States) have begun to incorporate folk medicines and practices from around the world into the treatment of their patients. For example, traditional Chinese medicine has become quite widely accepted, leading to the popularity of treatments like acupuncture and therapeutic massage. In addition, many modern drugs have been developed from traditional folk ingredients. For example, aspirin was developed from folk uses of the willow tree, and certain cancer drugs have evolved from folk traditions using the periwinkle flower. American folk medicine traditions have also influenced mainstream medical practice. These traditions have been best preserved in the Upland South, on some Indian reservations, and in the Mexican borderland.

The outlook of the Upland Southerner toward cures is well expressed in the comments of an eastern Tennessee mountaineer root digger who, in an interview with cultural geographer Edward Price, said that "the good Lord has put these yerbs here for man to make hisself well with. They is a yerb, could we but find it, to cure every illness." Root digging has been popularized to the extent that much of the produce of the Appalachians is now funneled to dealers, who serve a larger market outside the folk culture. Nonetheless, root digging remains at heart a folk enterprise, carried on in the old ways and requiring the traditionally thorough knowledge of the plant environment.

Although the attention to conservation varies from culture to culture, folk cultures' close ties to the land and local environment enhance the environmental perception of folk groups. This becomes particularly evident when they migrate. Typically, they seek new lands similar to the ones left behind. A good example can be seen in the migrations of Upland Southerners from the mountains of Appalachia between 1830 and 1930. As the Appalachians became increasingly populous, many Upland Southerners began looking elsewhere for similar areas to settle. Initially, they found an

environmental twin of the Appalachians in the Ozark-Ouachita Mountains of Missouri and Arkansas. Somewhat later, others sought out the hollows, coves, and gaps of the central Texas Hill Country. The final migration of Appalachian hill people brought some 15,000 members of this folk culture to the Cascade and Coast mountain ranges of Washington State between 1880 and 1930 (**Figure 2.20**).

People so close to nature tend to remain sensitive to very subtle environmental qualities. Nowhere is this sensitivity more evident than in the practice of "planting by the signs," found among folk farmers in the United States and elsewhere. Reliance on the movement and appearance of planets, stars, and the moon might seem absurd to the managers of huge corporate farms, but these beliefs and practices still exist among the members of folk cultures.

Nature in Popular Culture

Popular culture is less directly tied to the physical environment than folk culture, which is not to say that popular culture does not have an enormous impact on the environment. Urban dwellers generally do not draw their livelihoods from the land. They have no direct experience with farming, mining, or logging activities, though they could not live without the commodities produced from those activities. Gone is the intimate association between people and land known by our folk ancestors. Gone, too, is our direct vulnerability to many environmental forces, although the security is more apparent than real. Because popular culture is so tied to mass consumption, it can have enormous environmental impacts, such as the production

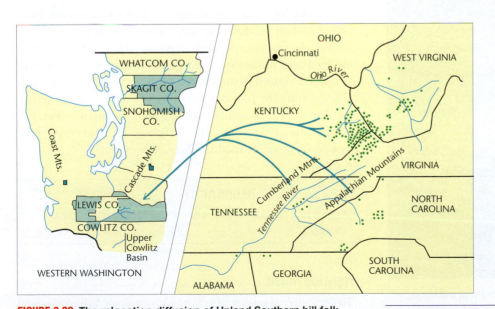

FIGURE 2.20 **The relocation diffusion of Upland Southern hill folk from Appalachia to western Washington.** Each dot represents the former home of an individual or family that migrated to the Upper Cowlitz River basin in the Cascade Mountains of Washington State between 1884 and 1937. Some 3000 descendants of these migrants lived in the Cowlitz area by 1940. (After Clevinger, 1938: 120; Clevinger, 1942: 4.)

THINKING GEOGRAPHICALLY What does the high degree of clustering of the sources of the migrants and subsequent clustering in Washington suggest about the processes of folk migrations? How should we interpret their choices of familiar terrain and vegetation for a new home? Why might members of a folk society who migrate choose a new land similar to the old one?

of air and water pollution and massive amounts of solid waste. Also, because popular culture fosters limited contact with and knowledge of the physical world, usually through recreational activities, our environmental perceptions can become quite distorted.

Popular culture makes heavy demands on ecosystems. This is true even in the seemingly benign realm of recreation. Recreational activities have increased greatly in the world's economically affluent regions. Many of these activities require machines, such as snowmobiles, off-road vehicles, and jet skis, that are powered by internal combustion engines and have numerous adverse ecological impacts ranging from air pollution to soil erosion. In national parks and protected areas worldwide, affluent tourists in search

of nature have overtaxed protected environments and wildlife and produced levels of congestion approaching those of urban areas (**Figure 2.21**).

Such a massive presence of people in our recreational areas inevitably results in damage to the physical environment. A study by geographer Jeanne Kay and her students in Utah revealed substantial environmental damage done by off-road recreational vehicles, including "soil loss and long-term soil deterioration." One of the paradoxes of the modern age and popular culture seems to be that the more we cluster in cities and suburbs, the greater our impact on open areas; we carry our popular culture with us when we vacation in such regions.

CULTURAL **INTERACTION** IN FOLK AND POPULAR CULTURES

Folk groups, or the remnants of folk groups, are almost never completely segregated from popular culture. The theme of cultural interaction allows us to see how groups such as the Amish can retain their folk character and yet be in almost daily contact with popular cultures—that is, how folk groups are integrated into the nonfolk world. A lively exchange is constantly under way between the folk and popular cultures. Perhaps most commonly, members of the folk culture absorb ideas filtering down from the popular culture, perhaps altering their way of life, but occasionally elements of the folk culture instead penetrate the popular society.

Traditional folk handicrafts and arts often fetch high prices among modern city dwellers, perhaps because they exhibit the quality, attention to detail, and uniqueness generally absent in factory-made goods. Sometimes these folk goods are revised in ways to make them even more marketable. Among the folk items that have successfully penetrated the popular culture are Irish "fisherman sweaters," Shaker furniture, and the brightly painted miniature wooden animals crafted in Oaxaca, Mexico.

REFLECTING ON **GEOGRAPHY**

Can a folk culture survive when it interacts with popular culture?

Hip-Hop Music

To see cultural interaction at work, consider urban hip-hop music, which has made a significant impact on American popular culture. Hip-hop music is derived, to a great degree, from elements of African-American music and jazz poetry. However, its roots can be traced farther back, to

FIGURE 2.21 **Traffic jam in Yosemite National Park at the height of the summer tourist season.** (Lonely Planet Images/Getty Images.)

THINKING GEOGRAPHICALLY What kind of diffusion would be involved in the spread of the desire to visit Yosemite (and other popular national parks)? What environmental impacts would result?

West African cultures from which many American slave families descended. For example, some experts believe that the practice of spoken rapping, used in much of hip-hop music, stems from "call and response" patterns used in certain African religious ceremonies. Today, hip-hop is a broad international musical style that contains elements of not only rap but also beat-boxing and DJing/scratching. There are also nonmusical cultural elements woven into hip-hop culture, such as graffiti art and break dancing.

Unlike some other musical forms that have roots in traditional folk cultures (like Appalachian music), hip-hop music has not remained confined to the societies in which it first developed. Instead, its popularity spread across the United States, leapfrogging quickly between urban centers beginning in the 1970s. This spread was encouraged by the style's progression from a performance-based form to a recorded one. During the 1980s, hip-hop began to diffuse throughout the world, with some countries, like Japan, developing large hip-hop subcultures. Since the 1980s, the rapping component of hip-hop has undergone many changes and has branched into multiple (and sometimes regional) styles such as East Coast, West Coast, Southern, and Alternative. It has also begun to merge with other musical styles such as pop and world music. As these mixtures occur, the themes of hip-hop music and rap lyrics also change, commenting on life and issues in diverse popular cultures in the United States and around the world.

From Difference to Convergence . . .

Globalization, which we introduced in Chapter 1, brings up an important question: Does globalization homogenize cultural difference? As we look for the answer in real-world examples, we find that the answer is less than straightforward.

Globalization is most directly and visibly at work in popular culture. Increased leisure time, instant communications, greater affluence for many people, heightened mobility, greater access to the Internet, and weakened attachment to family and place—all attributes of popular culture—have the potential, through interaction, to cause massive spatial restructuring. Most geographers long assumed that the result of such globalizing forces and trends, especially mobility and the electronic media, would be the homogenization of culture, wherein the differences among places are reduced or eliminated. This assumption is called the **convergence hypothesis;** that is, cultures are converging, or becoming more alike. In the geographical sense, this would yield placelessness, a concept discussed earlier in the chapter.

> **convergence hypothesis** A hypothesis holding that cultural differences among places are being reduced by improved transportation and communications systems and networks, leading to a homogenization of popular culture.

. . . And Difference Revitalized

Geographer Peter Jackson is a strong proponent of the position that cultural differences are not simply obliterated under the wave of globalization. For Jackson, globalization is not an all-powerful force. People in different places respond in different ways, rejecting outright some of what globalization brings while transforming and absorbing other aspects into local culture. Rather than one homogeneous globalized culture, Jackson sees multiple **local consumption cultures.**

> **local consumption cultures** Distinct consumption practices and preferences in food, clothing, music, and so forth, formed in specific places and historical moments.

Local consumption culture refers to the consumption practices and preferences—in food, clothing, music, and so on—formed in specific places and historical moments. These local consumption cultures often shape globalization and its effects. In some ways, globalization revitalizes local difference. That is, people reject or incorporate into their cultural practices the ideas and artifacts of globalization and in the process reassert place-based identities.

For example, Jackson suggests that the introduction of Cadbury's chocolate into China is more than simply another sign of globalization. He argues that the case "demonstrates the resilience of local consumption cultures to which transnational corporations must adapt." In cases where companies' products are negatively associated with their place of origin, such as exports from apartheid South Africa, the global ambitions of multinational companies can be thwarted. Local circumstances thus can make a difference to the outcomes of globalization.

Local resistance to globalization often takes the form of **consumer nationalism,** which occurs when local consumers avoid imported products and favor locally produced alternatives. India and China, in particular, have a long history of resisting outside domination through boycotts of imported goods. Jackson discusses a case in China in which Chinese entrepreneurs invented a local alternative to Kentucky Fried Chicken called Ronhua Fried Chicken Company. The company uses what it claims are traditional Chinese herbs in its recipe, delivering a product more suitable to Chinese cultural tastes. France also has a long history of objecting to foreign companies doing business on its soil.

> **consumer nationalism** Local consumers' favoring of nationally produced goods over imported goods as part of a nationalist political agenda.

Place Images

The same media that serve and reflect the rise of personal preference—the Internet, movies, television, photography, music, advertising, art, and others—often produce place images, a subject studied by geographers Brian Godfrey and Leo Zonn, among others. Place, portrayer, and medium interact to produce the image, which in turn colors our perception of and beliefs about places and regions we have never visited. The focus on place images highlights the role of the collective imagination in the formation and dissolution of culture regions. It also explores the degree to which the image of a region fits the reality on the ground. That is, in imagining a region or place, oftentimes certain regional characteristics are stressed while others are ignored.

The images may be inaccurate or misleading, but they nevertheless create a world in our minds that has an array of unique places and place meanings. Our decisions about tourism and migration can be influenced by these images. For example, through the media, Hawaii has become in the American mind a sort of earthly paradise peopled by scantily clad, eternally happy, invariably good-looking natives who live in a setting of unparalleled natural beauty and idyllic climate. People have always formed images of faraway places. Through the interworkings of popular culture, these images proliferate and become more vivid, if not more accurate.

FOLK AND POPULAR CULTURAL LANDSCAPES

Do folk and popular cultures look different? Do different cultures have distinctive cultural landscapes? The theme of cultural landscape reveals the important differences within and between cultures.

Folk Architecture

Every folk culture produces a highly distinctive landscape. One of the most visible aspects of these landscapes is **folk architecture.** These traditional buildings illustrate the theme of cultural landscape in folk geography.

> **folk architecture** Structures built by members of a folk society or culture in a traditional manner and style, without the assistance of professional architects or blueprints, using locally available raw materials.

Folk architecture springs not from the drafting tables of professional architects but from the collective memory of groups of traditional people (**Figure 2.22**). These buildings—whether dwellings, barns, churches, mills, or inns—are based not on blueprints but on mental images that change little from one generation to the next. Folk architecture is marked not by refined artistic genius or spectacular revolutionary design but rather by traditional, conservative, and functional structures. Material composition, floor plan, and layout are important ingredients of folk architecture, but numerous other characteristics help classify farmsteads and dwellings. The form or shape of the roof, the placement of the chimney, and even such details as the number and location of doors and windows can be important classifying criteria. E. Estyn Evans, a noted expert on Irish folk geography, considered roof form and chimney placement, among other traits, in devising an informal classification of Irish houses.

The house, or dwelling, is the most basic structure that people erect, regardless of culture. For most people in

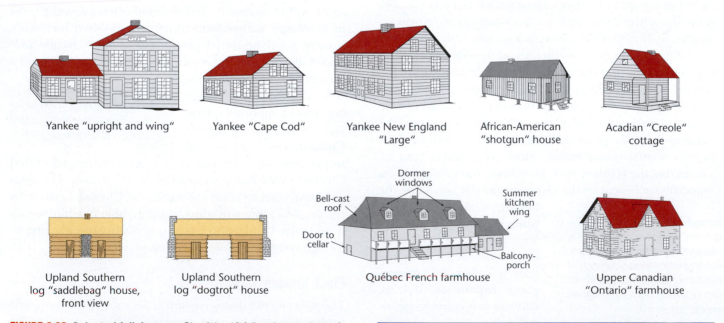

Yankee "upright and wing"
Yankee "Cape Cod"
Yankee New England "Large"
African-American "shotgun" house
Acadian "Creole" cottage

Upland Southern log "saddlebag" house, front view
Upland Southern log "dogtrot" house

Québec French farmhouse
— Dormer windows
— Bell-cast roof
— Summer kitchen wing
— Door to cellar
— Balcony-porch

Upper Canadian "Ontario" farmhouse

FIGURE 2.22 Selected folk houses. Six of the 13 folk culture regions of North America are represented (see Figure 2.1). (After Glassie, 1968, Kniffen, 1965.)

THINKING GEOGRAPHICALLY What features of these houses reflect adaptation to the environment of the folk region that developed them? What might be the origins of other features?

nearly all folk cultures, a house is the single most important thing they ever build. Folk cultures as a rule are rural and agricultural. For these reasons, it seems appropriate to focus on the folk house.

Folk Housing in North America

In the United States and Canada, folk architecture today is a *relict* form preserved in the cultural landscape. That is, it has survived from an earlier period or in a primitive form. For the most part, popular culture, with its mass-produced, commercially built houses, has so overwhelmed the folk traditions that few folk houses are built today, yet many survive in the refuge regions of American and Canadian folk culture (see Figures 2.1, 2.22).

Yankee folk houses are of wooden frame construction, and shingle siding often covers the exterior walls. They are built with a variety of floor plans, including the Yankee New England "large" house, a huge two-and-a-half-story house built around a central chimney and two rooms deep. As the Yankee folk migrated westward, they developed the upright-and-wing dwelling. These particular Yankee houses are often massive, in part because the cold winters of the region forced most work to be done indoors. By contrast, Upland Southern folk houses are smaller and built of notched logs. Many houses in this folk tradition consist of two log rooms, with either a double fireplace between, forming the saddlebag house, or a roofed breezeway separating the two rooms, a plan known as the dogtrot house (**Figure 2.23**). An example of an African-American folk dwelling is the shotgun house, a narrow structure only one room in width but two, three, or even four rooms in depth. Acadiana, a French-derived folk region in Louisiana, is characterized by the half-timbered Creole cottage, which has a central chimney and built-in porch. Scores of other folk house types

survive in the American landscape, although most such dwellings now stand abandoned and derelict.

Canada also offers a variety of traditional folk houses (see Figure 2.22). In French-speaking Québec, one of the common types consists of a main story atop a cellar, with attic rooms beneath a curved, bell-shaped (or bell-cast) roof. A balcony-porch with railing extends across the front, sheltered by the overhanging eaves. Attached to one side of this type of French-Canadian folk house is a summer kitchen that is sealed off during the long, cold winter. Many of the folk houses of Québec are built of stone. To the west, in the Upper Canadian folk region, one type of folk house occurs so frequently that it is known as the Ontario farmhouse. One and a half stories high, the Ontario farmhouse is usually built of brick and has a distinctive gabled front dormer window.

Examine **Figure 2.24** on page 50 and test your ability to identify each type of house in the figure.

The interpretation of folk architecture is by no means a simple process. Folk geographers often work for years trying to "read" such structures, seeking clues to diffusion and traditional adaptive strategies. The old problem of independent invention versus diffusion is raised repeatedly in the folk landscape, as **Figure 2.25** on page 51 illustrates. Precisely because interpretation is often difficult, however, geographers find these old structures challenging and well worth studying. Folk cultures rarely leave behind much in the way of written records, making their landscape artifacts all the more important in seeking explanations.

FIGURE 2.23 **A dogtrot house, typical of the Upland Southern folk region.** The distinguishing feature is the open-air passageway, or dogtrot, between the two main rooms. This house is located in central Texas. (Courtesy of Terry G. Jordan-Bychkov.)

THINKING GEOGRAPHICALLY How is this house related to the environment in the region that developed it?

(a)

(b)

(c)

(d)

FIGURE 2.24 Four folk houses in North America. (Courtesy of Terry G. Jordan-Bychkov.)

THINKING GEOGRAPHICALLY Using the sketches in Figure 2.22 and the related section of the text, determine the regional affiliation and type of each of these houses.

Folk Housing in Sub-Saharan Africa

Throughout East Africa and southern Africa, rural family homesteads take a common form. Most consist of a compound of buildings called a *kraal*, a term related to the English word *corral* and used across the region. The compound typically includes a main house (or houses in polygamous cultures), a detached building in the rear for cooking, and smaller buildings or enclosures for livestock.

All construction is done with local materials. Small, flexible sticks are woven in between poles that have been driven into the ground to serve as the frame. Then a mixture of clay and animal dung is plastered against the woven sticks, layer by layer, until it is entirely covered. Dried tall grasses are tied together in bundles to form the roof. In a recent innovation, rural people who can afford the expense have replaced the grass with corrugated iron sheets. In some societies the dwellings are round, in others square or rectangular. You can often tell when you have

entered a different culture region by the change in house types.

One of the most distinctive house types is found in the Ndebele culture region of southern Africa, which stretches from South Africa north into southern Zimbabwe. In the rural parts of the Ndebele region, people are farmers and livestock keepers and live in traditional kraals. What makes these houses distinctive is the Ndebele custom of painting brightly colored designs on the exterior house walls and sometimes on the walls and gates surrounding the kraal (**Figure 2.26**). The precise origins of this custom are unclear, but it seems to date to the mid-nineteenth century. Some suggest that it was an assertion of cultural identity in response to their displacement and domination at the hands of white settlers. Others point to a religious or sacred role.

What is clear is that the custom has always been the purview of women, a skill and practice passed down from mother to daughter. Many of the symbols and patterns

(a)

(b)

FIGURE 2.25 Two polygonal folk houses. (a) Buriat Mongol yurt in southern Siberia, near Lake Baikal. (b) Navajo hogan in New Mexico. The two dwellings, almost identical and each built of notched logs, lie on opposite sides of the world, among unrelated folk groups who never had contact with each other. Such houses do not occur anywhere in between. (a: Courtesy of Terry G. Jordan-Bychkov; b: Number 7/ Alamy.)

THINKING GEOGRAPHICALLY Is cultural diffusion or independent invention responsible? How might a folk geographer go about finding the answer?

are associated with particular families or clans. Initially, women used natural pigments from clay, charcoal, and local plants, which restricted their palette to earth tones of brown, red, and black. Today many women use commercial paints—expensive but longer lasting—to apply a range of bright colors, limited only by the imagination. Another new development is in the types of designs and symbols used. People are incorporating modern machines such as automobiles, televisions, and airplanes into their designs. In many cases, traditional paints and symbols are blended with the modern to produce a synthetic design of old and new. Ndebele house painting is developing and evolving in new directions, all the while continuing to signal a persistent cultural identity to all who pass through the region.

Landscapes of Popular Culture

Popular culture permeates the landscape of countries such as the United States, Canada, and Australia, including everything from mass-produced suburban houses to golf courses and neon-lit strips. So overwhelming is the presence of popular culture in most American settlement landscapes that an observer must often search diligently to find visual fragments of the older folk cultures. The popular landscape is in continual flux, for change is a hallmark of popular culture.

Few aspects of the popular landscape are more visually striking than the ubiquitous commercial malls and strips on urban arterial streets, which geographer Robert Sack calls *landscapes of consumption* (see Figure 2.6). In an Illinois college town, two other cultural geographers, John Jakle and Richard Mattson, made a study of the evolution of one such

FIGURE 2.26 Ndebele village in South Africa. Although the origins and meaning of Ndebele house painting are debated, there is no question that the paintings create a visually distinct cultural landscape. (Ariadne Van Zandbergen/ Alamy.)

THINKING GEOGRAPHICALLY What other cultures feature decorations on the exterior of their houses? Are the purposes purely decorative? Explain.

FIGURE 2.27 **The West Edmonton Mall, Alberta, Canada.** This enormous mall represents the growing infusion of spectacle and fantasy into the urban landscape. In addition to retail outlets, the mall includes an aquarium, an ice-skating rink and the submarine ride, shown here. (©James Marshall/Corbis.)

THINKING GEOGRAPHICALLY What role in the community does your local mall play? Is it designed as a getaway destination?

strip. During a 60-year span, the street under study changed from a single-family residential area to a commercial district. The researchers suggested a five-stage model of strip evolution, beginning with the single-family residential period, moving through stages of increasing commercialization, which drives owner-residents out, and culminating in stage 5, where the residential function of the street disappears and a totally commercial landscape prevails. Business properties expand so that off-street parking can be provided.

Public outcries over the ugliness of such strips are common. Even landscapes such as these are subject to interpretation, however, for the people who create them perceive them differently. For example, geographer Yi-Fu Tuan suggests that a commercial strip of stores, fast-food restaurants,

filling stations, and used-car lots may appear as visual blight to an outsider, but the owners or operators of the businesses are very proud of them and of their role in the community. Hard work and high hopes color their perceptions of the popular landscape.

Perhaps no landscape of consumption is more reflective of popular culture than the indoor shopping mall, numerous examples of which now dot both urban and suburban landscapes. Of these, the largest is West Edmonton Mall in the Canadian province of Alberta (**Figure 2.27**). Enclosing some 5.3 million square feet (493,000 square meters) and opened in 1986, West Edmonton Mall employs 23,500 people in more than 800 stores and services, accounts for nearly one-fourth of the total retail space in greater Edmonton, earned 42 percent of the dollars spent in local shopping centers, and experienced 2800 crimes in its first nine months of operation. Beyond its sheer size, West Edmonton Mall also boasts a water park, a sea aquarium, an ice-skating rink, a miniature golf course, a roller coaster, 21 movie theaters, and a 360-room hotel. Its "streets" feature motifs from such distant places as New Orleans, represented by a Bourbon Street complete with fiberglass ladies of the evening. Jeffrey Hopkins, a geographer who studied this mall, refers to this as a "landscape of myth and elsewhereness," a "simulated landscape" that reveals the "growing intrusion of spectacle, fantasy, and escapism into the urban landscape."

Leisure Landscapes

Another common feature of popular culture is what geographer Karl Raitz labeled **leisure landscapes.** Leisure landscapes are designed to entertain people on weekends and vacations; often they are included as part of a larger tourist experience. Golf courses and theme parks such as Disney World are good examples of such landscapes. **Amenity landscapes** are a related landscape form. These are regions with attractive natural features such as forests, scenic mountains, or lakes and rivers that have become desirable locations for retirement or vacation homes. One such landscape is in the Minnesota North Woods lake country, where, in a sampling of home ownership, geographer Richard Hecock found that fully 40 percent of all dwellings were not permanent residences but instead weekend cottages or vacation homes. These are often purposefully made rustic or even humble in appearance.

The past, reflected in relict buildings, has also been incorporated into the leisure landscape. Most often, collections of old structures are relocated to form "historylands,"

> **leisure landscape** A landscape that is planned and designed primarily for entertainment purposes, such as ski and beach resorts.
>
> **amenity landscape** Landscape that is prized for its natural and cultural aesthetic qualities by the tourism and real estate industries and their customers.

often enclosed by imposing chain-link fences and open only during certain seasons or hours. If the desired bit of visual history has perished, Americans and Canadians do not hesitate to rebuild it from scratch, undisturbed by the lack of authenticity—as, for example, at Jamestown, Virginia, or Louisbourg on Cape Breton Island, Nova Scotia. Normally, the history parks are put in out-of-the-way places and sanitized to the extent that people no longer live in them. Role-playing actors sometimes prowl these parks, pretending to live in some past era, adding "elsewhenness" to "elsewhereness."

Elitist Landscapes

A characteristic of popular culture is the development of social classes. A small elite group—consisting of persons of wealth, education, and expensive tastes—occupies the top economic position in popular cultures. The important geographical fact about such people is that because of their wealth, desire to be around similar people, and affluent lifestyle, they can and do create distinctive cultural landscapes, often over fairly large areas.

Daniel Gade, a cultural geographer, coined the term *elitist space* to describe such landscapes, using the French Riviera as an example. A photo illustrating the opulence of this elitist space can be found in Chapter 1 (Figure 1.9). Like other elitist landscapes, the Riviera visually produces distinctive cultural divides between the wealthy and the less affluent. These social-class divisions are equally evident in many evolving cultural landscapes around the world. **Figure 2.28** illustrates one contemporary elitist landscape in the oil- and tourism-rich United Arab Emirates (UAE). Due to the warm, sunny climate in this Middle Eastern state, as well as the availability of amenities like world-class golf courses and luxury shopping venues, the UAE has become a popular destination for celebrities and the ultra-wealthy. The grand scale and over-the-top opulence of the resorts, high-rise dwellings, and human-made private islands (all constructed to attract affluent people) have become part of a new elitist landscape. Other elitist landscapes are also taking shape in many of the world's tropical regions in the form of private island ownership by celebrities such as Johnny Depp and Mel Gibson and business moguls like Virgin's Richard Branson. These wealthy individuals participate in the creation of new elite spaces that are financially inaccessible to people in less affluent social classes. In this way, social divisions are made apparent and emphasized on the cultural landscape.

FIGURE 2.28 **The Emirates Palace Hotel in Abu Dhabi.** This resort offers million-dollar getaway packages for the world's elite. Partaking in "trophy tourism," guests who can afford these exclusive holidays are treated to private jet excursions, a personal butler, and other amenities. (Press Association/The Image Works.)

THINKING GEOGRAPHICALLY   Not all elitist spaces are highly visible on the everyday cultural landscape. Can you imagine why this would be?

FIGURE 2.29 **Gentleman farm in the Kentucky Bluegrass region near Lexington.** Here is "real" rural America as it should be (but never was). (Courtesy of Terry G. Jordan-Bychkov.)

THINKING GEOGRAPHICALLY Why would the people of Lexington find it in their economic interest to see that such landscapes are preserved?

America, too, offers elitist landscapes. An excellent example is the gentleman farm, an agricultural unit operated for pleasure rather than profit (**Figure 2.29**). Typically, affluent city people own gentleman farms as an avocation, and such farms help to create or maintain a high social standing for those who own them. Some rural landscapes in America now contain many such gentleman farms; perhaps most notable among these places are the inner Bluegrass Basin of north-central Kentucky, the Virginia Piedmont west of Washington, D.C., eastern Long Island in New York, and parts of southeastern Pennsylvania. Gentleman farmers engage in such activities as breeding fine cattle, racing horses, or hunting foxes.

Geographer Karl Raitz conducted a study of gentleman farms in the Kentucky Bluegrass Basin, where their concentration is so great that they constitute the dominant feature of the cultural landscape. The result revealed an idyllic scene, a rural landscape created more for appearance than for function. Raitz provided a list of visual indicators of Kentucky gentleman farms: wooden fences, either painted white or creosoted black; an elaborate entrance gate; a fine hand-painted sign giving the name of the farm and owner; a network of surfaced, well-maintained driveways and pasture roads; and a large, elegant house, visible in the distance from the public highway through a lawnlike parkland dotted with clumps of trees and perhaps a pond or two. So attractive are these estates to the eye that tourists travel the rural lands to view them, convinced they are seeing the "real" rural America, or at least rural America as it ought to be.

REFLECTING ON GEOGRAPHY
Can you think of other types of consumption, leisure, and elitist landscapes?

The American Popular Landscape: The Cult of Bigness?

Geographer David Lowenthal has attempted to analyze the cumulative visible impact of popular culture on the American countryside. Lowenthal identified the main characteristics of popular landscape in the United States, including the "cult of bigness"; the tolerance of present ugliness to achieve a supposedly glorious future; an emphasis on individual features at the expense of aggregates, producing a "casual chaos"; and the preeminence of function over form.

The American fondness for massive structures is reflected in edifices such as the Empire State Building, the Pentagon, the San Francisco–Oakland Bay Bridge, and Salt Lake City's Mormon Temple. Americans have dotted their cultural landscape with the world's largest of this or that, perhaps in an effort to match the grand scale of the physical environment, which includes such landmarks as the Grand Canyon, the towering redwoods of California, and the Rocky Mountains.

Americans, argues Lowenthal, tend to regard their cultural landscape as unfinished. As a result, they are "predisposed to accept present structures that are makeshift, flimsy, and transient," resembling "throwaway stage sets." Similarly, the hardships of pioneer life perhaps preconditioned Americans to value function more highly than beauty and form. In summary, American popular culture seems to have produced a built landscape that stresses bigness, utilitarianism, and transience. Sometimes these are opposing trends, as in the case of many of the massive structures previously cited, which are clearly built to last. Sometimes the trends mesh, as in the explosion of "big box" retail chains. These giant retail buildings are no more than oversized metal sheds that one can easily imagine being razed overnight to be replaced by the next big thing.

Key Terms

amenity landscape, 52	local consumption
consumer nationalism, 47	cultures, 47
convergence hypothesis, 47	material culture, 30
ethnomedicine, 44	mass culture, 32
folk, 30	nonmaterial culture, 30
folk architecture, 48	placelessness, 33
folk culture, 31	popular culture, 31
folk geography, 31	subculture, 30
leisure landscape, 52	vernacular culture region, 37

Folk and Popular Cultures on the Internet

You can learn more about the main categories of culture discussed in the chapter on the Internet at the following web sites:

American Memory, Library of Congress

http://memory.loc.gov

A project of the Library of Congress that presents a history of American popular culture, complete with documentation and maps.

Center for Folklife and Cultural Heritage

http://www.folklife.si.edu

A research and educational unit of the Smithsonian Institution promoting the understanding and continuity of diverse, contemporary grassroots cultures in the United States and around the world. The center produces exhibitions, films and videos, and educational materials.

Popular Culture Association

http://pcaaca.org

A multidisciplinary organization dedicated to the academic discussion of popular culture, where activities of the association are discussed.

Recommended Books on Folk and Popular Culture

Bronner, Simon. 2011. *Explaining Traditions: Folk Behavior in Modern Culture.* Lexington: The University of Kentucky Press. In this volume, Bronner discusses the ways in which folk traditions contribute to contemporary cultures, including the culture of the Internet. The book includes many examples of American folk culture behavior surrounding phenomena like storytelling, football, and the construction of roadside shrines for the dead.

Carney, George O. (ed.). 1998. *Baseball, Barns and Bluegrass: A Geography of American Folklife.* Boulder, CO.: Rowman & Littlefield. A wonderful collection of readings that, contrary to the title, span the gap between folk and popular culture.

Ensminger, Robert F. 1992. *The Pennsylvania Barn: Its Origin, Evolution, and Distribution in North America.* Baltimore: Johns Hopkins University Press. A common American folk barn, part of the rural cultural landscape, provides geographer Ensminger with visual clues to its origin and diffusion; a fascinating detective story showing how geographers "read" cultural landscapes and what they learn in the process.

Glassie, Henry. 1968. *Pattern in the Material Folk Culture of the Eastern United States.* Philadelphia: University of Pennsylvania Press. Glassie, a student of folk geographer Fred Kniffen, considers the geographical distribution of a wide array of folk culture items in this classic overview.

Jordan, Terry G., Jon T. Kilpinen, and Charles F. Gritzner. 1997. *The Mountain West: Interpreting the Folk Landscape.* Baltimore: Johns Hopkins University Press. Reading the folk landscapes of the American West, three geographers reach conclusions about the regional culture and how it evolved.

Skelton, Tracey, and Gill Valentine (eds.). 1998. *Cool Places: Geographies of Youth Cultures.* London: Routledge. The engaging essays in *Cool Places* explore the dichotomy of youthful lives by addressing the issues of representation and resistance in youth culture today. Using first-person vignettes to illustrate the wide-ranging experiences of youth, the authors consider how the media have imagined young people as a particular community with shared interests and how young people resist these stereotypes, instead creating their own independent representations of their lives.

Weiss, Michael J. 1994. *Latitudes and Attitudes: An Atlas of American Tastes, Trends, Politics, and Passions.* New York: Little, Brown. Using marketing data organized by postal zip codes, Weiss reveals the geographical diversity of American popular culture.

Zelinsky, Wilbur. 1992. *The Cultural Geography of the United States,* 2nd ed. Englewood Cliffs, NJ: Prentice-Hall. This revised edition of a sprightly classic book, originally published in 1973, reveals the cultural sectionalism in modern America in the era of popular culture, with attention also to folk roots.

Journals in Folk and Popular Culture

Journal of Popular Culture. Published by the Popular Culture Association since 1967, this journal focuses on the role of popular culture in the making of contemporary society. See in particular Volume 11, No. 4, 1978, a special issue on cultural geography and popular culture.

Material Culture: Journal of the Pioneer America Society. Published twice annually, this leading periodical specializes in the subject of the American rural material culture of the past. Volume 1 was published in 1969, and prior to 1984 the journal was called *Pioneer America.*

Answers to Thinking Geographically Questions

Figure 2.1: Buildings, designs, music, languages, and crafts often survive, sometimes in forms that are now produced in factories. They become nostalgia items for people who wish to remember former times and ways of life. Places with many remains of folk cultures may become tourist destinations.

Figure 2.2: It provides mechanisms for the manufacture of items of popular culture, as well as the communications technology that disseminates the information about it and the advertising to promote it.

Figure 2.3: The idea of the barn would have come with migrants from central Europe, who then modified it to fit their needs in their new homeland. The forebay is useful because hay is stored in the upper level, while animal pens are in the lower level. Farmers can throw the hay out the doors to animals in the barnyard without blocking the doors to the animal pens. This barn seems to have lost its role on a farm producing livestock and is used mainly for general storage.

Figure 2.4: Although the actual origins of the practice, which probably came from Africa, have been lost, it has been connected to the danger of snakes. The practice also eliminates the need to maintain vegetation by watering and mowing.

Figure 2.5: The apparatus is crudely constructed, with no superfluous parts. It was made of local materials, although the original design as brought by migrants demanded the use of scarce wood. It would probably never have been advertised, because each rancher would know how to construct a "beef wheel" if needed.

Figure 2.6: The scenes were taken in the following "unplaces": (a) McDonald's in Tokyo, (b) Breezewod, Pennsylvania, sometimes called "Town of Motels."

Figure 2.7: There are no signs for multinational companies and no billboards advertising anything. Houses are in traditional style, oriented to the topography.

Figure 2.8: You may think of other lifestyle clusters, especially if your home area is not included in the three regions mapped. You may also dispute the distribution of one or more of these regions, for example pointing out that the coast of California south of San Francisco is home to a great many groups besides upper-middle-class retirees.

Figure 2.9: These regions are renowned for their vineyards and wineries and have long cultural traditions of winemaking.

Figure 2.10: Answers will vary, but some other examples in the United States are Tex-Mex cuisine in the Southwest, seafood specialties in various coastal regions, Cajun food in Louisiana (and surrounding areas), boiled peanuts in South Carolina, and soul food in the South.

Figure 2.11: The patterns of fast-food consumption could be related to such factors as availability of fast-food restaurants in states with small populations like North Dakota. Regional differences in average income could also be a factor, as fast food is generally cheaper than many healthier food choices.

Figure 2.12: It would likely show that states named in rap/hip-hop lyrics tend to be those that contain major urban centers (such as New York, Los Angeles, and Atlanta), where rap/hip-hop music developed and enjoys large fan bases. Therefore, this music's popularity would not show a strongly regionalized pattern like that of country music.

Figure 2.13: Population in the area marked "West" is more sparsely distributed, so that towns are farther apart. All of this area was once part of the frontier, and thus it gained the perception of being "the West" as it was settled. Businesses use that perception to self-identify with an important element in American cultural history. Places with no regional perception are transition zones between culture regions and have received influences from several sources. An example is the area of western Pennsylvania–northern West Virginia–southeastern Ohio, which has influences from the Middle Atlantic, the South, and especially the Appalachian Mountains.

Figure 2.14: Although landscape features tend toward sameness, people tend to identify with a particular region and use it as a relative location. Mass media (for example, The Weather Channel) also use such regions as identifiers, solidifying the perception of their extent.

Figure 2.15: Contagious diffusion was first, and given the time period when the first fair was held (1810), that would have been the most available kind. In the far West, probably hierarchical diffusion played a role as fairs were established in major towns, especially county seats.

Figure 2.16: Answers will vary, but you may think of examples such as the diffusion of popular music styles like reggae from the Caribbean, newly popular competitive ballroom dances that diffused to the United States from older European dance styles, or popular reality television programming that diffused to the United States from European popular television.

Figure 2.17: In such countries, urban areas are frequently a major draw for migrants because they are perceived to have more wealth and job opportunities. Similarly, Western culture is perceived as more prestigious because the West is perceived as wealthy. McDonald's acquires its reputation by this association.

Figure 2.18: Besides the Pacific Ocean, the major barrier would be the eastward extent of ranching and thus of local cowboys and cowboy life. Rodeos would not resonate with local populations in the East in the same way.

Figure 2.19: The distribution seems to have been both independent invention and diffusion. Probably the blowgun was developed independently in the Africa/Pacific region and in the Americas. The distribution in the Indian and Pacific oceans matches that of Austronesian languages, suggesting transfer by long-distance traders. Among the data that might be used to investigate these questions would be other cultural transfers and the details of the blowguns' designs.

Figure 2.20: Migrations probably took place in family groups and in groups of people from the same valley in Appalachia. This could have been groups migrating together at the same time as well as chain migration. Not only would migrants find mountainous territory appealing as a reminder of the environment in which their culture developed, but they would have similar resources (wood, water) available. However, the choice of a place to settle would also depend on available land.

Figure 2.21: Both hierarchical diffusion through media and contagious diffusion (conversation with friends or relatives who have been there) would be involved. The large number of visitors adds to the environmental stress by its demands on water, food (which comes from outside the park), waste disposal, and the sheer impact of more feet and more vehicles in the park.

Figure 2.22: The Yankee houses have central fireplaces, providing heat for the long, cold winters. The African-American shotgun house probably originated in Africa and came to the lower Mississippi Valley via Haiti. The construction on piles allows for air circulation under the house, keeping it cooler in the hot summers, and also provides protection from moderate flooding.

Figure 2.23: The dogtrot house allows air circulation in the hot summers. The construction material is logs, which were abundantly available in the Upland South.

Figure 2.24: (a) Québec French farmhouse, Port Joli, Québec; (b) Yankee New England large house, New Hampshire; (c) Yankee upright and wing house, Massachusetts; (d) African-American shotgun house, Alleyton, Texas.

Figure 2.25: Although it is possible that the idea traveled by migration and the in-between cultures have discontinued the style, the probable explanation is independent invention. The techniques involved are relatively simple and do not require extensive learning from other people. Cultural geographers might look for other cultural transfers between the same regions and for other buildings that use similar techniques to these polygonal structures.

Figure 2.26: Painted scenes are found on houses in southern Germany. Some appear to be purely decorative, while others have religious themes. In North American popular culture, plastic eagles, hex signs, flags, and holiday decorations are common exterior decorations. They express holiday observances but also souvenirs of travel and to some extent the rules of taste regarding decoration in the community. Flags may express patriotism or loyalty to a sports team as well as aesthetics.

Figure 2.27: Answers will vary.

Figure 2.28: The wealthy often wish to preserve their privacy when taking advantage of elitist landscapes. Therefore, these areas are often developed in secluded places or feature walls, vegetation barriers, sea barriers, or other privacy defenses so that the media and the less wealthy do not have visual or physical access to them.

Figure 2.29: Aside from the valuable horses raised on these farms (not visible in the picture), the beauty of the landscape is a major tourist draw.

A street in Ho Chi Minh City, Vietnam. (Tony Burns/Shooting the World/Image Brief.)

Population Geography

3

One of the most important aspects of the world's human population is its demographic characteristics, such as age, gender, health, mortality, density, and mobility. In fact, many cultural geographers argue that familiarity with the spatial dimensions of demography provides a baseline for the discipline. Thus, **population geography,** or **geodemography,** provides an ideal topic to launch a substantive discussion of the human mosaic.

The essential demographic fact is that more than 7 billion people inhabit Earth today. Numbers alone tell only part of the story of the delicate balance between human populations and the resources on which we depend for our survival, comfort, and enjoyment. Think of the sort of lifestyle you may now enjoy or aspire to in the future. Does it involve driving a car? Eating meat regularly? Owning a spacious house with central heating and air conditioning? If so, you are not alone. In fact,

population geography
Geodemography; the study of the spatial and ecological aspects of population, including distribution, density per unit of land area, fertility, gender, health, age, mortality, and migration.

geodemography Population geography.

POPULATION DENSITY

FIGURE 3.1 Population density in the world. Try to imagine the diverse causal forces—physical, environmental, and cultural—that have been at work over the centuries to produce this complicated spatial pattern. It represents the most basic cultural geographical distribution of all. (Sources: Population Reference Bureau, *World Population Data Sheet*; Statistical Abstract of the United States.)

THINKING GEOGRAPHICALLY Is the northeastern part of the United States a fourth cluster?

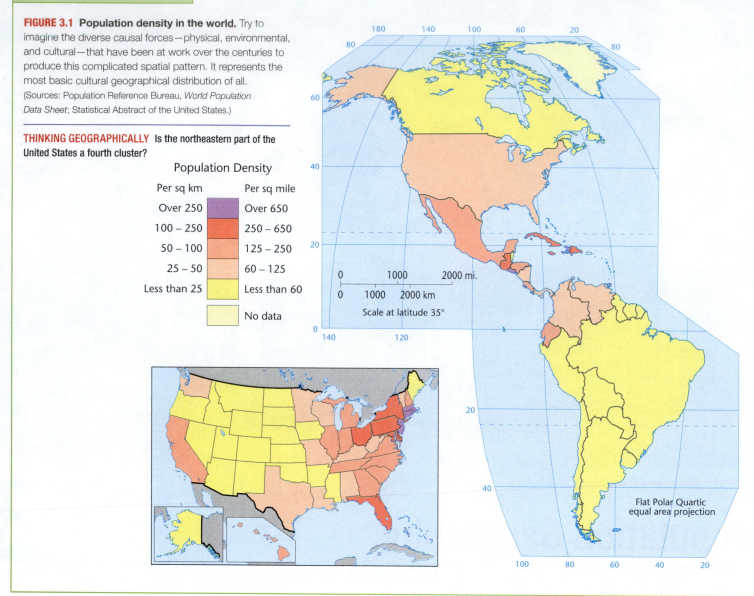

these "Western" consumption habits have become so widespread that they may now be more properly regarded as universal. Satisfying these demands requires using a wide range of nonrenewable resources—including fossil fuels, extensive farmlands, and fresh water—whose consumption ultimately limits the number of people Earth can support. Indeed, it has been argued that if Western lifestyles are adopted by a significant number of the globe's inhabitants, then our current population of 7 billion is already excessive and will soon deplete or contaminate Earth's life-support systems: the air, soil, and water we depend on for our very survival. Although we may think of our geodemographic choices, such as how many children we will have, as highly individual ones, when aggregated across whole groups they can have truly global repercussions.

As we do throughout this book, we approach our study using the five themes of cultural geography: culture region, diffusion, ecology, interaction, and landscape.

DEMOGRAPHIC **REGIONS**

In what ways do demographic traits vary regionally? How is the theme of culture region expressed in terms of population characteristics?

The principal characteristics of human populations—their densities, spatial distributions, age and gender structures, the ways they increase and decrease, and how rapidly population numbers change—vary enormously from place to place. Understanding the demographic characteristics of

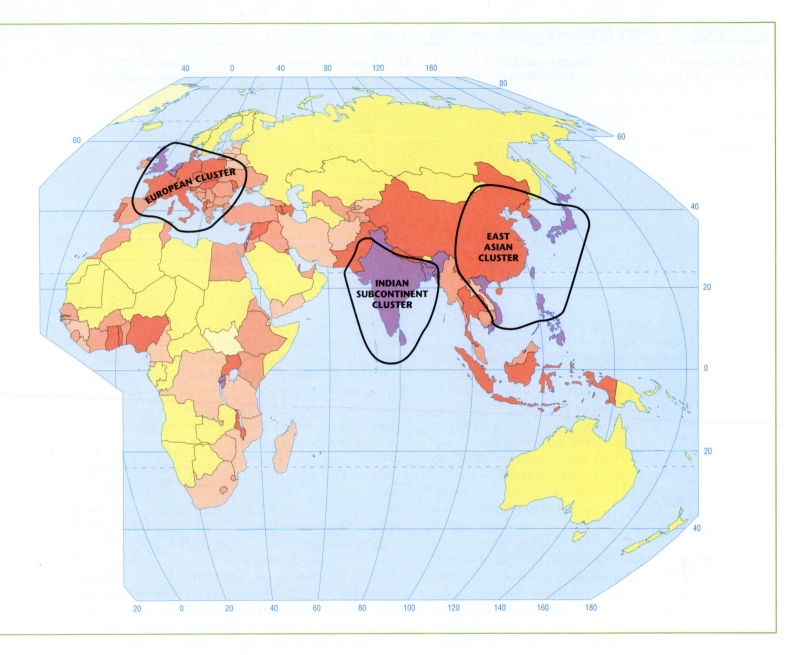

populations, and why and how they change over time, gives us important clues to their cultural characteristics.

Population Distribution and Density

If the over 7 billion inhabitants of Earth were evenly distributed across the land area, the **population density** would be about 121 persons per square mile (47 per square kilometer). However, people are very unevenly distributed, creating huge disparities in density. Iceland, for example, has approximately eight people per square mile (three per square kilometer), whereas Bangladesh has 2,709 people per square mile (1,046 per square kilometer) (**Figure 3.1**).

population density A measurement of population per unit area (for example, per square mile).

If we consider the distribution of people by continents, we find that 70.8 percent of the human race lives in Eurasia—Europe and Asia. North America (Canada, the United States, and Mexico) is home to only 6.6 percent of people, Africa to 15 percent, South America to 5.7 percent, and Australia and the Pacific islands to 0.5 percent. When we consider population distribution by country, we find that 19.3 percent of humans live in China (including Hong Kong and Macao), 17.7 percent in India, and only 4.5 percent in the third-largest nation in the world, the United States (**Table 3.1**, page 60). In fact, one in 50 humans lives in only one valley of one province of China: the Sichuan Basin.

Due to the variation in population density from one country to another, it is important to analyze population

TABLE 3.1	The World's 10 Most Populous Countries, 2011 and 2050		
Most Populous Countries, 2011	**Population in 2011 (in millions)**	**Most Populous Countries, 2050 (estimated)**	**Population in 2050 (estimated, in millions)**
China	1,346	India	1,692
India	1,241	China	1,313
United States	312	Nigeria	433
Indonesia	238	United States	423
Brazil	197	Pakistan	314
Pakistan	177	Indonesia	309
Nigeria	162	Bangladesh	226
Bangladesh	151	Brazil	223
Russia	143	Ethiopia	174
Japan	128	Philippines	150

(Source: Population Reference Bureau, *2011 World Population Data Sheet*)

data on a global scale. Figure 3.1 clearly shows certain patterns in population density in different regions of the world. On one end of the spectrum, there are thickly settled areas that have over 250 people per square mile (100 or more per square kilometer). On the other end of the spectrum, there are largely unpopulated areas that have fewer than 2 persons per square mile (less than 1 per square kilometer). Moderately settled areas and thinly settled areas fall between these two extremes. These categories of settlement create **formal demographic regions** based on the single trait of population density.

formal demographic region A demographic region based on the single trait of population density.

As Figure 3.1 shows, a fragmented crescent of densely settled areas stretches along the western, southern, and eastern edges of the huge Eurasian continent. Two-thirds of the human race is concentrated in this crescent, which contains three major population clusters: eastern Asia, the Indian subcontinent, and Europe. Outside of Eurasia, only scattered districts are so densely settled. Despite the image of a crowded world, thinly settled regions are much more extensive than thickly settled ones, and they appear on every continent. Thin settlement characterizes the northern sections of Eurasia and North America, the interior of South America, most of Australia, and a desert belt through North Africa and the Arabian Peninsula into the heart of Eurasia.

As geographers, we want to know more about population geography than simply population density. For example, what are people's standards of living, and are they related to population density? Some of the most thickly populated areas in the world have the highest standards of living—and even suffer from labor shortages (for example, the major industrial areas of western Europe and Japan). In other cases, thinly settled regions may actually be

severely overpopulated relative to their ability to support their population, a situation that is usually associated with marginal agricultural lands. Although 1000 persons per square mile (400 per square kilometer) is a sparse population for an industrial district, it is dense for a rural area. For this reason, **carrying capacity**—the population beyond which a given environment cannot provide support without becoming significantly damaged—provides a far more meaningful index of overpopulation than density alone. Often, however, it is difficult to determine carrying capacity until the region under study is near or over the limit. Sometimes the carrying capacity of one place can be expanded by drawing on the resources of another place. Americans, for example, consume far more food, products, and natural resources than do most other people in the world: 22 percent of the entire world's petroleum production, for instance, is consumed in the United States. The carrying capacity of the United States would be exceeded if it did not utilize the resources—including the labor—of much of the rest of the world.

carrying capacity The maximum number of people who can be supported in a given area.

A critical feature of population geography is the demographic changes that occur over time. Analyzing these gives us a dynamic perspective from which we can glean insights into cultural changes occurring at local, regional, and global scales. Populations change primarily in two ways: people are born and others die in a particular place, and people move into and out of that place. The latter refers to *migration*, which we will consider later in this chapter. For now we discuss births and deaths, which can be thought of as additions to and subtractions from a population. They provide what demographers refer to as natural increases and natural decreases.

CRUDE BIRTH RATE

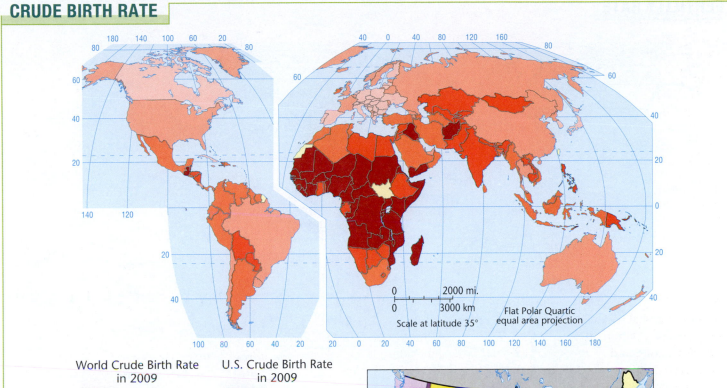

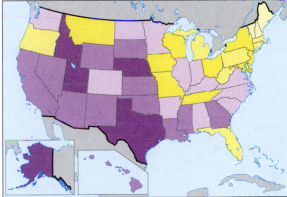

World Crude Birth Rate in 2009

- 33 or more
- 23 – 32
- 17 – 22
- 12 – 16
- Less than 12
- Not available

U.S. Crude Birth Rate in 2009

- 15.0 or more
- 14.0 – 14.9
- 13.3 – 13.9
- 12.5 – 13.2
- 11.0 – 12.4
- Less than 11.0

Crude Birth Rate = Number of births occurring during the year, per 1000 population

FIGURE 3.2 The crude birthrate is calculated as the number of births in a given year per 1000 population (at a midyear estimate). This measure varies greatly from country to country, with sub-Saharan Africa currently experiencing the highest crude birthrates of any sizable region of the world. (Sources: *Statistical Abstract of the United States*; The World Bank)

THINKING GEOGRAPHICALLY Do you think that crude birthrate data is consistently accurate from country to country around the world?

Patterns of Natality

Births can be measured by several methods. One way is to calculate the **crude birthrate**—the number of births per thousand people per year (**Figure 3.2**). Birthrates ranging from 10 to 20 births per thousand are considered low, while rates from 40 to 50 births per thousand are considered high. While high birthrates can present problems for governments attempting to care for populations with large numbers of children, low birthrates can also put stress on governments to provide

crude birthrate The annual number of births per thousand population.

for senior citizens, since there will be fewer people of working age to support this segment of the population.

Another revealing statistic is the **total fertility rate,** or **TFR,** which is measured as the average number of children born per woman during her reproductive lifetime, which is considered to be from 15 to 49 years of age. The TFR is a more useful measure than the birthrate because it focuses on the female segment of the population, reveals average family size, and gives an indication of future changes in the

total fertility rate (TFR) The number of children the average woman will bear during her reproductive lifetime (15 to 49 years old).

FERTILITY RATE

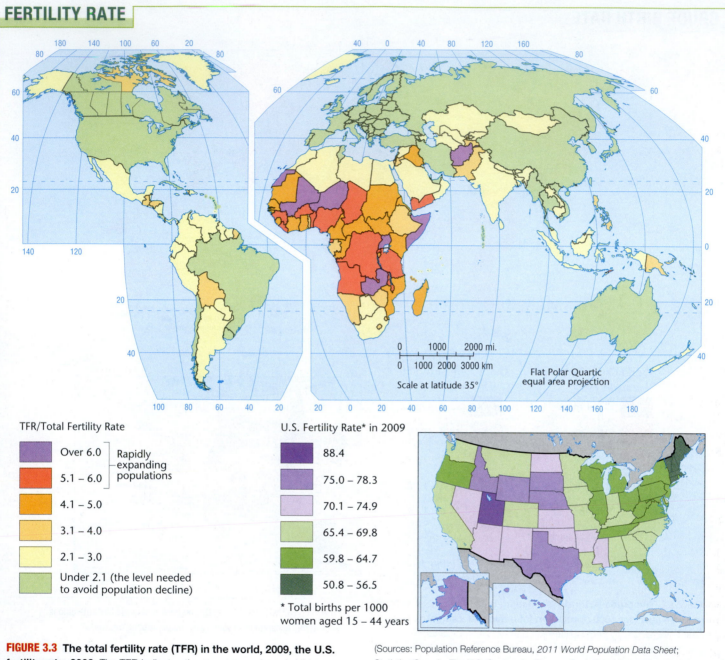

TFR/Total Fertility Rate

- Over 6.0 } Rapidly expanding populations
- 5.1 – 6.0
- 4.1 – 5.0
- 3.1 – 4.0
- 2.1 – 3.0
- Under 2.1 (the level needed to avoid population decline)

U.S. Fertility Rate* in 2009

- 88.4
- 75.0 – 78.3
- 70.1 – 74.9
- 65.4 – 69.8
- 59.8 – 64.7
- 50.8 – 56.5

* Total births per 1000 women aged 15 – 44 years

FIGURE 3.3 **The total fertility rate (TFR) in the world, 2009, the U.S. fertility rate, 2009.** The TFR indicates the average number of children a woman would have during her child-bearing years (ages 15-49), assuming that age-specific birthrates stay constant. A rate of 2.1 is needed to produce a stable population over the long run; below that, population will decline. Fast growth is associated with a TFR of 5.0 or higher. Within the United States, the highest fertility rate is found in the state of Utah, while the lowest rate occurs in Vermont.

(Sources: Population Reference Bureau, *2011 World Population Data Sheet*; Statistics Canada; The U.S. Centers for Disease Control and Prevention, *National Vital Statistics Reports*.)

THINKING GEOGRAPHICALLY Why do you think the TFR in many developing states is higher than that in more developed regions?

---

population structure. A TFR of 2.1 is needed to produce an eventually stabilized population—that is, one that does not increase or decrease. Once achieved, this condition is called **zero population growth.**

zero population growth A condition of population stability; zero population growth is achieved when an average of only 2.1 children per couple survive to adulthood, so that eventually the number of deaths equals the number of births.

The TFR varies markedly from one part of the world to another, revealing a vivid geographical pattern (**Figure 3.3**). In southern and eastern Europe, the average TFR is well under 2.0, so every country in these regions can eventually expect population decline. Bulgaria, for example, has a TFR of 1.5 and is expected to lose 15 percent of its population by 2050. By contrast, sub-Saharan Africa has

CRUDE DEATH RATE

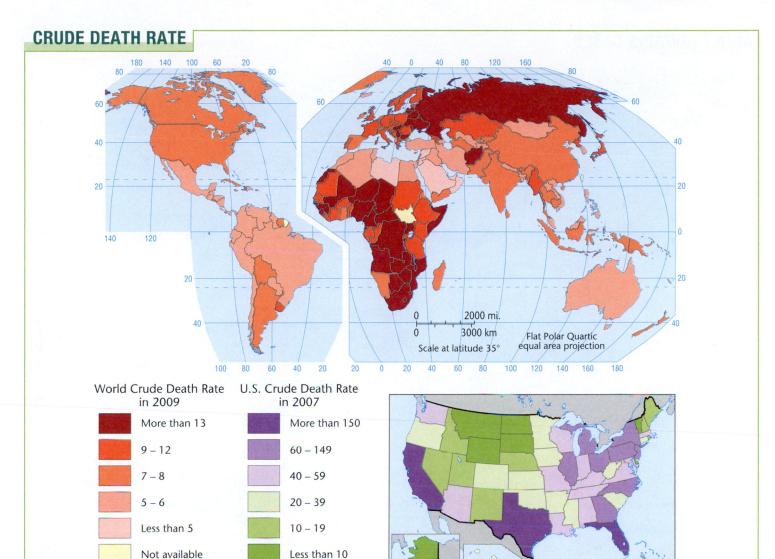

World Crude Death Rate in 2009

■	More than 13
■	9 – 12
■	7 – 8
■	5 – 6
■	Less than 5
■	Not available

U.S. Crude Death Rate in 2007

■	More than 150
■	60 – 149
■	40 – 59
■	20 – 39
■	10 – 19
■	Less than 10

Crude Death Rate = Number of deaths occurring during the year, per 1000 population

FIGURE 3.4 The world crude death rate, 2009, and the U.S. crude death rate, 2007. This is calculated as the number of deaths in a given year per thousand population (at a mid-year estimate). The national measures reflect the overall quality of health care in a country as well as episodic drops in population due to events such as war or famine.

(Sources: *Statistical Abstract of the United States*; The World Bank.)

THINKING GEOGRAPHICALLY Why do you think that the crude death rates for Florida and Texas are higher than in most other U.S. states?

the highest TFR of any region of the world, led by Niger, with a TFR of 7.0. Therefore, Niger can expect its population to increase by 112 percent by 2050. Elsewhere in the world, only Afghanistan in south-central Asia and the Arabian peninsula can rival the sub-Saharan African rates. Interestingly, according to the World Bank, during the past two decades, TFRs have begun to fall in all sub-Saharan African nations.

The Geography of Mortality

Another way to assess demographic change is to analyze the **crude death rate:** the deaths per year per thousand people (**Figure 3.4**).

crude death rate The annual deaths per thousand persons in the population.

In looking at the map of world crude death rates, we can observe patterns across different regions, with lower death rates occurring in developed states that can offer better and more widely available health care to citizens.

Of course, death is a natural part of the life cycle, and there is no way to achieve a death rate of zero. Geographically speaking, death comes in different forms. In the developed world, most people die of age-induced degenerative conditions, such as heart disease, or from maladies caused by industrial pollution of the environment. Many types of cancer fall into the latter category. By contrast, contagious diseases such as malaria, HIV/AIDS, and diarrheal diseases are a leading cause of death in poorer countries. Civil warfare, inadequate health services,

ADULT HIV/AIDS CASES

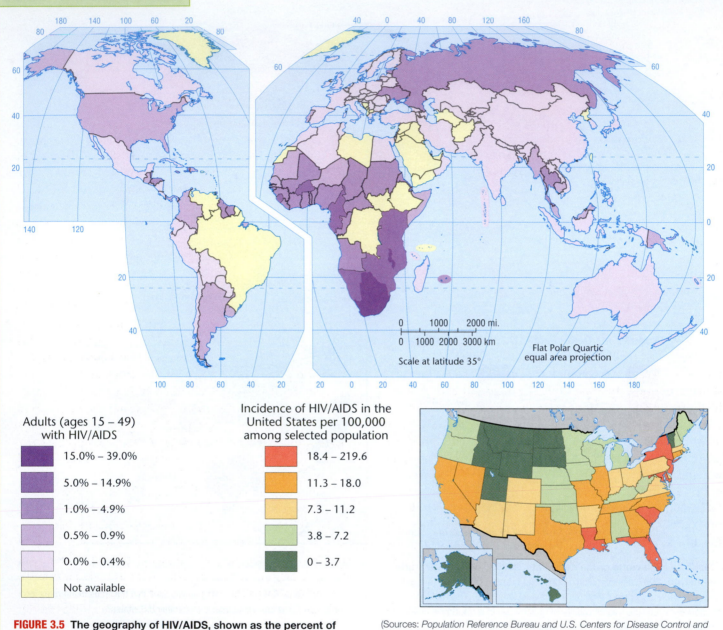

Adults (ages 15 – 49) with HIV/AIDS

- 15.0% – 39.0%
- 5.0% – 14.9%
- 1.0% – 4.9%
- 0.5% – 0.9%
- 0.0% – 0.4%
- Not available

Incidence of HIV/AIDS in the United States per 100,000 among selected population

- 18.4 – 219.6
- 11.3 – 18.0
- 7.3 – 11.2
- 3.8 – 7.2
- 0 – 3.7

FIGURE 3.5 The geography of HIV/AIDS, shown as the percent of adults aged 15 to 49 with HIV/AIDS in 2009. The quality of data gathering varies widely from one country to another and is particularly poor in Africa and most of Asia. The U.S. map shows the incidence of HIV/AIDS per 100,000 people.

(Sources: *Population Reference Bureau and U.S. Centers for Disease Control and Prevention.*)

THINKING GEOGRAPHICALLY The highest rates of HIV/AIDS are in Africa. What parts of Africa have lower rates? What cultural factors might explain this?

and the age structure of a country's population also affect its death rate.

The highest death rates occur in sub-Saharan Africa, the poorest world region and the one most afflicted by life-threatening diseases and civil strife. (**Figure 3.5** illustrates the geography of HIV/AIDS.) In general, death rates of more than 25 per thousand people are uncommon today. The world's highest death rate as of 2007—slightly more than 30 per thousand people—was found in Swaziland, in southern Africa, and is the result in great part because of the HIV/AIDS prevalence rate there, which is nearly 40 percent among the adult population. By contrast, the American tropics generally have rather low death rates, as does the desert belt across North Africa, the Middle East, and central Asia. In these regions, the predominantly young population depresses the death rate. Compared to Swaziland, Ecuador's death rate of only 5 per thousand seems quite low. Because of its older population, the average death rate in the European Union is 10 per thousand. Australia, Canada, and the United States, which continue to attract young immigrants, have lower death rates than most of Europe. Canada's death rate, for instance, is slightly less than 8 people per thousand.

Population Explosion?

One of the fundamental issues of the modern age is the **population explosion:** a dramatic increase in world population since 1900 (**Figure 3.6** on page 66). The crucial element triggering this explosion has been a steep decline in the death rate,

population explosion The rapid, accelerating increase in world population since about 1650 and especially since 1900.

particularly for infants and children, in most of the world, without an accompanying universal decline in fertility. At one time in traditional cultures, only two or three offspring in a family of six to eight children might live to adulthood, but when improved health conditions allowed more children to survive, the cultural norm encouraging large families persisted.

Until very recently, the number of people in the world has been increasing geometrically, doubling in shorter and shorter periods. It took from the beginning of human history until A.D. 1800 for the Earth's population to grow to 1 billion people, but from 1800 to 1930, it grew to 2 billion, and in only 45 more years it doubled again (**Figure 3.7** on page 66). Looking at the population explosion in another way, it is estimated that 61 billion humans have lived in the entire 200,000-year period since *Homo sapiens* originated. Of these, 7 billion (roughly 11 percent) are alive today. Approximately one of every 10 humans who ever lived on Earth is alive today. If we were to consider only those humans who survived into adulthood, the proportion alive today would come closer to one in five!

Some scholars foresaw long ago that an ever-increasing global population would eventually present difficulties. The most famous pioneer observer of population growth, the English economist and cleric Thomas Malthus, published *An Essay on the Principle of Population*—known as the "dismal essay"—in 1798. He believed that the human ability to multiply far exceeds our ability to increase food production. Consequently, Malthus maintained that "a strong and constantly operating check on population" will necessarily act as a natural control on numbers. Malthus regarded famine, disease, and war as the inevitable outcome of the human population's outstripping the food supply. He wrote, "Population, when unchecked, increases in a geometrical ratio. Subsistence only increases in an arithmetical ratio. A slight acquaintance with numbers will show the immensity of the first power in comparison of the second."

The word **Malthusian** entered the English language to describe the dismal future Malthus foresaw and the people who subscribe to Malthus's views. Being a cleric as well as an econo-

Malthusian One who holds the views of Thomas Malthus, who believed that overpopulation is the root cause of poverty, illness, and warfare.

mist, however, Malthus believed that if humans could voluntarily restrain the "passion between the sexes," they might avoid their otherwise miserable fate.

Was Malthus right? From the very first, his ideas were controversial. The founders of communism, Karl Marx and Friedrich Engels, blamed poverty and starvation on the evils of capitalist society. Taking this latter view might lead one to believe that the miseries of starvation, warfare, and disease are more the result of maldistribution of the world's wealth than of overpopulation. Indeed, severe food shortages within the past decade in several regions point to precisely this issue of inequitable food distribution and its catastrophic consequences.

Malthus did not consider that when faced with conundrums such as scarce food supplies, human beings are highly creative. This has led critics of Malthus and his modern-day followers to point out that while the global population has doubled three times since Malthus wrote his essay, food supplies have doubled five times. Scientific innovations such as the green revolution have led to food increases that far outpaced population growth (see Chapter 8). Other measures of well-being, including life expectancy, air quality, and average education levels, have all improved, too.

The fact is that at the beginning of the new millennium, the world's population is growing more slowly than before. The world's TFR has fallen to 2.8; one demographer has declared that "the population explosion is over"; and Figure 3.5 depicts a leveling-off of global population at about 11 billion by the year 2100. Relatively stable population totals worldwide, however, mask the population declines and aging of the population structure already under way in some regions of the world.

It is difficult to speculate about the state of the world's population beyond the year 2050. What is clear, however, is that the lifestyles we adopt will affect how many people Earth can ultimately support. In particular, whether wealthy countries continue to use the amount of resources they currently do, and whether developing countries decide to follow a Western-style route toward more and more resource consumption, will bear greatly on whether we have already overpopulated the planet.

The Demographic Transition

All industrialized, technologically advanced countries have low fertility rates and stabilized or declining populations,

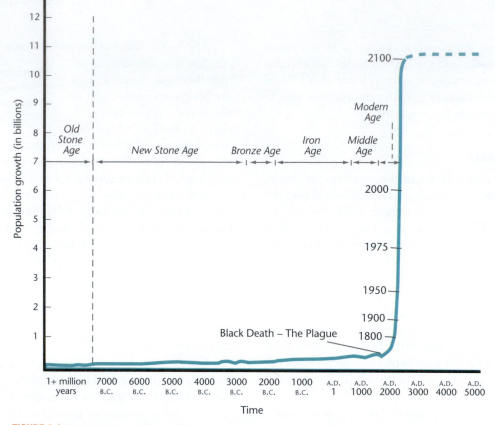

FIGURE 3.6 S-shaped world population curve. Is the global population explosion nearing its end? If this graph is right, the world's population will stabilize at nearly 11 billion by the year 2100. (Sources: Adapted from Population Reference Bureau and United Nations, *World Population Projections 2100*, 1998.)

THINKING GEOGRAPHICALLY What assumptions underlie the broken line at the right side of the graph?

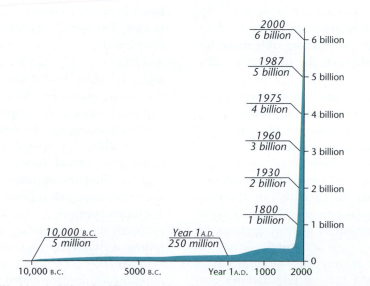

FIGURE 3.7 World population doubling time. This graph illustrates the ever-faster doubling times of the world's population. Whereas accumulating the first billion people took all of human history until about A.D. 1800, the next billion took slightly more than a century to add, the third billion took only 30 years, and the fourth took only 15 years. (Source: Adapted from Sustainablescale.org.)

THINKING GEOGRAPHICALLY Relate the shape of this growth to the demographic transition (Figure 3.9).

having passed through what is called the **demographic transition** (**Figure 3.8**). In preindustrial societies, birth and death rates were both high, resulting in almost no population growth. Because these were agrarian societies that depended on family labor, many children meant larger work-forces, thus the high birthrates. But low levels of public health and limited access to

demographic transition The movement from high birth and death rates to low birth and death rates.

health care, particularly for the very young, also meant high death rates. With the coming of the industrial era, medical advances and improvements in diet set the stage for a drop in death rates. Human life expectancy in industrialized countries soared from an average of 35 years in the eighteenth century to 75 years or more at present. Yet birthrates did not fall so quickly, leading to a population explosion as fertility outpaced mortality. In Figure 3.7, this is shown in late stage 2 and early stage 3 of the model. Eventually, a decline in the birthrate followed the decline in the death rate, slowing population growth. An important reason leading to lower fertility levels involves the high cost of children in industrial societies, particularly because childhood itself becomes a prolonged period of economic dependence on parents. Finally, in the postindustrial period, the demographic transition produced zero population growth or actual population decline (**Figure 3.9** on page 68).

Achieving lower death rates is relatively cost effective, historically requiring little more than the provision of safe drinking water and vaccinations against common infectious diseases. Less death tends to be uncontroversial and fast acting, demographically speaking. Getting birthrates to fall, however, can be far more difficult, especially for a government official who wants to be reelected. Birth control, abortion, and challenging long-held beliefs about family size can prove quite controversial, and political leaders may be reluctant to legislate for them. Indeed, the Chinese implementation of its one child per couple policy probably would never have been possible in a country with a democratically elected government. In addition, because it involves changing a cultural norm, the idea of smaller families can take three or four generations to take hold. Increasing educational levels for women is closely associated with falling fertility levels, as is access to various contraceptive devices (**Figure 3.10** on page 69).

The demographic transition is a model that predicts trends in birthrates, death rates, and overall population

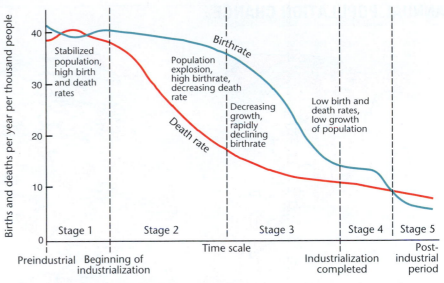
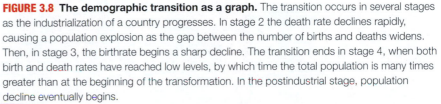

FIGURE 3.8 **The demographic transition as a graph.** The transition occurs in several stages as the industrialization of a country progresses. In stage 2 the death rate declines rapidly, causing a population explosion as the gap between the number of births and deaths widens. Then, in stage 3, the birthrate begins a sharp decline. The transition ends in stage 4, when both birth and death rates have reached low levels, by which time the total population is many times greater than at the beginning of the transformation. In the postindustrial stage, population decline eventually begins.

THINKING GEOGRAPHICALLY Why does the death rate decline before the birthrate?

levels in the abstract. It does a good job of describing population patterns over time in Europe, as well as in other wealthy regions. However, it has several shortcomings. First is the inexorable stage-by-stage progression implicit in the model. Have countries or regions ever skipped a stage or regressed? Certainly. The case of China shows how policy, in this case government-imposed restrictions on births, can fast-forward an entire nation to stage 4. War, too, can occasion a return to an earlier stage in the model by increasing death rates. For instance, Angola and Sierra Leone are two African countries with recent histories of conflict and with some of the highest death rates in the world: 24 per thousand and 22 per thousand, respectively. In other cases, wealth has not led to declining fertility. Indeed, the Population Reference Bureau has pointed to a "demographic divide" between countries where the demographic transition model applies well and others—mostly poorer countries or those with widespread conflict or disease—where birth and death rates do not necessarily follow the model's predictions.

Age Distributions

Some countries have overwhelmingly young populations. In a majority of countries in Africa, as well as some countries in Latin America and tropical Asia, close to half the population is younger than 15 years of age. In Uganda,

ANNUAL POPULATION CHANGE

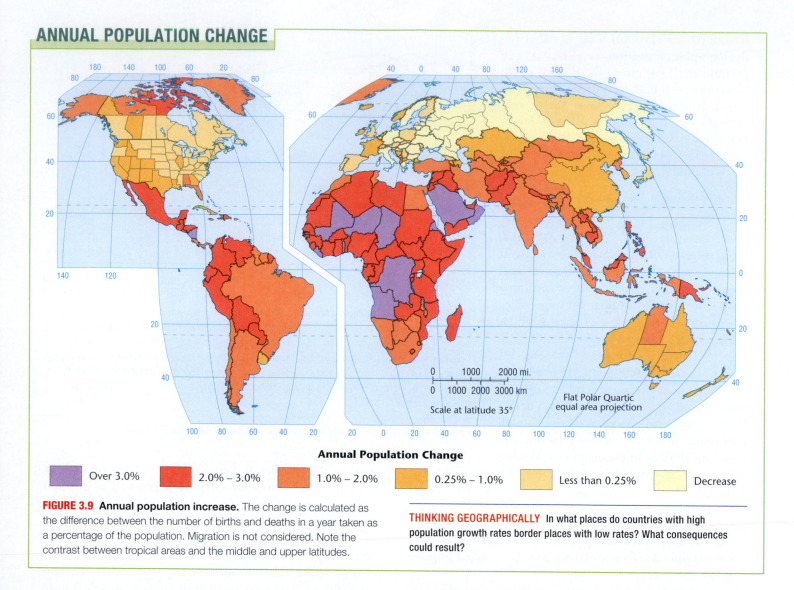

Annual Population Change

| | Over 3.0% | | 2.0% – 3.0% | | 1.0% – 2.0% | | 0.25% – 1.0% | | Less than 0.25% | | Decrease |

FIGURE 3.9 Annual population increase. The change is calculated as the difference between the number of births and deaths in a year taken as a percentage of the population. Migration is not considered. Note the contrast between tropical areas and the middle and upper latitudes.

THINKING GEOGRAPHICALLY In what places do countries with high population growth rates border places with low rates? What consequences could result?

for example, 51 percent of the population is younger than 15 years of age. In sub-Saharan Africa, 44 percent of all people are younger than 15. Other countries, generally those that industrialized early, have a preponderance of people aged 15 to 65. A growing number of affluent countries have remarkably aged populations. In Sweden, for example, fully 17 percent of the people have now passed the traditional retirement age of 65. Many other European countries are not far behind. A sharp contrast emerges when Europe is compared with Africa, Latin America, or parts of Asia, where the average person never even lives to age 65. In Mauritania, Niger, Afghanistan, Guatemala, and many other countries, only 2 to 3 percent of the people have reached that age.

Countries with disproportionate numbers of old or young people often address these imbalances in innovative ways. Italy, for example, has one of the lowest birthrates in the world, with a TFR of only 1.4. In addition, Italy's population is one of the oldest in the world; 18.6

percent of its population is age 65 or older. As a result of both its TFR and age distribution, Italy's population is projected to shrink by 10 percent between 2004 and 2050.

Given that the Italian culture does not embrace the institutionalization of the growing ranks of their elderly, and faced with the reality that more Italian women than ever work outside the home and thus few adult women are willing or able to stay at home full time to care for their elders, Italians have gotten creative. Elderly Italians can apply for adoption by families in need of grandfathers or grandmothers. One such man, Giorgio Angelozzi, recently moved in with the Rivas, a Roman family with two teenagers. Angelozzi said that Marlena Riva's voice reminded him of his deceased wife, Lucia, and this is what convinced him to choose the Riva family (adapted from D'Emilio, 2004).

Age structure also differs spatially within individual countries. For example, rural populations in the Unit-

GEOGRAPHY OF CONTRACEPTION

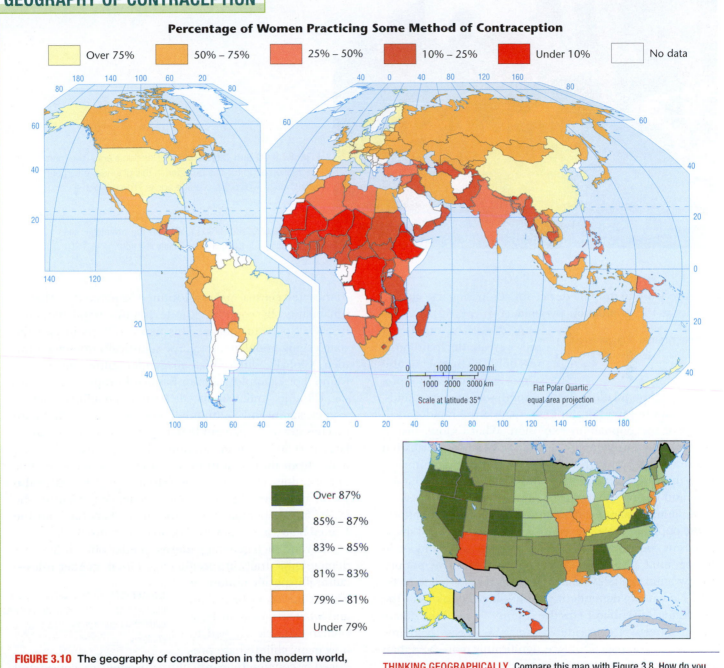

Percentage of Women Practicing Some Method of Contraception

Over 75% | 50% – 75% | 25% – 50% | 10% – 25% | Under 10% | No data

0   1000   2000 mi.
0 1000 2000 3000 km
Scale at latitude 35°

Flat Polar Quartic
equal area projection

Over 87%

85% – 87%

83% – 85%

81% – 83%

79% – 81%

Under 79%

FIGURE 3.10 **The geography of contraception in the modern world, as measured by the percentage of women using devices of any sort.** Contraception is much more widely practiced than abortion, but cultures differ greatly in their level of acceptance. Several devices are included. (Source: *World Population Data Sheet*.)

THINKING GEOGRAPHICALLY Compare this map with Figure 3.8. How do you explain the areas (for example, much of Latin America) where contraceptive use is relatively high and yet population growth rates are also high?

ed States and many other countries are usually older than those in urban areas. The flight of young people to the cities has left some rural counties in the midsection of the United States with populations whose median age is 45 years or older. Some warm areas of the United States have become retirement havens for the elderly; parts of Arizona and Florida, for example, have populations far above the average age. Communities such as Sun City near Phoenix, Arizona, legally restrict residence to the elderly (**Figure 3.11** on page 70). In Great Britain, coastal districts have a much higher proportion of elderly than does the interior, suggesting that many of the aged migrate to seaside locations when they retire.

FIGURE 3.11 Sun City, Arizona. Residents enjoy the many recreation opportunities provided in this planned retirement community, where the average age is 75 years old. (A. Ramey/PhotoEdit Inc.)

THINKING GEOGRAPHICALLY What are the advantages and disadvantages to living in a community like this?

A very useful graphic device for comparing age characteristics is the **population pyramid** (**Figure 3.12**). Careful study of such pyramids not only reveals the past progress of birth control but also allows geographers to predict future population trends. Youth-weighted pyramids, those that are broad at the base, suggest the rapid growth typical of the population explosion. What's more, broad-based population pyramids not only reflect past births but also show that future growth will rely on the momentum of all of those young people growing into their reproductive years and having their own family, regardless of how small those families may be in contrast to earlier generations. Those population pyramids with more of a cylindrical shape represent countries approaching population stability or in demographic decline. A quick look at a country's population pyramid can tell volumes about its past as well as its future. How many dependent people—the very old and the very young—live there? Has the country suffered the demographic effects of genocide or a massive disease epidemic (**Figure 3.13** on page 72)? Are significantly more boys than girls being born? These questions and more can be explored at a glance using a population pyramid.

> **population pyramid** A graph used to show the age and sex composition of a population.

Geography of Gender

Although the human race is divided almost evenly between females and males, geographical differences do occur in the **sex ratio,** the ratio between men and women in a population. Slightly more boys than girls are born, but infant boys have slightly higher mortality rates than do infant girls. Recently settled areas typically have more males than females. Examples of such areas are Alaska, northern Canada, and tropical Australia.

> **sex ratio** The numerical ratio of males to females in a population.

At the latest census, males constituted 52 percent of Alaska's inhabitants. By contrast, Rhode Island's population was nearly 52 percent female. This figure reflects in part the state's aging population, because statistically women outlive men. Globally, differentials in gender ratios can also be found. For example, states in Eastern Europe such as Russia, Ukraine, and Estonia all have populations wherein women significantly outnumber men. In Estonia, this figure reaches nearly 54 percent female. These imbalances are in large part due to the emigration of large numbers of young males from these countries in search of better economic opportunities elsewhere. The population pyramid is also useful in showing gender ratios. Note, for instance, the larger female populations in the upper bars for both the United States and Sun City, Arizona, in Figure 3.12.

Beyond such general patterns, gender often influences demographic traits in specific ways. Often, **gender roles**—culturally specific notions of what it means to be a man and what it means to be a woman—are closely tied to how many children are produced by couples. In many cultures, women are considered more womanly when they produce many offspring. Men are seen as more manly when they father many children. Because the raising of children often falls to women, the spaces that many cultures associate with women tend to be the private family spaces of the home. Public spaces such as streets, plazas, and the workplace, by contrast, are often associated with men. Some cultures go so far as to restrict where women and men may and may not go, resulting in a distinctive geography of gender. Falling fertility levels that coincide with higher levels of education for women, however, have resulted in numerous challenges to these cultural ideas of male and female spaces. As more and more women

> **gender role** What it means to be a man and what it means to be a woman in different cultural and historical contexts.

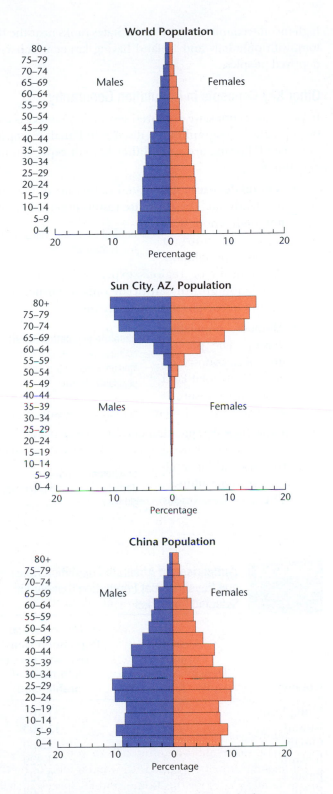

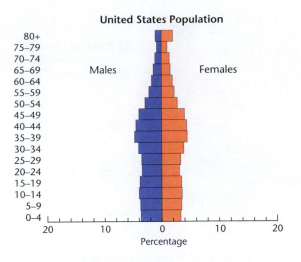

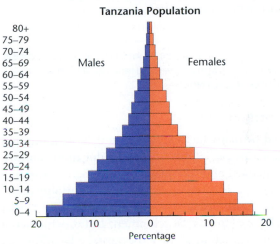

FIGURE 3.12 Population pyramids for the world and selected countries and communities. Tanzania displays the classic stepped pyramid of a rapidly expanding population, whereas the U.S. pyramid looks more like a precariously balanced pillar. China's population pyramid reflects the lowered numbers of young people as a result of that country's one-child policy. (Source: Population Reference Bureau.)

THINKING GEOGRAPHICALLY How do these pyramids help predict future population growth?

enter the workplace, for instance, ideas of where women should and should not go slowly become modified.

Standard of Living

Various demographic traits can be used to assess standard of living and analyze it geographically. **Figure 3.14** (on

page 73) is a simple attempt to map living standards using the **infant mortality rate:** a measure of how many children per thousand population die before reaching one year of age. Many experts believe that the infant mortality rate is the best single index of living standards because

infant mortality rate The number of infants per thousand live births who die before reaching one year of age.

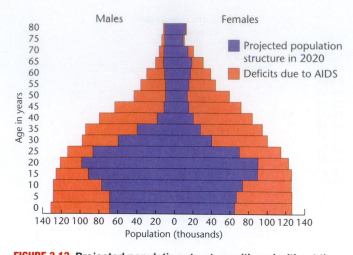

FIGURE 3.13 Projected population structure with and without the AIDS epidemic in Botswana in 2020. (Sources: U.S. Bureau of Census; Food and Agriculture Organization of the United Nations.)

THINKING GEOGRAPHICALLY Which are the widest bars in the blue pyramid in the 10- to 20-year cohorts? What does the rapid narrowing of the blue pyramid after age 20 indicate? What problems result?

it is affected by many factors: health, nutrition, sanitation, access to doctors, availability of clinics, education, ability to obtain medicines, and adequacy of housing. A vivid geographical pattern is revealed by the infant mortality rate.

Another good measure of quality of life is the United Nations HDI, or Human Development Index, which combines measures of literacy, life expectancy, education, and wealth (see Figure 1.12). The highest possible score is 1.000, and the two top-ranked countries in 2011 were Norway and Australia.

Examination of the HDI reveals some surprises. If all countries spent equally on those things that improve their HDI rankings, such as education and health care, then we would expect the wealthiest countries to place first on the list. According to the International Monetary Fund, the United States ranked fourth among 177 nations in wealth as measured by the gross domestic product, or GDP, per capita in 2006. Yet it ranked twelfth in the world on the HDI. Compare this to Barbados, which ranked forty-first by GDP per capita but came in much higher on the HDI at thirty-first. Why did Barbados rank higher in living standards than its monetary wealth would indicate, whereas the United States ranked lower? We would have to conclude that the government of Barbados places a relative priority on spending for education and health care, whereas the government of the United States does not.

Even more striking is the low standing of the United States when the Human Poverty Index, or HPI, is used. The HPI measures social and economic deprivation by examining the prevalence of factors such as low life expectancy, illiteracy, poverty, and unemployment. Among the world's

high-income countries, the United States ranks near the bottom, with only Italy and Ireland having larger numbers of deprived peoples.

Other Key Concepts in Population Geography

In previous sections, we explored several key concepts used by population geographers to discuss and analyze population. The following are some other key concepts you may encounter:

- **Natural decrease:** Calculated as the birthrate minus the death rate, this decrease results in fertility that is not high enough to replace a country or region's population. In countries or regions experiencing this trend, population growth can occur only when net in-migration outweighs natural decrease.

  natural decrease The birthrate minus the death rate.

- **Absolute population density:** The population of a country divided by its total land area. This measure is also known as arithmetic population density.

  absolute population density A country's population divided by its total land area. Also known as arithmetic population density.

- **Physiological population density:** The population of a country divided by its total *arable land* (land that is suitable and/or available for agriculture). This measure provides

  physiological population density A country's population divided by its total arable land.

TABLE 3.2 Comparison of Absolute Population Density and Physiological Population Density for Selected Countries

Country	Absolute Population Density (persons/square kilometer)	Physiological Population Density (persons/square kilometer of arable land)
Hong Kong	6,488	131,101
Kuwait	131	15,603
Oman	14	11,780
Iceland	3	4,229
Japan	339	2,924
Egypt	83	2,668
United Kingdom	250	1,077
China	140	943
United States	32	179
Canada	4	78

(Sources: Population Reference Bureau, *World Population Data Sheet*, 2011; *CIA World Factbook*, 2012.)

INFANT MORTALITY RATE

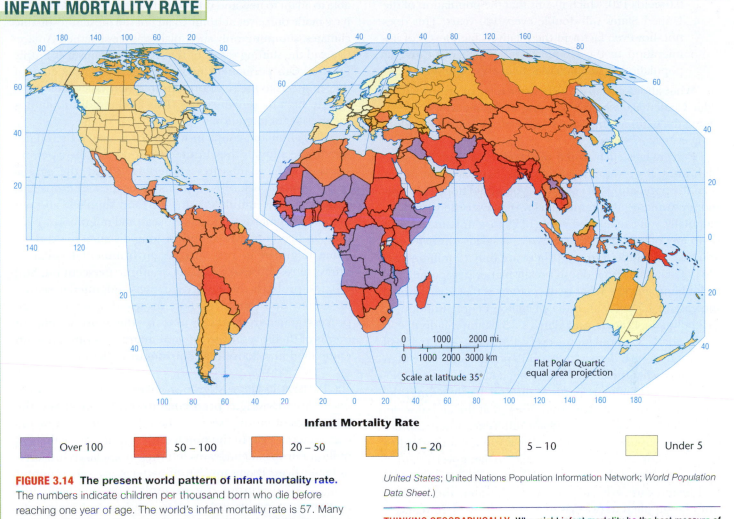

Infant Mortality Rate

Over 100	50 – 100	20 – 50	10 – 20	5 – 10	Under 5

FIGURE 3.14 **The present world pattern of infant mortality rate.** The numbers indicate children per thousand born who die before reaching one year of age. The world's infant mortality rate is 57. Many experts believe that this rate is the best single measure of living standards. (Sources: Population Reference Bureau; *Statistical Abstract of the United States*; United Nations Population Information Network; *World Population Data Sheet*.)

THINKING GEOGRAPHICALLY Why might infant mortality be the best measure of living standards?

insight into the relationship between a population and its available resources. **Table 3.2** provides a comparison of absolute population density and physiological population density for selected countries.

- **Dependency ratio:** Calculated as the number of dependent individuals (those under age 15 or over age 64) in a population divided by the number of productive individuals, expressed as a percentage. This measure provides insight into the pressure on the productive portion of a population to provide for its nonworking citizens.

dependency ratio The number of dependent individuals (those under age 15 or over age 64) in a population divided by the number of productive individuals, expressed as a percentage.

- **Population growth rate:** The increase in a country's population during a period (usually one year), expressed as a percentage of the population at the start of that period. This measure reflects the number of births and deaths during that year as well as the number of people migrating to and from that country. A country's population growth rate can be a negative number if death rates and out-migration have surpassed birthrates and in-migration for that year.

population growth rate The increase in a country's population during a period of time (usually one year), expressed as a percentage of the population at the start of that period.

- **Population doubling time:** The time it takes a population to double in size from any given numerical point. This length of time is determined by dividing the population growth rate by the number 72. For example, the natural annual growth of the United States in 2007 was 0.6 percent. Dividing 72 by

population doubling time The time it takes a population to double in size from any given numerical point. Determined by dividing the growth rate for a population into the number 72.

0.6 yields 120, which means that the population of the United States will double every 120 years. This does not, however, factor in the relatively high levels of immigration to the United States, which will cause its population to double faster than every 120 years.

What about countries with faster rates of growth? Consider Guatemala, which is growing at 2.8 percent per year. Doing the math, we find that Guatemala's population is doubling every 25.7 years!

Percentages such as 0.6 or 2.8 don't sound like such high rates. If we were discussing your bank account rather than the populations of countries, you would hope that your money would double more quickly than every 25 years! (Incidentally, you can apply the rule of 72 to your bank account or to any other figure that grows at a steady annual rate.) You may be tempted to say, "There are only 12 million people in Guatemala, so it doesn't really matter if its population is doubling quickly. What really matters is that at an annual increase of 1.7 percent, India's 1 billion people will double to 2 billion in 42 years, and China's 1.3 billion will double as fast as the U.S. population—every 120 years—but that will add another 1.3 billion to the world's population, more than four times what the United States will add by doubling!"

This assessment is partly right and partly wrong, depending on your vantage point. Viewed at the global scale, it does indeed make a significant difference when China's or India's population doubles. But if you are a resident of Guatemala or an official of the Guatemalan government, a doubling of your country's population every 25 years means that health care, education, jobs, fresh water, and housing must be supplied to twice as many people every 25 years.

DIFFUSION IN POPULATION GEOGRAPHY

How does demography relate to the theme of diffusion? After natural increases and decreases in populations, the second basic way that population numbers are altered is the result of the migration of people from one place to another, and it offers a straightforward illustration of relocation diffusion. When people migrate, they sometimes bring more than simply their culture; they can also bring disease. The introduction of diseases to new places can have dramatic and devastating consequences. Thus, the spread of diseases is also considered as part of the theme of diffusion because disease has a direct bearing on population geography.

Migration

Humankind is not tied to one locale. *Homo sapiens* most likely evolved in Africa, and ever since, we have proved remarkably able to adapt to new and different physical environments. We have made ourselves at home in all but the most inhospitable climates, shunning only such places as ice-sheathed Antarctica and the shifting sands of the Arabian Peninsula's Empty Quarter. Our permanent habitat extends from the edge of the ice sheets to the seashores, from desert valleys below sea level to high mountain slopes. This far-flung distribution is the product of migration.

Early human groups moved in response to the migration of the animals they hunted for food and the ripening seasons of the plants they gathered. Indeed, the *agricultural revolution*, whereby humans domesticated crops and animals and accumulated surplus food supplies, allowed human settlements to stop their seasonal migrations. Why, then, did certain groups opt for long-distance relocation? Some migrated in response to environmental collapse, others in response to religious or ethnic persecution. Still others—probably the majority, in fact—migrated in search of better opportunities. For those who migrate, the process generally ranks as one of the most significant events of their lives. Even ancient migrations often remain embedded in folklore for centuries or millennia.

The Decision to Move Migration takes place when people decide that moving is preferable to staying and when the difficulties of moving seem to be more than offset by the expected rewards. In the nineteenth century, more than 50 million European emigrants left their homelands in search of better lives. Today, migration patterns are very different (**Figure 3.15**). Europe, for example, now predominantly receives immigrants rather than sending out emigrants. International migration stands at an all-time high, with about 240 million people (3.1 percent of the world's population) living outside of the country of their birth. Much of this movement is due to labor migration associated with globalization. However, not all migrants are **voluntary migrants.** In fact, millions of individuals around the world today are **forced migrants** who have either been internally displaced within their own countries (27.5 million) or are **refugees** (15.4 million) fleeing political, ethnic or religious persecution, wars, famines, or other natural and environmental disasters. Interestingly, economic persecution does not fall under the definition of *refugee*. Forced migrations have occurred many times throughout history. The westward displacement of the Native American population of the United States; the dispersal of Jews from Palestine in

voluntary migrant A person leaving his or her place of residence in order to improve his or her quality of life, often for the purpose of employment.

forced migrant A migrant whose movements are coerced by threats to life and livelihood or by threats arising from natural or human-made causes

refugee A person fleeing from persecution in his or her country of nationality. The persecution can be religious, political, racial, or ethnic. A refugee is a forced migrant..

FIGURE 3.15 Migration flows today. (a) Major and minor international migration flows in the twenty-first century; (b) immigration flows to the United States in 2007. As the map illustrates, Asia and other parts of America are the leading source regions for immigrants to the United States. (Sources: [a] "Stalker's Guide to International Migration," www.pstalker.com/migration; [b] Created by Rusty Jones, Online GIS Certificate Program Student, University of West Florida, 2010.)

THINKING GEOGRAPHICALLY Why do you think that there are major migration flows from parts of Africa and Asia to Europe? What reasons can you give for the comparatively high number of migrants coming from Canada and Mexico to the United States?

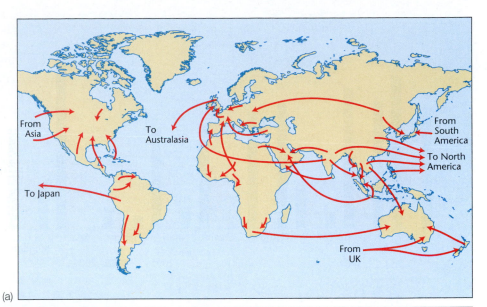

(a)

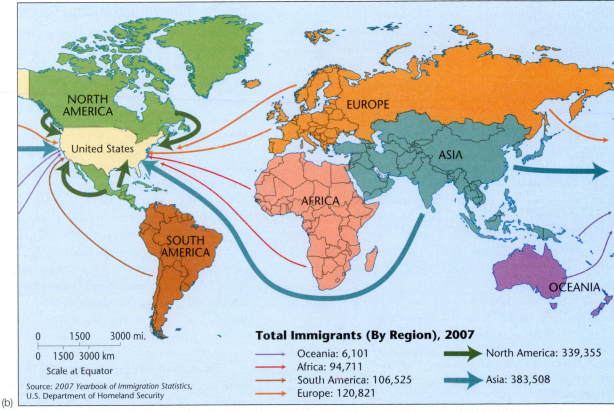

Total Immigrants (By Region), 2007

→ Oceania: 6,101	→ North America: 339,355
→ Africa: 94,711	→ Asia: 383,508
→ South America: 106,525	
→ Europe: 120,821	

0   1500   3000 mi.
0   1500 3000 km
Scale at Equator

Source: *2007 Yearbook of Immigration Statistics,* U.S. Department of Homeland Security

(b)

Roman times and from Europe in the mid-twentieth century; the export of Africans to the Americas as slaves; and the Clearances, or forced removal of farmers from Scotland's Highlands to make way for large-scale sheep raising, provide depressing examples.

Though the total rate of migration around the world has remained fairly stable in the past decade, rates of migration from country to country vary greatly. For example, the population of the state of Qatar in western Asia is composed of approximately 76 percent migrants, while the population of Indonesia has only 0.07 percent migrants. **Figure 3.16** on page 76 provides further data regarding the percentage of selected national populations that is composed of foreign-born individuals. The U.S. population is approximately 13 percent foreign-born. *Within* countries, the rate of domestic, or internal, migration varies greatly as well. In traditional societies, people have often chosen to remain close to the location of their

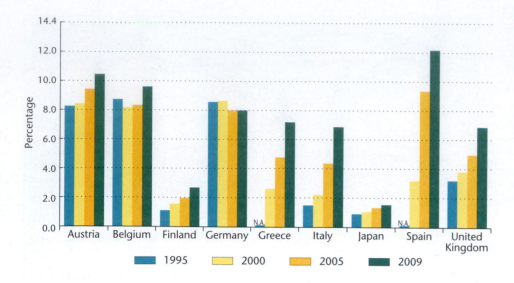

FIGURE 3.16 Foreign-born population as a percentage of the total population for selected countries, 1995–2009. Japan has quite a low percentage of foreign-born residents in its population, while Spain's foreign-born population has risen dramatically since 1995. (Source: Originally published on the Migration Policy Institute Data Hub, www.migrationpolicy.org/datahub.)

THINKING GEOGRAPHICALLY What factors do you think may influence migrants' decisions to move to one country versus another country?

birth throughout their lives. However, this trend has been changing in recent decades due to increasing urbanization in many developing states, because most opportunities for education and employment tend to be found in cities. These internal migrations to burgeoning twenty-first-century cities may be rewarded with higher wages or greater access to education, but they also entail many risks, particularly for certain ethnic minorities. Language barriers, lack of access to adequate health care and other social services, discrimination, and abuse of many kinds often await workers desperate to better their economic situation. Indeed, the United States has a long history of internal movement for the purpose of securing better economic opportunity and other personal reasons. In fact, more than 10 million Americans moved to other counties within the United States during 2008 alone.

Every migration, from the ancient dispersal of humankind out of Africa to the present-day movement toward urban areas, is governed by a host of **push-and-pull factors** that act to make the old home unattractive or unlivable and the new land attractive. Generally, push factors are the most important. After all, a basic dissatisfaction with the homeland is prerequisite to voluntary migration. The most important factor prompting migration throughout the thousands of years of human existence is economic. More often than not, migrating people seek greater prosperity through better access to resources, especially land. Both forced migrations and refugee movements, however, challenge the basic assumption of the push-and-pull model, which posits that human movement is the result of unforced choices.

push-and-pull factors Unfavorable, repelling conditions and favorable, attractive conditions that interact to affect migration and other elements of diffusion.

Disease Diffusion and Medical Geography

The branch of geography that deals most closely with health and disease-related topics is **medical geography.** Medical geographers seek to understand the various factors that affect the health of populations and to discover and map spatial patterns of disease and health-care provision. The foundations of medical geography and an interest in relationships between place and disease can be traced back as far as the time of Hippocrates in the third century B.C. Hippocrates observed, even at that time, that certain diseases seemed to occur in some places but not in others. In other words, Hippocrates recognized that diseases both move in spatially specific ways and occasion responses that are spatial in nature.

medical geography The branch of geography that deals with health and disease-related topics.

Throughout history, infectious disease has periodically decimated human populations. One has only to think of the vivid accounts of the Black Death episodes that together killed one-third of medieval Europe's population to get a sense of how devastating disease can be (**Figure 3.17**).

The spread of disease provides classic illustrations of both expansion and relocation diffusion. Some diseases are noted for their tendency to expand outward from their point of origin. Commonly borne by air or water, the viral pathogens for contagious diseases such as influenza and cholera spread from person to person throughout an affected area. Some diseases spread in a hierarchical diffusion fashion, whereby only certain social strata are exposed. The poor, for example, have always been much more exposed to the unsanitary conditions—rats, fleas, excrement, and crowding—that have occasioned disease epidemics. Other illnesses, particularly airborne diseases, affect all social and economic classes without regard for human hierarchies: their diffusion pattern is literally contagious.

FIGURE 3.17 **Fleeing from disease.** This woodcut, which was printed in 1630, depicts Londoners leaving the pestilent city in a cart. In 1665–1666, London experienced the so-called "Great Plague," an outbreak of bubonic plague that killed one-fifth of the city's population. (Bettmann/ CORBIS.)

THINKING GEOGRAPHICALLY How do modern outbreaks of disease influence travel?

Indeed, disease and migration have long enjoyed a close relationship. Diseases have spread and relocated thanks to human movement. Likewise, widespread human migrations have occurred as the result of disease outbreaks.

When humans engage in long-distance mobility, their diseases move along with them. Thus, cholera—a waterborne disease that had for centuries been endemic to the Ganges River region of India—broke out in Calcutta in the early 1800s. Because Calcutta was an important node in Britain's colonial empire, cholera quickly began to spread far beyond India's borders. Soldiers, pilgrims, traders, and travelers spread cholera throughout Europe, then from port to port across the world through relocation diffusion, making cholera the first disease of global proportions.

Early responses to contagious diseases were often spatial. Isolation of infected people from the healthy population—known as *quarantine*—was practiced. Sometimes only the sick individuals were quarantined, whereas at other times households or entire villages were shut off from outside contact. Whole shiploads of eastern European immigrants to New York City were routinely quarantined in the late nineteenth century. Another early spatial response was to flee the area of infection. During the time of the plague in fourteenth-century Europe, some healthy individuals abandoned their ill neighbors, spouses, and even children. Some of them formed altogether separate communities, whereas others sought merely to escape the city walls for the countryside.

Targeted spatial strategies could be enacted once it became known that some diseases spread through specific means. The best-known example is John Snow's 1854 mapping of cholera outbreaks in London, illustrated in Figure 3.18 on page 78, which allowed him to trace the source of the infection to one water pump. Thanks to his detective work and the development of a broader medical understanding of the role of germs in the spread of disease, cholera was understood as a waterborne disease that could be controlled by increasing the sanitary conditions of water delivery.

The threat of deadly disease is hardly a thing of the past. Today, medical geographers conduct important research used in the making of health-care policy, in the planning of health-care systems, and in the treatment of disease. This research can particularly benefit rural populations and those in developing regions that have relatively high rates of infant mortality, malnutrition, HIV/ AIDS, and other indicators of poor health care. It is hoped that mapping the spread of these contemporary diseases, understanding the pathways traveled by their carriers, and developing appropriate spatial responses can help prevent mass epidemics and disease-related panics (**Figure 3.19** on page 79).

Though the work of medical geographers is critical to improving quality of life for the world's citizens, unfortunately it is often impeded by an inability to collect accurate data in many regions. For example, due to linguistic and other cultural differences, geographers can seldom be sure that medical terminology, diagnostic techniques, and medical data collection techniques are standardized across regions and countries. In addition, the ability for an ill person to see a doctor varies greatly from region to region and country to country. This makes the incidence of disease and other health problems difficult to accurately measure and map. Despite these obstacles, medical geographers strive to produce maps and other forms of

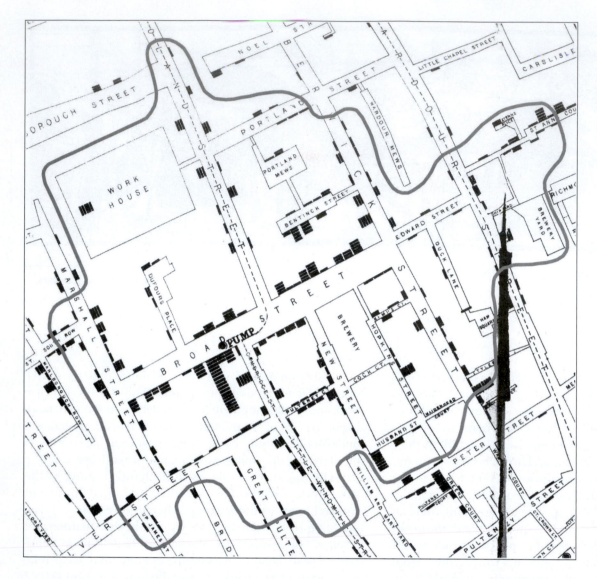

FIGURE 3.18 Mapping disease. This map was constructed by London physician John Snow. Snow was skeptical of the notion that "bad air" somehow carried disease. He interviewed residents of the Soho neighborhood to construct this map of cholera cases and used it to trace the outbreak to the contaminated Broad Street pump. Snow is considered to be a founder of modern epidemiology. (Courtesy of The John Snow Archive.)

THINKING GEOGRAPHICALLY What evidence could a geographer use to track the source of an airborne disease?

data that enable health-care professionals and governments assess and provide health-care needs.

POPULATION **ECOLOGY**

The reasons that human populations are so diverse are often cultural. Though the differences may have started out as adaptations to given physical conditions, when repeated from generation to generation these patterns become woven tightly into the cultural fabric of places. Thus, demographic practices such as living in crowded settlements or having large families may well have deep roots in both nature and culture.

Environmental Influence

Local population characteristics are often influenced in a possibilistic manner by the availability of resources. In the middle latitudes, population densities tend to be greatest where the terrain is level, the climate is mild and humid, the soil is fertile, mineral resources are abundant, and the sea is accessible. Conversely, population tends to thin out with excessive elevation, aridity, coldness, ruggedness of terrain, and distance from the coast.

Climatic factors influence where people settle. Most of the sparsely populated zones in the world have, in some respect, "defective" climates from the human viewpoint (see Figure 3.1). The thinly populated northern edges of Eurasia and North America are excessively cold, and the belt from North Africa into the heart of Eurasia matches the major desert zones of the Eastern Hemisphere. Humans remain creatures of the humid and subhumid tropics, subtropics, or midlatitudes and have not fared well in excessively cold or dry areas. Small populations of Inuit (Eskimo), Sami (Lapps), and other peoples live in some of the less hospitable areas of Earth, but these regions do not support large populations. Humans have proven remarkably

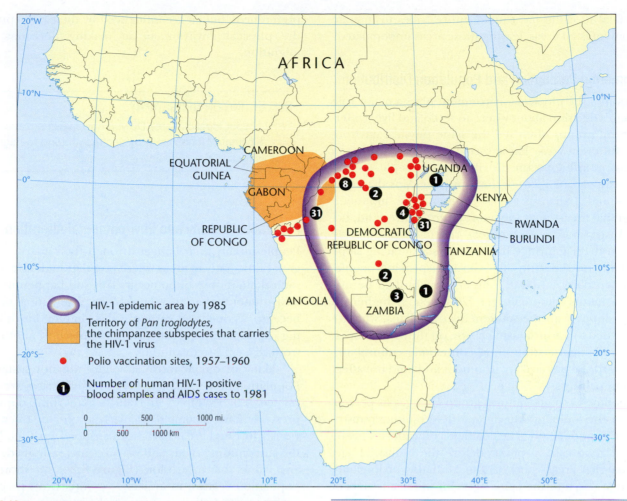

FIGURE 3.19 Early diffusion of the HIV-1 virus, which causes AIDS, involving an infected chimpanzee subspecies. (Sources: Gould, 1993; Hooper, 1999; Paul, 1994; Shannon, Pyle, and Bashshur, 1991: 49, 68, 73.)

THINKING GEOGRAPHICALLY How is the effect of long-distance truck transportation evident in the locations of the first HIV-1 positive blood samples in 1981?

adaptable, and our cultures contain strategies that allow us to live in many different physical environments, but perhaps, as a species, we have not entirely moved beyond the adaptive strategies that suited us so well to the climatic features of sub-Saharan Africa, where we began.

Humankind's preference for lower elevations is true especially for the middle and higher latitudes. Most mountain ranges in those latitudes stand out as sparsely populated regions. By contrast, inhabitants of the tropics often prefer to live at higher elevations, concentrating in dense clusters in mountain valleys and basins. For example, in tropical portions of South America, more people live in the Andes Mountains than in the nearby Amazon lowlands. The capital cities of many tropical and subtropical nations lie in mountain areas above 3000 feet (900 meters) in elevation. Living at higher elevations allows residents to escape the humid, hot climate and diseases of the tropic lowlands. In addition, these areas were settled because the fertile volcanic soils of these mountain valleys and basins were able to support larger populations in agrarian societies.

Many humans live near the sea. Eurasia, Australia, and South America resemble hollow shells, with the majority of the population clustered around the rim of each continent (see Figure 3.1). In Australia, half of the population lives in only five port cities, and most of the remainder is spread out over nearby coastal areas. This preference for living by the sea stems partly from the trade and fishing opportunities the sea offers. At the same time, continental interiors tend to be regions of climatic extremes. For example, Australians speak of the "dead heart" of their continent, an interior land of excessive dryness and heat.

Disease also affects population distribution. Some diseases attack valuable domestic animals, depriving people of food and clothing resources. Such diseases have an indirect effect on population density. For example, in parts of East Africa, a form of sleeping sickness attacks livestock. This particular disease is almost invariably fatal to cattle but not to humans. The people in this part of East Africa depend heavily on cattle, which provide food, represent wealth, and serve a religious function in some tribes. The spread of a disease fatal to cattle has caused entire tribes to migrate away from infested areas, leaving those areas unpopulated.

Environmental Perception and Population Distribution

Perception of the physical environment plays a major role in a group's decision about where to settle and live. Different cultural groups often see the same physical environment in different ways. These varied responses to a single environment influence the distribution of people. A good example appears in a part of the European Alps shared by German- and Italian-speaking peoples. The mountain ridges in that area—near the point where Switzerland, Italy, and Austria border each other—run in an east-west direction, so that each ridge has a sunny south-facing slope and a shady north-facing one. German-speaking people, who rely on dairy farming, long ago established permanent settlements some 650 feet (200 meters) higher on the *shady* slopes than the settlements of Italians, who are culturally tied to warmth-loving crops, on the *sunny* slopes. This example demonstrates how contrasting cultural attitudes toward the physical environment and land use affect settlement patterns.

Sometimes a single cultural group changes its perception of an environment over time, with a resulting redistribution of its population. The coalfields of western Europe provide a good case in point. Before the industrial age, many coal-rich areas—such as the Midlands of England, southern Wales, and the lands between the headwaters of the Oder (or Odra) and Vistula rivers in Poland—were only sparsely or moderately settled. The development of steam-powered engines and the increased use of coal in the iron-smelting process, however, created a tremendous demand. Industries grew up near the European coalfields, and people flocked to these areas to take advantage of the new jobs. In other words, once a technological development gave a new cultural value to coal, many sparsely populated areas containing that resource acquired large concentrations of people.

Recent studies indicate that much of the interregional migration in the United States today is prompted by a desire for a pleasant climate and other desirable physical environmental traits, such as beautiful scenery. Surveys of immigrants to Arizona revealed that its sunny, warm climate is a major reason for migration. An attractive environment provided the dominant factor in the growth of the population and economy of Florida. The most desirable environmental traits that serve as stimulants for American migration include (1) mild winter climate and mountainous terrain, (2) diverse natural vegetation that includes forests and a mild summer climate with low humidity, (3) lakes and rivers, and (4) nearness to the seacoast. Different age and cultural groups often express different preferences, but all are influenced by their perceptions of the physical environment in making decisions about migration.

Population Density and Environmental Alteration

People modify their habitat through their adaptive strategies. Particularly in areas where population density is high, radical alterations often occur. This can happen in fragile environments even at relatively low population densities because, as discussed earlier in this chapter, Earth's carrying capacity varies greatly from one place to another and from one culture to another.

Many of our adaptive strategies are not sustainable. Population pressures and local ecological crises are closely related. For example, in Haiti, where rural population pressures have become particularly severe, the trees in previously forested areas have been stripped for fuel, leaving the surrounding fields and pastures increasingly denuded and vulnerable to erosion (**Figure 3.20**). In short, overpopulation relative to available resources can precipitate environmental destruction—which, in turn, results in a downward cycle of worsening poverty, with an eventual catastrophe that is both ecological and demographic. Thus, many cultural ecologists believe that attempts to restore the balance of nature will not succeed until we halt or even reverse population growth, although they recognize that other causes are at work in ecological crises.

The worldwide ecological crisis is not solely a function of overpopulation. A relatively small percentage of the Earth's population controls much of the industrial technology and consumes a disproportionate percentage of the world's resources each year. Americans, who make up less than 4.5 percent of the global population, account for about 23 percent of the natural resources consumed globally each year. New houses built in the United States in 2009 were, on average, 46 percent bigger than those built in 1960, despite a shrinking average household size during that period. If everyone in the world had an average American standard of living, Earth could support only about 1.5 billion people—only 21 percent of the present global population. As the economies of large countries such as India and China continue to surge, the resource consumption of their populations is likely to rise as well because the human desire to consume appears to be limited only by the ability to pay for it.

FIGURE 3.20 **Overpopulation and deforestation.**
This aerial photograph depicts the border between
Haiti and the Dominican Republic. The two nations
share the Caribbean island of Hispaniola. Population
pressures in Haiti, on the left side of the photograph,
have led to deforestation; the Dominican Republic is
the greener area to the right. The political border is
also an environmental border. (NASA.)

THINKING GEOGRAPHICALLY What are the alternatives to
the massive removal of forests to obtain fuel and building
materials? What environmental impacts are inherent in
these alternatives?

CULTURAL **INTERACTION** AND POPULATION PATTERNS

**Can we gain additional understanding of de-
mographic patterns by using the theme of cultural interaction?** Are culture and demography intertwined?
Yes, in diverse and causal ways. Culture influences population density, migration, and population growth. Inheritance laws, food preferences, politics, differing attitudes
toward migration, and many other cultural features can
influence demography.

Cultural Factors

Many of the forces that influence the distribution of people are basic characteristics of a group's culture. For example, we must understand the preference for rice of
people living in Southeast Asia before we can try to interpret the dense concentrations of people in rural areas
there. The population in the humid lands of tropical and
subtropical Asia expanded as this highly prolific grain was
domesticated and widely adopted. Environmentally similar
rural zones elsewhere in the world, where rice is not a staple of the inhabitants' diet, never developed such great
population densities. Similarly, the introduction of the potato into Ireland in the 1700s allowed a great increase in
rural population, because it yielded much more food per
acre than did traditional Irish crops. Failure of the potato

harvests in the 1840s greatly reduced the Irish population,
through both starvation and emigration.

Geographers have found major cultural contrasts in attitudes toward population growth. Sustained fertility decline first took place in France in the nineteenth century,
but neighboring countries such as Germany, Italy, and the
United Kingdom did not experience the same decline.
France, the most populous of these four countries in the
1800s, became the least populous shortly after 1930. During this same period, millions of Germans, British people,
and Italians emigrated overseas, whereas relatively few
French left their homeland. Clearly, some factor in the culture of France worked to produce the demographic decline. What might the reason be?

Cultural groups also differ in their tendency to migrate. Religious ties bind some groups to their traditional
homelands, while other religious cultures place no stigma
on migration. In fact, some groups, such as the Irish until
recently, consider migration a way of life.

Culture can also condition a people to accept or reject
crowding. **Personal space**—the amount of space that individuals feel "belongs" to
them as they go about their
everyday business—varies
from one cultural group to
another. When Americans
talk to one another, they typically stand farther apart than,
say, Italians do. The large personal space demanded by

> **personal space** The amount of
> space that individuals feel "belongs" to
> them as they go about their daily
> business.

Americans may well come from a heritage of sparse settlement. Early pioneers felt uncomfortable when they first saw smoke from the chimneys of neighboring cabins. As a result, American cities sprawl across large areas, with huge suburbs dominated by separate houses surrounded by private yards. Most European cities are compact, and their residential areas consist largely of row houses or apartments.

Political and Economic Factors

The political mosaic is linked to population geography in many ways. Governmental policies, such as China's one-child policy, often influence the fertility rate, and forced migration is usually the result of political forces.

Governments can also restrict voluntary migration. Two independent countries, Haiti and the Dominican Republic, share the tropical Caribbean island of Hispaniola. Haiti is far more densely settled than the Dominican Republic. Government restrictions make migration from Haiti to the Dominican Republic difficult and help maintain the different population densities. If Hispaniola were one country, its population would be more evenly distributed over the island.

Governments also set in motion most *involuntary* migrations. These have become common in the past century or so, usually to achieve *ethnic cleansing*—the removal of unwanted minorities from nation-states. A twenty-first-century example of ethnic cleansing occurred in the Darfur region of Sudan, where decades of conflict between various rebel groups and the Muslim Sudanese government resulted in the deaths of over 1.5 million people and the displacement of over 2.5 million people from their homes. After more than eight years of armed disputes and widespread raids on villages in the region, a 2005 referendum decided that the southern region of the country (populated by a number of distinct ethnic groups) would split from the Arab north to form the new nation-state of South Sudan. This split officially occurred on July 9, 2011. The infant nation-state of South Sudan faces many challenges, as it is oil-rich but is also one of the more poorly developed nations on Earth. Most children in South Sudan below age 13 are not in school, and more than 80 percent of the nation's females are illiterate. Further, there are issues to be resolved between Sudan and its new southern neighbor involving citizenship rights, oil pipeline ownership rights, and many other political issues.

In addition to involuntary migrations of populations precipitated by governments, economic conditions often influence population density in profound ways. The industrialization of the past 200 years has caused the greatest voluntary relocation of people in world history. Within industrial nations, people have moved from rural areas to

cluster in manufacturing regions. Agricultural changes can also influence population density. For example, the complete mechanization of cotton and wheat cultivation in twentieth-century America allowed those crops to be raised by a much smaller labor force. As a result, profound depopulation occurred, to the extent that many small towns serving these rural inhabitants ceased to exist.

Population Control Programs Though the debates may occur at a global scale, most population control programs are devised and implemented at the national level. When faced with perceived national security threats, some governments respond by supporting pronatalist programs that are designed to increase the population. Labor shortages in Nordic countries have led to incentives for couples to have larger families. Most population programs, however, are antinatalist: they seek to reduce fertility. Needless to say, this is an easier task for a nonelected government to carry out because limiting fertility challenges traditional gender roles and norms about family size.

Although China is not the only country that has sought to limit its population growth, its so-called "one-child policy" provides the best-known modern example. Mao Zedong, the leader of the People's Republic of China from 1949 to 1976, did not initially discourage large families because he believed that "every mouth comes with two hands." In other words, he believed that a large population would strengthen the country in the face of external political pressures. He reversed his position in the 1970s, when it became clear that China faced resource shortages as a result of its burgeoning population. In 1979, the one-child-per-couple policy was adopted. With it, Chinese authorities sought not merely to halt population growth but ultimately to decrease the national population. Some families were exempt from this policy, including ethnic minority households and farming families whose first child was female. These exceptions were granted to prevent ethnic minority populations from dying out over time and to allow farming families to produce a male child to help with agricultural work and eventually to take over the family farm. Outside of these exceptions, during the three decades that the one-child policy has been in force, billboards and posters admonishing citizens that "one couple, one child" is the ideal family have been commonplace on the Chinese cultural landscape (**Figure 3.21**). Violators of the policy have faced huge monetary fines and ineligibility for new housing. They have also lost government benefits to the elderly, access to higher education for their children, and even their jobs. In addition to enforcing these penalties, the Chinese government has sought to slow population growth by encouraging late marriages (thus delaying the start of families). In response, between 1970 and 1980, the TFR in China plummeted from 5.9 births

(a)

(b)

FIGURE 3.21 Population control in the People's Republic of China. China has aggressively promoted a policy of "one couple, one child" in an attempt to relieve the pressures of overpopulation. These billboards convey the government's message. Violators—those with more than one child—are subject to fines, loss of job and old-age benefits, loss of access to better housing, and other penalties. (Courtesy of Terry G. Jordan-Bychkov.)

THINKING GEOGRAPHICALLY How effective would such billboards be in influencing people's decisions? Why is one of the signs in English?

per woman to only 2.7, then to 2.2 by 1990, 2.0 by 1994, 1.7 in 2007, and 1.5 in 2011.

As a result of these policies, China achieved one of the greatest short-term reductions of birthrates ever recorded, thus proving that cultural changes can be imposed from above, rather than simply by waiting for them to diffuse organically. In recent years, the Chinese population control program has been less rigidly enforced, as periods of economic growth eroded the government's control over the people. This relaxation allowed more couples to have two children instead of one; however, the rise of economic opportunity and migration to cities have led some couples to have smaller families voluntarily. Recently, the Chinese government has acknowledged that the country's socioeconomic trends, along with its rigid 30-year policy, have created significant demographic imbalances. For example, the gender ratio went from 108.5 boys to 100 girls in the early 1980s to 111 boys in 1990 to 116 in 2000, and it is now at nearly 120 boys for every 100 girls. This imbalance is largely due to millions of families over the past 30 years resorting to abortion (aided by widespread but illegal prenatal ultrasound scans) and female infanticide in order to ensure that their one child was a boy. (Throughout Chinese history, male children have represented continuity of the family lineage and protection and support for the parents in their old age. These ideas are still very much alive in China.)

As a result of worry over the implications of growing gender imbalances, the Chinese government has begun to consider whether to move to a two-child policy within the next five years. However, some opponents of the changes have argued that the result will be a sudden surge in the number of births, creating new social problems. Nevertheless, proposals are being discussed that may allow families whose first child is a girl to have a second child as early as 2015. Relaxing policies in this way could help ease the future burden of the Chinese working population to support an ever-growing elderly population, which will account for one-third of the country's people in the next three decades. It is also hoped that loosening the one-child policy will decrease the rising incidence of human trafficking in China as more and more men resort to "purchasing" their wives from "bachelors' villages" that have begun to spring up in remote areas of the country.

REFLECTING ON GEOGRAPHY
Would a global population control policy be desirable—or even feasible?

THE SETTLEMENT **LANDSCAPE**

What are the many ways that demographic factors are expressed in the cultural landscape? Population geographies are visible in the landscapes around us. The varied densities of human settlements and the shapes these take in different places provide clues to the intertwined cultural and demographic strategies enacted in response to diverse local factors. These clues are evident wherever people have settled, be it urban, suburban, or rural. Less visible—but no less important—factors are also at work on and throughout the cultural landscape. Politics, economics, and gender relations provide three examples of the larger power relations that can shape the cultural-demographic landscape. We open this final part of the chapter by examining rural settlement patterns, which illustrate how demographics, culture, and the landscape work together to create a variety of settlements.

Rural Settlement Patterns

Differing densities and arrangements of population are revealed, at the largest scale, by maps showing the distribution of dwellings. These differences in the cultural landscape can be illustrated by using the example of rural settlement types. Farm people differ from one culture to another, one place to another, in how they situate their dwellings, producing greatly varied rural cultural landscapes. They range from tightly clustered villages on the

one extreme to fully dispersed farmsteads on the other, as shown in **Figure 3.22**.

Farm Villages In many parts of the world, farming people group themselves together in clustered settlements called **farm villages.** These tightly bunched settlements vary in size from a few dozen inhabitants to several thousand. Contained in the village **farmstead** are the house, barn, sheds, pens, and garden. The fields, pastures, and meadows lie out in the country beyond the limits of the village, and farmers must journey out from the village each day to work the land.

> **farm village** Clustered, tightly bunched rural settlement inhabited by people who are engaged in farming.
>
> **farmstead** The center of farm operations, containing the house, barn, sheds, livestock pens, and garden.

Farm villages are the most common form of agricultural settlement in much of Europe, in many parts of Latin America, in the densely settled farming regions of Asia (including much of India, China, and Japan), and among the sedentary farming peoples of Africa and the Middle East.

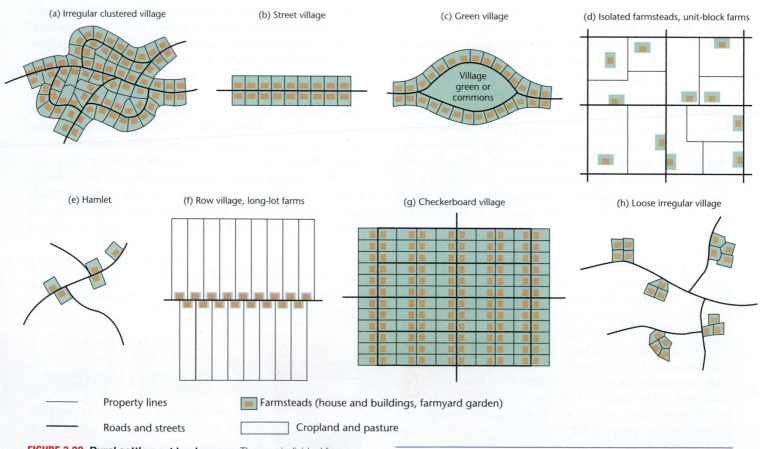

(a) Irregular clustered village
(b) Street village
(c) Green village — Village green or commons
(d) Isolated farmsteads, unit-block farms
(e) Hamlet
(f) Row village, long-lot farms
(g) Checkerboard village
(h) Loose irregular village

——— Property lines

——— Roads and streets

▢ Farmsteads (house and buildings, farmyard garden)

▢ Cropland and pasture

FIGURE 3.22 Rural settlement landscapes. The way individual farmers choose to locate their farmsteads leads to a general settlement pattern. In some areas, farmsteads are scattered and isolated. In areas where farmsteads are grouped, there are several possible patterns of clustering.

THINKING GEOGRAPHICALLY What are the advantages of patterns where farmers live in villages and travel to fields outside the village to work? What are the disadvantages?

(a)

(b)

FIGURE 3.23 **Two irregular clustered villages.** (a) The village on the left with adjacent irrigated grain fields is near Gonggar in southeastern Tibet. The village nestles against a hill, and much of it is built on land unfit for cultivation. (b) The village shown on the right is in northern Switzerland. Although halfway around the world from the village in Tibet, it also shows the irregular plan. (Courtesy of Terry G. Jordan-Bychkov.)

These compact villages come in many forms. Most are irregular clusterings—a maze of winding, narrow streets and a jumble of farmsteads (**Figures 3.22a** and **3.23**). Such irregular clustered farm villages developed organically over centuries, without any orderly plan to direct their growth. Other types of farm villages are very regular in their layout, revealing the imprint of planned design. The street village, the simplest of these planned types, consists of farmsteads grouped along both sides of a single central street, producing an elongated settlement (**Figures 3.22b** and **3.24**). Street villages are particularly common in eastern Europe and much of Russia. Another type, the green village, consists of farmsteads grouped around a central open place, or green, which forms a commons (see Figure 3.22c). Green villages occur throughout most of the plains areas of northern and northwestern Europe, and English immigrants laid out some such settlements in colonial New England. Also regular in layout is the checkerboard village, based on a gridiron pattern of streets meeting at right angles (see Figure 3.22g). Mormon farm villages in Utah are of this type, and checkerboard villages also dominate most of rural Latin America and northeastern China.

Why do so many farm people settle together in villages? Historically, the countryside was unsafe, threatened by roving bands of outlaws and raiders. Farmers could better defend themselves against such dangers by grouping together

FIGURE 3.24 **Street village beside the great Siberian river Lena in the Sakha Republic (Yakutia), part of Russia.** Farmsteads lie along a single street, creating an elongated settlement. One can distinguish Russian ethnic villages in Sakha by this form. In this way, the cultural landscape reveals the ethnicity of the inhabitants. (Courtesy of Terry G. Jordan-Bychkov.)

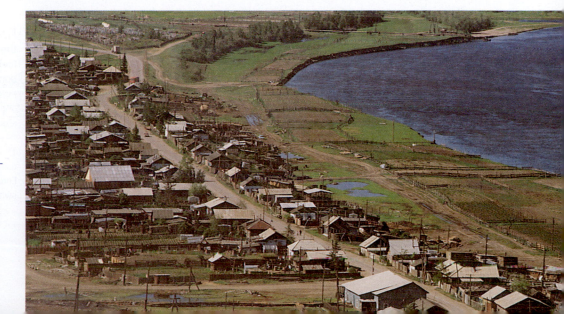

in villages. In many parts of the world, the populations of villages have grown larger during periods of insecurity and shrunk again when peace returned. Many farm villages occupy the most easily defended sites in their vicinity, what geographers call *strong-point settlements.*

In addition to defense, the quality of the environment helps determine whether people settle in villages. In deserts and in limestone areas where the ground absorbs moisture quickly, farmsteads are built around the few sources of water. Such *wet-point villages* cluster around oases or deep wells. Conversely, a superabundance of water—in marshes, swamps, and areas subject to floods—often prompts people to settle in villages on available dry points of higher elevation.

Various communal ties strongly bind villagers together. Farmers linked to one another by blood relationships, religious customs, communal landownership, or other similar bonds usually form clustered villages. Mormon farm villages in the United States provide an excellent example of the clustering force of religion. Communal or state ownership of the land—as in China and parts of Israel—encourages the formation of farm villages.

Isolated Farmsteads In many other parts of the world, the rural population lives on dispersed, isolated farmsteads, often some distance from their nearest neighbors (see Figure 3.22d). These dispersed rural settlements grew up mainly in Anglo-America, Australia, New Zealand, and South Africa—that is, in the lands colonized by emigrating Europeans. But even in areas dominated by village settlements—such as Japan, Europe, and parts of India—some isolated farmsteads appear (**Figure 3.25**).

The conditions encouraging dispersed settlement are precisely the opposite of those favoring village development. These include peace and security in the countryside, eliminating the need for defense; colonization by individual pioneer families rather than by socially cohesive groups;

private agricultural enterprise, as opposed to some form of communalism; and well-drained land where water is readily available. Most dispersed farmsteads originated rather recently, dating primarily from the colonization of new farmland in the past two or three centuries.

REFLECTING ON GEOGRAPHY

What challenges might a settlement pattern of isolated farmsteads present for the people who live there?

Historical Factors Shaping the Cultural-Demographic Landscape

The rural settlement forms described previously give geographers a chance to read the cultural landscape. In doing so, we must always be cautious, looking for the subtle as well as the overt, and not form conclusions too quickly.

For example, the Maya of the Yucatán Peninsula in Mexico reside in checkerboard villages, a rural settlement landscape that is both suggestive and potentially misleading (**Figure 3.26**). Before the Spanish conquest in the late 1400s, the Maya lived in wet-point villages of the irregular clustered type, situated alongside *cenotes*—natural sinkholes that provided water in a land with no surface streams. The Spaniards destroyed these settlements, replacing them with checkerboard villages. Wide, straight streets accommodated the wheeled vehicles of the European conquerors.

Superficially, the checkerboard landscape suggests the cultural victory of the Spaniards. A closer look, however, shows that, in fact, Mayan culture prevailed. Even today, many Maya make little use of wheeled vehicles in village life, and many of the Spaniards' streets merely make way for Mayan footpaths that wind among boulders and outcroppings of bedrock. Irregularities in the checkerboard, coupled with a casual distribution of dwellings, suggest

FIGURE 3.25 A truly isolated farmstead in the Vestfjörds district of northwestern Iceland. This type of rural settlement dominates almost all lands colonized by Europeans migrating overseas. Iceland was settled by Norse Vikings 1000 years ago. (Courtesy of Terry G. Jordan-Bychkov.)

THINKING GEOGRAPHICALLY How might the dominance of the isolated farmstead be related to European feudalism, in which only the wealthy owned land?

Street

Laid out as street, but used only as footpath; obstructions such as boulders make wheeled vehicles impossible

Foot trails to nearby cornfields

Dry rock walls enclosing farmsteads and outlying gardens

Mayan oval-shaped pole huts, with detached kitchen

Dwellings

Stores and corn mill

Town hall

} Flat-roofed, Spanish-style buildings

S School

FIGURE 3.26 **A hypothetical modern Mayan checkerboard farm village in Yucatán Province, Mexico.** Spanish influence—seen in the grid pattern, plaza, church, hacienda, and flat-roofed buildings—weakens with distance from the center, and the rigid checkerboard masks a certain irregularity of farmstead layout. A cenote is a large, deep sinkhole filled with water, and these natural pools served a major religious function among the Mayas before Christianity came. (Source: Composite of 1987 field observations by Terry G. Jordan-Bychkov in some 15 villages east and southeast of Mérida.)

THINKING GEOGRAPHICALLY What does the location of the Catholic church on the central plaza symbolize? The town hall?

Mayan resistance to the new geometry. Spanish-influenced architecture—flat-roofed houses of stone, the town hall, a church, and a hacienda mansion—remain largely confined to the area near the central plaza. The Catholic church stands on the very place where an ancient Mayan temple was. A block away, the traditional Mayan pole huts with thatched, hipped roofs prevail, echoed by cookhouses of the same design. Indian influence increases markedly with distance from the plaza.

The dooryard gardens surrounding each hut are full of traditional native plants—such as papayas, bananas, chili peppers, nuts, yucca, and maize—with only a few citrus trees to reveal Spanish influence. In the same yards, each carefully ringed with dry rock walls, as in pre-Columbian times, pigs descended from those introduced by the conquerors share the ground with the traditional turkeys of the Maya and apiaries for indigenous stingless bees. Occasionally, the Mayan language is heard, although Spanish prevails. So does Catholicism, but the absence of huts around the once-sacred cenote suggests a lingering pagan influence.

Gender and the Cultural-Demographic Landscape

Gender often interacts with other factors to influence geodemographic patterns and migrations. For example, together gender, race, and nationality can create situations in which women from specific countries are viewed as desirable immigrants. In nineteenth-century America, Irish female immigrants were considered to be highly reliable employees and often found work as domestic servants.

For several decades now, women from the Philippines have migrated to Hong Kong to work as domestic servants, doing chores for and looking after the children of the families who hire them. Wages in Hong Kong are much higher than back home in the Philippines, prompting the migration. However, working conditions can be far from ideal, involving hard physical labor and long hours. Reports of abuse of Filipina servants by their employers are numerous. If a servant is fired and cannot quickly find another job, she will be deported. The stress of long separations from husbands and children back in the Philippines has led to the breakup of families.

Some of these women are coerced into Asia's booming sex industry, as well. Geographer James Tyner has studied migration from the Philippines to Japan. Incredibly, 93 percent of this migration consists of female "entertainers." In part, poverty in the Philippines provided the push factor in this movement. Tyner also points to more complex pull factors, such as the stereotypical view held by Japanese men in which Filipinas are seen as highly desirable, exotic sex objects but also culturally inferior and thus "willing victims" who gladly become prostitutes. Thus, the gendered landscape becomes an attraction for human mobility other than work-related migration: in this case, sex tourism.

Landscapes and Demographic Change

Population change in a place can occur rapidly or more slowly over time. Rapid **depopulation** can come about as the result of sudden catastrophic events, such as natural disasters, disease epidemics, and warfare. Or depopulation can take place at a slower pace. For example, places may lose population over time because of the gradual out-migration of people in search of opportunities elsewhere, as a cumulative result of declining fertility rates, or as a response to climate change.

> **depopulation** A decrease in population that sometimes occurs as the result of sudden catastrophic events, such as natural disasters, disease epidemics, and warfare.

Populations can also grow in more or less rapid fashions. A place might experience an abrupt influx of people who have been displaced from other areas. Population increases commonly occur more gradually, as the result of demographic improvements such as longer life spans or lower infant mortality rates, or the accumulation of steady streams of immigrants to attractive areas over time.

Depopulation in one area and population increase in another are often linked processes. For instance, you could easily envision a scenario in which the same natural disaster victims leaving one place become the refugees suddenly flooding into a neighboring area. As a longer-term example, you could equally well imagine how broad economic changes or technological innovations could lead to regional population shifts whereby some areas become depopulated and others gain population.

Regardless of the reasons, population changes must be accommodated, and these adaptations are invariably reflected in the cultural landscape.

Bogotá Rising Dilapidated-looking landscapes can result from population decline, but they can also arise from rapid population influx. Many of the world's shantytowns exemplify the sort of chaotic landscape that can result from rapid population increases. Los Altos de Cazucá, a neighborhood in the Colombian capital of Bogotá, is but one example. In this case, the population influx comes mostly from people arriving in the capital after being displaced by armed conflict in the countryside.

The United Nations High Commissioner for Refugees (UNHCR) estimates that by 2009 there were 26 million internally displaced persons (IDPs) due to armed conflict worldwide. Colombia is one of the globe's hotspots for IDPs, alongside Sudan. Colombia's estimated 5 million

IDPs have been uprooted largely as the result of ongoing civil warfare between the paramilitary group known as the FARC (Revolutionary Armed Forces of Colombia) and government forces bent on eradicating them. The UN-HCR has identified this as the "worst humanitarian crisis in the Western Hemisphere."

Colombia's armed conflict has depopulated the rural areas where it occurs, forcing its victims into the cities. Terrorized, landless, and impoverished, the mainly indigenous and Afro-Colombian IDPs settle in the outskirts of cities like Colombia's capital, Bogotá.

Los Altos de Cazucá is one of the settlements inhabited by Colombia's internally displaced persons (**Figure 3.27**). Known as **shantytowns,** areas like Los Altos de Cazucá arise for various reasons, but they exist in all large

> **shantytown** A deprived area on the outskirts of a town consisting of a large number of crude dwellings.

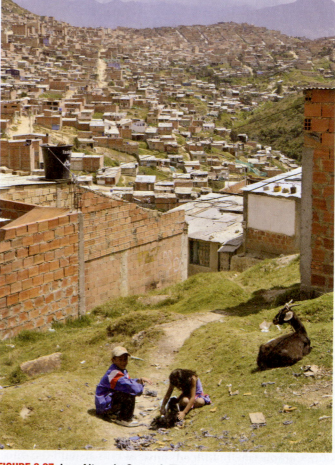

FIGURE 3.27 Los Altos de Cazucá. This is the shantytown on the outskirts of Colombia's capital city, Bogotá. Many of its approximately 50,000 residents are displaced people from other parts of the country. (imagebroker.net/SuperStock.)

THINKING GEOGRAPHICALLY Search the Internet for several alternative terms for *shantytown* used in Latin America.

FIGURE 3.28 Landscapes of poverty and wealth. This scene from Bogotá is a common Latin American urban landscape, with wealthy high-rises located next to impoverished shantytowns. (Victor Englebert.)

THINKING GEOGRAPHICALLY How is this landscape similar to or different from city landscapes in the United States?

cities throughout the developing world. Housing is constructed by the residents themselves, using found materials like cardboard, tin panels, and old tires. Some shantytowns are far from downtown areas where wealthy people reside and where jobs are, while others arc literally pressed up against wealthier neighborhoods (**Figure 3.28**).

Most shantytowns arise spontaneously to address population influx to cities that lack the resources to plan systematically for rapid growth. But are shantytowns temporary, disappearing as their residents become incorporated into city life? Interestingly, shantytowns gradually become part of the urban fabric. Indeed, most of the spatial expansion of Latin American cities like Bogotá occurs precisely thanks to growth of the shantytowns surrounding them. Over time, the dwellings are constructed of more-permanent materials such as concrete blocks. Roads are paved and running water installed. Power and phone lines are extended. The once-temporary areas slowly become visually,

economically, and culturally integrated into the permanent fabric of the city. New shantytowns to accommodate recent arrivals then arise beyond their borders.

Key Terms

absolute population density, 72	natural decrease, 72
carrying capacity, 60	personal space, 81
crude birthrate, 61	physiological population density, 72
crude death rate, 63	population density, 59
demographic transition, 67	population doubling time, 73
dependency ratio, 73	
depopulation, 88	population explosion, 65
farm village, 84	population geography, 57
farmstead, 84	population growth rate, 73
forced migrant, 74	population pyramid, 70
formal demographic region, 60	push-and-pull factors, 76
	refugee, 74
gender role, 70	sex ratio, 70
geodemography, 57	shantytown, 88
infant mortality rate, 71	total fertility rate (TFR), 61
Malthusian, 65	voluntary migrant, 74
medical geography, 76	zero population growth, 62

Population Geography on the Internet

You can learn more about population geography on the Internet at the following web sites:

Population Reference Bureau, Inc., Washington, D.C.
http://www.prb.org
This organization is concerned principally with the issues of overpopulation and standard of living. The *World Population Data Sheet* provides up-to-date basic demographic information at a glance, and the Datafinder section has a wealth of images on all aspects of global population for use in presentations and reports. Some of the maps in this chapter were adapted from PRB maps.

United Nations High Commissioner for Refugees
http://www.unhcr.org
This United Nations web site provides basic information about refugee situations worldwide. The site includes regularly updated maps showing refugee locations and populations and photos of refugee life.

U.S. Census Bureau Population Clocks
http://www.census.gov/main/www/popclock.html
Check real-time figures here for the population of the United States and the population of the world. The main web site, http://www.census.gov, hosts the most important and comprehensive data sets available on the U.S. population.

World Health Organization, Geneva, Switzerland
http://www.who.int/en
Learn about the group that distributes information on health, mortality, and epidemics as it seeks to improve health conditions around the globe. The Global Health Atlas allows you to create detailed maps from WHO data.

Worldwatch Institute, Washington, D.C.
http://www.worldwatch.org
This organization is concerned with the ecological consequences of overpopulation and the wasteful use of resources. It seeks sustainable ways to support the world's population and brings attention to ecological crises.

Recommended Books on Population Geography

Castles, Stephen, and Mark J. Miller. 1998. *The Age of Migration: International Population Movements in the Modern World,* 2nd ed. New York: Guilford. A global perspective on migrations, why they occur, and the effects they have on different countries in an age of an unprecedented volume of migration. Explores how migration has led to the formation of ethnic minorities in numerous countries as well as its impact on domestic politics and economics.

Johnson, Steven. 2006. *The Ghost Map: The Story of London's Most Terrifying Epidemic—And How It Changed Science, Cities, and the Modern World.* New York: Riverhead Books. In this gripping historical narrative, physician John Snow's tracing of the Soho cholera outbreak of 1854 is placed in the larger context of the history of science and urbanization.

Meade, Melinda S., and Robert J. Earickson. 2005. *Medical Geography,* 2nd ed. New York: Guilford. These surveys of the perspectives, theories, and methodologies that geographers use in studying human health provide a primary text that undergraduates can readily understand.

Newbold, K. Bruce. 2007. *Six Billion Plus: World Population in the Twenty-First Century,* 2nd ed. Lanham, MD: Rowman and Littlefield. The impact of increased global population levels across the next century is assessed with respect to interaction with environmental, epidemiological, mobility, and security issues.

Seager, Joni, and Mona Domosh. 2001. *Putting Women in Place: Feminist Geographers Make Sense of the World.* New York: Guilford. A highly readable account of why paying attention to gender is crucial to understanding the spaces in which we live and work.

Tone, Andrea. 2002. *Devices and Desires: A History of Contraceptives in America.* New York: Hill & Wang. This social history of birth control in the United States details the fascinating relationship between the state and the longstanding attempts of men and women to limit their fertility.

Journals in Population Geography

Gender, Place and Culture: A Journal of Feminist Geography. Published by the Carfax Publishing Co., P.O. Box 2025, Dunnellon, FL. 34430. Volume 1 appeared in 1994.

Population and Environment: A Journal of Interdisciplinary Studies. Volume 1 was published in 1996.

Answers to Thinking Geographically Questions

Figure 3.1: Yes. Northeastern United States also shows higher population density, although the areas of dense settlement are smaller than they are in the other three clusters.

Figure 3.2: There are likely to be variations in the quality of this data due to variations in health statistics reporting from country to country and because in less developed regions, births frequently occur outside of hospital settings and may therefore go unrecorded.

Figure 3.3: Answers will vary, but possible reasons for this differential are that children in developing states often contribute to the family income and care for elderly family members, so having more children is seen as advantageous. In more developed regions, people often choose to have fewer children because the government contributes to care for the elderly, young people typically do not work, and there are educational and lifestyle advantages for families with fewer children.

Figure 3.4: This elevated crude death rate is likely due to the fact that large numbers of senior citizens choose these states as retirement destinations.

Figure 3.5: North Africa has lower rates. These countries are predominantly Muslim, like Southwest Asia, which also has low rates.

Figure 3.6: The projection assumes that birth and death rates will remain stable, neither rising nor falling by very much.

Figure 3.7: The main reason for the rapid growth in the last 200 years or so is the decline in death rates, which has now affected the entire world. Birth rates typically drop later, allowing rapid population growth in the meantime.

Figure 3.8: While it is fairly straightforward to persuade people to take measures to prevent death, the number of children a family wants relates to deeply held values, which are much slower to change.

Figure 3.9: Among the places are Mexico and the United States. Papua New Guinea and Australia have very different rates, although they do not share a land boundary. In the case of Mexico and the United States, the high population growth rates in Mexico add pressure to northward migration.

Figure 3.10: In many societies, women use contraceptives only after having the number of children that their cultural values deem sufficient. Furthermore, the population growth rate also includes rising life expectancy. People are living longer, in addition to adding more children.

Figure 3.11: Physical facilities for the elderly (ramps, assist bars in bathrooms, and the like) are likely to be in place, and services for the elderly are readily available. Recreational facilities are conveniently located. Taxes are low, as they do not have to fund many child-oriented services such as schools. Disadvantages include a restricted diversity in age and frequent deaths of aging residents, which may be psychologically depressing. Life in such communities can seem very artificial.

Figure 3.12: The percentage of persons in the lowest age groups indicates the size of the childbearing population in the next couple of decades.

Figure 3.13: In the 10 to 20 age group, AIDS has not yet developed. The narrowing indicates high mortality among adults as AIDS sets in. The narrower bars in the youngest age groups are a result: AIDS-infected adults die before they have many children, or their AIDS-infected children die young. Many of the children who do survive are orphaned, and the prime workforce of the population is either ill or has died.

Figure 3.14: Infants are completely dependent upon the care of others and are vulnerable to communicable diseases. Therefore, their well-being depends on good nutrition, good medical care, sanitation, caregivers who have sufficient health and strength to give care, and proper temperature and protection from elements and hazards.

Figure 3.15: There are many reasons to migrate, but many migrants leave regions in Africa and Asia in search of better economic opportunities in Europe and to provide better education and health care to their children. They may also follow other family members who have migrated to Europe previously. Immigrants coming to the United States from Canada and Mexico would be able to remain closer to family and friends in their home countries after emigrating than if they had chosen to go to other regions of the world. Further, migrants from these countries would likely feel more familiar with American culture than with cultures in other migration destination regions of the world.

Figure 3.16: Answers will vary, but some factors that could influence migrants' decisions are economic opportunities in a particular country, the presence of family and/or friends who have migrated to that country earlier, or educational opportunities for themselves or their children.

Figure 3.17: Recent outbreaks of diseases like H1N1 influenza and SARS have led people to avoid travel to certain places where the disease has appeared. Government agencies even encourage such avoidance and may restrict or closely examine travelers who have been to such places. In 2005, for example, airline travelers were asked to fill out questionnaires to reveal possible exposure to SARS.

Figure 3.18: Mapping the incidence of the disease is still the first step, but from there, mapping possible sources of air pollution or releases of toxic airborne substances (instead of water pumps) would be the next step.

Figure 3.19: The locations of the samples are too far apart to be spread by simple contagious diffusion.

Figure 3.20: Fossil fuels are alternatives to burning wood, and synthetic building materials also use less timber. However, burning fossil fuels involves possible environmental destruction in their extraction (oil spills, mountaintop removal strip mining, mine waste), transportation (air pollution), and use (air pollution, production of coal ash).

Figure 3.21: The billboards are only part of the campaign and are reinforced by family planning personnel (medical and lay) who carefully monitor compliance and by enormous social pressure. English, as the language of the West, is regarded as high-status and the mark of modern ideas and prosperity.

Figure 3.22: Villages have the advantage of a greater possibility of mutual aid. From the point of view of an authoritarian government, a clustered village is easier to control. Provision of services can be more efficient. Disadvantages include time consumed in traveling to fields and lack of protection of the fields from thievery and predators.

Figure 3.23: The Tibetan village is carefully placed on land that cannot be used for farming. The land on which the Swiss village sits could be used as farmland.

Figure 3.24: Without building many roads, street villages ensure that everyone has access to the road for transportation.

Figure 3.25: A major attraction of migrating to new lands was the availability of farmland and thus a chance to become one of the elite landowning class.

Figure 3.26: The centrality of their location is meant to show their importance. These institutions were major elements in the power of Spanish rulers.

Figure 3.27: The terms *favelas*, *coloniadas*, and *barriadas* are often used for Latin American shantytowns.

Figure 3.28: Answers will vary, but many American cities have wealthy areas that are fairly close to more impoverished areas.

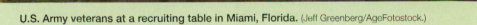

U.S. Army veterans at a recruiting table in Miami, Florida. (Jeff Greenberg/AgeFotostock.)

The Geography of Language

4

Language is one of the primary features that distinguishes humans from other animals. Many animals, including dolphins, whales, and birds, do indeed communicate with one another through patterned systems of sounds, movements, or scents and other scent- and taste-based chemicals. However, the complexity of human language, its ability to convey nuanced emotions and ideas, and its importance for our existence as social beings set it apart from the communication systems used by other animals.

In many ways, language is the essence of culture. It provides the single most common variable by which different cultural groups are identified and by which groups assert their unique identity. Language not only facilitates the cultural diffusion of innovations but also helps to shape the way we think about, perceive, and name our environment. **Language,** a mutually agreed-on system of symbolic communication that has a spoken and

language A mutually agreed-on system of symbolic communication that has a spoken and usually a written expression.

LANGUAGES

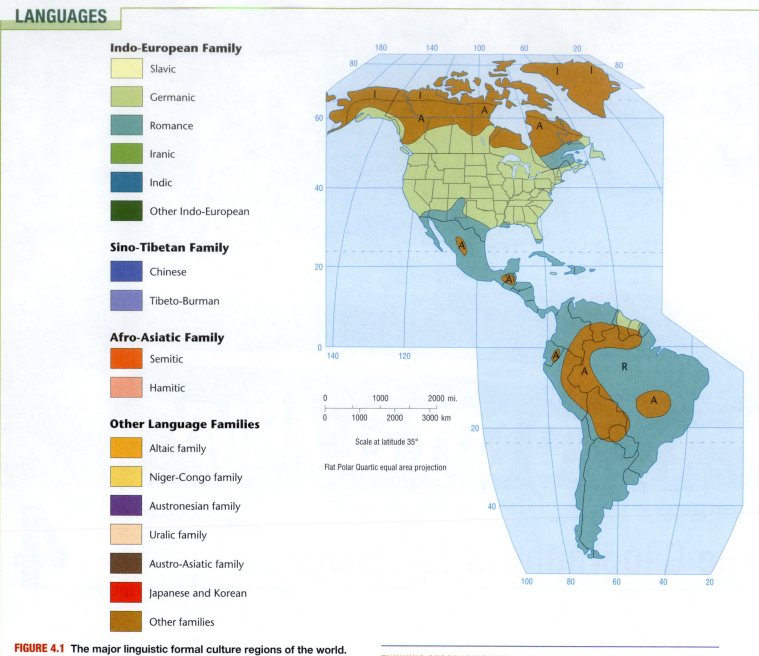

FIGURE 4.1 The major linguistic formal culture regions of the world.
Although there are thousands of languages and dialects in the world, they can be grouped into a few linguistic families. The Indo-European language family represents about half of the world's population. It was spread throughout the world in part through Europe's empire-building efforts.

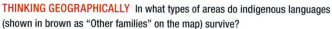

THINKING GEOGRAPHICALLY In what types of areas do indigenous languages (shown in brown as "Other families" on the map) survive?

usually a written expression, offers the main means by which learned belief systems, customs, and skills pass from one generation to the next.

Most cultural groups have their own distinctive form of speech, either a separate language or a dialect. Because languages vary spatially and tend to form spatial groupings, they reinforce the sense of place through the linguistic culture regions they form. Different types of diffusion have helped shape the contemporary linguistic map, as languages relocate through human movement or adapt to the

needs of their users. Clearly, the specific physical habitats in which languages evolve shape their vocabularies. Moreover, the environment can guide the migrations of linguistic groups or provide refuges for languages in retreat. The fact that language is entwined with all aspects of culture allows us to probe complex cultural-linguistic phenomena. Finally, the cultural landscape is shaped by such linguistic acts as naming, writing, and speaking. Power relations are reflected in the cultural landscape, in this case, by who is heard and who is silenced.

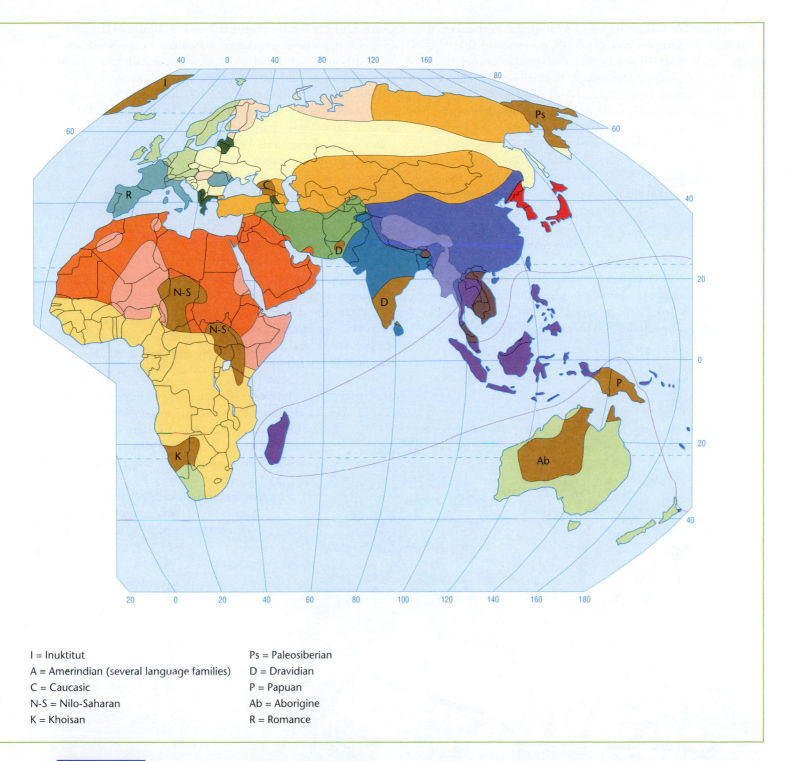

I = Inuktitut
A = Amerindian (several language families)
C = Caucasic
N-S = Nilo-Saharan
K = Khoisan

Ps = Paleosiberian
D = Dravidian
P = Papuan
Ab = Aborigine
R = Romance

LINGUISTIC CULTURE **REGIONS**

What is the geographical patterning of languages? Do various languages provide the basis for formal and functional culture regions? The spatial variation of speech is remarkably complicated, adding intricate patterns to the human mosaic (**Figure 4.1**). Because language is such a central component of culture, understanding the spatiality of language and how and why its patterns change over time provides a

particularly valuable window into cultural geography. The logical place to begin our geographical study of language is with the regional theme.

The Basics of Language

Separate languages are those that cannot be mutually understood. In other words, a monolingual speaker of one language cannot comprehend the speaker of another. **Dialects,** by contrast, are mutually

dialect A distinctive local or regional variant of a language that remains mutually intelligible to speakers of other dialects of that language.

comprehensible variant forms of a language. A speaker of English, for example, can generally understand that language's various dialects, whether the speaker comes from Australia, Scotland, or Mississippi. Nevertheless, a dialect is distinctive enough in vocabulary and pronunciation to label its speaker as hailing from one place or another or even from a particular city. About 6000 languages and many more dialects are spoken in the world today.

When different linguistic groups come into contact, a **pidgin** language, characterized by a very small vocabulary derived from the languages of the groups in contact, often results. Pidgins primarily serve the purposes of trade and commerce: they facilitate exchange at a basic level but do not have complex vocabulary or grammatical structure. An example is Tok Pisin, meaning "talk business." Tok Pisin is a largely English-derived pidgin spoken in Papua New Guinea, where it has become the official national language in a country where many native Papuan tongues are spoken. Although New Guinea pidgin is not readily intelligible to a speaker of Standard English, certain common words such as *gut bai* ("good-bye"), *tenkyu* ("thank you"), and *haumas* ("how much") reflect the influence of English. When pidgin languages acquire fuller vocabularies and become native languages of their speakers, they are called **creole** languages. Obviously, deciding precisely *when* a pidgin becomes a creole language has at least as much to do with a group's political and social recognition as it does with what are, in practice, fuzzy boundaries between language forms.

Another response to the need for speakers of different languages to communicate with each other is the elevation of one existing language to the status of a **lingua franca.** A lingua franca is a language of communication and commerce spoken across a wide area where it is not a mother tongue. Swahili enjoys lingua franca status in much of East Africa, where inhabitants speak a number of other regional languages and dialects. English is fast becoming a global lingua franca. Finally, regions that have linguistically mixed populations may be characterized by **bilingualism,** which is the ability to

> **pidgin** A composite language consisting of a small vocabulary borrowed from the linguistic groups involved in trade and commerce.

> **creole** A language derived from a pidgin language that has acquired a fuller vocabulary and become the native language of its speakers.

> **lingua franca** A language of communication and commerce used widely where it is not a mother tongue.
> **bilingualism** The ability to speak two languages fluently.

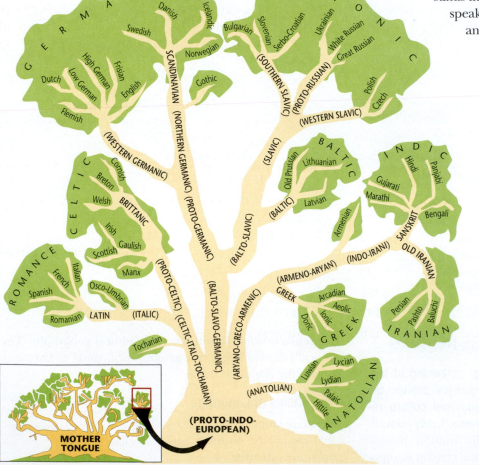

FIGURE 4.2 Linguistic family tree. Shown here are the relationships between the main branches of the linguistic family tree and a detailed image of one branch of that tree.

THINKING GEOGRAPHICALLY Why is a tree a useful metaphor for the relationship between languages? What are its weaknesses?

speak two languages with fluency. For example, along the U.S.–Mexican border, so many residents speak both English and Spanish (with varying degrees of fluidity) that bilingualism in practice—even if not in policy—means there is no need for a lingua franca.

Language Families

One way in which geolinguists often simplify the mapping of languages is by grouping them into **language families:** tongues that are related and share a common ancestor. Words are simply arbitrary sounds associated with certain meanings. Thus, when words in different languages are alike in both sound and meaning, they may well be related. Over time, languages interact with one another, borrowing words, imposing themselves through conquest, or organically diverging from a common ground. Languages and their interrelations can thus be graphically depicted as a tree with various branches (**Figure 4.2**). This classification makes the complicated linguistic mosaic a bit easier to comprehend.

language family A group of related languages derived from a common ancestor.

Indo-European Language Family The largest and most widespread language family is Indo-European, which is spoken on all of the continents and is dominant in Europe, Russia, North and South America, Australia, and parts of southwestern Asia and India (see Figure 4.1). Romance, Slavic, Germanic, Indic, Celtic, and Iranic are all Indo-European

subfamilies. These subfamilies are in turn divided into individual languages. For example, English is a Germanic Indo-European language. Seven Indo-European tongues, including English, are among the 10 most spoken languages in the world as classified by numbers of native speakers (**Table 4.1**).

Comparing the vocabularies of various Indo-European tongues reveals their kinship. For example, the English word *mother* is similar to the Polish *matka,* the Greek *meter,* the Spanish *madre,* the Farsi *madar* in Iran, and the Sinhalese *mava* in Sri Lanka. Such similarities demonstrate that these languages have a common ancestral tongue.

Sino-Tibetan Family Sino-Tibetan is another of the major language families of the world and is second only to Indo-European in numbers of native speakers. The Sino-Tibetan region extends throughout most of China and Southeast Asia (see Figure 4.1). The two language branches that make up this group, Sino- and Tibeto-Burman, are believed to have a common origin some 6000 years ago in the Himalayan Plateau; speakers of the two language groups subsequently moved along the great Asian rivers that originate in this area. *Sino* refers to China and in this context indicates the various languages spoken by more than 1.2 billion people in China. Han Chinese (Mandarin) is spoken in a variety of dialects and serves as the official language of China. The nearly 400 languages and dialects that make up the Burmese and Tibetan branch of this language family border the Chinese language region on the south and

| TABLE 4.1 | The 10 Leading Languages in Numbers of Native Speakers* | | |

Language	Family	Speakers (in millions)	Main Areas Where Spoken
Chinese (all dialects)	Sino-Tibetan	1,213	China, Taiwan, Singapore
Spanish	Indo-European	329	Spain, Latin America, southwestern United States
English	Indo-European	328	British Isles, Anglo-America, Australia, New Zealand, South Africa, Philippines, former British colonies in tropical Asia and Africa
Arabic (all dialects)	Afro-Asiatic	221	Middle East, North Africa
Hindi, Urdu	Indo-European	182	Northern India, Pakistan
Bengali	Indo-European	181	Bangladesh, eastern India
Portuguese	Indo-European	178	Portugal, Brazil, southern Africa
Russian	Indo-European	144	Russia, Kazakhstan, parts of Ukraine, other former Soviet republics
Japanese	Japanese and Korean	122	Japan
German (standard)	Indo-European	90	Germany, Austria, Switzerland, Luxembourg, eastern France, northern Italy

*Native speakers *means the language is their mother tongue.*

(Source: Lewis, M. Paul [ed.], 2009. *Ethnologue: Languages of the World,* 16th ed. Dallas, TX: SIL International. Online version: http://www.ethnologue.com/.)

west. Other East Asian languages, such as Vietnamese, have been heavily influenced by contact with the Chinese and their languages, although it is not clear that they are linguistically related to Chinese at all.

Afro-Asiatic Family The third major language family is the Afro-Asiatic. It consists of two major divisions: Semitic and Hamitic. The Semitic languages cover the area from the Arabian Peninsula and the Tigris-Euphrates river valley of Iraq westward through Syria and North Africa to the Atlantic Ocean. Despite the considerable size of this region, there are fewer speakers of the Semitic languages than you might expect because most of the areas that Semites inhabit are sparsely populated deserts. Arabic is by far the most widespread Semitic language and has the greatest number of native speakers, about 221 million. Although many different dialects of Arabic are spoken, there is only one written form.

Hebrew, which is closely related to Arabic, is another Semitic tongue. For many centuries, Hebrew was a "dead" language, used only in religious ceremonies by millions of Jews throughout the world. With the creation of the state of Israel in 1948, a common language was needed to unite the immigrant Jews, who spoke the languages of their many different countries of origin. Hebrew was revived as the official national language of what otherwise would have been a **polyglot**, or multilanguage, state (**Figure 4.3**). Amharic, a third major Semitic tongue, today claims 18 million speakers in the mountains of East Africa.

> **polyglot** A mixture of languages.

FIGURE 4.3 Sign in Arab Quarter of Nazareth. Many of Israel's cities are home to diverse populations. This child-care center sign reflects Israel's polyglot population, with its English, Arabic, and Hebrew wording. (Source: Courtesy of Patricia L. Price.)

THINKING GEOGRAPHICALLY What examples of signs in more than one language can you find in your community?

Smaller numbers of people who speak Hamitic languages share North and East Africa with the speakers of Semitic languages. Like the Semitic languages, these tongues originated in Asia but today are spoken almost exclusively in Africa by the Berbers of Morocco and Algeria, the Tuaregs of the Sahara, and the Cushites of East Africa.

Other Major Language Families Most of the rest of the world's population speak languages belonging to one of six remaining major families. The Niger-Congo language family, also called Niger-Kordofanian, dominates Africa south of the Sahara Desert. The greater part of the Niger-Congo culture region belongs to the Bantu subgroup. Both Niger-Congo and its Bantu constituent are fragmented into a great many different languages and dialects, including Swahili. The Bantu and their many related languages spread from what is now southeastern Nigeria about 4000 years ago, first west and then south in response to climate change and new agricultural techniques (**Figure 4.4**).

Flanking the Slavic Indo-Europeans on the north and south in Asia are the speakers of the Altaic language family, including Turkic, Mongolic, and several other subgroups. The Altaic homeland lies largely in the inhospitable deserts, tundra, and coniferous forests of northern and central Asia. Also occupying tundra and grassland areas adjacent to the Slavs is the Uralic family. Finnish and Hungarian are the two most widely spoken Uralic tongues, and both enjoy the status of official languages in their respective countries.

One of the most remarkable language families in terms of distribution is the Austronesian. Representatives of this group live mainly on tropical islands stretching from Madagascar, off the east coast of Africa, through Indonesia and the Pacific islands to Hawaii and Easter Island. This longitudinal span is more than half the distance around the world. The north-south, or latitudinal, range of this language area is bounded by Hawaii and Taiwan in the north and New Zealand in the south. The largest single language in this family is Malay-Indonesian, with 58 million native speakers, but the most widespread is Polynesian.

Japanese and Korean, with about 200 million speakers combined, probably form another Asian language family. The two perhaps have some link to the Altaic family, but even their kinship to each other remains controversial and unproven.

In Southeast Asia, the Vietnamese, Cambodians, Thais, and some tribal peoples of Malaysia and parts of India speak languages that constitute the Austro-Asiatic family. They occupy an area into which Sino-Tibetan, Indo-European, and Austronesian languages have all encroached.

Language's Shifting Boundaries

Dialects, as well as the language families discussed previously, reveal a vivid geography. Their boundaries—what

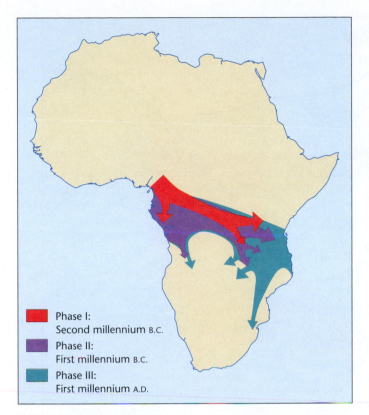

FIGURE 4.4 Bantu expansionism. The Bantu and their many languages spread from what is today southeastern Nigeria to the west and south. (Source: Adapted from Mark Dingemanse/Wikipedia.)

Phase I:
Second millennium B.C.

Phase II:
First millennium B.C.

Phase III:
First millennium A.D.

THINKING GEOGRAPHICALLY Compare this map with Figure 4.7. How far did the resulting Niger-Congo family extend? What language seems to have survived in a restricted region?

tú

vos

tú/vos

Not applicable

FIGURE 4.5 Dialect boundaries in Latin America. Spanish speakers in the Americas use either *vos* or *tú* as the second-person singular pronoun. They are dialects of Spanish: both are correct, linguistically speaking, but the *vos* form is older. Some regions use *vos* and *tú* interchangeably. (Source: Adapted from Pountain, 2005.)

THINKING GEOGRAPHICALLY Why do you think standard Spanish textbooks in the United States teach *tú* as the second-person singular form?

separates them from other dialects and languages—shift over time, both spatially and in terms of what elements they contain or discard.

Geolinguists map dialects by using **isoglosses,** which indicate the spatial borders of individual words or pronunciations. The dialect boundaries between Latin American Spanish speakers using *tú* and those using *vos* are clearly defined in some areas, as shown on the map in **Figure 4.5**. The choice of *tú* or *vos* for the second-person singular carries with it a cultural indication. *Vos* represents a usage closer to the original Spanish but is considered by many in the Spanish-speaking world to be rather archaic and, in fact, has died out in Spain itself. In other regions, particularly throughout Argentina, *vos* has long been used by the media; often, it is considered to reflect the "standard" dialect and usage for the area in which it is used. Because certain words or dialects can fall out of fashion or simply become overwhelmed by an influx of new speakers, isogloss boundaries are rarely clear or stable over time. Indeed, in Central America, the media are increasingly using *vos*—long used in conversation in the region—thus ele-

isogloss The border of usage of an individual word or pronunciation.

vating *vos* to a more official status covering a larger territory. Because of this, geolinguists often disagree about how many dialects are present in an area or exactly where isogloss borders should be drawn. The language map of any place is a constantly shifting kaleidoscope.

The dialects of American English provide another good example. By the time of the American Revolution at least three major dialects, corresponding to major culture regions, had developed in the eastern United States: the Northern, Midland, and Southern dialects (**Figure 4.6** on page 100). As the three subcultures expanded westward, their dialects spread and fragmented. Nevertheless, they retained much of their basic character, even beyond the Mississippi River. These culture regions have unusually stable boundaries. Even today, the "r-less" pronunciation of words such as *car* ("cah") and *storm* ("stohm"), characteristic of the East Coast Midland regions, is readily discernible in the speech of its inhabitants.

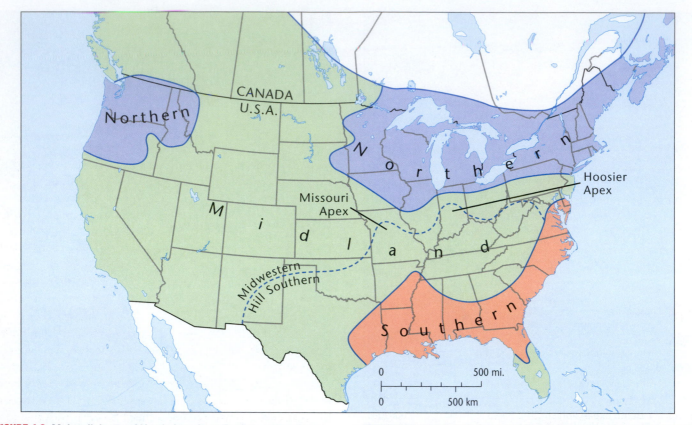

FIGURE 4.6 Major dialects of North American English, with a few selected subdialects. These dialects had developed by the time of the American Revolution and have remained remarkably stable over time. (Sources: After Carver, 1986; Kurath, 1949.)

THINKING GEOGRAPHICALLY What might have caused the Hoosier Apex and the Missouri Apex of Hill Southern speech to penetrate Indiana and Missouri, respectively?

Although we are sometimes led to believe that Americans are becoming more alike as a national culture overwhelms regional ones, the current status of American English dialects suggests otherwise. Linguistic divergence is still under way, and dialects continue to mutate on a regional level, just as they always have. Local variations in grammar and pronunciation proliferate, confounding the proponents of standardized speech and defying the homogenizing influence of the Internet, television, and other mass media.

Shifting language boundaries involve content, as well as spatial reach, and this too changes over time. Today, for example, some of the unique vocabulary of American English dialects is becoming old-fashioned. For instance, the term *icebox*, which was literally a wooden box with a compartment for ice that was used to cool food, was used widely throughout the United States to refer to the precursor of the refrigerator. Although the modern electric refrigerator is ubiquitous in the United States today, some people, particularly those of older generations and in the South, still use the term *icebox*. Many young people, by contrast, no longer say the entire word *refrigerator*, shortening it instead to *fridge*.

As illustrated by the birth of the new word *fridge*, slang terms are quite common in most languages, and American English is no exception. **Slang** refers to words and phrases that are not a part of a standard, recognized vocabulary for a given language but are nonetheless used and understood by some or most of its speakers.

Often, subcultures (for example, youth, street gangs, and users of certain Internet social networking or gaming sites), have their own slang that is used within that community but is not readily understood by nonmembers.

Slang words tend to be used for a period and then are discarded as newer terms replace them. For example, in the 1980s in the United States, *awesome* and *rad(ical)* were used to refer to things that were desirable or fantastic. These slang terms were popularized largely as a result of movies and other media that idealized the California culture of groups such as surfers and "Valley girls" and diffused it nationwide. Although many people today still recognize these words, the current generation of young people is adopting new slang terms emerging from hip-hop culture and other subgroups within the American population. Their children, in turn, will likely use yet an-

> **slang** Words and phrases that are not part of a standard, recognized vocabulary for a given language but that are nonetheless used and understood by some of its speakers.

other set of slang words to express themselves in the future.

In the United States today, many descendants of Spanish speakers have adapted their speech to include words and variants of words in both Spanish and English in a dialect known as Spanglish. Similarly, French and English have been blended together by certain cultural groups within the United States, Canada, and former French colonies like Cameroon to produce Franglais, or Franglish. This combination has been particularly notable among native English speakers and immigrant bilingual language learners in French-speaking regions of Canada like the province of Québec. The provincial government of Québec has officially discouraged such "incorrect" uses of the French language, but the new language blend appears to be persisting despite these admonishments.

LINGUISTIC **DIFFUSION**

How did the mosaic of languages and dialects come to exist? How is the spatial patterning of language changing today? Different types of cultural diffusion have helped shape the linguistic map. Relocation diffusion has been extremely important because languages spread when groups, in whole or in part, migrate from one area to another. Some individual tongues or entire language families are no longer spoken in the regions where they originated, and in certain other cases the linguistic hearth is peripheral to the present distribution (compare Figure 4.1 on page 94 and **Figure 4.7** on page 102). Today, languages continue to evolve and change based on the shifting locations of peoples and on their needs, as well as on outside forces.

Indo-European Diffusion

According to a widely accepted new theory, the earliest speakers of the Indo-European languages lived in southern and southeastern Turkey, a region known as Anatolia, about 9000 years ago. According to the so-called **Anatolian hypothesis,** the initial diffusion of these Indo-European speakers was facilitated by the innovation of plant domestication. As sedentary farming was adopted throughout Europe, a gradual and peaceful expansion and diffusion of Indo-European languages occurred. As these people dispersed and lost contact with one another, different Indo-European groups gradually developed variant forms of the language, causing fragmentation of the language family.

The Anatolian hypothesis has been criticized by scholars who note that specific terms related to animals, particu-

> **Anatolian hypothesis** A theory of language diffusion holding that the movement of Indo-European languages from the area in contemporary Turkey known as Anatolia followed the spread of plant domestication technologies.

larly horses, as opposed to agriculture, link Indo-European languages to a common origin. The so-called **Kurgan hypothesis** places the rise of Indo-European languages in the central Asian steppes only 6000 years ago, positing that the spread of Indo-European languages was both swifter and less peaceful than believed by those who subscribe to the Anatolian hypothesis. No one theory has been definitively proven to be correct.

> **Kurgan hypothesis** A theory of language diffusion holding that the spread of Indo-European languages originated with animal domestication; originated in the Central Asian steppes; and was later, more violent, and swifter than proponents of the Anatolian hypothesis maintain.

What is more certain is that in later millennia, the diffusion of certain Indo-European languages—in particular Latin, English, and Russian—occurred in conjunction with the territorial spread of great political empires. In such cases of imperial conquest, relocation diffusion and expansion diffusion were not mutually exclusive. Relocation diffusion occurred as a small number of conquering elite came to rule an area. The language of the conqueror, implanted by relocation diffusion, often gained wider acceptance through expansion diffusion. Typically, the conqueror's language spread hierarchically—adopted first by the more important and influential persons and by city dwellers. The diffusion of Latin with Roman conquests, and Spanish with the conquest of Latin America, occurred in this manner.

Austronesian Diffusion

One of the most impressive examples of linguistic diffusion is that of the Austronesian languages, 5000 years ago, from a presumed hearth in the interior of Southeast Asia that was completely outside the present Austronesian culture region. From here, it is theorized, speakers of this language family first spread southward into the Malay Peninsula (see Figure 4.7). Then, in a process lasting perhaps several thousand years and requiring remarkable navigational skills, they migrated through the islands of Indonesia and sailed in tiny boats across vast, uncharted expanses of ocean to New Zealand, Easter Island, Hawaii, and Madagascar. If agriculture was the technology permitting Indo-European diffusion, sailing and navigation provided the key to the spread of the Austronesians.

Most remarkable of all was the diffusionary achievement of the Polynesian people, who form the eastern part of the Austronesian culture region. Polynesians occupy a triangular realm consisting of hundreds of Pacific islands, with New Zealand, Easter Island, and Hawaii at the three corners (**Figure 4.8** on page 103). The Polynesians' watery leap of 2500 miles (4000 kilometers) from the South Pacific to Hawaii, a migration in outrigger canoes against prevailing winds into a new hemisphere with different navigational stars, must rank as one of the greatest achievements in seafaring. No humans had previously found the

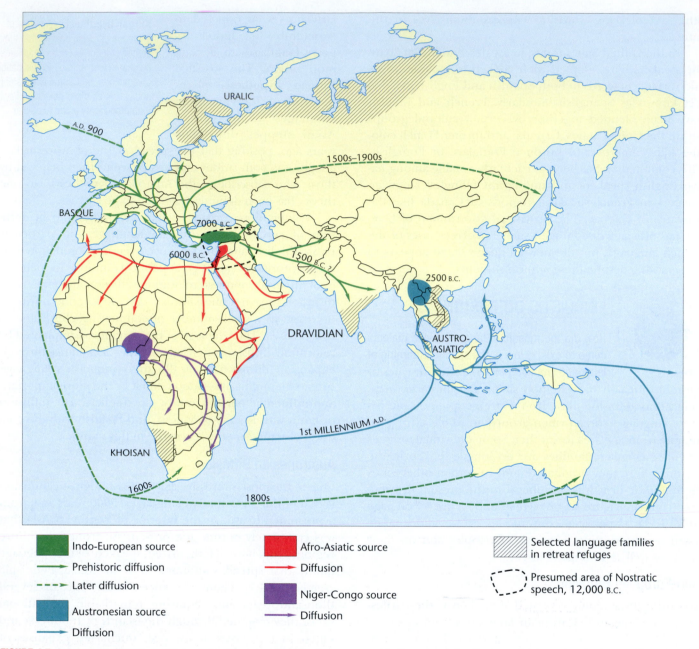

FIGURE 4.7 Origin and diffusion of four major language families in the Eastern Hemisphere. The early diffusion of Indo-European speech to the west and north probably occurred in conjunction with the diffusion of agriculture from the Middle Eastern center, as did the early spread of the Afro-Asiatic family. As these and other groups advanced, certain linguistic families retreated to refuges in remote places, where they hold out to the present day. Sources, dates, and routes are mostly speculative. (Sources: Kranta, 1988; Renfrew, 1989.)

THINKING GEOGRAPHICALLY Why would language diffuse along with agriculture?

isolated Hawaiian Islands, and the Polynesian sailors had no way of knowing ahead of time that land existed in that quarter of the Pacific.

The relocation diffusion that produced the remarkable present distribution of the Polynesian languages has long been the subject of controversy. How, and by what means, could a traditional people have achieved the diffusion? Geolinguists Michael Levison, Gerard Ward, and John Webb answered these questions by developing a computer model

incorporating data on winds, ocean currents, vessel traits and capabilities, island visibility, and duration of voyage. Both *drift* voyages, in which the boat simply floats with the winds and currents, and *navigated* voyages were considered.

After performing more than 100,000 voyage simulations, they concluded that the Polynesian triangle had probably been entered from the west, from the direction of the ancient Austronesian hearth area, in a process of island-hopping—that is, migrating from one island to another

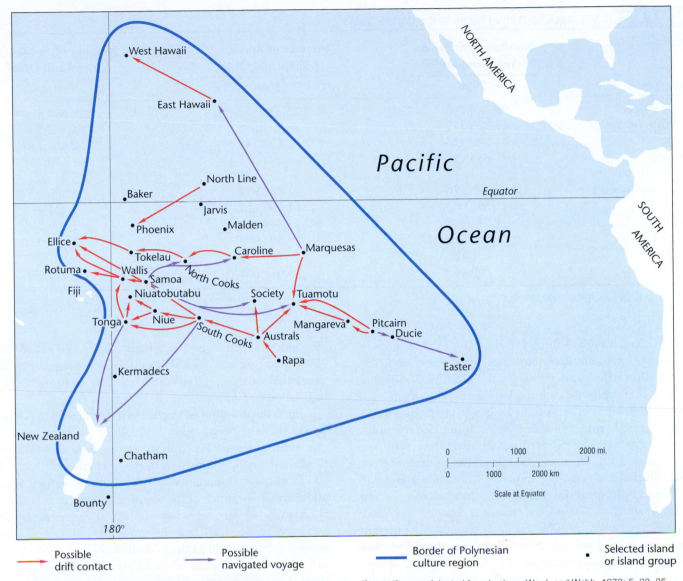

| Possible drift contact | Possible navigated voyage | Border of Polynesian culture region | • Selected island or island group |

FIGURE 4.8 **Probabilities of selected Polynesian drift and navigation voyages in the Pacific Ocean.** According to a computer model, the outer arc of Polynesia, represented by Hawaii, Easter Island, and New Zealand, could have been reached only by navigated voyages. The earliest known Polynesian pottery shards were found in Tonga by archaeologists David Burley and William Dickinson, suggesting that the first distinctively Polynesian culture originated

there. (Source: Adapted from Levison, Ward, and Webb, 1973: 5, 33, 35, 43, 61.)

THINKING GEOGRAPHICALLY The Maori, indigenous Polynesian people in New Zealand, link their identity to specific "canoes," or groups of ancestors, as they arrived in New Zealand. How does this system of identity support the idea that voyages to New Zealand were navigated rather than drift?

one visible in the distance. The core of eastern Polynesia was probably reached in navigated voyages, but once this was attained, drift voyages easily explain much of the internal diffusion. According to Levison and his colleagues, a peripheral region, the "outer arc from Hawaii through Easter Island to New Zealand," could be reached only by means of intentionally navigated voyages: truly astonishing and daring feats (see Figure 4.8).

In carrying out their investigation, Levison and his colleagues employed the themes of region (present distribution of Polynesians) and cultural ecology (currents, winds,

visibility of islands) to help describe and explain the workings of a third theme, cultural diffusion.

Language Proliferation: One or Many?

Using techniques that remain controversial, certain linguists are probing the origin and diffusion of languages, seeking elusive prehistoric tongues. Some scholars believe that an ancestral speech called *Nostratic*, spoken in the Middle East 12,000 to 20,000 years ago, was central to six modern language families: Indo-European, Uralic, Altaic, Afro-Asiatic, Caucasic, and Dravidian (see Figures 4.2 and 4.7).

TABLE 4.2	Languages Spoken at Home in the United States		
Language	Number of Speakers 5 Years and Older 1990	Number of Speakers 5 Years and Older 2000	Number of Speakers 5 Years and Older 2010
English only	198,600,798	215,423,557	228,699,523
Spanish, Spanish Creole	17,339,172	28,101,052	35,468,501
Chinese	1,249,213	2,022,143	2,600,150
Tagalog	843,251	1,224,241	1,513,734
French (including Patois, Cajun)	1,702,176	1,643,838	1,305,503
Vietnamese	507,069	1,009,627	1,251,468
German	1,547,099	1,383,442	1,109,216
Korean	626,478	894,063	1,039,021
Russian	241,798	706,242	881,723
Arabic	355,150	614,582	845,396

(Source: U.S. Census, 1990; 2000; 2010.)

These linguists seek nothing less than the original linguistic hearth area, almost certainly in Africa, where complex speech first arose and from which it diffused. Skeptics counter that similarities among languages arose from coincidence or from speakers of one language borrowing words from others through trade and other cultural interactions.

Are the forces of modernization working to produce, through cultural diffusion, a single world language? If so, what will that language be? English? Worldwide about 328 million people speak English as their mother tongue and perhaps another 350 million speak it well as a second, learned language. Adding other reasonably competent speakers who can get by in English, the world total reaches about 1.5 billion, more than for any other language. What's more, the Internet is one of the most potent agents of diffusion, and its language, overwhelmingly, is English.

English earlier diffused widely with the British Empire and U.S. imperialism, and today it has become the de facto language of globalization. Consider the case of India, where the English language imposed by British rulers was retained (after independence) as the country's language of business, government, and education. It provided some linguistic unity for India, which had 800 indigenous languages and dialects. This is why today many of India's nearly 1 billion people speak English well enough to provide customer support services over the telephone for clients in the United States. Even so, many people resent its use and wish India to be rid of this hated linguistic colonial legacy once and for all (see Figure 4.21, page 115). Although English is not likely to be driven out of India any time soon, it is true that the spoken English of India has drifted away from Standard British English. The same is true for the English of

Singapore, which is now a separate language called Singlish. Many other regional English-based languages have developed, languages that could not be understood readily in London or Chicago.

But is the diffusion of English to the entire world population likely? Will globalization and cultural diffusion produce one world language? Probably not. More likely, the world will be divided largely among 5 to 10 major languages.

Despite the dominance of English in the United States and throughout much of the world in the twenty-first century, many other languages are spoken in American homes every day (**Table 4.2**). Examining the prominence of these various languages within the United States provides valuable insight into the presence of its different immigrant populations. Given that international migration trends fluctuate over time, it is not surprising that languages (other than English) spoken in American homes fluctuate from decade to decade. For example, by examining Table 4.2, we can see that the number of speakers of French and German in the United States has steadily declined since 1990. This trend will likely continue in future decades, as most families in the United States today that are of French or German descent can be traced to immigrants who came to the country during the eighteenth, nineteenth, and early twentieth centuries. Twenty-first century generations of these families have typically acquired English and begun using it as their first language. In contrast, Table 4.2 shows that speakers of Chinese and Arabic are on the increase in the United States since 1990. These trends reflect contemporary migration flows from Asia and the Middle East, as immigrants from these regions have more recently been drawn to the United States in search of economic and educational opportunities.

LINGUISTIC **ECOLOGY**

What relationships exist between language and the physical environment? Language interacts with the environment in two basic ways. First, the specific physical habitats in which languages evolve help shape their vocabularies. Second, the environment can guide the migrations of linguistic groups or provide refuges for languages in retreat. The following section, from the viewpoint of possibilism—the notion that the physical environment shapes but does not fully determine cultural phenomena—illustrates how the physical environment influences vocabulary and the distribution of language.

Environment and Vocabulary

Humankind's relationship to the land played a strong role in the emergence of linguistic differences, even at the level of vocabulary. For example, the Spanish language—which originated in Castile, Spain, a dry and relatively barren land rimmed by hills and high mountains—is especially

FIGURE 4.9 A scene in the desert of the western United States. "Mountains," yes, but what kind of mountains? (See Table 4.3.) The English language cannot describe such a place adequately because it is the product of a very different physical landscape, humid and cool. As a result, the ability of English speakers to name dryland environmental features will be less precise in such places. (David Muench/Corbis.)

THINKING GEOGRAPHICALLY How many words in Table 4.3 can you find that English has adopted in its descriptive language or in place names?

TABLE 4.3 Some Spanish Words Describing Mountains and Hills

Spanish Word	English Meaning
candelas	Literally "candles"; a collection of *peñas*
ceja	Steep-sided breaks or escarpments separating two plains of different elevations
cerro	A single eminence, intermediate in size between English *hill* and *mountain*
cordillera	A mass of mountains, as distinguished from a single mountain summit
cuchilla	Literally "knife"; the comblike secondary crests that project at right angles from the sides of a *sierra*
cumbre	The highest elevation or peak within a *sierra* or *cordillera*; a summit
eminencia	A mountainous or hilly protuberance
loma	A hill in the midst of a plain
mesa	Literally "table"; a flat-topped eminence
montaña	Equivalent to English *mountain*
pelado	A barren, treeless mountain
pelon	A bare conical eminence
peña	A needlelike eminence
picacho	A peaked or pointed eminence
sandia	Literally "watermelon"; an oblong, rounded eminence
sierra	An elongated mountain mass with a serrated crest

(Source: Hill, 1986.)

rich in words describing rough terrain, allowing speakers of this tongue to distinguish even subtle differences in the shape and configuration of mountains, as **Table 4.3** reveals. Similarly, Scottish Gaelic possesses a rich vocabulary to describe types of topography; this terrain-focused vocabulary is an attribute of all of the Celtic languages spoken by hill peoples. In the Romanian tongue, also born of a rugged landscape, words relating to mountainous features emphasize use of that terrain for livestock herding. English, by contrast, which developed in the temperate wet coastal plains of northern Europe, is relatively deficient in words describing mountainous terrain (**Figure 4.9**). However, English abounds with words describing flowing streams and wetlands: typical physical features found in northern Europe. This vocabulary transferred well to the temperate East Coast of the United States. In the rural American South alone, one finds *river, creek, branch, fork, prong, run, bayou,* and *slough*. This vocabulary indicates that the area is a well-watered land with a dense network of streams.

Clearly, then, language serves an adaptive strategy. Vocabularies are highly developed for those features of the environment that involve livelihood. Without such detailed vocabularies, it would be difficult to communicate sophisticated information relevant to the community's livelihood, which in most places is closely bound to the physical landscape.

The Environment Helps Shape Language Areas and Guide Migration

Environmental barriers and natural routes have often guided linguistic groups onto certain paths. The wide distribution of the Austronesian language group, as we have seen, was profoundly affected by prevailing winds and water currents in the Pacific and Indian oceans. The Himalayas and the barren Deccan Plateau deflected migrating Indo-Europeans entering the Indian subcontinent into the rich Ganges-Indus river plain. Even today in parts of India, according to Charles Bennett, the Indo-European/Dravidian "language boundary seems to approximate an ecological boundary" between the black soils of the plains and the thinner, reddish Deccan soils.

Because such physical barriers as mountain ridges can discourage groups from migrating from one area to another, they often serve as linguistic borders as well. In parts of the Alps, speakers of German and Italian live on opposite sides of a major mountain ridge. Portions of the mountain rim along the northern edge of the Fertile Crescent in the Middle East form the border between Semitic and Indo-European tongues. Linguistic borders that follow such physical features generally tend to be stable, and they often endure for thousands of years. By contrast, language borders that cross plains and major routes of communication are often unstable.

The Environment Provides Refuge

The environment also influences language insofar as inhospitable areas provide protection and isolation. Such areas often provide minority linguistic groups refuge from aggressive neighbors and are accordingly known as **linguistic refuge areas.** Rugged hilly and mountainous areas, excessively cold or dry climates, dense forests, remote islands, and extensive marshes and swamps can all offer protection to minority language groups. For one thing, unpleasant environments rarely attract conquerors. Also, mountains tend to isolate the inhabitants of one valley from those in adjacent ones, discouraging contact that might lead to linguistic diffusion.

Examples of these linguistic refuge areas are numerous. The rugged Caucasus Mountains and nearby ranges in central Eurasia are populated by a large variety of peoples and languages (**Figure 4.10**). In the Rocky Mountains of northern New Mexico, an archaic form of Spanish survives, largely as a result of isolation that ended only in the early 1900s. Similarly, the Alps, the Himalayas, and the highlands of Mexico form fine-grained linguistic mosaics, thanks to the mountains that provide both isolation and

linguistic refuge area An area protected by isolation or inhospitable environmental conditions in which specific languages or dialects have survived.

protection for multitudinous languages. The Dhofar, a mountain tribe in Oman, preserves Hamitic speech, a language family otherwise vanished from all of Asia. Bitterly cold tundra climates of the far north have sheltered Uralic and Inuktitut speakers, and a desert has shielded Khoisan speakers from Bantu invaders. In short, rugged, hostile, or isolated environments protect linguistic groups that might otherwise be eclipsed by more dominant languages.

Still, environmental isolation is no longer the vital linguistic force it once was. Fewer and fewer places are so isolated that they remain little touched by outside influences. Today, inhospitable lands may offer linguistic refuge, but it is no longer certain that they will in the future. Even an island in the middle of the vast Pacific Ocean does not offer reliable refuge in an age of airplanes, satellite communications, and global tourism. Similarly, marshes and forests provide refuge only if they are not drained and cleared by those who wish to use the land more intensively. The nearly 10,000 Gullah-speaking descendants of African slaves have long nurtured their distinctive African-influenced culture and language, in part because they reside on the Sea Islands of South Carolina, Georgia, and North Florida. Today, the development of these islands for tourism and housing for wealthy nonlocals threatens the survival of the Gullah culture and language, as has the out-migration of Gullah youth in search of better economic opportunities. The reality of the world is no longer isolation, but rather contact.

CULTURAL-LINGUISTIC INTERACTION

Language is intertwined with all aspects of culture. The theme of cultural interaction permits us to probe some of these complex links between speech and other cultural phenomena. The complicated linguistic map cannot be understood without reference to the comparative social, demographic, political, and technological characteristics of the groups in question. At root, linguistic cultural interaction often reflects the dominance of one group over another, a dominance based in culture.

Religion and Language

Cultural interaction creates situations in which language is linked to a particular religious faith or denomination, a linkage that greatly heightens cultural identity. Perhaps Arabic provides the best example of this cultural link. It spread from a core area on the Arabian Peninsula with the expansion of Islam. Had it not been for the evangelical success of the Muslims, Arabic would not have diffused so widely. The other Semitic languages also correspond to particular religious groups. Most Hebrew-speaking people are Jewish, and the Amharic speakers in Ethiopia tend to

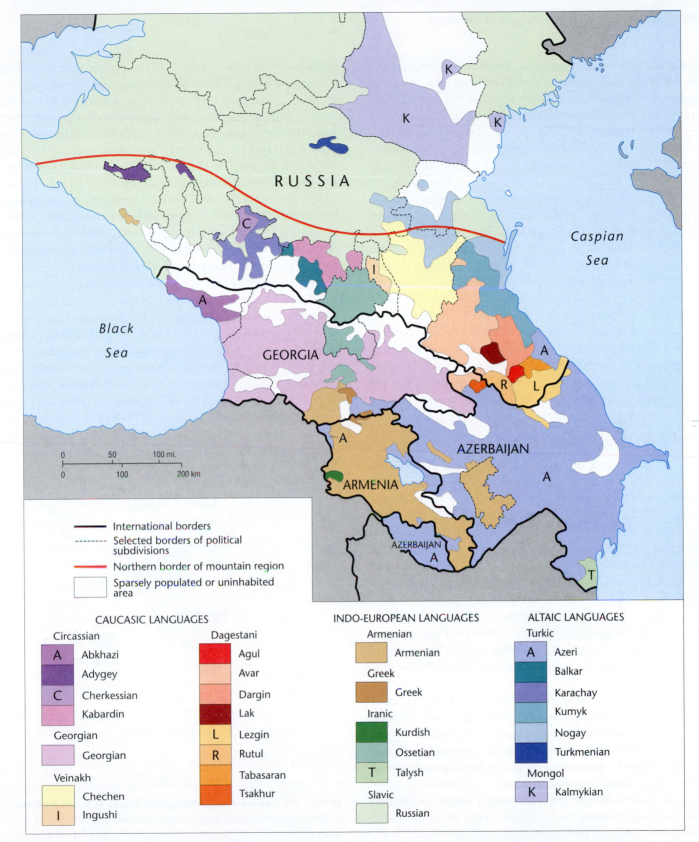

CAUCASIC LANGUAGES

Circassian
- A Abkhazi
- Adygey
- C Cherkessian
- Kabardin

Georgian
- Georgian

Veinakh
- Chechen
- I Ingushi

Dagestani
- Agul
- Avar
- Dargin
- Lak
- L Lezgin
- R Rutul
- Tabasaran
- Tsakhur

INDO-EUROPEAN LANGUAGES

Armenian
- Armenian

Greek
- Greek

Iranic
- Kurdish
- Ossetian
- T Talysh

Slavic
- Russian

ALTAIC LANGUAGES

Turkic
- A Azeri
- Balkar
- Karachay
- Kumyk
- Nogay
- Turkmenian

Mongol
- K Kalmykian

Map legend:
- International borders
- Selected borders of political subdivisions
- Northern border of mountain region
- Sparsely populated or uninhabited area

FIGURE 4.10 The environment is a linguistic refuge in the Caucasus Mountains. The rugged mountainous region between the Black and Caspian seas—including parts of Armenia, Russia, Georgia, and Azerbaijan—is peopled by a great variety of linguistic groups, representing three major language families. Mountain areas are often linguistic mosaics because the rough terrain provides refuge and isolation. For more information about this fascinating and diverse region, see Wixman, 1980.

THINKING GEOGRAPHICALLY Why has the Caucasus been a region of considerable political conflict since the breakup of the Soviet Union?

be Coptic, or Eastern, Christians. Indeed, we can attribute the preservation and recent revival of Hebrew to its active promotion by Jewish nationalists who believe that teaching and promoting Hebrew to diasporic Jews facilitates unity.

Certain languages have acquired a religious status. Latin survived mainly as the ceremonial language of the Roman Catholic Church and Vatican City. In non-Arabic Muslim lands, such as Iran, where people consider themselves Persians and speak Farsi, Arabic is still used in religious ceremonies. Great religious books can also shape languages by providing them with a standard form. Martin Luther's translation of the Bible led to the standardization of the German language, and the Koran is the model for written Arabic. Because they act as common points of frequent cultural reference and interaction, great religious books can also aid in the survival of languages that would otherwise become extinct. The early appearance of a hymnal and the Bible in Welsh aided the survival of that Celtic tongue, and Christian missionaries in diverse countries have translated the Bible into local languages, helping to preserve them. In Fiji, the appearance of the Bible in one of the 15 local dialects elevated it to the dominant native language of the islands.

Technology, Language, and Empire

Technological innovations affecting language range from the basic practice of writing down spoken languages to the sophisticated information superhighway provided by the Internet. Technological innovations have in the past facilitated the spread and proliferation of multiple languages, but more recently they have encouraged the tendency of only a few languages—especially English—to dominate all others. Particular language groups achieve cultural dominance over neighboring groups in a variety of ways, often with profound results for the linguistic map of the world. Technological superiority is usually involved. Plant and animal domestication—the technology of the "agricultural revolution"—aided the early diffusion of the Indo-European language family.

An even more basic technology was the invention of writing, which appears to have developed as early as 5300 years ago in several hearth areas, including in Egypt, among the Sumerians in what is today Iraq, and in China. Writing helped civilizations develop and spread, giving written languages a major advantage over those that remained spoken only. Written languages can be published and distributed widely, and they carry with them the status of standard, official, and legal communication.

Written language facilitates record keeping, allowing governments and bureaucracies to develop. Thus, the languages of conquerors tend to spread with imperial expansion. The imperial expansion of Britain, France, the Netherlands, Belgium, Portugal, Spain, and the United States across the globe altered the linguistic practices of

millions of people (**Figure 4.11**). This empire building superimposed Indo-European tongues on the map of the tropics and subtropics. The areas most affected were Asia, Africa, and the Austronesian island world. A parallel case from the ancient world is China, also a formidable imperial power that spread its language to those it conquered. During the Tang dynasty (A.D. 618–907), Chinese control extended to Tibet, Mongolia, Manchuria (in contemporary northeastern China), and Korea. The 4000-year-old written Chinese language proved essential for the cohesion and maintenance of its far-flung empire. Although people throughout the empire spoke different dialects or even different languages, a common writing system lent a measure of mutual intelligibility at the level of the written word.

Even though imperial nations have, for the most part, given up their colonial empires, the languages they transplanted overseas survive. As a result, English still has a foothold in much of Africa, South Asia, the Philippines, and the Pacific islands. French persists in former French and Belgian colonies, especially in northern, western, and central Africa; Madagascar; and Polynesia (**Figure 4.12**). In most of these areas, English and French function as the languages of the educated elite, often holding official status. They are also used as a lingua franca of government,

SPAIN | PORTUGAL

—— Present border of Brazil (Portuguese as official language)

Language spoken:
- Portuguese
- Spanish
- English
- Dutch
- French

N

Approximate line of demarcation Treaty of Tordesillas, 1494

0   800   1600 km
0   600   1000 mi.

FIGURE 4.11 **The mesh of language and empire in South America.** Latin America was colonized by Spanish speakers and Portuguese speakers as well as by speakers of French, Dutch, and English in northern areas. The Treaty of Tordesillas, signed by Spain and Portugal in 1494, established the political basis for the present linguistic pattern in South America. Portugal was awarded the eastern part of the continent, and Spain, the rest. (Source: Courtesy of Terry G. Jordan-Bychkov.)

THINKING GEOGRAPHICALLY How do you account for the extensions of Portuguese west of the treaty line?

FIGURE 4.12 French and Polynesian combined. French, the colonial language of the empire, shares this sign on the isle of Bora Bora in French Polynesia with the native variant of the Polynesian tongue. Until recently, French rulers allowed no public display of the Polynesian language and tried to make the natives adopt French. (Courtesy of Terry G. Jordan-Bychkov.)

THINKING GEOGRAPHICALLY On a global scale, what advantages would people on Bora Bora find in knowing French?

commerce, and higher education, helping hold together states with multiple native languages.

Transportation technology also profoundly affects the geography of languages. Ships, railroads, and highways all serve to spread the languages of the cultural groups that build them, sometimes spelling doom for the speech of less technologically advanced peoples whose lands are suddenly opened to outside contacts. The Trans-Siberian Railroad, built about a century ago, spread the Russian language eastward to the Pacific Ocean. The Alaska Highway, which runs through Canada, carried English into Native American refuges. The construction of highways in Brazil's remote Amazonian interior threatens the native languages of that region.

REFLECTING ON GEOGRAPHY

Can we view the Internet as a principal transportation route responsible for spreading English throughout the world today? If so, what does the spread of Internet access to areas formerly isolated by their physical landscape mean for the survival of linguistic minorities?

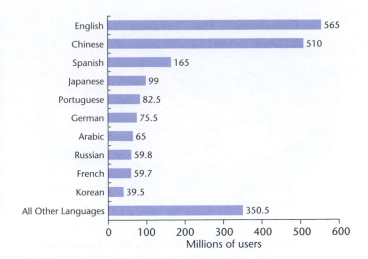

FIGURE 4.13 The 10 most prevalent languages on the Internet as of 2011, measured by total users in each language. English is the third most widely spoken language on Earth, after Chinese and Spanish. English is the most widely spoken *second* language in the world. It dominates the Internet as well. (Source: Adapted from http://internetworldstats.com/stats7.htm.)

THINKING GEOGRAPHICALLY If Chinese and Spanish have more speakers, why would English dominate the Internet?

Another example is the predominance of English on the Internet, which can be understood as a contemporary information highway (**Figure 4.13**). What will happen when other languages begin to challenge the dominance of English on the Internet? This will inevitably happen sooner or later, although whether English will be surpassed by another language is anyone's guess. For example, although only 23.7 percent of today's Internet users speak Chinese, from 2000 to 2011, there was 1,478.7 percent growth in the number of Chinese speakers on the Internet. If this trend continues, and when—not if—the 60.8 percent of Chinese speakers who do not now use the Internet begin to log on, we can expect the prevalence of Chinese on the Internet to continue expanding significantly.

Language and Cultural Survival

Because language is the primary way of expressing culture, if a language dies out, there is a good chance that the culture of its speakers will too. Some languages, like animal species, are classified as endangered or extinct. Endangered languages are those that are not being taught to children by their parents and are not being used actively in everyday matters. Some linguists believe that more than half of the world's roughly 6000 languages are endangered. Ethnologue, an online language resource (see Linguistic Geography on the Internet at the end of the chapter for this web site address), considers 417 world languages to be

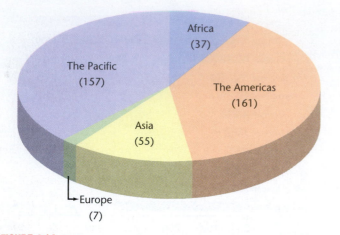

FIGURE 4.14 Distribution of the world's nearly extinct languages. This chart shows the regional distribution of the world's nearly extinct languages, or those languages that have only a few elderly speakers still living. Nearly extinct languages account for 6 percent of the world's total existing languages. (Source: Adapted from Ethnologue, "Nearly Extinct Languages.")

THINKING GEOGRAPHICALLY How do you account for the small number of nearly extinct languages in Europe?

nearly extinct. Languages that have only a few elderly speakers still living fall into this category. Today the Americas and the Pacific regions together account for more than three-quarters of the world's nearly extinct languages, thanks to their many and varied indigenous language groups. In Argentina, for example, only five families speak Vilela, and only seven or eight speakers of Tuscarora remain in Canada. When these speakers die, it is likely that their language will die out with them.

As **Figure 4.14** shows, almost 40 percent of the world's nearly extinct languages are found in the Americas. They include a wealth of Native American languages that are slowly becoming suffocated by English, Spanish, and Portu-

guese. Other **language hotspots**—places with unique, misunderstood, or endangered tongues—are located around the globe (**Figure 4.15**). Two of the world's five most vulnerable regions—northern Australia, central South America, North America's upper Pacific coast, eastern Siberia, and Oklahoma—are located in the United States.

Related to linguistic extinction is the existence of so-called remnant languages that survive, typically in small linguistic islands that are surrounded by the dominant language. One example is Khoisan, found in the Kalahari Desert of southwestern Africa and characterized by distinctive clicking sounds. The pockets of Khoisan seen in Figure 4.1 survived after the expansion of the Bantu, discussed earlier (see Figure 4.4). Other remnant languages include Dravidian, spoken by hundreds of millions of people in southern India, adjacent northern Sri Lanka, and a part of Pakistan, as well as Australian Aborigine, Papuan, Caucasic, Nilo-Saharan, Paleosiberian, Inuktitut, and a variety of Native American language families. In a few cases, individual minor languages represent the sole survivors of former families. Basque, spoken in the borderland between Spain and France, is such a survivor, unrelated to any other language in the world.

language hotspot A place on Earth that is home to a unique, misunderstood, or endangered language.

Texting and Language Modification

Though English dominates the Internet (see Figure 4.13), much of what comes across our computer and cell phone screens isn't a readily recognizable form of English. As with the diffusion of spoken English to far-flung regions of the British Empire, the English language that is spread via electronic correspondence is subject to significant modification. E-mailing, instant messaging, and text

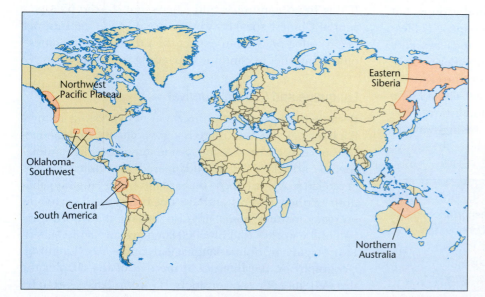

FIGURE 4.15 Global language hotspots. The Enduring Voices Project and the National Geographic Society have teamed up to document endangered languages and thereby attempt to prevent language extinction. (Source: Anderson and Harrison, 2007.)

THINKING GEOGRAPHICALLY Some areas with concentrations of indigenous peoples are not language hotspots. Why do you think this is the case?

FIGURE 4.16 **Sign in Apia, Western Samoa.** Samoan, a Polynesian language belonging to the Austronesian family, becomes part of the linguistic landscape of Apia, the capital of independent Western Samoa in the Pacific Ocean. When this area was still a British colony, such a visual display of the native language would not have been permitted. (Courtesy of Terry G. Jordan-Bychkov.)

THINKING GEOGRAPHICALLY What part of this sign is international and understandable to speakers of any language? Why would it have been included?

messaging Standard English on cell phones requires a lot of typing, and text is notoriously deficient in conveying emotions when compared to the spoken word. For these reasons, users often use abbreviations and symbols to decrease the number of keystrokes used, to add emotional punctuation to their correspondence, and to make electronic communication hard to monitor by those who don't understand the language—particularly parents and teachers.

English is an *alphabetic* writing system; its letters represent discrete sounds that must be strung together to form a word's complete sound. English-speaking texters quickly learn to take shortcuts around the lengthiness inherent to alphabetic writing. The simplest and most commonly used shortcut involves using abbreviations instead of whole words to convey common phrases. POS (parent over shoulder), AFK (away from keyboard), VBG (very big grin), LOL (laughing out loud), and GMTA (great minds think alike) are some examples of this technique. Another shortcut involves using characters that sound like words. For example, the number "8" can substitute for the word or sound "ate," so "h8" is "hate," and "i 8" is "I ate." As your instructor for this class will likely confirm, such abbreviations and symbols have even worked their way into term papers at the college level, much to the consternation or delight of language scholars.

In texting, the use of symbols such as ☺ and ♥, which are called *pictograms* (word pictures), is similar to Chinese writing. Their popularity has resulted in a vast lexicon of pictograms as well as symbol groupings, called *emoticons*, used to convey entire words, ideas, and emotions in a compact and often humorous form. Here are some examples:

> :)   (user is smiling or joking)
>
> :@   (user is screaming or cursing)
>
> :#   (well, shut my mouth!)

Although the meaning of these symbols is understood by speakers of many languages because of their ubiquity, non-English languages also employ their own symbol combinations. In Korean, for instance, ^^ is used instead of :) to convey a smiling face and -_- is used instead of :(to depict a sad face. In Chinese, the number 5 is pronounced in a way that resembles crying, so "555" is the Chinese texter's way of conveying sadness.

LINGUISTIC **LANDSCAPES**

In what ways are languages visible and, as a result, part of the cultural landscape?

Road signs, billboards, graffiti, placards, and other publicly displayed writings not only reveal the locally dominant language but also can be a visual index to bilingualism, linguistic oppression of minorities, and other facets of linguistic geography (**Figure 4.16**). Furthermore, differences in writing systems render some linguistic landscapes illegible to those not familiar with these forms of writing (**Figure 4.17** on page 112).

Messages

Linguistic landscapes send messages, both friendly and hostile. Often these messages have a political content and deal with power, domination, subjugation, or freedom. In Turkey, for example, until recently Kurdish-speaking minorities were not allowed to broadcast music or television programs in Kurdish, to publish books in Kurdish, or even to give their children Kurdish names. Because Turkey wishes to join the European Union, these minority language restrictions have come under intense outside scrutiny. In 2002, Turkey reformed its legal restrictions to allow the Kurdish language to be used in daily life but not in public education. The Canadian province of Québec, similarly, has tried to eliminate English-language signs. French-speaking

FIGURE 4.17 Linguistic landscapes can be hard to read for those who are not familiar with the script used for writing. For many English-speaking monoglots, who are visually accustomed to the Latin alphabet, the linguistic landscape of countries such as Korea appears illegible. (Courtesy of Terry G. Jordan-Bychkov.)

THINKING GEOGRAPHICALLY How does the use of a script like this create unity within the area that uses it? How does it create barriers to diffusion between its region and the rest of the world?

immigrants settled Québec, and its official language is French, in contrast to Canada's policy of bilingualism in English and French. By 2003, in most regions of Ireland, formerly English-only place name signs have been modified to reflect the country's dual official languages of Irish (*Gaeilge*) and English. Irish place names are listed first, with the more recent English place names listed second. In specially designated Irish culture preservation regions of the country (known as *Gaeltachts*), only *Gaeilge* appears on place name signs, while English has no legal status at all. The suppression of minority languages and moves to reinstate them in the landscape offer an indication of the social and political status of minority populations more generally.

Other types of writing, such as gang-related graffiti, can denote ownership of territory or send messages to others that they are not welcome (**Figure 4.18**). Only those who understand the specific gang symbols used will be able to decipher the message. Misreading such writing can have dangerous consequences for those who stray into unfriendly territory. In this way, gang symbols can be understood as a dialect that is particular to a subculture and transmitted through symbols or a highly stylized script.

Toponyms

Language and culture also intersect in the names that people place on the land, whether they are given to settlements, terrain features, streams, or various other aspects of their surroundings. These place names, or **toponyms,** often directly reflect the spatial patterns of language, dialect, and ethnicity. Toponyms become part of the cultural landscape when they appear on signs and placards.

> **toponym** A place name, usually consisting of two parts, the generic and the specific.

FIGURE 4.18 Graffiti is used to mark gang territory. This wall in the Polanco neighborhood of Guadalajara, Mexico, is covered with graffiti. Gangs use stylized scripts that are often unintelligible to nonmembers to mark their territory. (Source: Courtesy of Patricia L. Price.)

THINKING GEOGRAPHICALLY How is gang use of stylized script in their graffiti similar to the effect of languages written in unique kinds of script?

Toponyms can be very revealing, because, as geographer Stephen Jett said, they often provide insights into "linguistic origins, diffusion, habitat, and environmental perception." Many place names consist of two parts—the generic and the specific. For example, in the American place names Huntsville, Harrisburg, Ohio River, Newfound Gap, and Cape Hatteras, the specific segments are *Hunts-*, *Harris-*, *Ohio*, *Newfound*, and *Hatteras*. The generic parts, which tell what *kind* of place is being described, are *-ville*, *-burg*, *River*, *Gap*, and *Cape*.

Types of Toponyms Geographers classify toponyms into several types when studying their origins and the cultures that conceived them. **Table 4.4** provides examples of some common types of toponyms found on the cultural landscape. One type is a **descriptive toponym,** which describes some physical feature or environmental characteristic of a place—for example, the *Rocky* Mountains. Sometimes, rather than being named for a physical feature, places are named to honor a famous person. An example of this type of **commemorative toponym** is the Hudson River, which was named for explorer Henry Hudson. Another type of place name is a **manufactured toponym,** or one that has simply been made up. An example of this form of place naming occurred in Truth or Consequences, Arizona, in the 1950s, when a popular radio program by the same name challenged towns to rename themselves after the program. The first that did so won the right to have the radio show

> **descriptive toponym** A place name that describes a physical feature or environmental characteristic of a place.
>
> **commemorative toponym** A place name that honors a famous or important person.
>
> **manufactured toponym** A place name that is made up, often by early founders of a settlement or an influential member of a community.

TABLE 4.4	Types of Toponyms	
Type of Toponym	**Origin**	**Examples**
Commemorative	Honors a famous or important person	• Monrovia, Liberia (U.S. President James Monroe) • Seattle, Washington (Suquamish Indian Chief Si'ahl [Seattle]) • Illinois (the Illini Indians)
Commendatory	Praises some physical or environmental characteristic	• Pleasant Valley, Greenland • Sun City, Arizona • Paradise, Texas
Descriptive	Describes a physical feature or environmental characteristic	• Rocky Mountains • Great Falls, Montana • Land's End, England
Incident related	Recalls an historic event	• Battle Creek, Michigan • Fourth of July Mountain, Washington • Massacre Lake, Nevada
Manufactured	Made-up or coined	• Tesnus, Texas ("sunset" spelled backward) • Reklaw, Texas (for Ann Walker, with "Walker" spelled backward) • Truth or Consequences, New Mexico (after a 1950s radio show as part of a contest)
Mistaken	Traceable to an historic error in identification or translation	• The West Indies (neither part of the Indies nor west of them) • Texas (mistakenly translated from the word *teychas*, a Caddoan Indian word for "friend") • Lasker, North Carolina (mistranslation of *Alaska,* for which the town is named)
Possessive	Indicates an historic claim to ownership or control of a place	• Castro Valley, Pennsylvania • Johnson City, Tennessee • Hall's Store, Texas
Shift	Relocated from another place, often settlers' homeland	• New Leipzig, North Dakota (Germany) • Lancaster, Pennsylvania (England) • New Mexico

broadcast live from their town. In this way, a town once called Hot Springs became known as Truth or Consequences and has remained so ever since.

Generic Toponyms of the United States

Generic toponyms are of greater potential value to the cultural geographer than specific names because they appear again and again throughout a culture region. There are literally thousands of generic place names, and every culture or subculture has its own distinctive set of them. They are particularly valuable both in tracing the spread of a culture and in reconstructing culture regions of the past. Sometimes generic toponyms provide information about changes people wrought long ago in their physical surroundings.

generic toponym The descriptive part of many place names, often repeated throughout a cultural area.

The three dialects of the eastern United States (see Figure 4.6)—Northern, Midland, and Southern—illustrate the value of generic toponyms in cultural geographical detective work. For example, New Englanders, speakers of the Northern dialect, often used the terms *Center* and *Corners* in the names of the towns or hamlets. Outlying settlements frequently bear the prefix *East, West, North,* or *South,* with the specific name of the township as the suffix. Thus, in Randolph Township, Orange County, Vermont, we find settlements named Randolph Center, South Randolph, East Randolph, and North Randolph. A few miles away lies Hewetts Corners.

These generic usages and duplications are peculiar to New England, and we can locate areas settled by New Englanders as they migrated westward by looking for such place names in other parts of the country. A trail of "Centers" and name duplications extending westward from New England through upstate New York and Ontario and into the upper Midwest clearly indicates their path of migration and settlement (**Figure 4.19**). Toponymic evidence of New England exists in areas as far afield as Walworth County, Wisconsin, where Troy, Troy Center, East Troy, and Abels Corners are clustered; in Dufferin County, Ontario,

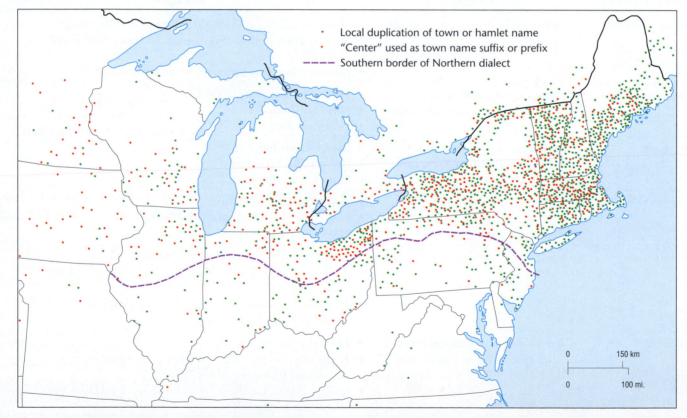

FIGURE 4.19 Generic place names reveal the migration of Yankee New Englanders and the spread of the Northern dialect. Two of the most typical place name characteristics in New England are the use of *Center* in the names of the principal settlements in a political subdivision and the tendency to duplicate the names of local towns and villages by adding the prefixes *East, West, North,* and *South* to the subdivision's name. As the concentration of such place names suggests, these two Yankee traits originated in Massachusetts, the first New England colony.

Note how these toponyms moved westward with New England settlers but thinned out rapidly to the south, in areas not colonized by New Englanders.

THINKING GEOGRAPHICALLY Based on this trait, where was initial settlement by Yankees most common? What other landscape features might you find in those places?

where one finds places such as Mono Centre; and even in distant Alberta, near Edmonton, where the toponym Michigan Centre doubly suggests a particular cultural diffusion. Similarly, we can identify Midland American areas by such terms as *Gap, Cove, Hollow, Knob* (a low, rounded hill), and *-burg*, as in Stone Gap, Cades Cove, Stillhouse Hollow, Bald Knob, and Fredericksburg. We can recognize southern speech by such names as *Bayou, Gully,* and *Store* (for rural hamlets), as in Cypress Bayou, Gum Gully, and Halls Store.

Toponyms and Cultures of the Past

Place names often survive long after the culture that produced them vanishes from an area, thereby preserving traces of the past. Australia abounds in Aborigine toponyms, even in areas from which the native peoples disappeared long ago (**Figure 4.20**). No toponyms are more permanently established than those identifying physical geographical features, such as rivers and mountains. Even the most absolute conquest, exterminating an aboriginal people, usually does not entirely destroy such names. Quite the contrary, in fact. Geographer R. D. K. Herman speaks of anticonquest, in which the defeated people finds its toponyms venerated and perpetuated by the conqueror, who at the same time denies the people any real power or cultural influence. The abundance of Native American toponyms in the United States provides an example. India, however, has

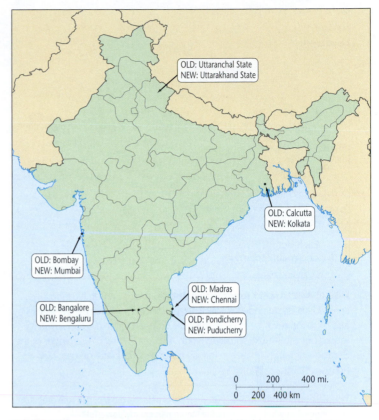

FIGURE 4.21 **India's postcolonial toponym shift.** More than 50 years after the English colonizers "quit" India, their colonial place names are being swept from the map, too. (Source: Adapted from Sappenfield, 2006.)

THINKING GEOGRAPHICALLY What other countries or cities changed their names after independence?

recently decided to revert to traditional toponyms; many Indian place names had been Anglicized under British colonial rule (**Figure 4.21**).

In Spain and Portugal, seven centuries of Moorish rule left behind a great many Arabic place names (**Figure 4.22** on page 116). An example is the prefix *guada-* on river names (as in Guadalquivir and Guadalupejo). The prefix is a corruption of the Arabic *wadi*, meaning "river" or "stream." Thus, Guadalquivir, corrupted from Wadi-al-Kabir, means "the great river." The frequent occurrence of Arabic names in any particular region or province of Spain reveals the remnants of Moorish cultural influence in that area. Many such names were brought to the Americas through Iberian conquest, so that Guadalajara, for example, appears on the map as an important Mexican city.

New Zealand, too, offers some intriguing examples of the subtle messages that can be conveyed by archaic toponyms. The native Polynesian people of New Zealand are the Maori. As cultural geographer Hong-key Yoon has observed, the survival rate of Maori names for towns varies according to the size of its population. The smaller the

FIGURE 4.20 **An Australian Aborigine specific toponym joined to an English generic name, near Omeo in Victoria state, Australia.** Such signs give a special, distinctive look to the linguistic landscape and speak of a now-vanished culture region. (Courtesy of Terry G. Jordan-Bychkov.)

THINKING GEOGRAPHICALLY What similar kinds of toponyms can you name in your region?

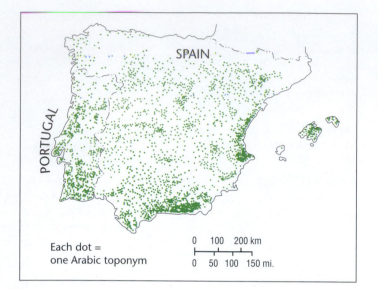

SPAIN

PORTUGAL

Each dot =
one Arabic toponym

0   100   200 km

0   50 100 150 mi.

FIGURE 4.22 **Arabic toponyms in Iberia.** Arabic, a Semitic language, spread into Spain and Portugal with the Moors more than 1000 years ago. A reconquest by Romance Indo-European speakers subsequently rooted out the Arabic-speaking North Africans in Iberia. A reminder of the Semitic language survives in Iberian toponyms, or place names. (Source: Houston, 1967.)

THINKING GEOGRAPHICALLY In what parts of Iberia did Moorish rule probably last longest? In what parts were the Moors particularly powerful?

town, the more likely it is to bear a Maori name. The four largest New Zealand cities all have European names, but of the 20 regional centers, with populations of 10,000 to 100,000, 40 percent have Maori names. Almost 60 percent of the towns with fewer than 10,000 inhabitants bear Maori toponyms. Similarly, whereas only 20 percent of New Zealand's provinces have Maori names, 56 percent of the counties do. Nearly all streams, hills, and mountains retain Maori names. The implication is that the British settlement of New Zealand was largely an urban phenomenon.

Key Terms

Anatolian hypothesis, 101	language, 93
bilingualism, 96	language family, 97
commemorative toponym, 113	language hotspot, 110
	lingua franca, 96
creole, 96	linguistic refuge area, 106
descriptive toponym, 113	manufactured toponym, 113
dialect, 95	pidgin, 96
generic toponym, 114	polyglot, 98
isogloss, 99	slang, 100
Kurgan hypothesis, 101	toponym, 112

Linguistic Geography on the Internet

You can learn more about linguistic geography on the Internet at the following web sites:

Dictionary of American Regional English
http://dare.wise.edu
Discover a reference web site that describes regional vocabulary contrasts of English in the United States and includes numerous maps.

Enduring Voices
http://www.nationalgeographic.com/mission/enduringvoices/
This flash map allows you to explore the world's language hotspots, those regions that are home to the most linguistic diversity, the highest levels of linguistic endangerment, and the least-studied tongues.

Ethnologue
http://www.ethnologue.com
This site provides information on how languages change over time as well as on endangered and nearly extinct languages.

Language Log
http://itre.cis.upenn.edu/~myl/languagelog/
Search the fascinating posts on this language-themed blog, run by University of Pennsylvania phonetician Mark Liberman and featuring guest linguists. Themes range from aversion to the word *moist* to the peculiar naming of coffee cup sizes at Starbucks to insulting words.

UNESCO Atlas of the World's Languages in Danger
http://www.unesco.org/culture/languages-atlas/
Explore UNESCO's flagship activity in safeguarding endangered languages by browsing this interactive atlas. Statistics regarding number of speakers and the degree of vitality are available for approximately 6000 languages spoken around the world—half of which are expected to disappear by the end of this century.

Recommended Books on the Geography of Language

Carver, Craig M. 1986. *American Regional Dialects: Word Geography.* Ann Arbor: University of Michigan Press. One of the best overall presentations of American English dialects from the standpoint of vocabulary.

Cassidy, Frederic C. (ed.). 1985–2002. *Dictionary of American Regional English.* 4 vols. Cambridge, MA.: Harvard University Press. A massive compilation of words used only regionally within the United States, with maps showing distributions.

Kurath, Hans. 1949. *Word Geography of the Eastern United States.* Ann Arbor: University of Michigan Press. The classic study that gave rise to the geographical study of American English dialects.

Moseley, Christopher, and R. E. Asher (eds.). 1994. *Atlas of the World's Languages.* London: Routledge. A wonderfully detailed color map portrait of the world's complex linguistic mosaic; thumb through it at your library and you will come to appreciate how complicated the patterns and spatial distributions of languages remain, even in the age of globalization.

Ostler, Nicholas. 2005. *Empires of the Word: A Language History of the World*. New York: Harper-Collins. This fascinating book explores the spread and evolution of languages through conquest, with many maps accompanying the text.

Journal in the Geography of Language

World Englishes. Published by the International Association for World Englishes, the journal documents the fragmentation of English into separate languages around the world. Edited by Margie Berns and Daniel R. Davis.

Answers to Thinking Geographically Questions

Figure 4.1: Most areas of "other families" are mountainous, cold, or desert lands.

Figure 4.2: A tree and its branches show divergence as people became separated over time. However, languages continue to influence each other despite being on separate branches, especially under conditions of modern mass communications.

Figure 4.3: More and more stores are erecting signage in English and Spanish. In some neighborhoods and regions, other languages are also used—for example, French near the border of Québec and Vietnamese in neighborhoods with residents who immigrated from there. International airports often use signage in multiple languages, supplemented by pictographs.

Figure 4.4: The Niger-Congo family extends from the west coast of Africa to the southern tip of the continent, with the exception of the area dominated by Germanic languages as a result of colonization. Khoisan is a preexisting language now confined to a part of Namibia and Botswana.

Figure 4.5: *Tú* is the form used in Mexico, the closest Spanish-speaking country to the United States. The dialect for Spanish education in the United States is Mexican, although the pronunciation taught is Castilian, from Spain.

Figure 4.6: Both are related to migration of Upland Southerners into nearby areas of the Midwest, seeking farmland and, later, jobs.

Figure 4.7: Both language and agriculture are culture traits. People learning farming would be instructed through speech, so they would learn the language as well as the methods. Languages of peoples without agriculture would not contain the words needed to explain the process, so those words would be learned from the people teaching the concept.

Figure 4.8: Navigated voyages would entail somewhat more determination and planning, which would tend to promote specific memories and traditions of the voyage as opposed to the more random contact by simply drifting in the currents.

Figure 4.9: Among the most common words are *cordillera, loma, mesa, montaña,* and *sierra*.

Figure 4.10: Many political boundaries do not follow the pattern of linguistic groups, which are also usually ethnic and sometimes religious groups. Note, for example, the exclave of Armenian in Azerbaijan.

Figure 4.11: The extension in the north shows the Portuguese speakers following the Amazon River into the interior. In the south, Portuguese speakers moved inland seeking mineral wealth and ranchland. In neither case did the Spanish regard those lands as worth contesting. Their interests lay in the mountains with their highly civilized (and thus suitable for labor) Inca and rich silver mines.

Figure 4.12: Because French is much more widely spoken in the world, residents of Bora Bora would be better able to conduct foreign trade, diplomacy, and business if they knew French.

Figure 4.13: English-speaking people had access to the Internet years before the Chinese did. They were in the forefront of development of computer technology and the Internet, so their language became the standard. English is also spoken over a much broader geographic area than Chinese, most of whose speakers live in China.

Figure 4.14: In almost all of Europe, languages spoken by a very small number of people have already been replaced by national, standard, and regional tongues spoken by a sizable population. In other words, the kinds of languages that are nearly extinct in the Americas, Africa, Asia, and the Pacific have already died out in Europe.

Figure 4.15: Some indigenous peoples already speak national or world languages and have already lost their original languages.

Figure 4.16: The pictograph indicating "no smoking" is international. It would convey its message to anyone; no one would have an excuse not to understand its meaning.

Figure 4.17: Because languages develop among communities that communicate, there is a feeling of a shared communication and meaning among those who use the script. However, communicating with those who do not know the script becomes even more difficult as transliteration as well as translation is necessary.

Figure 4.18: In both cases, the script reinforces the community that developed it and can read it, while creating barriers to diffusion against users of other scripts.

Figure 4.19: Western New York State and northeastern Ohio show the greatest concentration outside New England. New England styles of houses (upright-and-wing, New England large, and Cape Cod) would be common, along with towns laid out around a green.

Figure 4.20: Answers will depend on where you live, but two examples would be Appalachian Mountains and Susquehanna River, which are Native American names.

Figure 4.21: Many countries and cities in Africa changed names. For example, Southern Rhodesia became Zimbabwe and its capital of Salisbury was changed to Harare.

Figure 4.22: The east coast, the far south, and the western part (modern Portugal) show the greatest density of Moorish names and consequently were probably areas of longest-lasting rule and power.

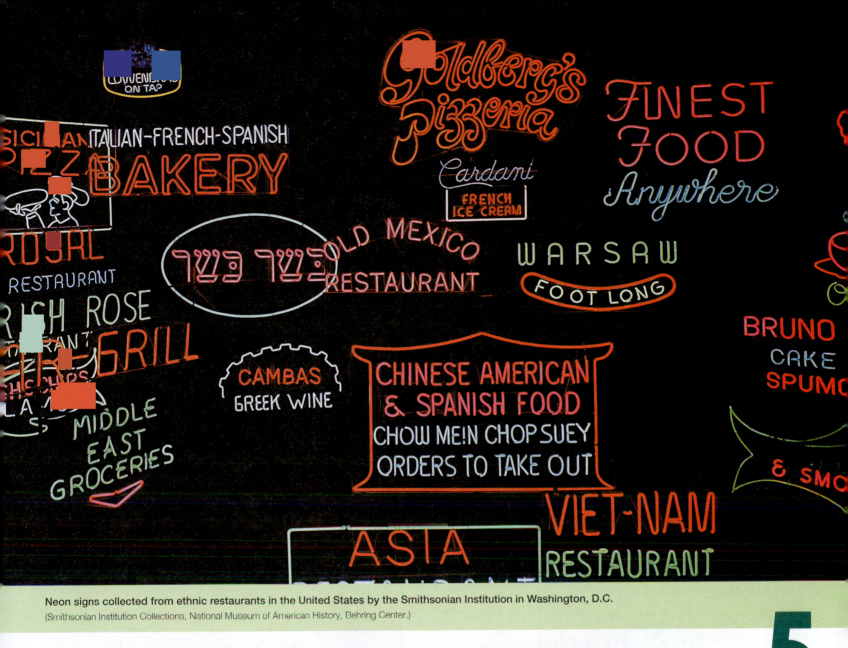

Neon signs collected from ethnic restaurants in the United States by the Smithsonian Institution in Washington, D.C.
(Smithsonian Institution Collections, National Museum of American History, Behring Center.)

Geographies of Race and Ethnicity

5

One of the enduring stories that the people of the United States proudly tell themselves is that "ours is an immigrant nation." This story is displayed during annual festivals celebrating the mosaic of ethnic traditions in countless cities, towns, and villages across the nation. For example, the midwestern town of Wilber, settled by Bohemian immigrants beginning about 1865, bills itself as "The Czech Capital of Nebraska" and annually invites visitors to attend a National Czech Festival. Celebrants are promised Czech foods, such as *koláce, jaternice,* poppy seed cake, and *jelita*; Czech folk dancing; "colored Czech postcards and souvenirs" imported from Europe; and handicraft items made by Nebraska Czechs (bearing an official seal and trademark to prove authenticity). Thousands of visitors attend the festival each year. Without leaving Nebraska, these tourists can move on to Norwegian Days at Newman Grove, the Greek Festival at Bridgeport, the Danish Grundlovs Fest in Dannebrog, German Heritage Days at McCook, the Swedish Festival at Stromsburg,

(a)

(b)

FIGURE 5.1 **Ethnic pride.** (a) The town of Stromsburg, Nebraska. Proud of its Swedish heritage, Stromsburg holds a "Swedish Festival" each year in June. (b) Hispanic Heritage Festival in Nebraska. Hispanic immigrants have expanded the range of ethnic pride festivals throughout the Midwest. (Steve Skjold/Alamy.)

THINKING GEOGRAPHICALLY Besides generating tourist income, what purposes do such festivals serve?

the St. Patrick's Day Celebration at O'Neill, several Native American powwows, and assorted other ethnic celebrations (**Figure 5.1**).

Today, Nebraska is still a magnet for immigrants, but since the 1990s, the state's new arrivals have been overwhelmingly non-European. In particular, Mexican immigrants employed in Nebraska's meat-processing industry find destinations such as Nebraska and other upper midwestern states attractive. In general, immigrants to the United States today are more likely to come from Asia or Latin America than from Europe, and they are changing the face of ethnicity in the United States (**Figures 5.2** and **5.3**). Indeed, ethnicity is a central aspect of the cultural geography of most places, forming one of the brightest motifs in the human mosaic.

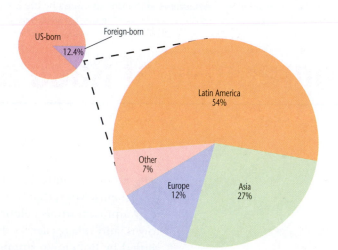

FIGURE 5.2 **The foreign-born population of the United States.** The smaller chart shows that 12.4 percent, or approximately 37.6 million, of the total U.S. population in 2010 (304.3 million) was born abroad. The larger chart shows that the majority of these people came from Latin America and Asia. (Source: U.S. Bureau of the Census, 2010.)

THINKING GEOGRAPHICALLY What characteristics of Asia and Latin America might have contributed to such large numbers of people migrating to the United States?

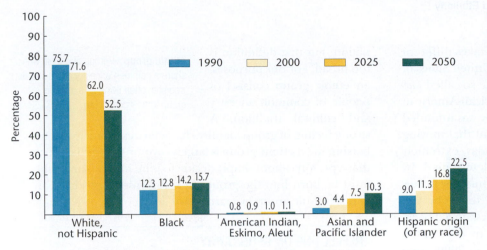

FIGURE 5.3 **Percent of the U.S. population by race and origin, 1990–2050.** This histogram illustrates the changing mosaic of race and ethnicity in the United States. The non-Hispanic white population is projected to steadily decline through the mid-part of the twenty-first century, while Asian groups and those citing Hispanic origin are expected to significantly increase as a percentage of the national population. (Source: U.S. Bureau of the Census, 2010.)

THINKING GEOGRAPHICALLY What reasons can you suggest for the significant projected increase in the percentage of Hispanic-origin individuals (as compared to other population groups) in the United States through the year 2050?

FIGURE 5.4 *Castas* **painting.** These paintings were common in colonial Mexico. They depict the myriad racial and ethnic combinations perceived to arise from intermixing among Europeans, indigenous peoples, and African slaves in the New World. They provided a way for those in power to keep track of the confusing racial hierarchy they had created. (Schalkwijk/Art Resource, NY.)

THINKING GEOGRAPHICALLY Why were such distinctions important when this painting was made? Why did Europeans find these matters so confusing at that time?

WHAT ARE RACE AND ETHNICITY?

The word *race* is often used interchangeably with **ethnicity**, but the two have very different meanings, and one must be careful in choosing between the two terms. **Race** can be understood as a genetically significant difference among human populations. A few biologists today support the view that human populations do form racially distinct groupings, arguing that race explains phenomena such as the susceptibility to certain diseases. In contrast, many social scientists (and many biologists) have noted the fluidity of definitions of *race* across time and space, suggesting that race is a social construct rather than a biological fact (see **Figures 5.4** and **5.5**).

ethnicity See *ethnic group*.

race A classification system that is sometimes understood as arising from genetically significant differences among human populations, or visible differences in human physiognomy, or as a social construction that varies across time and space.

FIGURE 5.5 **Mug shots of people from different races.** Phenotype variations—visible bodily differences such as facial features, skin color, and hair texture—are considered indicative of "race" by the U.S. Federal Bureau of Investigation (FBI). "Race" is one of several visible characteristics, which also include tattoos, scars, height, and weight, often used by law enforcement to identify suspects. These are photos from the FBI's Most Wanted list. (Courtesy of the Federal Bureau of Investigation.)

THINKING GEOGRAPHICALLY How is this idea of race as defined by visible differences different from race as it is commonly understood in the United States?

Because race is a social construction, it takes different forms in different places and times. In the United States of the early twentieth century, for example, the so-called *one-drop rule* meant that anyone with any African-American ancestry at all was considered black. This law was intended to prevent interracial marriage. It also meant that moving out of the category *black* was, and still is today, extremely difficult because one's racial status is determined by ancestry. Yet the notion that "black blood" somehow makes a person completely black is challenged today by the growing numbers of young people with diverse racial backgrounds. The rise in interracial marriages in the United States and the fact that—for the first time in 2000—one could declare multiple races on the Census form mean that more and more people identify themselves as "racially mixed" or "biracial" instead of feeling they must choose only one facet of their ancestry as their sole identity. Thus, golfer Tiger Woods calls himself a "Cablinasian" to describe his mixed Caucasian, black, American Indian, and Asian background. Race and ethnicity arise from multiple sources: your own definition of yourself; the way others see you; and the way society treats you, particularly legally. All three of these are subject to change over time. Still, President Barack Obama, whose mixed-race ancestry has led him to identify with both his black and his white relatives, struggled during the 2008 presidential campaign with the claims that he was at once "too black" and "not black enough" to be a viable presidential candidate.

Studies of genetic variation have demonstrated that there is far more variability within so-called racial groups than between them, which has led most scholars to believe that all human beings are, genetically speaking, members of only one race: *Homo sapiens sapiens*. In fact, many social scientists have dropped the term *race* altogether in favor of *ethnicity*. This is not to say, however, that **racism,** the belief

> **racism** The belief that certain individuals are inferior because they are born into a particular ethnic, racial, or cultural group. Racism often leads to prejudice and discrimination, and it reinforces relationships of unequal power between groups.

that certain individuals are inferior because they are born into a particular ethnic, racial, or cultural group, does not exist. Indeed, it *does* exist, and often leads to attitudes of prejudice or acts of discrimination. It also reinforces relationships of unequal power between groups and/or individuals from differing ethno-cultural backgrounds. In this chapter, we use the verb *racialize* to refer to the processes whereby these socially constructed differences are understood—usually, but not always, by the powerful majority—to be impervious to assimilation. Across the world, hatreds based in racism are at the root of the most incendiary conflicts imaginable.

What exactly is an *ethnic* group? The word *ethnic* is derived from the Greek word *ethnos*, meaning a people or nation, but that definition is too broad. For our purposes, an **ethnic group** consists of people of common ancestry and cultural tradition. A

> **ethnic group** A group of people who share a common ancestry and cultural tradition, often living as a minority group in a larger society.

strong feeling of group identity characterizes ethnicity. Membership in an ethnic group is largely involuntary, in the sense that a person cannot simply decide to join; instead, he or she must be born into the group. In some cases, outsiders can join an ethnic group by marriage or adoption.

> ## REFLECTING ON **GEOGRAPHY**
> Are ethnic groups always minorities, or do majority groups also have an ethnicity? What different sorts of traits might a majority group use to define its identity, as compared to a minority group? Is it possible for a racialized minority group to racialize other groups, or even itself?

Different ethnic groups may base their identities on different traits. For some, such as Jews, ethnicity primarily means religion; for the Amish, it is both folk culture and religion; for Swiss-Americans, it is national origin; for German-Americans, it is ancestral language; for African-Americans, it is a shared history stemming from slavery. Religion, language, folk culture, history, and place of origin can all help provide the basis of the sense of "we-ness" that underlies ethnicity.

As with race, ethnicity is a notion that is at once vexingly vague and hugely powerful. The boundaries of ethnicity are often fuzzy and shift over time. Moreover, ethnicity often serves to mark minority groups as different, yet majority groups also have an ethnicity, which may be based on a common heritage, language, religion, or culture. Indeed, an important aspect of being a member of the majority is the ability to decide how, when, and even if one's ethnicity forms an overt aspect of one's identity. Finally, some scholars question whether ethnicity even exists as intrinsic qualities of a group, suggesting instead that the perception of ethnic difference arises only through contact and interaction.

Apropos of this last point is the distinction between *immigrant* and *indigenous* (sometimes called *aboriginal*) groups. Many, if not most, ethnic groups around the world originated when they migrated from their native lands and settled in a new country. In their old home, they often belonged to the host culture and were not ethnic, but when they were transplanted by relocation diffusion to a foreign land, they simultaneously became a minority and ethnic. Han Chinese are not ethnic in China (**Figure 5.6**), but if they come to North America they are. Indigenous ethnic groups that continue to live in their ancient homes become ethnic when they are absorbed into larger political states. The Navajo, for example, reside on their

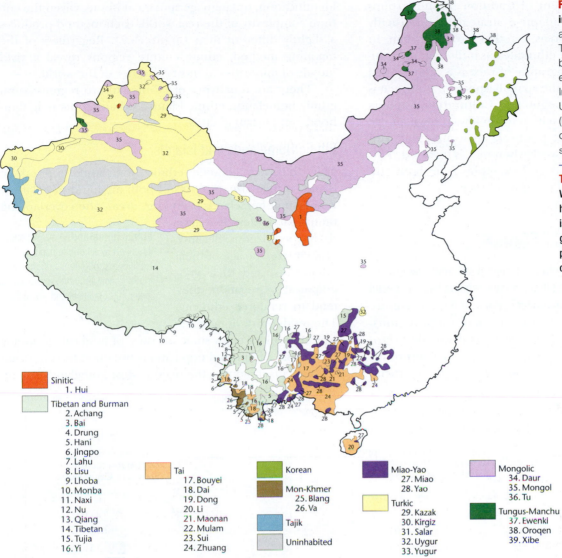

FIGURE 5.6 Ethnic minorities in China. Most ethnic groups are Turkic, Mongolic, Tai, Tibetan, or Burman in speech, but the rich diversity extends even to the Tajiks of the Indo-European language family. Unshaded areas are Han (Mandarin) Chinese, the host culture. (Source: Adapted and simplified from Carter et al., 1980.)

THINKING GEOGRAPHICALLY

Which of these ethnic regions are homelands and which are ethnic islands? Why are China's ethnic groups concentrated in sparsely populated peripheries of the country?

Sinitic
1. Hui

Tibetan and Burman
2. Achang
3. Bai
4. Drung
5. Hani
6. Jingpo
7. Lahu
8. Lisu
9. Lhoba
10. Monba
11. Naxi
12. Nu
13. Qiang
14. Tibetan
15. Tujia
16. Yi

Tai
17. Bouyei
18. Dai
19. Dong
20. Li
21. Maonan
22. Mulam
23. Sui
24. Zhuang

Korean

Mon-Khmer
25. Blang
26. Va

Tajik

Uninhabited

Miao-Yao
27. Miao
28. Yao

Turkic
29. Kazak
30. Kirgiz
31. Salar
32. Uygur
33. Yugur

Mongolic
34. Daur
35. Mongol
36. Tu

Tungus-Manchu
37. Ewenki
38. Oroqen
39. Xibe

traditional and ancient lands and became ethnic only when the United States annexed their territory. The same is true of Mexicans who, long resident in what is today the southwestern United States, found themselves labeled ethnic minorities when the border was moved after the 1848 U.S.–Mexican War. "We did not jump the border, the border jumped us!" is a common local comeback to this sudden shift in ethnic status.

This is not to say that ethnic minorities remain unchanged by their host culture. **Acculturation** often occurs, meaning that the ethnic group adopts enough of the ways of the host society to be able to function economically and socially. Stronger still is **assimilation,** which implies a complete blending with the host

culture and may involve the loss of many or all distinctive ethnic traits. Intermarriage is perhaps the most effective way of encouraging assimilation. Many students of American culture have long assumed that all ethnic groups would eventually be assimilated into the American melting pot, but relatively few ethnic groups have been; instead they use acculturation as their way of survival. The past three decades, in fact, have witnessed a resurgence of ethnic identity across the globe.

Ethnic geography is the study of the spatial aspects of ethnicity. Ethnic groups are the keepers of distinctive cultural traditions and the focal points of various kinds of social interaction. They are the basis or source not only of group identity but also of friendships, marriage partners, recreational outlets, business success, and political power. These interactions can offer cultural

acculturation The adoption by an ethnic group of enough of the ways of the host society to be able to function economically and socially.

assimilation The complete blending of an ethnic group into the host society, resulting in the loss of many or all distinctive ethnic traits.

ethnic geography The study of the spatial aspects of ethnicity.

security and reinforcement of tradition. Ethnic groups often practice unique adaptive strategies and usually occupy clearly defined areas, whether rural or urban. In other words, the study of ethnicity has built-in geographical dimensions. The geography of race is a related field of study that focuses on the spatial aspects of how race is socially constructed and negotiated. Cultural geographers who study race and ethnicity tend to always have an eye on the larger economic, political, environmental, and social power relations at work when race is involved. This chapter draws on insights and examples from both ethnic geography and the geography of race.

ETHNIC **REGIONS**

How are ethnic groups distributed geographically? Do ethnic culture regions have a special spatial character? Formal ethnic culture regions exist in most countries. To map these regions, geographers rely on data as diverse as probable origin of surnames in telephone directories and census totals for answers to questions on ancestry, primary racial identification, or language spoken at home. Given the cultural complexity of the real world, each method produces a slightly different map (**Figure 5.7**). Regardless of the mapping method, ethnic culture regions reveal a vivid mosaic of minorities in most countries of the world.

There are four types of ethnic culture regions: rural ethnic homelands, ethnic islands, urban ethnic neighborhoods, and urban ghettos.

Ethnic Homelands and Islands

The difference between ethnic homelands and ethnic islands is their size, in terms of both area and population. Rural **ethnic homelands** cover large areas, often overlapping municipal borders, and have sizable populations. Because of their size, the age of their inhabitants, and their geographical segregation, they tend to reinforce ethnicity. The residents of homelands typically seek or enjoy some measure of political autonomy or self-rule. Homeland populations usually exhibit a strong sense of attachment to the region. Most homelands belong

> **ethnic homeland** A sizable area inhabited by an ethnic minority that exhibits a strong sense of attachment to the region and often exercises some measure of political and social control over it.

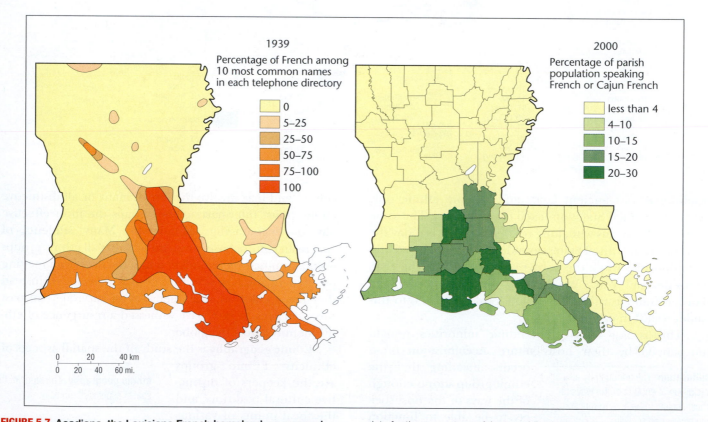

FIGURE 5.7 Acadiana, the Louisiana French homeland, as mapped by two different methods. The 1939 map was compiled by sampling the surnames in telephone directories. The 10 most common names in each directory were determined, and the percentage of these 10 that were of French origin was recorded. When no telephone directories were available, surnames on mailboxes were used. The 2000 map is based on census data for the percentage of those residing in the parish who spoke French or Cajun French. (Sources: After Meigs, 1941: 245; U.S. Bureau of the Census, 2000.)

THINKING GEOGRAPHICALLY What differences are apparent in the two maps? Do you think they are the result of the way of compiling the data or of the passage of time? Why?

to indigenous ethnic groups and include special, venerated places that serve to symbolize and celebrate the region—shrines to the special identity of the ethnic group. In its fully developed form, the homeland represents the most powerful of geographical entities, one combining the attributes of both formal and functional culture regions. By contrast, **ethnic islands** (sometimes called *folk islands*) are small

ethnic island A small ethnic area in the rural countryside; sometimes called a folk island.

dots in the countryside, usually occupying an area smaller than a county and serving as home to several hundred to several thousand people. Because of their small size and isolation, they do not exert as powerful an influence as homelands do.

North America includes a number of viable ethnic homelands (**Figure 5.8**), including Acadiana, the Louisiana French homeland now increasingly identified with the Cajun people and also recognized as a vernacular region; the Hispano or

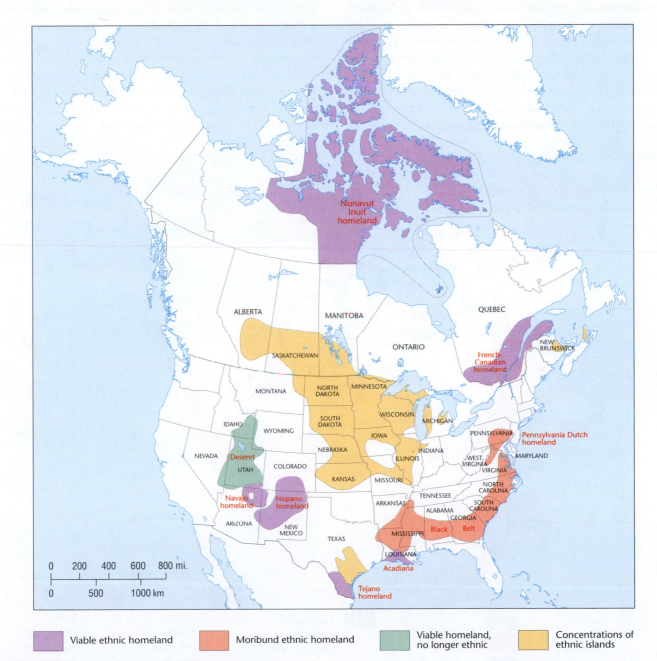

Viable ethnic homeland ■ **Moribund ethnic homeland** ■ **Viable homeland, no longer ethnic** ■ **Concentrations of ethnic islands**

FIGURE 5.8 Selected ethnic homelands in North America, past and present, and concentrations of rural ethnic islands. The Hispano homeland is also referred to as the Spanish-American homeland. With the return migration of African-Americans from northern industrial cities such as Chicago to rural southern areas, the moribund Black Belt homeland might soon be enjoying a second life. Note: Nunavut became a

functioning political unit within Canada in 1999. (Sources: Arreola, 2002; Carlson, 1990; Meinig, 1965; Nostrand and Estaville, 2001; U.S. Bureau of the Census, 2000.)

THINKING GEOGRAPHICALLY What landscape features are likely to remain even if an ethnic homeland becomes moribund?

Spanish-American homeland of highland New Mexico and Colorado; the Tejano homeland of south Texas; the Navajo reservation homeland in Arizona, Utah, and New Mexico; and the French-Canadian homeland centered on the valley of the lower St. Lawrence River in Québec. Some geographers would also include Deseret, a Mormon homeland in the Great Basin of the intermontane West. Some ethnic homelands have experienced decline and decay. These include the Pennsylvania Dutch (German) homeland, weakened to the point of extinction by assimilation, and the southern Black Belt, diminished by the collapse of the plantation-sharecrop system and the resulting African-American relocation to northern urban areas. Mormon absorption into the American cultural mainstream has eroded Mormon ethnic status, whereas nonethnic immigration has diluted the Hispano homeland. At present, the most vigorous ethnic homelands are those of the French-Canadians and south Texas Mexican-Americans.

If ethnic homelands succumb to assimilation and their people are absorbed into the host culture, at the very least a geographical residue, or **ethnic substrate,** remains. The resulting culture region, though no longer ethnic, nevertheless retains some distinctiveness, whether in local cuisine, dialect, or traditions. Thus, it differs from surrounding regions in a variety of ways. In seeking to explain its distinctiveness, geographers often discover an ancient, vanished ethnicity. For example, the Italian province of Tuscany owes both its name and some of its uniqueness to the Etruscan people, who ceased to be an ethnic group 2000 years ago, when they were absorbed into the Latin-speaking Roman Empire. More recently, the massive German presence in the American heartland (**Figure 5.9**), now largely nonethnic, helped shape the

> **ethnic substrate** Regional cultural distinctiveness that remains following the assimilation of an ethnic homeland.

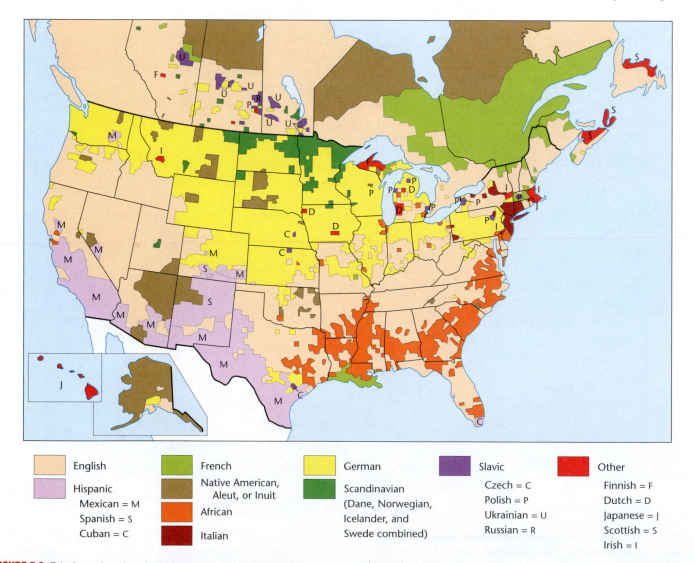

English

Hispanic
Mexican = M
Spanish = S
Cuban = C

French

Native American, Aleut, or Inuit

African

Italian

German

Scandinavian (Dane, Norwegian, Icelander, and Swede combined)

Slavic
Czech = C
Polish = P
Ukrainian = U
Russian = R

Other
Finnish = F
Dutch = D
Japanese = J
Scottish = S
Irish = I

FIGURE 5.9 Ethnic and national-origin groups in North America. Notice how the border between Canada and the United States generally also forms a cultural boundary. Several ethnic homelands appear on this map, as do many ethnic islands. Areas shown as Scandinavian are those where the total of all Scandinavian origins combined exceeds the origins of any other group. (Sources: Allen and Turner, 1988: 210; Census of Canada, 1991; Dawson, 1936: iv.; U.S. Bureau of the Census, 1990.)

THINKING GEOGRAPHICALLY Why are most Germans and Scandinavians found in the North?

cultural character of the Midwest, which can be said to have a German ethnic substrate.

Ethnic islands are much more numerous than homelands or substrates, peppering large areas of rural North America, as Figures 5.8 and 5.9 suggest. Ethnic islands develop because, in the words of geographer Alice Rechlin, "a minority group will tend to utilize space in such a way as to minimize the interaction distance between group members," facilitating contacts within the ethnic community and minimizing exposure to the outside world. People are drawn to rural places where others of the same ethnic background are found. Ethnic islands survive from one generation to the next because most land is inherited. Moreover, land is typically sold within the ethnic group, which helps to preserve the identity of the island. Social stigma is often attached to those who sell land to outsiders. Even so, the smaller size of ethnic islands makes their populations more susceptible to acculturation and assimilation.

Ethnic Neighborhoods and Racialized Ghettos

Formal ethnic culture regions also occur in cities throughout the world, as minority populations initially create, or are consigned to, separate ethnic residential quarters. Two types of urban ethnic culture regions exist. An **ethnic neighborhood** is a voluntary community where people of common ethnicity reside by choice. Such neighborhoods are, in the words of Peter Matwijiw, an Australian geographer, "the results of preferences shown

ethnic neighborhood A voluntary community where people of like origin reside by choice.

by different ethnic groups . . . toward maintaining group cohesiveness." An ethnic neighborhood has many benefits: common use of a language other than that of the majority culture, nearby kin, stores and services specially tailored to a certain group's tastes, the presence of employment that relies on an ethnically based division of labor, and institutions important to the group—such as churches and lodges—that remain viable only when a number of people live close enough to participate in their activities often. Miami's orthodox Jewish population clusters in Miami Beach–area neighborhoods in part because the proximity of synagogues and kosher food establishments makes religious observance far easier than it would be in a neighborhood that did not have a sizable orthodox Jewish population.

The second type of urban ethnic region is a **ghetto.** Historically, the term dates from thirteenth-century medieval Europe, when Jews lived in segregated, walled communities called ghettos (**Figure 5.10**). Ethnic residential quarters have, in fact, long

ghetto Traditionally, an area within a city where an ethnic group lives, either by choice or by force. Today in the United States, the term typically indicates an impoverished African-American urban neighborhood.

been a part of urban cultural geography. In ancient times, conquerors often forced the vanquished native people to live in ghettos. Religious minorities usually received similar treatment. Islamic cities, for example, had Christian districts. If one abides by the origin of the term, ghettos neither need to be composed of racialized minorities nor be impoverished. Typically, however, the term *ghetto* is used in the United States today to signal an impoverished urban African-American neighborhood. A related term, *barrio*, refers to an

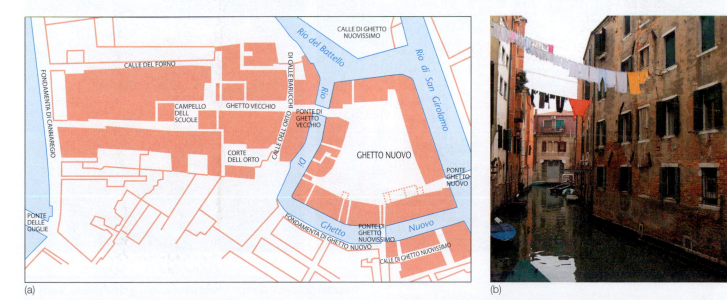

FIGURE 5.10 Venetian ghetto. In the sixteenth century, Venice's Jewish population lived in a segregated, walled neighborhood called a ghetto. On the left is a map of this early ghetto. Though most of Venice's Jews do not reside in the ghetto today, many attend religious services there, and the ghetto continues to be at the heart of Venetian Jewish life. (a: Adapted from the Jewish Museum of Venice; b: Lusoltaly/Alamy.)

THINKING GEOGRAPHICALLY What evidence is there that the ghetto expanded over time?

impoverished urban Hispanic neighborhood. Ghettos and barrios are as much functional culture regions as formal ones.

Coinciding with the urbanization and industrialization of North America, ethnic neighborhoods became typical in the northern United States and in Canada about 1840. Instead of dispersing throughout the residential areas of the city, immigrant groups clustered together. To some degree, ethnic groups that migrated to cities came from different parts of Europe from those who settled in rural areas. Whereas Germany and Scandinavia supplied most of the rural settlers, the cities attracted those from Ireland and eastern and southern Europe. Catholic Irish, Italians, and Poles, along with Jews from eastern Europe, became the main urban ethnic groups, although lesser numbers of virtually every nationality in Europe came to the cities of North America. These groups were later joined by French-Canadians, southern African-Americans, Puerto Ricans, Appalachian whites, Native Americans, Asians, and other non-European groups.

Regardless of their particular history, the neighborhoods created by ethnic migrants tend to be transitory. As a rule, urban ethnic groups remain in neighborhoods while undergoing acculturation. As a result, their central-city ethnic neighborhoods experience a life cycle in which one group is replaced by another, later-arriving one. We can see this process in action in the succession of groups that resided in certain neighborhoods and then moved on to more desirable areas. The list of groups that passed through one Chicago neighborhood from the nineteenth century to the present provides an almost complete history of American migratory patterns. First came the Germans and Irish, who were succeeded by the Greeks, Poles, French-Canadians, Czechs, and Russian Jews, who were soon replaced by the Italians. The Italians, in turn, were replaced by Chicanos and a small group of Puerto Ricans. As this succession occurred, the established groups had often attained enough economic and cultural capital to move to new areas of the city. In many cities, established ethnic groups moved to the suburbs. Even when ethnic groups relocated from inner-city neighborhoods to the suburbs, residential clustering survived (**Figure 5.11**). The San Gabriel Valley, about 20 miles (32 kilometers) from downtown Los Angeles, has developed as a major Chinese suburb. These

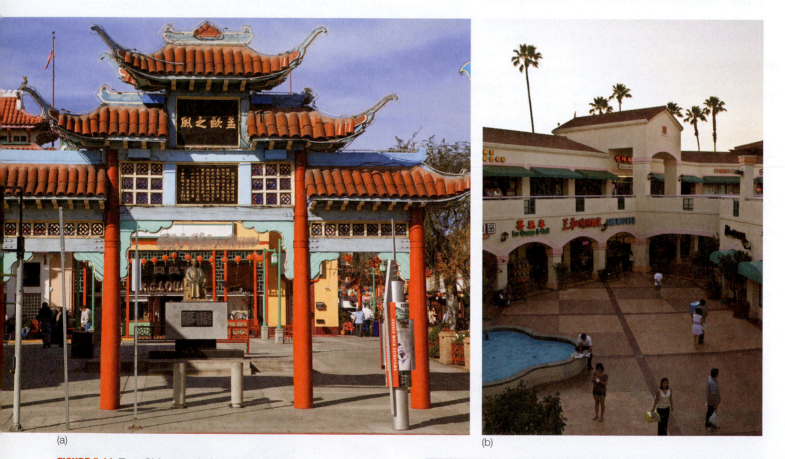

(a)

(b)

FIGURE 5.11 Two Chinese ethnic neighborhoods. The image on the left depicts Los Angeles's traditional urban Chinatown, which is a popular tourist destination as well. The image on the right, from the San Gabriel Valley outside Los Angeles, illustrates the suburban location and flavor of new ethnoburbs. (a: © Richard Cummins/Robert Harding World Imagery/Corbis; b: Ron Lim.)

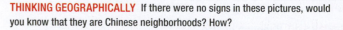

THINKING GEOGRAPHICALLY If there were no signs in these pictures, would you know that they are Chinese neighborhoods? How?

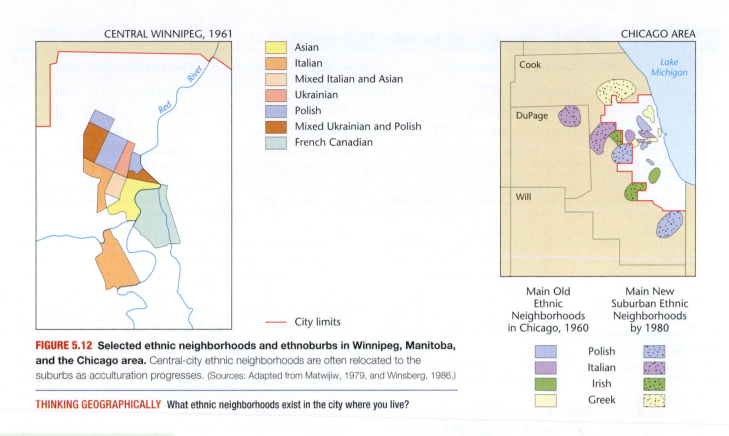

FIGURE 5.12 **Selected ethnic neighborhoods and ethnoburbs in Winnipeg, Manitoba, and the Chicago area.** Central-city ethnic neighborhoods are often relocated to the suburbs as acculturation progresses. (Sources: Adapted from Matwijiw, 1979, and Winsberg, 1986.)

THINKING GEOGRAPHICALLY What ethnic neighborhoods exist in the city where you live?

ethnoburb A suburban ethnic neighborhood, sometimes home to a relatively affluent immigrant population.

suburban ethnic neighborhoods can house relatively affluent immigrant populations and are called **ethnoburbs** (**Figure 5.12**).

example, 85% of Mexican immigrants lived in just 10 U.S. states in 2006. This unevenness is expected to continue through future decades, as new immigrants will often be drawn to areas where they have familial and social networks

Recent Shifts in Ethnic Mosaics

In the United States, immigration laws have changed during the past 45 years, shifting in 1965 from the quota system based on national origins to one that allowed a certain number of immigrants from the Eastern and Western hemispheres, as well as giving preference to certain categories of migrants, such as family members of those already residing in the United States. These changes, and the rising levels of undocumented immigration, have led to a growing ethnic variety in North American cities.

Latin America and Asia, rather than Europe, are now the principal sources of immigrants to North America. In fact, Latin Americans now constitute 53% of the U.S. foreign-born population, with Mexicans alone (including naturalized citizens and noncitizens) accounting for about 30% of the U.S. immigrant population. This number has increased from only 8% in 1970. Due to economic *push factors*, the number of Mexican (documented and undocumented) immigrants to the United States is expected to continue to increase through 2050, with Mexicans remaining the largest immigrant group within the country. As shown in **Figure 5.13**, the distribution of Latin American immigrants within the United States is highly uneven. For

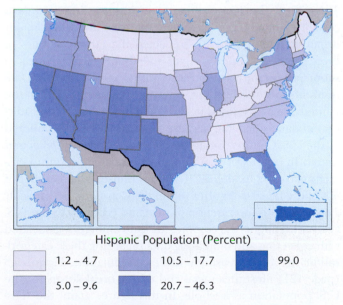

FIGURE 5.13 **Hispanic population by state.** Hispanic or Latino population as a percentage of the total population by state. *Hispanic* is an ethnic category that can encompass several racial designations. (Source: U.S. Bureau of the Census, 2010.)

THINKING GEOGRAPHICALLY How do you explain the pattern on this map?

TABLE 5.1	U.S. Cities of 100,000 or More with the Highest Ethnic Concentrations		
	Largest Concentration	**Second-Largest Concentration**	**Third-Largest Concentration**
African-American	Detroit, MI (82.7%)	Jackson, MS (79.4%)	Miami Gardens, FL (76.3%)
White	Hialeah, FL (92.6%)	Fargo, ND (90.2%)	Arvada, CO (89.8%)
Hispanic	East Los Angeles, CA (97.1%)	Laredo, TX (95.6%)	Hialeah, FL (94.7%)
Asian	Daly City, CA (55.6%)	Honolulu, HI (54.8%)	Fremont, CA (50.6%)

The figures for African-Americans, whites, and Asians are for that category alone, not in combination with other races. The figures for Hispanics can include any racial designation. In U.S. Census documents, the designation *white* overlaps, as do all other official racial categories, with the term *Hispanic* or *Latino*, which was introduced in the 1980 Census as a category of ethnicity, separate and independent of race. Hispanic and Latino Americans make up a racially diverse group, and as a whole compose the largest minority in the country. *(Source: U.S. Bureau of the Census, 2010.)*

awaiting them. Further, immigrants often settle in **gateway cities** and regions that are located close to their native country. For example, California and the American southwest (bordering Mexico) and Florida (nearest Cuba and other Caribbean countries) have been popular destinations for many newcomers to the United States in recent decades.

gateway city A city that acts as a port of entry, early settlement area, and distribution center for immigrants to a country because of its relative proximity to the homeland.

In some gateway cities, Hispanic or Latino immigrants now constitute large segments of new "minority majority" populations. For example, the population of Los Angeles is now more than 50 percent nonwhite (excluding Hispanic-white). Other cities like Las Vegas and San Diego also now have similarly nonwhite majorities. **Table 5.1** lists the cities in the United States with the highest ethnic concentrations. This table makes it clear that the attraction of immigrants to certain gateway cities (and their suburbs) and other major urban centers has greatly impacted the ethnic mosaics of these places. It is interesting to note that Hialeah, Florida (a suburb of Miami), is the most "white" city of over 100,000 residents, but it is also the third-most "Hispanic" city in the United States. As noted in the table, this *apparent* contradiction in data is possible because all racial categories denoted by the U.S. Census overlap the *ethnic* categories *Hispanic* and *Latino*. In other words, a person of Hispanic or Latino ethnicity can elect to identify herself or himself as a member of *any* of the racial categories within the Census or as ethnically Hispanic. This intersection of race and ethnicity in the U.S. Census highlights the complexities and ambiguity associated with collecting and tabulating data regarding these categorizations in an increasingly diverse population. Figure 5.3 (page 121) illustrates racial and ethnic breakdowns for the U.S. population as a whole. In this figure, again, the "Hispanic origin" population includes people of any racial classification. We can see from this figure that when only "non-Hispanic whites" are counted, the population of the United States has become less "white" and more "Hispanic" since 1990. The projections in Figure 5.3 show that these trends are set to continue at least through the year 2050.

The second-largest source region for contemporary immigrants to the United States is Asia. Asians constitute nearly 28% of all immigrants in the United States today; they accounted for just over 5% in 1960. China, the Philippines, and India are currently the largest source countries for Asian immigrants to the United States. As with immigrants of Hispanic origin, the distribution of Asian immigrants within the United States (as well as Canada) is uneven. In the United States, the West Coast is home to nearly 50 percent of the Asian population, mostly in California, with another concentration occurring in the urban corridor that stretches from New York City to Boston and further concentrations occurring in Illinois and Texas (**Figure 5.14**). In Hawaii, people of Japanese ancestry form the largest national-origin group, with 57 percent of the state's population claiming Asian (or multiracial Asian) ancestry. Because they are relatively close to Asia, many West Coast cities in both the United States and Canada have acted as gateway cities to Asian immigrants over

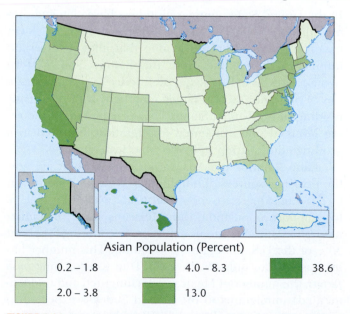

Asian Population (Percent)

0.2 – 1.8	4.0 – 8.3	38.6
2.0 – 3.8	13.0	

FIGURE 5.14 Asian population by state. People indicating "Asian" alone as a percentage of the total population by state. (Source: U.S. Bureau of the Census, 2010.)

THINKING GEOGRAPHICALLY How do you explain the pattern you see here?

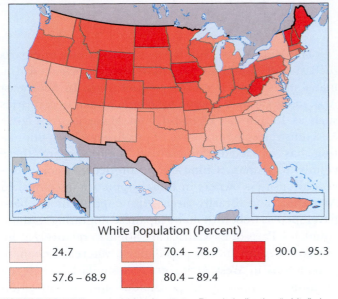

White Population (Percent)

24.7	70.4 – 78.9	90.0 – 95.3
57.6 – 68.9	80.4 – 89.4	

FIGURE 5.15 **White population by state.** People indicating "white" alone as a percentage of the total population by state. (Source: U.S. Bureau of the Census, 2010.)

THINKING GEOGRAPHICALLY How do you explain the patterns on this map?

the past several decades. For example, Vancouver has absorbed many Asian immigrants for decades, but shortly preceding the 1997 return of Hong Kong to the People's Republic of China (and continuing since that time), many Chinese immigrants dramatically increased these numbers. People of Chinese origin now constitute the third-largest ethnic-origin group in the city, accounting for more than 19 percent of the current population.

As ethnic populations continue to concentrate in gateway cities, a large swath of the United States has remained largely non-Hispanic white (**Figure 5.15**). **Table 5.2** shows the 10 most-identified ancestries among this segment of the U.S. population. However, these ancestries do not take into consideration the 54.4% of residents who chose to identify

TABLE 5.2 Top 10 Most-Identified Ancestries as a Percentage of the Total U.S. Population*

Ancestry Group	Percentage of Total U.S. Population
German	16.5
Irish	12.0
English	9.0
Italian	5.9
Polish	3.3
French (except Basque)	3.1
Scottish	1.9
Dutch	1.6
Norwegian	1.5
Swedish	1.4

*Does not include those individuals who identified themselves as Other or American or who chose not to classify or report their ancestry. (Source: U.S. Bureau of the Census, 2009.)

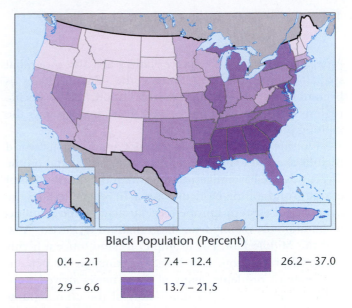

Black Population (Percent)

0.4 – 2.1	7.4 – 12.4	26.2 – 37.0
2.9 – 6.6	13.7 – 21.5	

FIGURE 5.16 **Black population by state.** People indicating "black or African-American" alone as a percentage of the total population by state. (Source: U.S. Bureau of the Census, 2010.)

THINKING GEOGRAPHICALLY How do you explain the patterns on this map?

themselves as American or Other or simply declined to report their ethnicity in the U.S. Census. This large percentage is another reflection of the increasingly complex nature of U.S. ethnicity. Fewer and fewer people are able (or wish) to identify themselves by a single ancestry. Where ancestry *is* identified, it is increasingly less European when compared to figures from earlier periods in U.S. history.

In examining other segments of the U.S. population, **Figures 5.16** and **5.17** show that African-Americans,

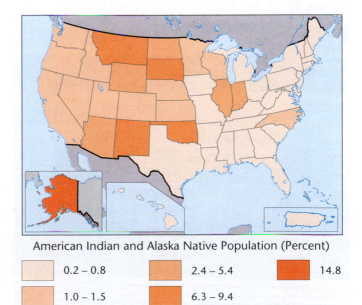

American Indian and Alaska Native Population (Percent)

0.2 – 0.8	2.4 – 5.4	14.8
1.0 – 1.5	6.3 – 9.4	

FIGURE 5.17 **American Indian and Alaska Native population by state.** People indicating "American Indian or Alaska Native" alone as a percentage of the total population by state. (Source: U.S. Bureau of the Census, 2010.)

THINKING GEOGRAPHICALLY How do you explain the patterns on this map?

American Indians, and Alaska Natives are also somewhat spatially concentrated. Although the United States is becoming increasingly diverse as a nation, *within* specific regions of the country the United States is not always the melting pot it is portrayed to be.

Outside the United States and Canada, many world regions are also experiencing increasing racial and ethnic diversity, as well as similar unevenness in the distribution of ethnic groups within particular areas. In recent decades the countries of western Europe have attracted large numbers of immigrants, mainly from Eastern Europe, Africa, Asia, and the Middle East. For example, Ireland, once a highly monocultural nation, now has approximately 11 percent foreign-born residents, many of them newly arrived from Poland, Nigeria, and Pakistan. Many of these immigrants are concentrated in Dublin and other major cities where employment opportunities have historically been most abundant for newcomers.

Similarly significant increases in ethnic diversity have occurred in the populations of France (nearly 6 percent foreign-born), Iceland (nearly 8 percent foreign-born), and Germany (nearly 9 percent foreign-born). These dramatic demographic shifts have sparked heated controversies between various factions within the governments and citizenries regarding immigration laws, access to employment, and the provision of state benefits to foreign-born residents. The concentration of large numbers of immigrants in European urban centers has raised tensions in cities such as Paris, leading to public demonstrations and outbreaks of violence on both sides of the ongoing immigration debate.

The Middle East, South America, Australia, and parts of East and Southeast Asia have also experienced dramatic shifts in their ethnic compositions. In Southeast Asia, many Chinese people have immigrated to countries such as Indonesia, Thailand, and Malaysia. As **Figure 5.18** illustrates, the presence of this foreign-born population has impacted the cultural landscapes of these countries, as many Chinese entrepreneurs become successful business owners in their adopted homelands. In Argentina and Brazil, large ethnic populations have developed as well, with many Japanese migrating to these countries. Parts of East Africa are home to relatively affluent South Asian Indian populations, whereas Lebanese populations in West Africa enjoy a similarly privileged position. Increasing globalization and urbanization throughout the world appear likely to ensure that the trend toward ethnic diversity will continue in many world regions in coming decades.

CULTURAL **DIFFUSION** AND ETHNICITY

How do the various types of cultural diffusion— relocation, hierarchical, and contagious—help us understand the complicated geographical patterns of ethnicity? Do ethnic homelands, islands, ghettos, and neighborhoods result from different types of diffusion? When groups move, how do cultural patterns reassemble in new places? As noted in the introduction to this chapter, it is often through migration that a group formerly in the majority becomes, in a new land, different from the mainstream and is thus labeled as ethnic or racialized. The many motives for migration can have different results in terms of where a group chooses to go, which parts of their original culture relocate and which do not, and who may become their neighbors in their new home.

Migration and Ethnicity

Much of the ethnic pattern in many parts of the world is the result of relocation diffusion. The migration process itself often creates ethnicity, as people leave countries where they belonged to a nonethnic majority and become a minority in a new home. *National Geographic*'s Genographic Project uses DNA samples from volunteers worldwide to substantiate the

FIGURE 5.18 Store owned by a prosperous ethnic Chinese retailer on the island of Bora Bora in French Polynesia. The population of the island is overwhelmingly Polynesian. Only 0.8 percent of the people are Europeans, 6.6 are "Demis" (a mixture of white and Polynesian), and less than 0.5 percent are Asian. Yet this store and many others in the archipelagoes of the Pacific are owned by persons of Chinese heritage. (Courtesy of Terry G. Jordan-Bychkov.)

THINKING GEOGRAPHICALLY Why don't Polynesians own such stores?

claim that all humans descended from a group of Africans who began to migrate out of Africa about 60,000 years ago. The contemporary global mosaic of ethnic populations can be traced to their specific journeys. Today's voluntary migrations have also produced much of the ethnic diversity in the United States and Canada, and the involuntary migration of political and economic refugees has always been an important factor in ethnicity worldwide and is becoming ever more so in North America.

Chain migration may be involved in relocation diffusion. In chain migration, an individual or small group decides to migrate to a foreign country. This decision typically arises from negative conditions in the home area, such as political persecution or lack of employment, and the perception of better conditions in the receiving country. Often ties between the sending and receiving areas are preexisting, such as those formed when military bases of receiving countries are established in sending countries. The first emigrants, or "innovators," may be natural leaders who influence others, particularly friends and relatives, to accompany them in the migration. The word spreads to nearby communities, and soon a sizable migration is under way from a fairly small district in the source country to a comparably small area or neighborhood in the destination country (**Figure 5.19**). In village after village, the first emigrants often rank high in the local social order, so hierarchical diffusion also occurs. That is, the *decision* to migrate spreads by a mixture of hierarchical and contagious diffusion, whereas the actual migration itself represents relocation diffusion.

chain migration The tendency of people to migrate along channels, over a period of time, from specific source areas to specific destinations.

Involuntary migration (sometimes called *forced migration*) also contributes to ethnic diffusion and the formation of ethnic culture regions. African slavery constituted the most demographically significant involuntary migration in human history and has strongly shaped the ethnic mosaic of the Americas. Refugees from Cambodia and Vietnam formed ethnic groups in North America, as did Guatemalans and Salvadorans fleeing political repression in Central America. Often, such forced migrations result from policies of **ethnic cleansing,** in which countries expel or massacre minorities outright to produce cultural homogeneity in their populations. For example, upon achieving independence, Croatia systematically expelled its Serb minority. In 1994 in Rwanda, the government and military, controlled by the majority ethnic group, the Hutu, systematically killed up to 1 million members of a rival (minority) ethnic group called the Tutsi. Following this approximately 100-day genocide, the Tutsi minority regained control of the country. Fearing retaliation for the genocide, more than 2 million Hutu refugees then fled to neighboring countries like the Democratic Republic of the Congo, Burundi, Tanzania, and Uganda (**Figure 5.20** on page 134). Thousands of these individuals subsequently died of diseases common to refugee camps, like cholera and dysentery.

involuntary migration The forced displacement of a population, whether by government policy (such as a resettlement program), warfare or other violence, ethnic cleansing, disease, natural disaster, or enslavement. Also called forced migration.

ethnic cleansing The removal of unwanted ethnic minority populations from a nation-state through mass killing, deportation, or imprisonment.

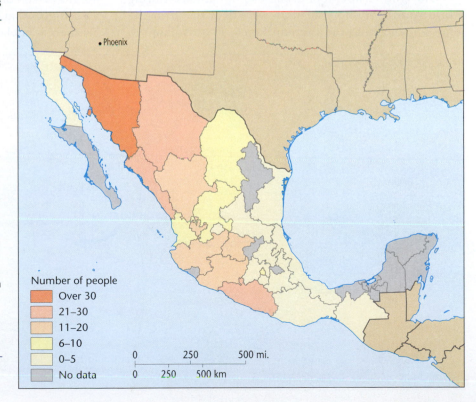

FIGURE 5.19 Origins of migrants. This map depicts the Mexican state of origin for the residents of Garfield, an ethnic neighborhood in Phoenix, Arizona, that is composed mostly of Mexican and Mexican-American residents. Note that most of the residents trace their homeland to only a few northern Mexican states; immigrants from Mexico's more southern states, such as Oaxaca and Veracruz, are relatively rare. (Source: Adapted from Oberle and Arreola, 2008.)

Number of people
- Over 30
- 21–30
- 11–20
- 6–10
- 0–5
- No data

THINKING GEOGRAPHICALLY How does this pattern demonstrate distance decay?

FIGURE 5.20 A refugee camp near the provincial capital of Goma, Democratic Republic of the Congo (DRC). Squalid conditions in refugee camps often lead to mass deaths due to diseases such as cholera, which results from contaminated water supplies and inadequate waste disposal. These deaths compound the horror of genocides, or ethnic cleansings, that necessitate the camps' existence. (ALAIN WANDIMOYI/EPA/Landov.)

THINKING GEOGRAPHICALLY In addition to the health hazards mentioned, what difficulties and dangers might refugees from ethnic cleansing face in their new host countries?

Another modern example of ethnic cleansing occurred in the Darfur region of Sudan beginning in the late 1990s. For more than eight years, Arab nomadic groups within the country and the Arab Sudanese government conflicted with several non-Arab ethnic groups, with the Arab groups conducting village raids and engaging in armed disputes with the non-Arabs. More than 1.5 million people died as a direct result of the conflict, while starvation and disease caused the death and displacement of millions more. The southern region of the country ultimately seceded to form the new nation of South Sudan, but the effects of this conflict leave unresolved tensions between the infant country and its neighbor, Sudan.

Outright warfare, too, displaces populations. During World War II, Polish populations were deported from the lands annexed into the German Reich, in an attempt to "Germanize" this region. In all, some 2 million Poles were expelled from their homes. In Afghanistan, over a 30-year period of warfare, 10 million people have fled the country. **Figure 5.21** shows that, though many Afghan refugees have returned during that time, over 2.8 million remain in Iran and Pakistan, with nearly 100,000 still in other countries. In addition, nearly half a million Afghans are internally displaced persons (IDPs) in their own country. Attempts to find long-term solutions for Afghan refugees are complicated by the continued flow of Afghans into Pakistan and Pakistan's increasing unwillingness to continue hosting them. Additionally, voluntary returns to Afghanistan have decreased dramatically since 2007 due to increased conflict within the country and a realization by refugees that there are few opportunities for livelihood in their homeland. Overall, the decades of Afghan displacement now constitute one of the world's most protracted refugee crises.

FIGURE 5.21 Afghans displaced by conflict. This map shows the war-related displacement of more than 3 million Afghans who are involuntary migrants. These individuals may face discrimination and poverty as unwanted minority populations in Iran, Pakistan, and other countries in which they have taken refuge. (Source: http://afghanistan101.blogspot.com.)

THINKING GEOGRAPHICALLY What difficulties do you think governments and humanitarian aid organizations face in making accurate estimates of the number of people displaced in a conflict such as that in Afghanistan?

return migration A type of ethnic diffusion that involves the voluntary movement of a group of migrants back to its ancestral or native country or homeland.

channelization A migration process in which a specific source location becomes linked to a particular destination, so that neighbors in the old place become neighbors in the new place.

Return migration represents another type of ethnic diffusion and involves the voluntary movement of a group back to its ancestral homeland or native country. The large-scale return since 1975 of African-Americans from the cities of the northern and western United States to the Black Belt ethnic homeland in the South is one of the most notable such movements now under way. This type of ethnic migration is also channelized. **Channelization** is a process in which a specific source region becomes linked to a particular destination, so that neighbors in the old place become neighbors in the new place as well. Geographers James Johnson and Curtis Roseman found that 7 percent of African-Americans in Los Angeles County, California, moved away between 1985 and 1990, including many who went to the American South. Indeed, the 1990s witnessed the largest return migration of African Americans to the American South ever, from all parts of the United States. This acts to revitalize the southeastern African American homeland depicted as "moribund" in Figure 5.8.

Similarly, many of the 200,000 or so expatriate Estonians, Latvians, and Lithuanians left Russia and other former Soviet republics to return to their newly independent Baltic home countries in the 1990s, losing their ethnic status in the process. Clearly, migration of all kinds turns the ethnic mosaic into an ever-changing kaleidoscope.

REFLECTING ON GEOGRAPHY

Why might African-Americans have begun return migration to the South after 1975, and why did this movement accelerate in the 1990s?

Simplification and Isolation

When groups migrate and become ethnic in a new land, they have, in theory at least, the potential to introduce the totality of their culture by relocation diffusion. Conceivably, they could reestablish every facet of their traditional way of life in the area where they settle. However, ethnic immigrants never successfully introduce the totality of their culture. Rather, profound **cultural simplification** occurs. As geographer Cole Harris noted, "Europeans established overseas drastically simplified versions of European society." This happens, in part, because of chain migration: only

cultural simplification The process by which immigrant ethnic groups lose certain aspects of their traditional culture in the process of settling elsewhere, creating a new culture that is less complex than the old.

fragments of a culture diffuse overseas, borne by groups from particular places migrating in particular eras. In other words, some simplification occurs at the point of departure. Moreover, far more cultural traits are implanted in the new home than actually survive. Only selected traits are successfully introduced, and others undergo considerable modification before becoming established in the new homeland. In other words, absorbent barriers prevent the diffusion of many traits, and permeable barriers cause changes in many other traits, greatly simplifying the migrant cultures. In addition, choices that did not exist in the old home become available to immigrant ethnic groups. They can borrow novel ways from those they encounter in the new land, invent new techniques better suited to the adopted place, or modify existing approaches as they see fit. Most immigrant ethnic groups resort to all these devices to varying degrees.

The displacement of a group and its relocation to a new homeland can have widely differing results. The degree of isolation an ethnic group experiences in the new home helps determine whether traditional traits will be retained, modified, or abandoned. If the new settlement area is remote and contacts with outsiders are few, diffusion of traits from the sending area is more likely. Because contacts with groups in the receiving area are rare, little borrowing of traits can occur. Isolated ethnic groups often preserve in archaic form cultural elements that disappear from their ancestral country. That is, they may, in some respects, change less than their kinfolk back in the mother country.

Language and dialects offer some good examples of this preservation of the archaic. Germans living in ethnic islands in the Balkan region of southeastern Europe preserve archaic South German dialects better than do Germans living in Germany itself, and some medieval elements survive in the Spanish spoken in the Hispano homeland of New Mexico. The highland location of Taiwanese aboriginal peoples helped maintain their archaic Formosan language, belonging to the Austronesian family, despite attempts by Chinese conquerors to acculturate them to the Han Chinese culture and language.

ETHNIC ECOLOGY

How do ethnic groups interact with their habitats? Is there a special bond between ethnic groups and the land they inhabit that helps to form their self-identity? Do ethnic groups find shelter in certain habitats? Are the areas inhabited by racialized groups targets of environmental racism? Ethnicity is very closely linked to cultural ecology. The possibilistic interplay between people and physical environment is often evident in the pattern of ethnic culture regions, in ethnic migration, and in ethnic persistence or survival.

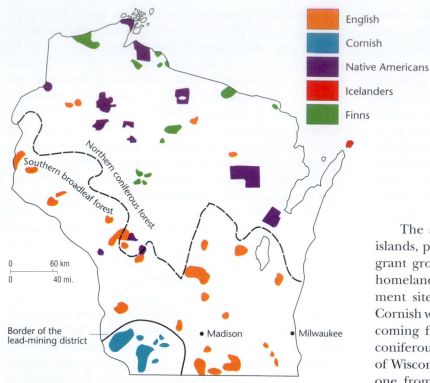

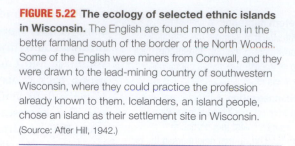

FIGURE 5.22 **The ecology of selected ethnic islands in Wisconsin.** The English are found more often in the better farmland south of the border of the North Woods. Some of the English were miners from Cornwall, and they were drawn to the lead-mining country of southwestern Wisconsin, where they could practice the profession already known to them. Icelanders, an island people, chose an island as their settlement site in Wisconsin. (Source: After Hill, 1942.)

THINKING GEOGRAPHICALLY Why are the Finns concentrated in the northern coniferous forest region? The Native Americans?

Ethnicity and race are sometimes also linked to the environment in harmful ways, particularly when the places they inhabit become the targets of pollution.

Cultural Preadaptation

For those ethnic groups created by migration or relocation diffusion, the concept of cultural preadaptation provides an interesting approach. **Cultural preadaptation** involves a set of adaptive traits possessed by a group in advance of migration that gives them the ability to survive and a competitive advantage in occupying the new environment. Most often, preadaptation occurs in groups migrating to a place environmentally similar to the one they left behind. The adaptive strategy they had pursued before migration works reasonably well in the new home.

The preadaptation may be accidental, but in some cases the immigrant ethnic group deliberately chooses a destination area that physically resembles their former home. In Africa, the Bantu expansion mentioned in Chapter 4 was probably initially driven by climate change and expansion of the Sahara (see Figure 4.4, page 99). The Bantu spread south and west in search of forested lands similar to those they had previously inhabited. Their progress southward was finally inhibited because their agricultural techniques and cattle were not adapted to the drier Mediterranean climate of southern Africa.

cultural preadaptation A set of adaptive traits and skills possessed in advance of migration by a group, giving it survival ability and competitive advantage in occupying the new environment.

The state of Wisconsin, dotted with scores of ethnic islands, provides some fine examples of preadapted immigrant groups that sought environments resembling their homelands. Particularly revealing are the choices of settlement sites made by the Finns, Icelanders, English, and Cornish who came to Wisconsin (**Figure 5.22**). The Finns—coming from a cold, thin-soiled, glaciated, lake-studded, coniferous forest zone in Europe—settled the North Woods of Wisconsin, a land similar in almost every respect to the one from which they had migrated. Icelanders, from a bleak, remote island in the North Atlantic, located their only Wisconsin colony on Washington Island, an isolated outpost surrounded by the waters of Lake Michigan. The English, accustomed to good farmland, generally founded ethnic islands in the better agricultural districts of southern and southwestern Wisconsin. Cornish miners from the Celtic highlands of Cornwall in southwestern England sought out the lead-mining communities of southwestern Wisconsin, where they continued their traditional occupation.

Such ethnic niche-filling has continued to the present day. Cubans have clustered in southernmost Florida, the only part of the United States mainland to have a tropical savanna climate identical to that in Cuba, and many Vietnamese have settled as fishers on the Gulf of Mexico, especially in Texas and Louisiana, where they could continue their traditional livelihood. Yet historical and political patterns, as well as the factors driving chain migration discussed earlier, are also at work in these contemporary patterns of ethnic clustering. They act to temper the influence of the physical environment on ethnic residential selection, making it only one of many considerations.

This deliberate site selection by ethnic immigrants represents a rather accurate environmental perception of the new land. As a rule, however, immigrants tend to perceive the ecosystem of their new home as more like that of their abandoned native land than is actually the case. Their perceptions of the new country emphasize the similarities and minimize the differences. Perhaps the search for similarity results from homesickness or an unwillingness to admit that migration has brought them to a largely alien land. Perhaps growing to adulthood in a particular kind of physical environment inhibits one's ability to perceive a different ecosystem accurately. Whatever the reason, the distorted perception occasionally caused problems for ethnic farming groups. A period of trial and error was often necessary to come to terms with the New World environment. Sometimes crops that thrived in the old homeland proved poorly suited to the new setting. In such cases, **cultural maladaptation** is said to occur.

> **cultural maladaptation** Poor or inadequate adaptation that occurs when a group pursues an adaptive strategy that, in the short run, fails to provide the necessities of life or, in the long run, destroys the environment that nourishes it.

Habitat and the Preservation of Difference

Certain habitats may act to shelter and protect ethnically or racially distinct populations. The high altitudes and rugged terrain of many mountainous regions make these areas relatively inaccessible and thus can provide refuge for minority populations while providing a barrier to outside influences. As we saw in Chapter 4 on language, mountain-dwellers sometimes speak archaic dialects or preserve their unique tongues thanks to the refuge provided by their habitat. In more general terms, the ways of life—including language—that are associated with ethnic distinctiveness are often preserved by a mountainous location.

The ethnic patchwork in the Caucasus region, for example, persists thanks in no small measure to the mountainous terrain found there. As **Figure 5.23** shows, distinct groups occupy the rugged landscape of valleys, plateaus, peaks, foothills, steppes, and plains in this region. Religions and languages overlap in complex ways with ethnic identities and are made even more confounding by the geopolitical boundaries in this region, which do not necessarily follow the contours of the Caucasian ethnic mosaic. Thus, although this is one of the most ethnically diverse places on Earth, the Caucasus region has seen more than its fair share of conflict as well, much of it ethnically motivated.

Islands, too, can provide a measure of isolation and protection for ethnically or racially distinct groups. The Gullah, or Geechee, people are descendants of African slaves brought to the United States in the eighteenth and nineteenth centuries to work on plantations. Many of their nearly 10,000 descendants today inhabit coastal islands off South Carolina, Georgia, and north Florida. Their island location has allowed many of their original African cultural

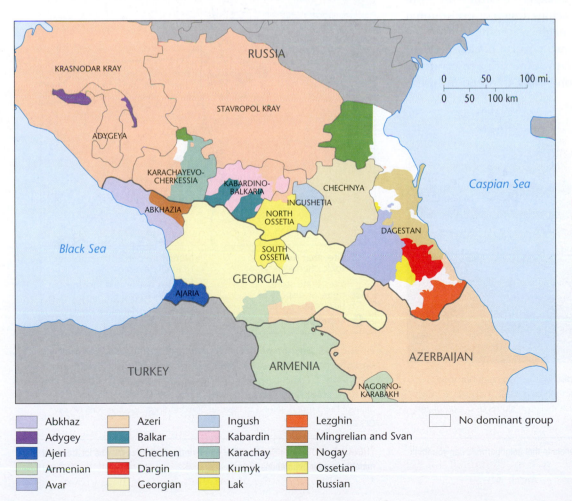

FIGURE 5.23 Ethnic pluralities in the Caucasus. The Caucasus Mountains, located in Southwest Asia, are home to one of the world's most ethnically diverse populations. Because ethnic territories often overlap, this map depicts pluralities. The populations shown comprise at least 40 percent of that place's ethnic population, and in most cases also constitute a majority population, although some of the more heterogeneous areas have no dominant ethnic population. The racialized term *Caucasian* is derived from this area's name, although, in fact, it has little to do with the people actually living in this region. (Source: Adapted from O'Loughlin et al., 2007.)

THINKING GEOGRAPHICALLY Why is this region one of almost perpetual conflict?

roots to be preserved, so much so that the Gullah are sometimes called the most African-American community in the United States. Today, as the tourist industry sets its sights on developing these coastal islands, and as Gullah youth migrate elsewhere in search of opportunity, the survival of this unique culture is in question.

Environmental Racism

Not only people but also the places where they dwell can be labeled by society as minority. It is a fact that, although belonging to a racial or ethnic minority group does not necessarily lead to poverty, the two often go hand-in-hand. In spatial terms, being poor frequently means having the last and worst choice of where to live. In cities, that means impoverished and racialized minorities often reside in run-down, ecologically precarious, or peripheral places that no one else wants to inhabit (**Figure 5.24**). In rural areas, racialized and indigenous minorities work the smallest and least-fertile lands, or—more commonly—they work the plots of others, having been dispossessed of their own lands.

Environmental racism refers to the likelihood that a racialized minority population inhabits a polluted area. As Robert Bullard asserts, "Whether by conscious design or institutional neglect, communities of color in urban ghettos, in rural 'poverty pockets,' or on economically impoverished Native American reservations face some of the worst environmental devastation in the

> **environmental racism** The targeting of areas where ethnic or racial minorities live with respect to environmental contamination or failure to enforce environmental regulations.

nation." In the United States, Hispanics, Native Americans, and blacks are more likely than whites to reside in places where toxic wastes are dumped, polluting industries are located, or environmental legislation is not enforced. Inner-city populations, which in many U.S. cities are disproportionately made up of minorities, occupy older buildings contaminated with asbestos and toxic lead-based paints. Farm workers, who in the United States are disproportionately Hispanic, are exposed to high levels of

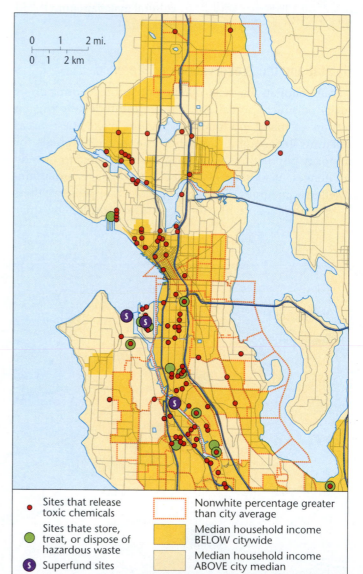

Sites that release toxic chemicals

Sites that store, treat, or dispose of hazardous waste

S Superfund sites

Nonwhite percentage greater than city average

Median household income BELOW citywide

Median household income ABOVE city median

FIGURE 5.25 Environmental racism in the Seattle area. This map illustrates how toxic sites are distributed unevenly amongst Seattle communities. Within the city, nonwhites and low-income persons live in the most polluted environments, while white residents and the more affluent have much safer and cleaner surroundings. (Source: King County GIS Center.)

FIGURE 5.24 Arab slum in Delhi, India. This ethnic neighborhood is home to low-income Arabs, who are an ethnic minority population in India. Nearly 50 percent of Delhi's residents live in slums or unauthorized settlements. (Viviane Dalles/REA/Redux.)

THINKING GEOGRAPHICALLY What parallels to this neighborhood can you think of in your community or country?

THINKING GEOGRAPHICALLY What other kinds of facilities may be located near minority or poor populations?

pesticides, fertilizers, and other agricultural chemicals. Native American reservation lands are targeted as possible sites for disposing of nuclear waste.

As **Figure 5.25** depicts, in the Seattle area there is an overlap of nonwhite (and low-income) populations with hazardous waste storage, treatment, and disposal facilities, as well as facilities that release toxic chemicals into the surrounding environment. This association is difficult to explain away as simply a random coincidence. Rather, it appears that contaminating industries and activities purposely located in neighborhoods composed of largely nonwhite residents and residents with incomes below the median level.

Environmental racism extends to the global scale. Industrial countries often dispose of their garbage, industrial waste, and other contaminants in poor countries. Industries headquartered in wealthy nations often outsource the dangerous and polluting phases of production to countries that do not have—or do not enforce—strict environmental legislation. Whether areas inhabited by poor, ethnic, or racialized minorities are deliberately targeted for unhealthy conditions, or whether they are simply the unfortunate recipients of the effects of larger detrimental processes, the connections among race, ethnicity, and the environment pose central questions for cultural geographers.

ETHNIC CULTURAL **INTERACTION**

How is ethnic identity linked to other aspects of culture? If ethnicity possesses a vital link to ecology, it is also firmly integrated into the fabric of culture. The very concept of ethnicity can find a basis in the theme of cultural interaction. For example, ethnicity plays a role in what members of an ethnic group eat, what religious faith they practice, how they vote, whom they marry, how they earn a living, and in what ways they spend their leisure time. In the process, an identity emerges. The complicated pattern of ethnic homelands, ghettos, and neighborhoods influences the spatial distribution of diverse cultural phenomena (**Figure 5.26**).

Over 250 Finns in county population, 1920

Incidence of male stomach cancer significantly higher than national average, by county

FIGURE 5.26 **Finnish ethnic areas and the incidence of stomach cancer in the northern United States.** (Sources: Allen and Turner, 1987; Shannon and Pyle, 1992.)

THINKING GEOGRAPHICALLY What ethnic behavior on the part of Finnish-Americans might explain this striking spatial correlation?

Ethnicity and Business Activity

Differential ethnic preferences give rise to distinct patterns of purchasing goods and services. These patterns, in turn, are reflected in the types of businesses and services available in different ethnic neighborhoods in a city. Geographer Keith Harries made a detailed study of businesses in the Los Angeles urban area in the early 1970s, comparing Anglo-American, African-American, and Mexican-American neighborhoods. He found that East Los Angeles Chicano neighborhoods had unusually large numbers of food stores, eating and drinking places, personal services, and repair shops. These Hispanic areas had, in fact, three times as many food stores as Anglo neighborhoods. In large part, this pattern reflected the dominance of small corner grocery stores and the fragmentation of food sales among several kinds of food stores, such as *tortillerias*. Abundant small barbershops provide one reason personal service establishments ranked so high.

African-American South Los Angeles ranked highest in personal service businesses, and vacant stores ranked second. Eating and drinking places were the third most numerous. In contrast to the Chicano east part of town, the south had relatively few bars but a large number of liquor stores and liquor departments in grocery stores and drugstores. Secondhand shops were very common, but there were no antique or jewelry stores. A distinctive African-American personal service enterprise, the shoeshine parlor, was found only in South Los Angeles.

Anglo neighborhoods ranked high in professional and financial service establishments, such as doctors' and lawyers' offices and banks. These services were much less common in the non-Anglo neighborhoods. Furniture, jewelry, antique, and apparel stores were more numerous among the Anglos, as were full-scale restaurants.

REFLECTING ON GEOGRAPHY

Do you think similar patterns to those described above persist in Los Angeles today? Why or why not?

Contrasts similar to those observed by Harries in the urban scene can also be found in rural areas and small towns. An example can be taken from a classic study by geographer Elaine Bjorklund of an ethnic island in southwestern Michigan settled in the mid-nineteenth century by Dutch Calvinists (**Figure 5.27**). Their descendants adhered

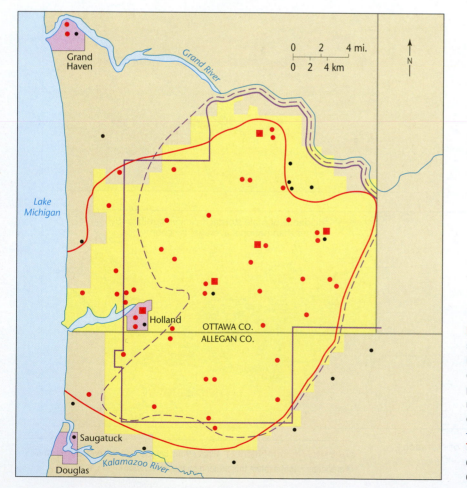

Legend:
— Exclusively Dutch-Reformed churches within this area (rural)
— No Sunday business within this area
-- No taverns within this area

• Dutch-Reformed church
■ Dutch-Reformed school
• Other church

▨ Incorporated places
▨ Area in which Dutch-Reformed people own half or more of the land

FIGURE 5.27 The impact of ethnicity in southwestern Michigan, about 1960. An ethnic island of Dutch Reformed (Calvinist) immigrants was established here in the 1840s and has survived to the present. The Calvinists kept taverns, movie theaters, non-Calvinist churches, and Sunday business activity out of their area.

THINKING GEOGRAPHICALLY Why do these traits, though causally related, not have exactly the same geographical distribution?

to a strict moral code and tended to regard the non-Dutch Reformed world outside their ethnic island as sinful and inferior. This adherence to the precepts of the Calvinist Reformed Church was clearly the main manifestation of their ethnicity, because the Dutch language had died out in the area. The impact of the Calvinist code of behavior on business activity was apparent in various ways. As recently as 1960, no taverns, dance halls, or movie theaters existed there except in the city of Holland, and no business activity was permitted on Sunday.

Ethnicity and Type of Employment

Closely related to type of business is type of employment. In many urban ethnic neighborhoods, specific groups gravitated early to particular kinds of jobs. These job identities, never rigid, were stronger in the decades immediately following immigration than they are today because of advancing acculturation, but some notable examples

can still be found. In some cases, the identification of ethnic groups and job types is strong enough to produce stereotyped images in the American popular mind, such as Irish police, Chinese launderers, and Korean grocers.

The contrast in ethnic activity in the restaurant trade is striking. Certain groups proved highly successful in marketing versions of their traditional cuisines to the population at large. In particular, the Chinese, Mexicans, and Italians succeeded in this venture (**Figure 5.28**). Each dominates a restaurant region in North America far larger than its ethnic homeland, island, or neighborhood.

In Boston, the Irish once provided most of the laborers in the warehouse and terminal facilities near the central business district; Italians dominated the distribution and marketing of fresh foods; Germans gravitated toward the sewing machine and port supply trades; and Jews found employment in merchandising and the manufacture of ready-made clothing. A more recent example is the immigration of Basques

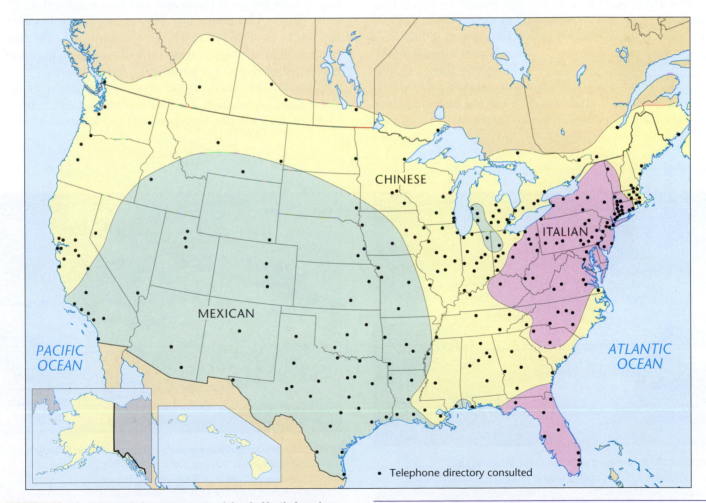

FIGURE 5.28 Dominant ethnic restaurant cuisine in North America. In each region, the cuisine indicated is that of the leading type of ethnic restaurant. All ethnic cuisines were considered in compiling the map, based on telephone Yellow Pages listings. (Source: After Zelinsky, 1985: 66.)

THINKING GEOGRAPHICALLY Why do Mexican foodways balloon northward faster than the main diffusion of Mexican-Americans? Why does Italian cuisine prevail so far south of the major concentration of Italian-Americans?

from Spain to serve as professional jai alai players in the cities of southern Florida, where their ancient ethnic ball game has become a major medium of legal gambling.

REFLECTING ON **GEOGRAPHY**

What is the future of ethnicity and race? Will the potent forces of globalization erase ethnic differences and wipe out racism? Or do ethnicity and race provide enduring constructs that predate—and will outlast—the global era?

ETHNIC **LANDSCAPES**

What traces do ethnicity and race leave on the landscape? Ethnic and racialized landscapes often differ from mainstream landscapes in the styles of traditional architecture, in the patterns of surveying the land, in the distribution of houses and other buildings, and in the degree to which they "humanize" the land. Often the imprint is subtle, discernible only to those who pause and look closely. Sometimes it is quite striking, flaunted as an **ethnic flag**: a readily visible marker of ethnicity on the landscape that strikes even the untrained eye (**Figure 5.29**). Sometimes the distinctive markers of race

ethnic flag A readily visible marker of ethnicity on the landscape.

and ethnicity are not visible at all but rather are audible, tactile, olfactory, or—as we explore here with culinary landscapes—tasty.

Urban Ethnic Landscapes

Ethnic cultural landscapes often appear in urban settings, in both neighborhoods and ghettos. A fine example is the brightly colored exterior mural typically found in Mexican-American ethnic neighborhoods in the southwestern United States (**Figure 5.30**). These began to appear in the 1960s in Southern California, and they exhibit influences rooted in both the Spanish and the indigenous cultures of Mexico, according to geographer Daniel Arreola. A wide variety of wall surfaces, from apartment house and store exteriors to bridge abutments, provide the space for this ethnic expression. The subjects portrayed are also wide ranging, from religious motifs to political ideology, from statements about historical wrongs to ones about urban zoning disputes. Often they are specific to the site, incorporating well-known elements of the local landscape and thus heightening the sense of place and ethnic turf. Inscriptions can be in either Spanish or English, but many Mexican murals do not contain a written message, relying instead on the sharpness of image and vividness of color to make an impression.

Usually, the visual ethnic expression is more subtle. Color alone can connote and reveal ethnicity to the trained eye. Red, for example, is a venerated and auspicious color

FIGURE 5.29 An ethnic flag in the cultural landscape. This maize granary, called a *cuezcomatl*, is unique to the indigenous population of Tlaxcala state, Mexico. The structure holds shelled maize. (Courtesy of Terry G. Jordan-Bychkov.)

THINKING GEOGRAPHICALLY Why would a facility for storage of corn remain an indicator of indigenous ethnicity?

FIGURE 5.30 Mexican-American exterior mural, Barrio Logan, San Diego, California. This mural bears an obvious ideological-political message and helps create a sense of this place as a Mexican-American neighborhood. See also Arreola, 1984. (Courtesy of Terry G. Jordan-Bychkov.)

THINKING GEOGRAPHICALLY What Mexican and Mexican-American symbols can you pick out in the mural?

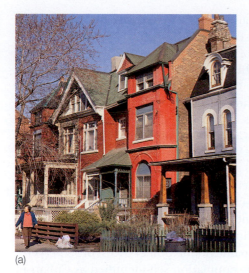

FIGURE 5.31 **Two urban ethnic landscapes.** (a) Houses painted red reveal the addition of a Toronto residential block to the local Chinatown. (b) The use of light blue trim, the Greek color, coupled with the planting of a grape arbor at the front door of a dwelling in the Astoria district of Queens, New York City, marks the neighborhood as Greek. (Courtesy of Terry G. Jordan-Bychkov.)

THINKING GEOGRAPHICALLY Would such ethnic expression be permitted in a new residential development with a neighborhood or homeowners' association set up by the developer? Why?

to the Chinese, and when they established Chinatowns in Canadian and American cities, red surfaces proliferated (**Figure 5.31**). Light blue is a Greek ethnic color, derived from the flag of their ancestral country. Not only that, but Greeks also avoid red, which is perceived as the color of their ancient enemy, the Turks. Green, an Irish Catholic color, also finds favor in Muslim ethnic neighborhoods in countries as far-flung as France and China because it is the sacred color of Islam (see Chapter 7).

Indeed, urban ethnic and racial landscapes are visible in cities around the world. This is true even when urban planners try their best to prevent the emergence of such landscapes. When Brazil's capital was moved from Rio de Janeiro to Brasília in 1960, for example, very few pedestrian-friendly public spaces were incorporated into the urban plan of architects Lúcio Costa and Oscar Niemeyer. Rather, residences were concentrated in high-rise apartments called *superquadra;* streets were designed for high-speed motor traffic only; and no smaller plazas, cafés, or sidewalks were included—all of which discouraged informal socialization (**Figure 5.32**). As historian James C. Scott notes in his discussion of Brasília, the lack of human-scale public gathering places was intentional. "Brasília was to be an exemplary city, a center that would transform the lives of the Brazilians who lived there—from their personal habits and household organization to their social lives, leisure, and work. The

FIGURE 5.32 **The planned urban landscape of Brasília.** As you can see from this aerial view, Brasília is a city of straight roadways, high-rise residential *superquadra*, and green spaces in between. Brasília is a city that was literally built from scratch in an area of cleared jungle in Brazil's tropical scrubland interior. This allowed the planners to follow a modern urban design blueprint down to the last detail. Residents of Brasília complain that its larger-than-life public spaces, high-speed vehicle traffic, and impersonal housing blocks inhibit socialization and make Brasília a difficult place to live. (Cassio Vasconcellos/SambaPhoto/Getty Images.)

THINKING GEOGRAPHICALLY What message were the plan and architecture meant to convey?

goal of making over Brazil and Brazilians necessarily implied a disdain for what Brazil had been." But because the housing needs of those building the city and serving the government workers had not been planned, Brasília soon developed surrounding slums that did not follow an orderly layout. And because the desires of wealthier residents were not met by the uniform *superquadra* apartments, unplanned but luxurious residences and private clubs were also built. By 1980, 75 percent of Brasília's population lived in unanticipated settlements.

The Re-creation of Ethnic Cultural Landscapes

As mentioned previously in this chapter, migration is the principal way that groups that were not ethnic in their homelands become ethnic or racialized, through relocation to a new place. As we have seen, ethnic groups may choose to relocate to places that remind them of their old homelands. Once there, they set about re-creating some—although, because of cultural simplification, not all—particular landscapes in their new homes.

Cuban-American immigrants, for example, have reconstructed aspects of their prerevolutionary Cuban homeland in Miami. In the early 1960s, the first wave of Cuban refugees came to the United States on the heels of Castro's communist revolution on the island. Some went north, to places such as Union City, New Jersey, where today there is a sizable Cuban-American population. Others, however, came (or eventually relocated) to Miami.

Once a predominantly Jewish neighborhood called Riverside, what is today called the Little Havana neighborhood became the heart of early Cuban immigration. Shops along the main thoroughfare, Calle Ocho (or Southwest Eighth Street), reflect the Cuban origins of the residents: grocers such as Sedanos and La Roca cater to the Spanish-inspired culinary traditions of Cuba; cafés selling small cups of strong, sweet *café Cubano* exist on every block; and famed restaurants such as Versailles are social gathering points for Cuban-Americans.

Because ethnicity and its expression on the landscape are fluid and ever-changing, the landscape of Calle Ocho reflects the current demographic changes under way in Little Havana. Although the neighborhood is still a Latino enclave, with a Hispanic population in excess of 95 percent, not even half of its population self-identified as Cuban in the 2010 Census. More than one-quarter of Little Havana's residents are recent arrivals from Central American countries, particularly Nicaragua and Honduras, whereas more and more Argentineans and Colombians are arriving—like the Cubans before them—on the heels of political and economic chaos in their home countries. Today, you are just as likely to see a Nicaraguan *fritanga* restaurant as a Cuban coffee shop along Calle Ocho. This has prompted some to suggest that the neighborhood's name be changed from Little Havana to The Latin Quarter.

Another example of a re-created ethnic landscape can be found in Washington, D.C. Beginning in the 1970s, large numbers of Ethiopian emigrants, fleeing a repressive government and impoverished conditions, relocated from their homeland to the United States. With changes in American immigration restrictions and the introduction of new refugee programs during the 1980s and 1990s, flows of Ethiopian migrants to the United States continued. Disproportionate numbers of these Ethiopian migrants (when compared to other American cities) settled in the greater Washington, D.C., area. Once there, they began to re-create many aspects of the cultural landscape of their homeland in their newly adopted country. For example, **Figure 5.33** depicts a Washington, D.C., neighborhood where a concentration of Ethiopian businesses and restaurants sprang up and grew in the area of Ninth and U streets. At the request of Ethiopian residents in this neighborhood, it has even been signposted "Little Ethiopia" by District officials. Culinary tours to Washington, D.C., now often include stops in this area so that visitors may partake of the many specialty foods on offer and

FIGURE 5.33 The "Little Ethiopia" neighborhood of Washington, D.C. The sounds of Ethiopian music and smells of heavily spiced cooking fill the air in this vibrant neighborhood populated by large numbers of Ethiopian immigrants, demonstrating that cultural landscapes can be constructed from any of the five senses. (Jack Rogers/Image Brief.)

THINKING GEOGRAPHICALLY In what other ways might the influx of Ethiopian immigrants be seen or sensed on the cultural landscape of Washington, D.C.?

shop in the neighborhood's vibrant food markets, clothing stores, and music shops.

It is not merely the presence of Ethiopian flags, fashions, and signage in native languages on brightly painted storefronts that signify the presence of a re-created cultural landscape in Little Ethiopia, however. Geographer Elizabeth Chacko has studied the making of ethnic places in Washington, D.C., and found that Ethiopian migrants have also re-created aspects of their native culture by participating in entrepreneurial endeavors in their new home. An entrepreneurial spirit is much admired among Ethiopians, and they have impressed this aspect of their culture on the landscape of their new home. The entrepreneurial spirit is particularly evidenced by one city block of Little Ethiopia, which is now home to more than 25 businesses started and run by Ethiopian migrants. In fact, many Ethiopian families own more than one business in the neighborhood and have also expanded their entrepreneurship outward into other areas of Washington, D.C. A telephone directory of Ethiopian-owned businesses in the District now includes over 900 listings. In the twenty-first century, the Ethiopian entrepreneurial spirit continues to diffuse with such contagion that a second (unofficial) Little Ethiopia has now sprung up across the Potomac River in Alexandria, Virginia.

Ethnic Culinary Landscapes

"Tell me what you eat, and I'll tell you who you are." This oft-quoted phrase arises from the connections between identity and **foodways,** or the customary behaviors associated with food preparation and consumption

foodway Customary behavior associated with food preparation and consumption.

that vary from place to place and from ethnic group to ethnic group. As geographers Barbara and James Shortridge point out, "Food is a sensitive indicator of identity and change." Immigration, intermarriage, technological innovation, and the availability of certain ingredients mean that modifications and simplifications of traditional foods are inevitable over time. An examination of culinary cultural landscapes reveals that, although landscapes are typically understood to be visual entities, they can be constructed from the other four senses as well: touch, smell, sound, and—in this case—taste.

Constituting part of the rich Southern culinary landscape in the United States is a regionalized foodway tradition known as Cajun cooking. The practices associated with Cajun cooking began as far back as the 1700s. At that time, American Indians, Arcadians (an exiled group from Canada), and other immigrant populations coming to the South (including Germans, French, English, Creoles, Africans, and Mexicans) began to blend their

cooking practices and traditional ingredients to develop new Cajun foodways. These immigrant groups were often poor and had to make do with inexpensive and easily accessible ingredients. As a result, they began the practice of adding spices to these otherwise bland or simple foods to produce the characteristic Cajun kick that has become so popular throughout the South and many other parts of the United States today. **Figure 5.34** depicts one popular Cajun dish, crawfish pie, which blends locally caught crawfish and crabs with rice, beans, and spicy Cajun sausage.

Another example of a regional foodway that has become a fixture in the contemporary American cultural landscape is traceable to traditional Mexican cooking practices. In many regions of Mexico, corn was traditionally ground daily to make fresh tortillas, which were used as the staple "bread" component in most meals. However, the grinding of corn is strenuous and time-consuming. Therefore, over time, corn tortillas began to be replaced with wheat tortillas, which are easier to make. As Mexican immigrants began to spread through the southwestern United States and other areas, they took these modified foodways with them. As a result of this diffusion, we now find wheat tortillas embedded in many U.S. fast foods (such as wraps) and home-cooked dishes.

Geographer Daniel Arreola has plotted the Taco-Burrito and the Taco-Barbeque isoglosses in southwestern

FIGURE 5.34 **A Cajun specialty.** Crawfish pie is a Cajun specialty combining several ingredients that were readily accessible to and affordable for early immigrants and native populations in southern Louisiana. Today, Cajun foodways have grown in popularity with non-Southern diners and have diffused throughout many other parts of the world as well. (Mario Mejia.)

THINKING GEOGRAPHICALLY What other American foodways can you think of that blend food practices and traditional ingredients from two or more cultures?

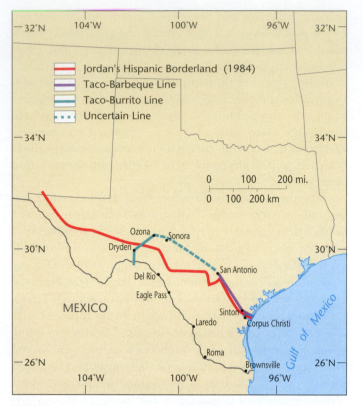

FIGURE 5.35 The Taco-Burrito and Taco-Barbeque lines. The transition between different styles of Mexican, and Mexican and European, foodways is depicted on this map of Texas. Note that the lines coincide with the borders of the Hispanic homeland as defined by Terry Jordan. In this map, Daniel Arreola builds on that work. (Sources: After Arreola, 2002: 175; Jordan, 1984.)

THINKING GEOGRAPHICALLY Thinking of restaurants in your area serving Mexican food, what kinds of tacos or burritos are most common? Is there a distance-decay force at work?

Texas in **Figure 5.35**. These lines show the transition between different Mexican culinary influences (tacos versus burritos) and between Mexican and European foodways (tacos versus barbeque), with the barbeque sandwich preferred by the German, Czech, Scandinavian, Anglo, and black Texan populations to the north and east of the line.

Key Terms

acculturation, 123
assimilation, 123
chain migration, 133
channelization, 135
cultural maladaptation, 137
cultural preadaptation, 136
cultural simplification, 135
environmental racism, 138
ethnic cleansing, 133
ethnic flag, 142
ethnic geography, 123
ethnic group, 122
ethnic homeland, 124

ethnic island, 125
ethnic neighborhood, 127
ethnic substrate, 126
ethnicity, 121
ethnoburb, 129
foodway, 145
gateway city, 130
ghetto, 127
involuntary migration, 133
race, 121
racism, 122
return migration, 135

Ethnic Geography on the Internet

You can learn more about ethnic geography and the geography of race on the Internet at the following Web sites:

2010 Census of the United States

http://www.census.gov
Go to the American FactFinder section of the web page. Here you will find a wonderful selection of maps that show themes (thematic maps) generated from census data.

Ethnic Geography Specialty Group, Washington, D.C.

http://www.uwec.edu/geography/ethnic/
This web site provides information about a specialty group within the Association of American Geographers, whose membership includes nearly all U.S. specialists in the study of ethnic geography.

Food: Past and Present

http://www.teacheroz.com/food.htm
Here you will find a list of links to hundreds of web pages that deal with all aspects of food: diets of historical and contemporary cultures, recipes, and foodways of regions and ethnic groups in the United States.

Racialicious

http://www.racialicious.com
Feel the need to know the top 10 trends in race and pop culture? Want to follow celebrity gaffes, politicians' missteps, and questionable media representations? Log on to Racialicious, a no-holds-barred blog about the intersection of race and pop culture. The site is mediated by Carmen Van Kerchove, the cofounder and president of New Demographic, a consulting firm that "helps organizations overcome diversity fatigue by facilitating relaxed, authentic, and productive conversations about race and racism."

Recommended Books on Ethnic Geography

Berry, Kate A., and Martha L. Henderson (eds.). 2002. *Geographical Identities of Ethnic America: Race, Space, and Place.* Reno: University of Nevada Press. Eighteen different experts give their views on American ethnic geography, explaining how place shapes ethnic and racial identities and, in turn, how these groups create distinctive spatial patterns and ethnic landscapes.

Jordan, Terry G., and Matti E. Kaups. 1989. *The American Backwoods Frontier: An Ethnic and Ecological Interpretation.* Baltimore: Johns Hopkins University Press. The authors use the concepts of cultural preadaptation and ethnic substrate to reveal how the forest colonization culture of the American eastern woodlands developed and helped shape half a continent.

McKee, Jesse O. (ed.). 2000. *Ethnicity in Contemporary America: A Geographical Appraisal,* 2nd ed. Lanham, MD.: Rowman & Littlefield. This clear and thoughtful text offers a geographical analysis of U.S. immigration patterns and the development of selected ethnic minority groups, focusing especially on their origin, diffusion, socioeconomic characteristics, and settlement patterns within the United States; many well-known geographers contributed chapters.

Miyares, Ines M., and Christopher A. Airriess. 2007. *Contemporary Ethnic Geographies in America.* Lanham, MD.: Rowman & Littlefield. An edited collection featuring chapters authored by renowned contemporary ethnic geographers on a variety of topics ranging from Central American soccer leagues to Muslim immigrants from Lebanon and Iran; also includes a geographer's view of ethnic festivals, with which we began this chapter.

Nostrand, Richard L., and Lawrence E. Estaville, Jr. (eds.). 2001. *Homelands: A Geography of Culture and Place Across America.* Baltimore: Johns Hopkins University Press. A collection of essays on an array of North American ethnic homelands, together with in-depth treatment of the geographical concept of homeland.

Rehder, John B. 2004. *Appalachian Folkways.* Baltimore: Johns Hopkins University Press. An exploration of the folk culture of the Appalachian region of the United States, including its distinctive settlement history, folk architecture, cuisine, speech, and belief systems.

Schein, Richard (ed.). 2006. *Landscape and Race in the United States.* London and New York: Routledge. Contributions by leading geographers studying the cultural geographies of race provide an updated and critical insight into this growing subfield.

Answers to Thinking Geographically Questions

Figure 5.1: They help build group solidarity. The Hispanic festival, by incorporating an American flag, also helps portray the immigrants as loyal, patriotic Americans.

Figure 5.2: Both Asia and Latin America have had rapid increases in population in the late twentieth century. In both areas, cultures have close family ties, leading to families bringing their relatives after one member immigrates. The number of Asian immigrants was also increased by refugees from warfare in Southeast Asia in the 1970s.

Figure 5.3: The percentage of the U.S. population that cites Hispanic origin is likely to increase in the next few decades for several reasons. The economies of countries such as Mexico and many Central American states are projected to continue to shrink in coming decades, offering citizens of those countries fewer and fewer opportunities for livelihood. Therefore, increasing numbers of these individuals are likely to seek economic opportunity in the (nearby) United States. Further, individuals of Hispanic origin already living in the United States tend to be, on average, younger than members of some other racial and ethnic groups within the country. Therefore, they are more likely to be of childbearing age in the coming decades, further increasing the Hispanic-origin population in comparison to groups experiencing lower birthrates.

Figure 5.4: Many aspects of social and economic position depended on "race." Europeans had not previously been in contact with many peoples of nonwhite races to the degree that they were when they settled in the Americas.

Figure 5.5: Race in the United States has traditionally been defined by ancestry, not necessarily appearance.

Figure 5.6: The large areas of the Tibetans, Turkic peoples, Mongols, and Hui would be homelands. There are numerous ethnic islands in the south, such as those of Miao-Yao peoples. In the Northeast are islands of Tungus-Manchu peoples. The peripheral location of the ethnic areas relates to China's having absorbed these areas as it expanded through its history from its origins in the Yellow River (Huang He) valley.

Figure 5.7: The 1939 map shows an extension of the Cajun-speaking area toward the northwest and an area north of Lake Pontchartrain. It also shows more Cajun speakers in the eastern part of the Mississippi Delta. Differences in the 2000 map may be related to changes in the population, but they may also reflect the use of English by people with Cajun surnames, so that they would not report themselves as Cajun-speaking.

Figure 5.8: Characteristic types of houses, barns, and houses of worship may remain. Place names are another fairly permanent part of an ethnic homeland. Although field patterns were greatly influenced by the survey system, other aspects of fields may have been developed by the ethnic group.

Figure 5.9: When most Germans and Scandinavians immigrated in the middle to late nineteenth century, they were seeking either good farmland or mine or factory employment. Both were more likely to be found in the North. The South had little industry, and most land was already claimed. Scandinavians were also willing to settle in cold climates.

Figure 5.10: A large section is labeled "Ghetto Nuovo" and a street in the lower right corner is "Calle di Ghetto Nuovissimo." *Nuovo* means new in Italian.

Figure 5.11: There is nothing about the buildings themselves to show that they are Chinese. The traditional Chinatown was probably built by others and then acquired by Chinese immigrants. Although Chinese entrepreneurs may have built the suburban shopping center, they used popular architectural styles in the area.

Figure 5.12: Answers will vary.

Figure 5.13: The Southwest is closest to Latin America and was under Spanish, then Mexican, rule before it became part of the United States. Puerto Rico, likewise, was a Spanish colony. Puerto Ricans have migrated to the major cities of the Northeast in search of jobs, and Cubans migrated in large numbers to Florida after Fidel Castro came to power in Cuba. Both Mexicans and Puerto Ricans have settled in Chicago. Hispanics are spreading out in the Southeast and Midwest, primarily in response to job opportunities.

Figure 5.14: California and Hawaii are physically closest to Asia; California began attracting Asian immigrants during the Gold Rush. The Northeast has attracted Asians looking for good jobs, especially in technology. Chicago, in Illinois, is also a draw for this reason, as is Washington State. Nevada is a migration destination for many Americans because of the growth of jobs in general.

Figure 5.15: States with the highest percentages of white population tend to be in the North, where settlement was primarily by European immigrants; other, later immigrants have not settled in large numbers. Places like northern New England, Iowa, North Dakota, and Wyoming lack large cities and large numbers of factory jobs that attract later nonwhite migrants. West Virginia is well known as a place of poverty with few jobs other than coal mining. New York, New Jersey, Illinois, and Michigan have lower white percentages because they have attracted blacks to factory jobs.

Figure 5.16: The major region of black population is still composed of the former Confederate states where slavery was practiced. Northward migration for factory jobs is also evident in the larger percentages in New Jersey, New York, Michigan, and Illinois, and to a lesser extent, California and Nevada, which are farther away from the major concentration of blacks.

Figure 5.17: In defining "American Indian and Alaska Native," the U.S Bureau of the Census includes people whose origins are from the original people of North, South, and Central America (excluding Native Hawaiians and other Pacific Islander groups). These populations are most prevalent in several midwestern, southwestern and Great Plains states such as New Mexico, Arizona, Montana, North and South Dakota, and Oklahoma, which have comparatively large areas of land designated as American Indian reservations. Alaska is also home to many native groups, including Alaska Athabascan, Eskimo, Aleut, and Tlingit-Haida.

Figure 5.18: Polynesians have no tradition of retailing, and theirs is a communal society that shares wealth. If a Polynesian were to open a store, all of his or her relatives would have the right to come and take merchandise for free, sharing the wealth. The store would fail within a month. The way was left open for the Chinese, who have a very different culture.

Figure 5.19: The states that contributed the most migrants are close by; more distant states are hardly represented. The exception is Baja California, where migrants would have other opportunities, such as moving to California.

Figure 5.20: Answers will vary, but you may consider the following challenges and conditions that refugees frequently encounter: racism and discrimination; poverty and unemployment; lack of access to education and other services; and separation from, and loss of contact with, family members in their home country.

Figure 5.21: Many refugees, in fleeing war-torn regions, take up residence illegally in neighboring countries that may not be officially hospitable to them. Therefore, these individuals are likely to keep their presence unknown to officials if possible. Furthermore, movements into and out of designated refugee camps and other sanctioned settlement areas are likely difficult to monitor due to chaotic conditions.

Figure 5.22: The Finns found the forested region similar to the land they had left. The Native Americans survived there because few other groups wanted such infertile land.

Figure 5.23: Political boundaries do not necessarily accommodate ethnic differences, as in the case of the enclave Nagorno-Karabakh in Azerbaijan. Divided ethnic groups may want to join together, as in the case of North and South Ossetia, and groups may desire independence from rule by different ethnic powers, as in the case of Chechnya.

Figure 5.24: Many ethnic neighborhoods began as areas of poor, substandard housing, for example, the Lower East Side of New York City. Ethnic neighborhoods tended to develop in areas that wealthier residents did not want.

Figure 5.25: Waste disposal, sewage treatment plants, and industries that cause noise, fumes, or odors.

Figure 5.26: It is suggested that the Finns' fondness for smoked fish may help explain their high cancer rate.

Figure 5.27: Culture regions are based on several traits, each of which may have a slightly different distribution and boundaries.

Figure 5.28: Americans have traveled in Mexico and in Italy and brought back a fondness for those types of cuisine, which nonmembers of the ethnic groups concerned have learned to reproduce (usually in a modified fashion). Both Mexican and Italian foodways have been adopted by major corporations for chain brands.

Figure 5.29: Corn (maize) is the traditional staple food of the indigenous peoples, while the Spanish prefer wheat, their traditional staple food.

Figure 5.30: Examples include the flag of César Chávez's farm-labor movement, the Virgin of Guadalupe, the Mexican flag, and indigenous clothing styles.

Figure 5.31: Such neighborhoods tend to have very strict regulations regarding exterior decoration in order to maintain an image of upper-class mainstream America. Bright red paint and grape arbors would probably be against these regulations.

Figure 5.32: The major message to be conveyed was that Brazil was a modern, forward-looking country.

Figure 5.33: Because the flow of Ethiopian migrants continued over an extended period of time, and because large numbers within this total flow of individuals relocated to the Washington, D.C., area, it is likely that there are many noticeable imprints on the local cultural landscape. For example, there are likely to be clothing stores specializing in Ethiopian fashion; salons catering to African hairstyles and styling techniques; and services in native languages targeted toward Ethiopian clients and businesses, such as accounting firms and legal services.

Figure 5.34: Answers will vary, but you may think of Polynesian cuisine in Hawaii, which blends many traditionally East Asian ingredients and preparations with those of several South Pacific island cultures. You may also suggest Tex-Mex cuisine in the American Southwest, which blends traditional Mexican foodways with those of American cowboy culture.

Figure 5.35: Students will relate different experiences depending on their location. The closer they live to Mexico, the more likely they are to report corn tortillas or at least more authentic Mexican tortillas. While large numbers of Mexican immigrants have resulted in large numbers and widely spread Mexican restaurants, they vary greatly in the authenticity of the food they serve.

Post–9/11 border security: the boundary between Israeli and Arab-Palestinian lands and the U.S.–Mexico border.
(Left: Debbie Hill/UPI/Landov; right: David R. Frazier/DanitaDelimont.com.)

Political Geography

6

From the breakup of empires to regional differences in voting patterns, from the drawing of international boundaries to congressional redistricting in the U.S. electoral system, from the resurgence of nationalism to separatist violence, human political behavior is inherently geographical. As geographer Gearóid Ó Tuathail has said, **political geography** "is about power, an ever-changing map revealing the struggle over borders, space, and authority."

> **political geography** The geographic study of politics and political matters.

POLITICAL **CULTURE** REGIONS

How is political geography revealed in culture regions? The theme of culture region is essential to the study of political geography because an array of both formal and functional political regions exist. Among these, the most important and influential is the *state*.

INDEPENDENT COUNTRIES

Abbreviations

A	AUSTRIA
AL	ALBANIA
B	BELGIUM
BA	BOSNIA-HERZEGOVINA
BF	BURKINA FASO
BG	BULGARIA
BOTS	BOTSWANA
BY	BELARUS
CH	SWITZERLAND
CZ	THE CZECH REPUBLIC
D	GERMANY
EG	EQUATORIAL GUINEA
EST	ESTONIA
ET	EAST TIMOR
GBI	GUINEA BISSAU
H	HUNGARY
HR	CROATIA
IC	IVORY COAST
L	LUXEMBOURG
LT	LITHUANIA
LV	LATVIA
MK	MACEDONIA
NL	NETHERLANDS
RCA	CENTRAL AFRICAN REPUBLIC
RL	LEBANON
RO	ROMANIA
RU	RUSSIA
SK	SLOVAKIA
SLO	SLOVENIA
S	SERBIA
M	MONTENEGRO
TC	TURKISH CYPRUS
TM	TURKMENISTAN
UAE	UNITED ARAB EMIRATES
WAG	THE GAMBIA
WAL	SIERRA LEONE
ZW	ZIMBABWE

Most of the countries on this map have homepages on the Worldwide Web. Visit The Human Mosaic *online to learn more about them.*

A World of States

The fundamental political-geographical fact is that Earth is divided into nearly 200 independent countries or states, creating a diverse mosaic of functional regions (**Figure 6.1**). The state is a political institution that has taken a variety of forms over the centuries, ranging from Greek city-states to Chinese dynastic states to European feudal states. When we talk about states today, however, we mean something very specific and historically recent. **States** are independent political units with a centralized authority that makes claims to sole jurisdiction over a bounded territory. Within that territory, the central authority

state An independent political unit with a centralized authority that makes claims to sole jurisdiction over a bounded territory, within which a central authority controls and enforces a single system of political and legal institutions; often used synonymously with "country."

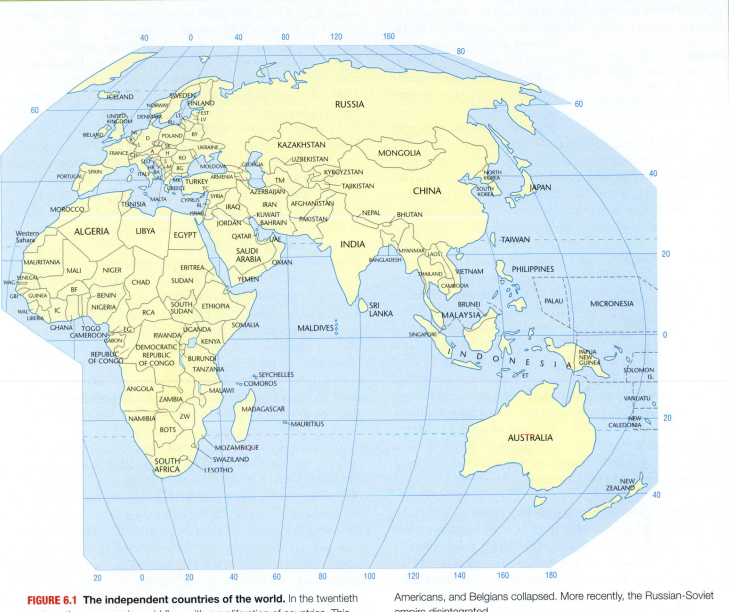

FIGURE 6.1 **The independent countries of the world.** In the twentieth century, the map was in rapid flux, with a proliferation of countries. This process began after World War I with the breakup of such empires as those of Austria-Hungary and Turkey, then intensified after World War II when the overseas empires of the British, French, Italians, Dutch, Americans, and Belgians collapsed. More recently, the Russian-Soviet empire disintegrated.

THINKING GEOGRAPHICALLY What changes might take place next?

controls and enforces a single system of political and legal institutions. Importantly, in the modern international system, states recognize each other's **sovereignty.** That is, virtually every state recognizes every other state's right to exist and control its own affairs within its territorial boundaries.

sovereignty The right of individual states to control political and economic affairs within their territorial boundaries without external interference.

Closer inspection of Figure 6.1 reveals that some parts of the world are fragmented into many states, whereas others exhibit much greater unity. The United States occupies about the same amount of territory as Europe, but the latter is divided into 45 independent countries. The continent of Australia is politically united, whereas South America has 12 independent entities and Africa has more than 50.

The modern state is a tangible geographical expression of one of the most common human tendencies: the need to belong to a larger group that controls its own piece of Earth, its own territory. So universal is this trait that scholars coined the term **territoriality** to describe it. Most geographers view territoriality as a learned cultural response. Robert Sack, for example, regards territoriality as a cultural strategy that uses power to control an area and communicate that control, thereby subjugating the inhabitants and acquiring resources. He argues that the precise marking of borders is a practice unique to modern Western culture. The modern territorial state, he claims, emerged rather recently in sixteenth-century Europe and diffused around the globe through European colonialism. **Colonialism** is the building and maintaining of colonies in one territory by people based elsewhere; it is the forceful appropriation of a territory by a distant state, often involving the displacement of indigenous populations to make way for colonial settlers.

Political territoriality, then, is a thoroughly cultural-geographical phenomenon. The sense of collective identity that we call *nationalism* (see page 155) springs from a learned or acquired attachment to region and place. Geography and national identity cannot be separated.

territoriality   A learned cultural response, rooted in European history, that produced the external bounding and internal territorial organization characteristic of modern states.

colonialism   The building and maintaining of colonies in one territory by people based elsewhere; the forceful appropriation of a territory by a distant state, often involving the displacement of indigenous populations to make way for colonial settlers.

Distribution of National Territory   An important geographical aspect of the modern state is the shape and configuration of the national territory. Theoretically, the more compact the territory, the easier is national governance. Circular or hexagonal forms maximize compactness, allow short communication lines, and minimize the amount of border to be defended. Of course, no country actually enjoys this ideal degree of compactness, although some—such as France and Brazil—come close (**Figure 6.2**).

An **enclave** is a piece of territory surrounded by, but not part of, a country. For example, Lesotho is an enclave that is completely surrounded by South Africa. **Exclaves** are parts of a national territory separated from the main body of the country to which they belong by the territory of another (e.g., the Kaliningrad District in Figure 6.2). Exclaves are particularly undesirable if a hostile power holds the intervening territory because defense of such an isolated area is difficult and makes substantial demands

enclave   A piece of territory surrounded by, but not part of, a country.

exclave   A piece of national territory separated from the main body of a country by the territory of another country.

on national resources. Moreover, an exclave's inhabitants, isolated from their compatriots, may develop separatist feelings, thereby causing additional problems.

Pakistan provides a good example of the national instability created by exclaves. Pakistan was created in 1947 as two main bodies of territory separated from each other by almost 1000 miles (1600 kilometers) of territory in northern India. West Pakistan had the capital and most of the territory, but East Pakistan was home to most of the people. West Pakistan hoarded the country's wealth, exploiting East Pakistan's resources but giving little in return. Ethnic differences between the peoples of the two sectors further complicated matters. In 1971, a quarter of a century after its founding, Pakistan broke apart. The distant exclave seceded and became the independent country of Bangladesh (see Figure 6.1).

Even when a national territory is geographically united, instability can develop if the shape of the state is awkward. Narrow, elongated countries, such as Chile, The Gambia, and Norway, can be difficult to administer, as can nations consisting of separate islands (see Figures 6.1 and 6.2). In these situations, transportation and communications are difficult, causing administrative problems. Several major secession movements threaten the multi-island country of Indonesia; one of these—in East Timor—recently succeeded. Similarly, the three-island country of Comoros, in the Indian Ocean, is troubled today by a separatist movement on Anjouan Island.

The shape and configuration of a country can make it harder or easier to administer, but it does not determine its stability. We can think of exceptions for all of the territorial forms in Figure 6.2. For example, Alaska is an exclave of the United States, but it is unlikely to produce a secessionist movement similar to that of Bangladesh. Conversely, the Democratic Republic of the Congo and Nigeria both have compact territories and both have suffered through secessionist wars. Geography is an important factor in national governance, but it is not the only factor.

Boundaries   Political territories have different types of boundaries. Until fairly recent times, many boundaries were not sharp, clearly defined lines but instead were referred to as zones called *marchlands*. Today, the nearest equivalent to the marchland is the **buffer state,** an independent but small and weak country lying between two powerful, potentially belligerent countries. Mongolia, for example, is a buffer state between Russia and China; Nepal occupies a similar position between India and China (see Figure 6.1). If one of the neighboring countries assumes control of the buffer state, it loses some or much of its independence and becomes a **satellite state.**

buffer state   An independent but small and weak country lying between two powerful countries.

satellite state   A small, weak country dominated by one powerful neighbor to the extent that some or much of its independence is lost.

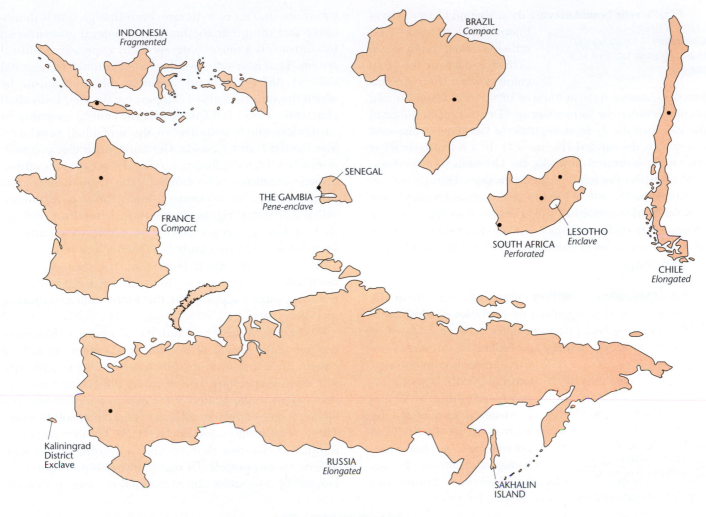

- Capital city (in South Africa, functions divided among three cities)

FIGURE 6.2 Differences in the distribution of national territory. The map, drawn from Eurasia, Africa, and South America, shows wide contrasts in territorial shape. France and, to a lesser extent, Brazil approach the ideal hexagonal shape, but Russia is elongated and has an exclave in the Kaliningrad District, whereas Indonesia is fragmented into a myriad of islands. The Gambia intrudes as a pene-enclave into the heart of Senegal, and South Africa has a foreign enclave, Lesotho. Chile must overcome extreme elongation.

THINKING GEOGRAPHICALLY What problems can arise from elongation, enclaves, fragmentation, and exclaves?

Most modern boundaries are lines rather than zones, and we can distinguish several types. **Natural boundaries** follow some feature of the natural landscape, such as a river or mountain ridge. Examples of natural boundaries are numerous; for example, the Pyrenees lie between Spain and France, and the Rio Grande serves as part of the border between Mexico and the United States. **Ethnographic boundaries** are drawn on the basis of some cultural trait, usually a particular language spoken or religion practiced. The border that divides India from predominantly Islamic Pakistan is one case. **Geometric boundaries** are regular, often perfectly straight lines drawn without regard for physical or cultural features of the area. The U.S.–Canada border west of the Lake of the Woods (about 93° west longitude) is a geometric boundary, as are most county, state, and province borders in the central and western United States and Canada. Not all boundaries are easily categorized; some boundaries are of mixed type, composites of two or more of the types listed.

natural boundary A political border that follows some feature of the natural environment, such as a river or mountain ridge.

ethnographic boundary A political border that follows some cultural border, such as a linguistic or religious border.

geometric boundary A political border drawn in a regular, geometric manner, often a straight line, without regard for environmental or cultural patterns.

Finally, **relic boundaries** are those that no longer exist as international borders. Nevertheless, they often leave behind a trace in the local cultures. With the reunification of Germany in the autumn of 1990, the old Iron Curtain border between the former German Democratic Republic in the east and the Federal Republic of Germany in the west was quickly dismantled (**Figure 6.3**). In a remarkably short time span, measured in weeks, the Germans reopened severed transport lines and created new ones, knitting the enlarged country together. Even so, remnants and reminders of the old border remained, as it continued to function as a provincial boundary within Germany. Furthermore, it still separated two parts of the country with strikingly different levels of prosperity.

> **relic boundary**   A former political border that no longer functions as a boundary.

Spatial Organization of Territory

States differ greatly in the way their territory is organized for purposes of administration. Political geographers recognize two basic types of spatial organization: **unitary** and **federal.** In unitary countries, power is concentrated centrally, with little or no provincial authority. All major decisions come from the central government, and policies are applied uniformly throughout the national territory. France and

> **unitary state**   An independent state that concentrates power in the central government and grants little authority to the provinces.
>
> **federal state**   An independent country that gives considerable powers and even autonomy to its constituent parts.

China are unitary in structure, even though one is democratic and the other totalitarian. A federal government, by contrast, is a more geographically expressive political system. That is, it acknowledges the existence of regional cultural differences and provides the mechanism by which the various regions can perpetuate their individual characters. Power is diffused, and the central government surrenders much authority to the individual provinces. The United States, Canada, Germany, Australia, and Switzerland, though exhibiting varying degrees of federalism, provide examples. The trend in the United States has been toward a more unitary, less federal government, with fewer states' rights. By contrast, federalism remains vital in Canada, representing an effort to counteract French-Canadian demands for Québec's independence. That is, by emphasizing federalism, the central government allows more latitude for provincial self-rule, thus reducing public support for the more radical option of secession.

Whether federal or unitary, a country functions through some system of political subdivisions. In federal systems, these subdivisions sometimes overlap in authority, with confusing results. For example, the Native American reservations in the United States occupy a unique and ambiguous place in the federal system of political subdivisions. These semiautonomous enclaves are legally sanctioned political territories that theoretically only indigenous Americans can possess. In reality, ownership is often fragmented by non-Native American land holdings within the

FIGURE 6.3 A boundary disappears. In Berlin, the view toward the Brandenburg Gate changed radically between 1989 and 1990, when the Berlin Wall was destroyed and Germany reunited. (Courtesy of Terry G. Byckhov.)

THINKING GEOGRAPHICALLY   Do you think the walls between Israel and the West Bank and between the United States and Mexico will someday be torn down? Why or why not?

reservations. Although not completely sovereign, Native Americans do have certain rights to self-government that differ from those of surrounding local authorities. For example, many reservations can (and do) build casinos on their land even though gambling may be illegal in surrounding political jurisdictions. Reservations do not fit neatly into the American political system of states, counties, townships, precincts, and incorporated municipalities.

Centrifugal and Centripetal Forces Although physical factors such as the spatial organization of territory, degree of compactness, and type of boundaries can influence an independent country's stability, other forces are also at work. In particular, cultural factors often make or break a country. The most viable independent countries, those least troubled by internal discord, have a strong feeling of group solidarity among their population. Group identity is the key.

In the case of the modern state, the primary source of group identity is **nationalism.** Nationalism is the idea that the individual derives a significant part of his or her social identity from a sense of belonging to a nation. We can trace the origins of modern nationalism to the late eighteenth century and the emergence

> **nationalism** The sense of belonging to and self-identification with a national culture.

of the modern nation-state. Political geographer John Agnew cautions us, however, that the meaning and form of nationalism are unstable, making it difficult to generalize. One version is *state nationalism,* wherein the nation-state is exalted and individuals are called to sacrifice for the good of the greater whole. The twentieth-century history of state nationalism in Europe is marked by two horrific world wars in which millions of lives were sacrificed. Referring to World War I, Agnew points out that the "war itself was also the outcome of a mentality in which the individual person had to sacrifice for the good of the greater whole: the nation-state." Another form is *sub-state nationalism,* in which ethnic or linguistic minority populations seek to secede from the state or to alter state territorial boundaries to promote cultural homogeneity and political autonomy.

Geographers refer to factors that promote national unity and solidarity as **centripetal forces.** By contrast, anything that disrupts internal order and furthers the destruc-

> **centripetal force** Any factor that supports the internal order of a country.
> **centrifugal force** Any factor that disrupts the internal order of a country.

tion of the country is called a **centrifugal force.** Many states encourage centripetal forces that help fuel nationalistic sentiment. Such things as an official national language, national history museums, national parks and monuments, and sometimes even a national religion are actively promoted and supported by the state. We consider an array of centripetal and centrifugal forces later, in the section on cultural ecology.

International Political Bodies

In addition to independent countries and their governmental subdivisions, the third major type of political functional region is the **international organization.** This term is often loosely used to describe varying types of organizations that attempt to promote cooperation and coordination amongst their member states. However, key functional differences can be used to more specifically classify international organizations as *intergovernmental organizations* or *supranational organizations.* Examples of both types of organizations are shown in **Figure 6.4** on page 156. **Supranationalism** exists when countries voluntarily give up some portion of their sovereignty to gain the advantages of a closer political, economic, and cultural association with their neighbors. The only organization that is currently recognized as truly supranational is

> **international organization** A general term used to describe varying types of organizations that attempt to promote cooperation and coordination among their member states.

> **supranationalism** Occurs when states willingly relinquish some degree of sovereignty in order to gain the benefits of belonging to a larger political-economic entity.
> **European Union (EU)** A supranational organization composed of 27 European nations.

the **European Union (EU).** The EU successfully grew from a core of six countries in the 1950s to its present membership of 27 states. **Figure 6.5** on page 157 illustrates this expansion over the course of seven stages of accession. Despite recent economic problems in the EU, including threats to withdraw from the union itself, the EU is likely to continue to grow in the future, as many nonmember European states express interest in joining the union.

At its inception, the EU was merely a customs union designed to lower or remove tariffs hindering trade, but it gradually took on political and cultural roles as well. An underlying motivation to form the EU was to weaken the power of member countries so that they could never again wage war against one another—a response to the devastation of Europe in two world wars. Today, EU law dictates that member states elect representatives to a European Parliament that supersedes their national governments in a broad range of policy making. However, some member states retain a degree of authority and autonomy in choosing whether or not to comply with EU measures. Most of the EU uses a single monetary currency, the euro, and most international borders within the EU are open, requiring no passport checks.

In addition to establishing measures that ease international trade and travel, the EU has attempted to standardize a range of social norms related to the tolerance of religious and ethnic difference, human rights, and gender relations. However, due to the ability of some member states to choose noncompliance with these (and other) EU measures, they have not, as yet, been adopted universally throughout the union.

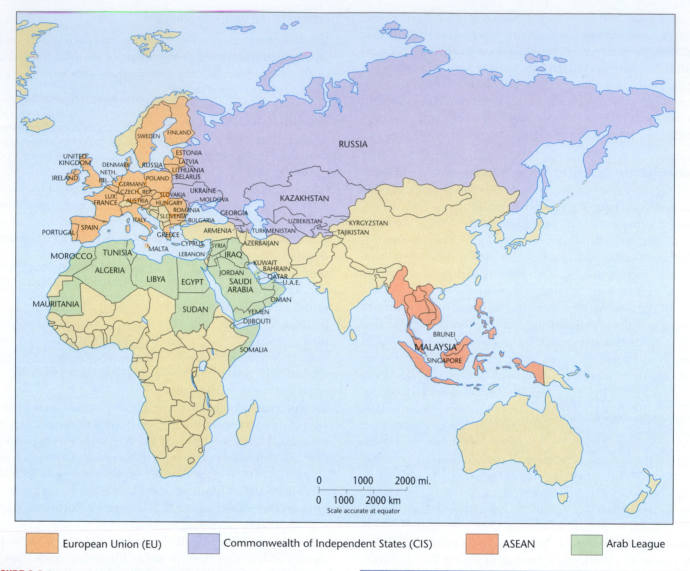

| | European Union (EU) | | Commonwealth of Independent States (CIS) | | ASEAN | | Arab League |

FIGURE 6.4 **International and supranational political organizations in the Eastern Hemisphere.** These organizations vary greatly in purpose and cohesion. ASEAN stands for the Association of Southeast Asian Nations, and its purposes are both economic and political.

THINKING GEOGRAPHICALLY What might this map indicate about globalization?

Resembling supranational organizations, but different from them in key ways, are **intergovernmental organizations.** These organizations do not require member states to relinquish sovereignty in order to benefit from cooperation between states within the association, and they exhibit a lesser degree of cohesion than supranational organizations. Examples of intergovernmental organizations include the Caribbean Community (CARICOM) and the Central American Integration System

intergovernmental organization An international organization that does not require member states to relinquish sovereignty in order to benefit from cooperation between states in the association.

(SICA) (**Table 6.1**, page 158). Some intergovernmental organizations, like the Union of South American Nations (UNASUR), aspire to become true supranational organizations one day and thus increasingly take on integrated functions. However, most remain simply cooperative associations with certain shared goals, like the Arab League and the Association of Southeast Asian Nations (ASEAN). Some intergovernmental organizations serve primarily as **regional trading blocs,** which are designed to reduce trade barriers among

regional trading bloc Entity formed by an agreement made among geographically proximate countries to reduce trade barriers and to better compete with other regional markets.

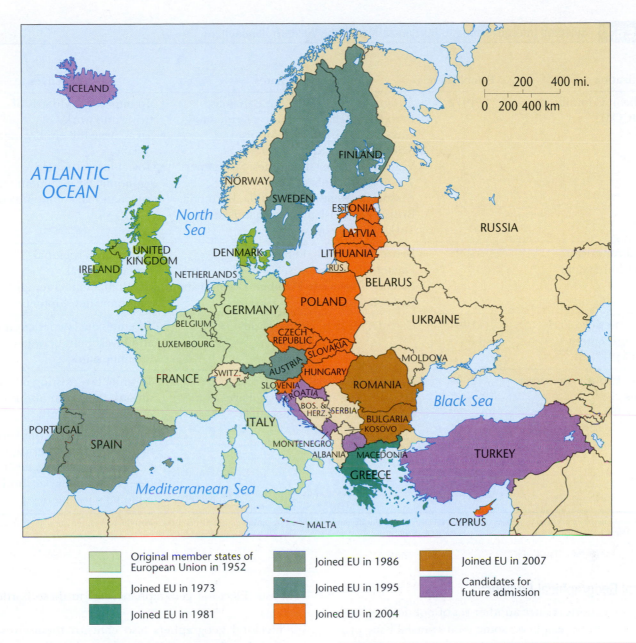

Legend:
- Original member states of European Union in 1952
- Joined EU in 1973
- Joined EU in 1981
- Joined EU in 1986
- Joined EU in 1995
- Joined EU in 2004
- Joined EU in 2007
- Candidates for future admission

FIGURE 6.5 **The Expansion of the European Union (EU).** The EU has continually expanded since its formation in 1951 (originally as the European Coal and Steel Community). The union reached its current size of 27 member countries with the accession of Romania and Bulgaria in 2007.

THINKING GEOGRAPHICALLY What reasons can you suggest for the accession of several Eastern European states to the EU in 2004 and 2007?

its members and to better compete with other regional markets. One such organization is the **North American Free Trade Association (NAFTA).** This organization comprises Canada, Mexico, and the United States, and it promotes the freer flow of goods and services across member state borders. There has been talk of NAFTA expanding its role to become a union of North American states, but only time will tell if supranational status will ultimately be established within this organization.

North American Free Trade Association (NAFTA) An intergovernmental organization composed of Canada, Mexico, and the United States, designed to serve as a trading bloc.

REFLECTING ON GEOGRAPHY

What will motivate the existing and prospective member countries of the European Union to continue to sacrifice aspects of their independence to create a stronger union?

Organization	Year Established	Member States	Key Objectives
Caribbean Community (CARICOM)	1972	Antigua and Barbuda, the Bahamas, Barbados, Belize, Dominica, Grenada, Guyana, Haiti, Jamaica, Montserrat, Saint Lucia, St. Kitts and Nevis, St. Vincent and the Grenadines, Suriname, Trinidad and Tobago (**Associate Members**: Anguilla, Bermuda, British Virgin Islands, Cayman Islands, Turks and Caicos Islands)	• To improve standards of living and work • To achieve full employment of labor • To achieve sustained economic development and international competitiveness
Central American Integration System (SICA)	1993	Belize, Costa Rica, El Salvador, Guatemala, Honduras, Nicaragua, Panama (**Associate Member**: Dominican Republic)	• To establish regional peace, political freedom, democracy and economic development
North American Free Trade Agreement (NAFTA)*	1994	Canada, Mexico, United States	• To eliminate barriers to trade and investment between member states
Union of South American Nations (UNASUR)	2004	Argentina, Bolivia, Chile, Colombia, Ecuador, Guyana, Paraguay, Peru, Suriname, Uruguay, Venezuela	• To unite two existing customs unions, Mercosur and the Andean Community • Ultimately to form the United States of South America, with a common currency, tariff-free common market, and regional parliament

TABLE 6.1 Selected Intergovernmental Organizations in the Western Hemisphere

*NAFTA has discussed incorporating greater integration measures in the future but currently is not equivalent in function or scope to the other international organizations shown in Table 6.1.

Electoral Geographical Regions

Voting in elections creates another set of political regions. A free vote of the people on some controversial issue provides one of the purest expressions of culture, revealing attitudes on religion, ethnicity, and ideology. Geographers can devise formal culture regions based on voting patterns, giving rise to the subspecialty known as **electoral geography.**

Mapping voting tendencies over many decades shows deep-rooted, formal electoral behavior regions. In Europe, for example, some districts and provinces have a long record of rightist sentiment, and many of these lie toward the center of Europe. Peripheral areas, especially in the east, are often leftist strongholds (**Figure 6.6**). Every country where free elections are permitted has a similarly varied electoral geography. In other words, cumulative voting patterns typically reveal sharp and pronounced regional

electoral geography The study of the interactions among space, place, and region and the conduct and results of elections.

contrasts. Electoral geographers refer to these borders as *cleavages.*

Electoral geographers also concern themselves with functional regions, in this case the voting district or precinct. Their interests are both scholarly and practical. For example, following each census, political redistricting takes place in the United States. In redistricting, new boundaries are drawn for congressional districts to reflect the population changes since the previous census. The goal is to establish voting areas of more or less equal population and to increase or reduce the number of districts depending on the amount and direction of change in total population. These districts form the electoral basis for the U.S. House of Representatives. State legislatures are based on similar districts. Geographers often assist in the redistricting process.

The pattern of voting precincts or districts can influence election results. If redistricting remains in the hands of legislators, then the majority political group or party will

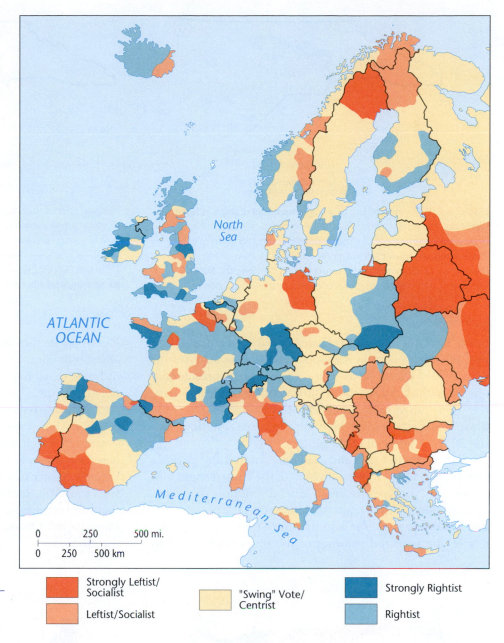

FIGURE 6.6 The electoral geography of Europe. A conservative-rightist core contrasts with a socialist-leftist periphery. Data are based on elections held in the period from 1950 to 1995. In the formerly communist countries, the record of free elections only began in 1990. (Source: Jordan-Bychkov and Jordan, 2002: 226.)

THINKING GEOGRAPHICALLY How does the communist era continue to express itself?

Legend:
- Strongly Leftist/Socialist
- Leftist/Socialist
- "Swing" Vote/Centrist
- Strongly Rightist
- Rightist

often try to arrange the voting districts geographically in such a way as to maximize and perpetuate its power. Cleavage lines will be crossed to create districts that have a majority of voters favoring the party in power or some politically important ethnic group. This practice is called **gerrymandering** (**Figure 6.7** on page 160), and the resultant voting districts often have awkward, elongated shapes. Gerrymandering can be accomplished by one of two methods. One is to draw district boundaries so as to concentrate all of the opposition party into one district, thereby creating an unnecessarily large majority while also

gerrymandering The drawing of electoral district boundaries in an awkward pattern to enhance the voting impact of one constituency at the expense of another.

ensuring that it cannot win elsewhere. A second is to draw the district boundaries so as to dilute the opposition's vote so that it does not form a simple majority in any district.

Red States, Blue States, Purple States

In recent years, various commentators in academia and the popular media have used maps of national election results to suggest that the United States is divided into distinct culture regions. We can trace this argument to the post-election analysis of the controversial 2000 presidential election, when for the first time news media adopted a universal color scheme for mapping voter preferences. States where the majority of voters favored the Democratic presidential candidate were assigned the color blue, whereas

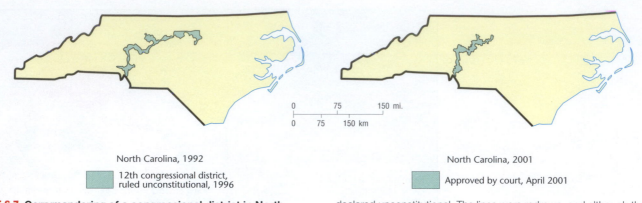

North Carolina, 1992

[ ] 12th congressional district,
ruled unconstitutional, 1996

North Carolina, 2001

[ ] Approved by court, April 2001

FIGURE 6.7 Gerrymandering of a congressional district in North Carolina. The North Carolina Twelfth District was created in 1992 to ensure an African-American majority so that an additional minority candidate could be elected to Congress. Note the awkward shape of these districts, often a sign of gerrymandering. In 1996, the North Carolina district and others gerrymandered for racial purposes were declared unconstitutional. The lines were redrawn, and although the district still looks very much gerrymandered, the courts approved the revised version. (Source: The New York Times.)

THINKING GEOGRAPHICALLY Besides ensuring a particular racial makeup of the winners, what other goals might be involved in gerrymandering?

those favoring the Republican candidate were assigned red. When cartographers mapped the state-by-state results of the 2000 election, as well as subsequent presidential elections, the solid blocks of starkly contrasting colors gave the appearance of a deeply divided country. **Figure 6.8** illustrates the "red state/blue state" pattern resulting from the 2012 presidential election.

The terms *red state* and *blue state* entered the popular lexicon in reference to the division of the United States into regions of "conservative" and "liberal" cultures that these maps implied. These maps, so common in the popular media, are overly simplistic and therefore misleading. For example, the solid red coloring of the states represented in

Figure 6.8 implies that much of the United States is culturally and politically conservative. However, this map gives inordinate visual importance to states of larger areal extent, no matter their population, and exaggerates the appearance of sharp political and geographic divisions within the country. This exaggeration results from the U.S. electoral system, in which a state is designated as "red" or "blue" based on a simple majority in that state's popular vote. In Figure 6.8 you can see that in the 2012 presidential election, Governor Mitt Romney took almost half the U.S. states (24 total) but received only 206 electoral votes. Incumbent President Barack Obama received 332 electoral votes and won the election.

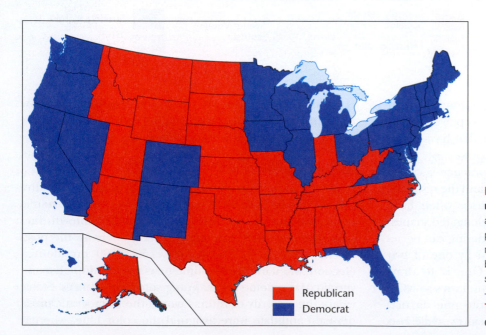

Republican
Democrat

FIGURE 6.8 The 2012 presidential election results. In their result maps, news media have adopted a standard color-coding scheme for political parties: red for Republican, blue for Democrat. It is now common to see maps of the United States being politically and culturally classified into "red state" and "blue state" regions.

THINKING GEOGRAPHICALLY In 2012, which parts of the country were "red"? Which were "blue"?

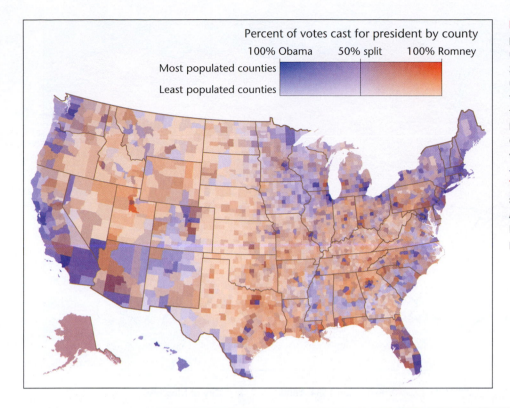

Percent of votes cast for president by county

100% Obama   50% split   100% Romney

Most populated counties

Least populated counties

FIGURE 6.9 Purple America and the presidential election of 2012. Rather than using stark color contrasts to represent the 2012 presidential election results, this map uses color shading based on the popular vote for each county in the United States. Counties that voted heavily for President Obama are bluer, while those that voted more heavily for Governor Romney are redder. Counties that were close to evenly split are purple.

THINKING GEOGRAPHICALLY How does the use of shading affect our understanding of a sharply divided American electorate that appears so prominent in Figure 6.8? What other kinds of geographic voting patterns emerge?

The limitations of the red state/blue state designation have led to the introduction of the term *purple state* to signify the mix of votes in a particular state. Virginia is one example of a purple state. Historically a Republican stronghold, Virginia has been carried (by a narrow margin) by the Democratic presidential candidate in the two most recent presidential elections. If we apply the purple state idea (that is, shadings of color rather than stark contrasts between just two colors) to county-level election results, we find that the simplistic division of the United States into liberal and conservative regions begins to break down. As **Figure 6.9** shows, there is a great deal of county-by-county variation within states across the United States.

POLITICAL **DIFFUSION**

Do political ideas, institutions, and countries expand and spread by means of cultural diffusion? Absolutely! Political events and developments can trigger massive human migrations, sometimes spanning continents. However, political boundaries also act as barriers to the movement of people, ideas, knowledge, and resources.

Movement Between Core and Periphery

Many independent states sprang fully grown into the world, but some expanded outward when powerful political entities emerged from a small nucleus called a **core area.** These core political powers enlarged their territories by annexing adjacent lands, often over many centuries. Generally, core areas possess a particularly attractive set of resources for human life and culture. Larger numbers of people cluster there than in surrounding districts, especially if the area has some measure of natural defense against aggressive neighboring political entities. This denser population, in turn, may produce enough wealth to support a large army, which then provides the base for further expansion outward from the core area. Resources, people, and capital investment begin to flow between the core and peripheral territories, usually resulting in a further consolidation of the core's wealth and power.

In states formed in this way, the core area typically remains the country's single most important district, housing the capital city and the cultural and economic heart of the nation. The core area is the node of a functional region. France expanded to its present size from a small core area around the capital city of Paris. China grew from a nucleus in the northeast, and Russia originated in the small principality of Moscow (**Figure 6.10** on the next page). The United States grew westward from a core between Massachusetts and Virginia on the Atlantic coastal plain, an area that still has the national capital and the densest population in the country.

core area The territorial nucleus from which a country grows in area and over time, often containing the national capital and the main center of commerce, culture, and industry.

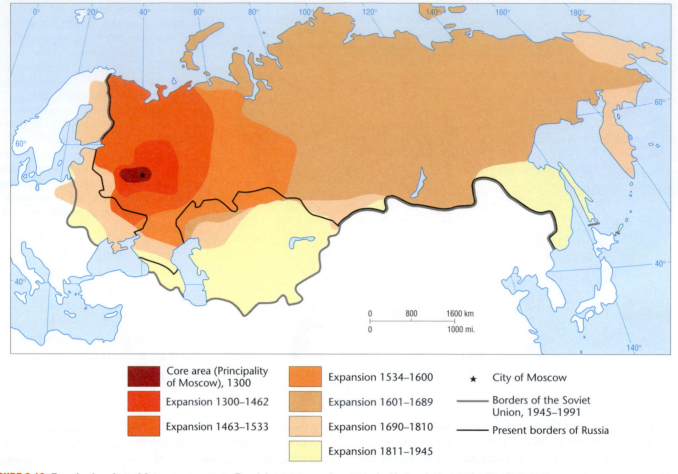

FIGURE 6.10 Russia developed from a core area. Russia's vast expansion started with the single principality of Moscow.

THINKING GEOGRAPHICALLY Can you think of reasons expansion to the east was greater than expansion to the west? What environmental goals might have motivated Russian expansion?

The evolution of independent countries in this manner produces the *core-periphery* configuration, described in Chapter 1 as typical of both functional and formal regions. Although the core dominates the periphery, a certain amount of friction exists between the two. Peripheral areas generally display pronounced, self-conscious regionality and occasionally provide the settings for secession movements. Even so, countries that diffused from core areas are, as a rule, more stable than those created all at once to fill a political void. The absence of a core area can blur or weaken citizens' national identity and make it easier for various provinces to develop strong local or even foreign allegiances. Belgium and the Democratic Republic of the Congo offer examples of countries without political core areas. In the case of the Congo, this situation partly accounts for the history of secessionist conflicts and internecine wars since the country gained independence from colonial rule in 1960.

Countries with multiple, competing core areas are potentially the least stable of all. This situation often develops when two or more independent countries are united. The main threat is that one of the competing cores will form the center of a separatist movement and break apart the country. In Spain, Castile and Aragon united in 1479, but tensions in the union remain more than five centuries later—in part because the old core areas of the two former countries, represented by the cities of Madrid and Barcelona, continue to compete for political control and cultural influence. That these two cities symbolize two language-based cultures, Castilian and Catalan, compounds the division.

Diffusion and Political Innovation

One of the consequences of colonialism was the creation of new patterns of mobility among the conquered populations. One pattern in particular contained the seeds of colonialism's demise. Colonial rule in Africa depended on the establishment of a relatively small cadre of educated African civil servants. The colonial rulers selected those whom they considered the best and the brightest (and most politically cooperative) among their African subjects and sent them to the best universities in Europe. In addition to learning the skills required of colonial functionaries, they absorbed the ideals of Western political philosophy, such as the right of self-determination, independent self-rule, and individual rights. Educated Africans returned to their countries armed with these ideas and organized new nationalist movements to oppose colonial rule.

The consequences can be seen in the spread of political independence in Africa. In 1914, only two African countries—Liberia and Ethiopia—were fully independent of European colonial or white minority rule, and even Ethiopia later fell temporarily under Italian control. Influenced by developments in India and Pakistan, the Arabs of North Africa began a movement for independence. Their movement gained momentum in the 1950s and swept southward across most of the African continent between 1960 and 1965. Many of Africa's great nationalist leaders of this period obtained university degrees in England, Scotland, France, and other European countries. By 1994, African nationalists had helped spread the idea of independence into all remaining parts of the continent, eventually reaching the Republic of South Africa, formerly under white minority rule (**Figure 6.11**).

Despite its rapid spread, diffusion of African self-rule occasionally encountered barriers. Portugal, for example, clung tenaciously to its African colonies until 1975, when a change in government in Lisbon reversed a 500-year-old policy, allowing the colonies to become independent. In colonies containing large populations of European settlers, independence came slowly and usually with bloodshed. France, for example, sought to

hold on to Algeria because many European colonists had settled there. The country nonetheless achieved independence in 1962, but only after years of violence. In Zimbabwe, a large population of European settlers refused London's orders to move toward majority rule, resulting in a bloody civil war and delaying the country's independence until 1979.

On a quite different scale, political innovations also spread within independent countries. American politics abounds with examples of cultural diffusion. A classic case is the spread of suffrage for women, a movement that culminated in 1920 with the ratification of a constitutional amendment (**Figure 6.12** on the next page). Opposition to women's suffrage was strongest in the U.S. South, a region that later exhibited the greatest resistance to ratification of the Equal Rights Amendment and displayed the most reluctance to elect women to public office.

Federal statutes permit, to some degree, laws to be adopted in the individual functional subdivisions. In the United States and Canada, for example, each state and province enjoys broad lawmaking powers, vested in the legislative bodies of these subdivisions. The result is often a patchwork legal pattern that reveals the processes of cultural diffusion at work. A good example is the clean air

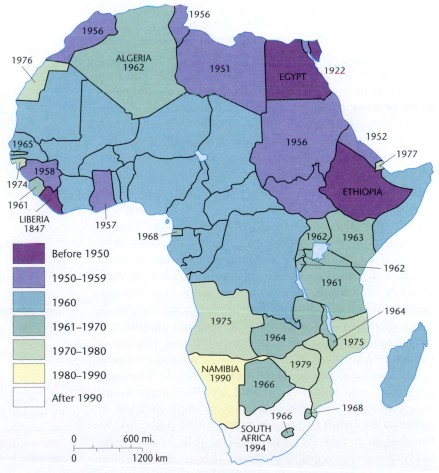

FIGURE 6.11 Independence from European colonial or white majority rule spread rapidly through Africa in the 1950s and 1960s. Prior to the 1950s, there were only three independent countries in Africa. Between 1951 and 1968, most others had attained self-rule. The remaining few gained freedom over the next three decades, so that by 1994 independence had spread from the Mediterranean to the Cape of Good Hope. Many of the ideas for national liberation were spread by African leaders educated in Europe.

THINKING GEOGRAPHICALLY Why do you think it took a few countries so much longer than most to gain independence?

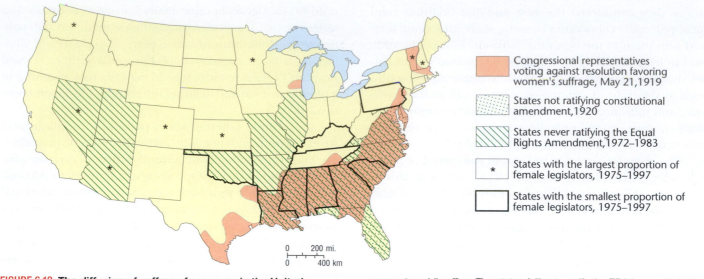

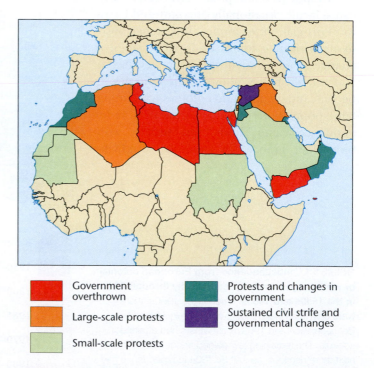

FIGURE 6.12 The diffusion of suffrage for women in the United States and of the Equal Rights Amendment. The suffrage movement achieved victory through a constitutional amendment in 1920. Both the suffrage movement and the campaign for an Equal Rights Amendment (ERA) for women failed to gain approval in the Deep South, an area that also lags behind most of the remainder of the country in the election of women to public office. The states failing to ratify the ERA lay mostly in the same area. The ERA movement did not succeed, in contrast to the earlier suffrage movement. (Source: Adapted in part from Paulin and Wright, 1932.)

THINKING GEOGRAPHICALLY What might be the barriers to diffusion in the Deep South?

movement, which began in California with the initiation of state legislation regulating automotive and industrial emissions. It later spread to other states and provided the model for clean air legislation at the federal level.

In recent decades, patterns in the initiation and organization of political change have begun to supersede state, provincial, and even national boundaries due to the increasing influence of the Internet. An important example of this shift in the diffusion of political activism and change began in December 2010 in Tunisia, where students and other dissatisfied citizens protested perceived state attempts to repress the population and to censor the Internet. These protests sparked a wave of similar demonstrations in many other Middle Eastern states, with the series of protests collectively coming to be known as the *Arab Spring*. The majority of these protests incorporated civil-resistance techniques such as strikes, marches, and rallies. These events were largely organized, and subsequently documented, primarily through social media on the Internet. The Internet allowed not only instantaneous communication among protestors but also the ability to raise awareness and garner support within specific countries and around the world. The ousting of national leaders in several Middle Eastern countries and significant governmental changes enacted in other states within the region attest to the power of the Internet as a tool for organizing people from diverse backgrounds for political purposes (**Figure 6.13**).

FIGURE 6.13 The Arab Spring. The Arab Spring movement resulted in the dismantling of several repressive Middle Eastern governments as well as the initiation of significant changes in many other states across the region.

THINKING GEOGRAPHICALLY Which type(s) of diffusion do you think were involved in the growth and spread of the Arab Spring?

The Internet and social media played a significant role in another mass political movement known as *Occupy Wall Street (OWS)*. In July 2011, taking inspiration from the Arab Spring, the anticonsumerism magazine *Adbusters* posted a suggestion on Twitter for a protest march in Lower Manhattan. This protest was intended to highlight perceived mass social injustices against the 99 percent of the population struggling with student debt, rising health-care costs, and foreclosure by the 1 percent of greedy and corrupt capitalists and government officials. In the months following this tweet, hundreds of thousands of protesters used Twitter, Facebook, YouTube, and other social networking sites to spread the OWS message globally. The capacity to stream live video on the Internet in conjunction with these social networking sites propelled the OWS movement to unprecedented global visibility. The Internet allowed organizers of and participants in the movement to get up-to-the-minute information about Occupy protests around the world; it also provided them with virtual places to talk about the images they were seeing. In this way, OWS established itself as a "leaderless movement" composed of individuals from many different political persuasions and national backgrounds.

Politics and Migration

Very often, political decisions or events provide the motivation for migration, both voluntary and forced. A contemporary example of political migration occurred in the Gambella region of western Ethiopia beginning in 2008. At that time, the state began to relocate tens of thousands of indigenous peoples living in this rich agricultural region to less fertile agricultural lands in other regions of the country. This migration was officially voluntary, with the government promising to compensate indigenous landowners and to provide health care and education in the new villages. However, many of the people who complied with the relocation program have faced food and water shortages and a lack of basic services in their new home. Cases of endemic hunger and starvation in these settlements have even been reported. Rather than face these conditions, many people fled to refugee camps in nearby Kenya.

The Ethiopian government has officially denied that the Gambella relocation is connected to the leasing of the vacated fertile lands at lucrative rates to foreign investors and wealthy Ethiopian businesspeople. Instead, state officials maintain that the program is designed to find alternative livelihoods for citizens who were not properly farming the fertile lands to grow cash crops. These differing claims have drawn the attention of international human rights watchdog groups, which have called for the suspension of the program until investigations of alleged threats, assaults, and arrests can be conducted. Further, some human-rights organizations fear that humanitarian funding to Ethiopia has been used, in part, to finance the Gambella displacements and to suppress growing political dissent. No matter how these disputes play out, it can be stated with certainty that political decisions to motivate, or forcibly conduct, mass migrations leave lasting imprints on the physical and cultural landscape.

POLITICAL ECOLOGY

What is the relationship between politics and the environment? How people use the land and natural resources is profoundly influenced by politics. Whether a particular habitat is conserved or degraded often has much to do with the structures of a country's land laws, tax codes, and agricultural policies. However, increasingly, politics is being defined by changing environmental conditions. National policies about environmental protection, guerrillas seeking a secure base for their operations, and the natural defense provided for an independent country by a surrounding sea all reveal an intertwining of environment and politics. How governments respond to ecological crises like the loss of biodiversity, pollution, and climate change has become an important political issue. Let's examine this complex two-way interaction between politics and the environment.

Chain of Explanation

When geographers Piers Blaikie and Harold Brookfield used the term *political ecology*, they were interested in trying to understand how political and economic forces affect people's relationships to the land. They suggested that focusing on proximate or immediate causes (for example, the farmer dumping pesticides in a river or the poor peasant cutting a patch of tropical forest) provided an inadequate and misleading explanation of human-environment relations. As an alternative, they developed the idea of a *chain of explanation* as a method for identifying ultimate causes. The chain of explanation begins with the individual land managers, the people with direct responsibility for land-use decisions—the farmers, timber cutters, firewood gatherers, or livestock keepers. The chain of explanation then moves up in spatial scale, tracing the land managers' economic, cultural, and political relationships from the local to the national and, ultimately, the global scale.

One of the primary areas addressed by the chain-of-explanation approach is the character of the state, particularly the way that national land laws, natural resource policies, tax codes, and credit policies influence land-use decision making. For example, if a state assesses high taxes on land improvements, such as terracing and channeling, its tax policies actually create disincentives for land managers to implement soil conservation measures. Conversely, if a state provides cheap loans to land managers to build such structures, its credit policies encourage soil conservation.

There are many examples of state influences on individual land-use decisions, leading Blaikie and Brookfield to argue that one cannot fully explain the causes of environmental problems without analyzing the role of the state.

Geopolitics and Folk Fortresses

Spatial variations in politics and the spread of political phenomena are often linked to terrain, soils, climate, natural resources, and other aspects of the physical environment. The term **geopolitics** was originally coined to describe the influence of geography and the environment on political entities. Conversely, established political authority can be a powerful instrument of environmental modification, providing the framework for organized alteration of the landscape and for environmental protection.

geopolitics The influence of geography and the habitat on political entities.

Before modern air and missile warfare, a country's survival was aided by some sort of natural protection, such as surrounding mountain ranges, deserts, or seas; bordering marshes or dense forests; or outward-facing escarpments. Political geographers named such natural strongholds **folk fortresses.** The folk fortress might shield an entire country or only its core area. In either case, it is a valuable asset. Surrounding seas have helped protect the British Isles from invasion for the past 900 years. In Egypt, desert wastelands to the east and west insulated the fertile, well-watered Nile Valley core. In the same way, Russia's core area was shielded by dense forests, expansive marshes, bitter winters, and vast expanses of sparsely inhabited lands. France—centered on the plains of the Paris Basin and flanked by mountains and hills such as the Alps, Pyrenees, Ardennes, and Jura along its borders—provides another good example (**Figure 6.14**).

folk fortress A stronghold area with natural defensive qualities, useful in the defense of a country against invaders.

Expanding countries often regard coastlines as the logical limits to their territorial growth, even if those areas belong to other peoples, as the drive across the United States from the East Coast to the Pacific Ocean in the first half of the nineteenth century made clear. U.S. expansion was justified by the doctrine of *manifest destiny*, which is based on the belief that the Pacific shoreline offered the logical and predestined western border for the country. A similar doctrine led Russia to expand in the directions of the Mediterranean and Baltic seas and the Pacific and Indian oceans.

The Heartland Theory

Discussions of environmental influence, manifest destiny, and Russian expansionism lead naturally to the **heartland theory** of Halford Mackinder. Propounded in the early twentieth century and based

heartland theory A 1904 proposal by Mackinder that the key to world conquest lay in control of the interior of Eurasia.

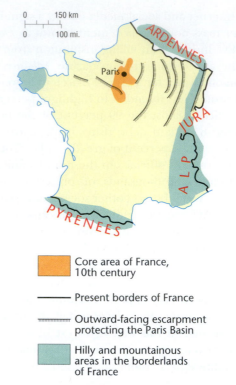

FIGURE 6.14 The distribution of landforms in France. Terrain features such as ridges, hills, and mountains offer protection. Outward-looking escarpments form a folk fortress that protected the core area and capital of France until as recently as World War I. Hill districts and mountain ranges lend stability to French boundaries in the south and southeast.

Legend:
- Core area of France, 10th century
- Present borders of France
- Outward-facing escarpment protecting the Paris Basin
- Hilly and mountainous areas in the borderlands of France

THINKING GEOGRAPHICALLY Would such landforms be effective today? Why?

on environmental determinism, the heartland theory addresses the balance of power in the world and the possibility of world conquest based on natural habitat advantage. It held that the Eurasian continent was the most likely base from which to launch a successful campaign for world conquest.

In examining this huge landmass, Mackinder discerned two environmental regions: the **heartland,** which lies remote from the ice-free seas, and the **rimland,** the densely populated coastal fringes of Eurasia in the east, south, and west (**Figure 6.15**). Far from the sea, the heartland was invulnerable to the naval power of rimland empires, but the cavalry and infantry of the heartland could spill out through diverse natural gateways and invade the rimland region. Mackinder thus reasoned that a unified heartland power could conquer the maritime countries with relative ease. He believed that the East European Plain would be the likely base for unification. As Russia had already unified

heartland The interior of a sizable landmass, removed from maritime connections; in particular, the interior of the Eurasian continent.

rimland The maritime fringe of a country or continent; in particular, the western, southern, and eastern edges of the Eurasian continent.

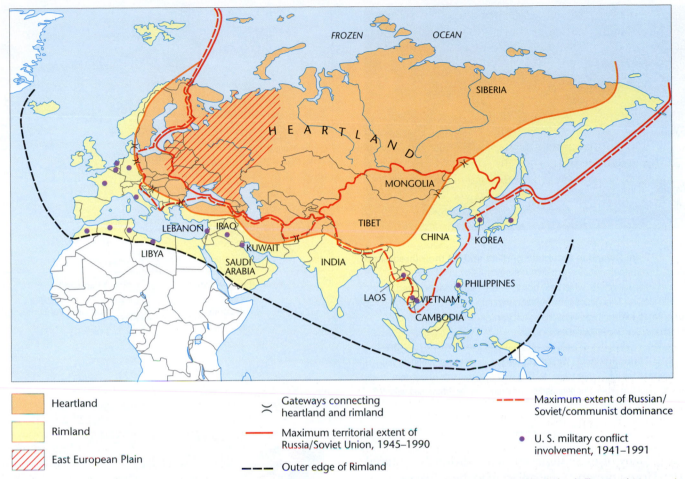

🟧	Heartland	⌇	Gateways connecting heartland and rimland	- - -	Maximum extent of Russian/ Soviet/communist dominance
🟨	Rimland	—	Maximum territorial extent of Russia/Soviet Union, 1945–1990	●	U. S. military conflict involvement, 1941–1991
▨	East European Plain	– – –	Outer edge of Rimland		

FIGURE 6.15 **Heartland versus rimland in Eurasia.** For most of the twentieth century, the heartland, epitomized by the Soviet Union and communism, was seen as a threat to the United States and the rest of the world, a notion based originally on the environmental deterministic theory of the political geographer Halford Mackinder. Control of the East European Plain would permit rule of the entire heartland, which in turn would be the territorial base for world conquest. During the cold war (1945–1990), the United States and its rimland allies sought to counter this perceived menace by a policy of containment—resisting every expansionist attempt by the heartland powers. (Sources: After Mackinder, 1904; Spykman, 1944.)

THINKING GEOGRAPHICALLY Has any force replaced the "heartland" in modern geopolitics?

this region at the time, Mackinder in effect predicted that the Russians would pursue world conquest.

Following Russia's communist revolution in 1917, the leaders of rimland empires and the United States employed a policy of containment. This policy, in no small measure, found its origin in Mackinder's theory and resulted in numerous wars to contain what was then considered a Russian-inspired conspiracy of communist expansion. Overlooked all the while were the fallacies of the heartland theory, particularly its reliance on the discredited doctrine of environmental determinism. In the end, Russia proved unable to hold together its own heartland empire, much less conquer the rimland and the world.

Warfare and Environmental Destruction

Of course, many political actions have an ecological impact, but perhaps none are as devastating as warfare.

Scorched earth—the systematic, intentional destruction of resources—has been a favored practice of retreating armies for millennia. However, perhaps even more devastating to local environments in war-torn regions are the *unintentional* effects wrought by war. For example, widespread deforestation and wildlife habitat destruction are often caused by the long-term use of heavy military equipment and explosives in delicate physical environments such as rainforests and deserts. This damage disrupts ecosystems for many years afterward, causing erosion, water-quality issues, and diminished food-production capacities.

Damage to fragile habitats has also been caused by the resettlement of refugees displaced by conflict. For example, for many years following the end of the Rwandan genocide of 1994, thousands of displaced refugees flowed out of refugee camps in search of land on which to resettle in order to feed and house their families in an already overpopulated

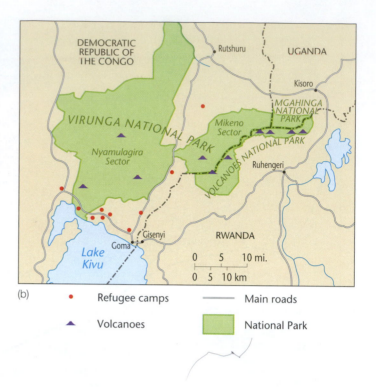

FIGURE 6.16 **(a) Rwandan mountain gorillas and (b) their threatened habitat.** The map shows the proximity of several Rwandan refugee camps to national parks, which are home to the highly endangered mountain gorilla. This geographic relationship has proven extremely detrimental to efforts to prevent the extinction of mountain gorillas. (a: © Ralph H. Bendjebar/DanitaDelimont.com.)

● Refugee camps	—— Main roads
▲ Volcanoes	▮ National Park

THINKING GEOGRAPHICALLY What obstacles to habitat preservation do you think organizations might face when working with governments in war-torn regions to establish protective boundaries and programs for threatened wildlife?

country. Many of these impoverished individuals ventured into vital forest preserve areas adjacent to the refugee camps, as there was little land available elsewhere (**Figure 6.16**). These preserved forests are home to highly endangered mountain gorilla populations that have suffered significant preservation setbacks as a result of human encroachment into their habitat.

Warfare can also cause damage to key regional infrastructure, which can lead to damage of local ecosystems. For example, damage to water pipes from bomb blasts or bullets can result in bacterial contamination and water leaks that reduce the amount of safe drinking water available for local residents. Water shortages due to damaged water systems can also lead to inadequate irrigation capacity on lands already damaged by heavy vehicles or rendered hazardous by the presence of landmines. Other infrastructure—such as local waste treatment, storage, and disposal facilities—can also be damaged by warfare, leading to soil and water contamination that impacts local residents and the environment in which they live long after the war ends. Clearly, warfare (especially modern, high-tech warfare) can be environmentally catastrophic. From an ecological standpoint, it does not matter who started or won a war. Everyone loses when such destruction occurs, given the global interconnectedness of the life-supporting ecosystem.

POLITICAL-**CULTURAL** INTERACTION

How are politics intertwined with the other diverse aspects of culture? Although we learn a great deal from studying how the physical environment and political phenomena interact, we gain an even broader perspective by examining the ties between politics and culture. The growth of independent countries, voting patterns, and other topics that interest political geographers are largely explained in cultural terms. In addition, political decisions often have far-reaching effects on the distribution of such cultural elements as economy, land use, and migration. Indeed, the political organization of territory, both past and present, is revealed to some degree in almost every facet of culture.

The Nation-State

The link between political and cultural patterns is epitomized by the **nation-state,** created when a nation—a people of common heritage, memories, myths, homeland, and culture; speaking the same language; and/or sharing a particular religious faith—achieves independence as a separate country. Nationality is culturally based in the nation-state, and the country's raison d'être lies in that

> **nation-state** An independent country dominated by a relatively homogeneous culture group.

cultural identity. The more the people have in common culturally, the more stable and potent is the resultant nationalism. Examples of modern nation-states include Germany, Sweden, Japan, Greece, Armenia, and Finland (**Figure 6.17**).

Scholars generally trace the nation-state model to Europe and the European settler colonies in the Americas of the late eighteenth century. Political philosophers of the time argued that *self-determination*—the freedom to rule one's own country—was a fundamental right of all peoples. The ideal of self-determination spread, and over the course of the nineteenth century, a globalized system of nearly 100 nation-states emerged. Many of these nation-states, such as Portugal, Holland, and France, ruled overseas territories as colonies of the mother country. Under the European Empire, self-determination did not apply to colonies. However, the United Nations incorporated the principle of national self-determination in its charter in 1945. This principle was adopted by independence movements in European colonies around the globe and led ultimately to decolonization.

Ethnic Separatism

Many independent countries—the large majority, in fact—are not nation-states but instead contain multiple national,

ethnic, and religious groups within their boundaries. India, Spain, and South Africa provide examples of older multiethnic countries. **Figure 6.18** on page 170 illustrates this point by dividing up South Africa linguistically.

Many, though not all, multiethnic countries came into being in the second half of the twentieth century. These were former colonies in Africa, South Asia, and Southeast Asia whose boundaries were a product of colonialism. European colonial powers drew political boundaries without regard to the territories of indigenous ethnic or tribal groups. These boundaries remained in place when the countries gained the right of self-determination. Although these states are often culturally diverse, they are sometimes plagued by internal ethnic conflict. What's more, members of a single, territorially homogeneous ethnic group may find themselves divided among different states by culturally arbitrary international borders.

Many independent countries function as nation-states because the political power rests in the hands of a dominant, nationalistic cultural group, whereas sizable ethnic minority groups reside in the national territory as second-class citizens. This creates a centrifugal force disrupting the country's unity. Many of the newest nation-states carved out of the former Soviet Union and Yugoslavia, such as

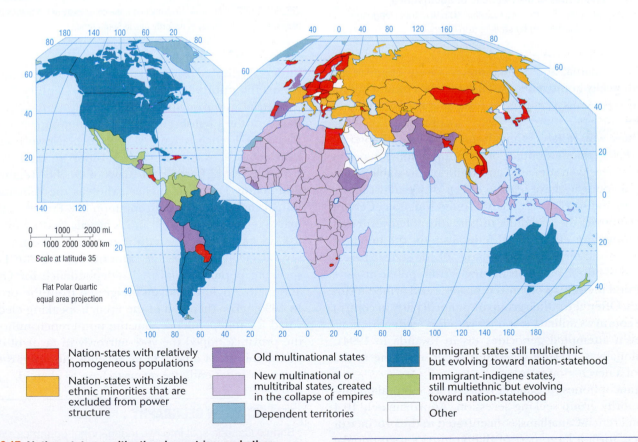

🟥 Nation-states with relatively homogeneous populations	🟪 Old multinational states	🟦 Immigrant states still multiethnic but evolving toward nation-statehood
🟧 Nation-states with sizable ethnic minorities that are excluded from power structure	🟪 New multinational or multitribal states, created in the collapse of empires	🟩 Immigrant-indigene states, still multiethnic but evolving toward nation-statehood
	🟦 Dependent territories	⬜ Other

FIGURE 6.17 Nation-states, multinational countries, and other types. This classification, as is true of all classifications, is arbitrary and debatable.

THINKING GEOGRAPHICALLY How would you change the classification, and why?

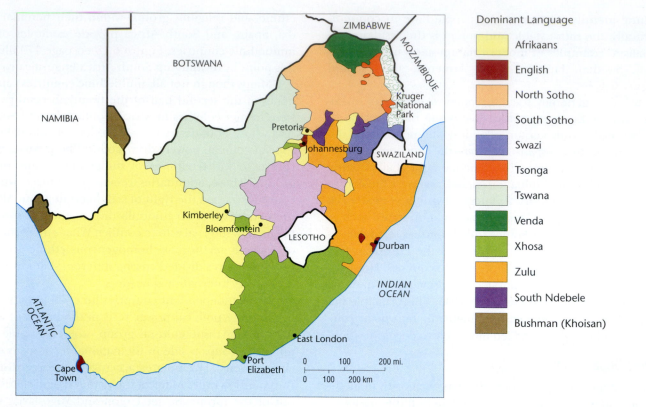

FIGURE 6.18 Languages of the Republic of South Africa, a multinational state. The mixture includes 10 native tribal languages and two languages introduced by settlers from Europe.

THINKING GEOGRAPHICALLY How might the complexity of languages have played a role in the delay of majority rule in South Africa?

Estonia, Armenia, and Serbia, are relatively linguistically and ethnically homogeneous. These states' cultural homogeneity represents a centripetal force that supports national unity.

Those who inhabit ethnic homelands (see Chapter 5) often seek greater autonomy or even full independence as nation-states. Even some old and traditionally stable multinational countries have felt the effects of separatist movements, including Canada and the United Kingdom. Certain other countries discarded the unitary form of government and adopted an ethnic-based federalism, in the hopes of preserving the territorial boundaries of the state. The expression of ethnic nationalism ranges from public displays of cultural identity to organized protests and armed insurgencies. Often the ethnic group or political party in control of the country's military responds with forced deportations and even attempted genocides (as in Rwanda in 1994). Occasionally, successful secessions occur, resulting in the birth of a new nation-state.

Francophones in Canada represent a cultural-linguistic minority group seeking secession. Approximately 10.4 million French-Canadians, concentrated in the province of Québec, form a large part of that country's total population of 34.8 million. Descended from French colonists who immigrated in the 1600s and 1700s, these Canadians lived under English or Anglo-Canadian rule and domination from 1760 until well into the twentieth century. Even the provincial government of Québec long remained in the hands of the English. A political awakening eventually allowed the French to gain control of their own homeland province, and as a result Québec differs in many respects from the rest of Canada. The laws of Québec retain a predominantly French influence, whereas the remainder of Canada adheres to English common law. French is the primary language of Québec and is heavily favored over English in provincial law, education, and government. In several elections, a sizable minority among the French-speaking population favored independence for Québec, and many Anglo-Canadians emigrated from the province. In 1995, more than half of the French-speaking electorate voted for independence, but the non-French minority in the province tipped the vote narrowly in favor of continued union with Canada. More recently, however, the campaign for independence seems to have weakened.

REFLECTING ON GEOGRAPHY

Should Canada split into two independent countries? What would be the advantages and disadvantages for an independent Québec if this split occurred?

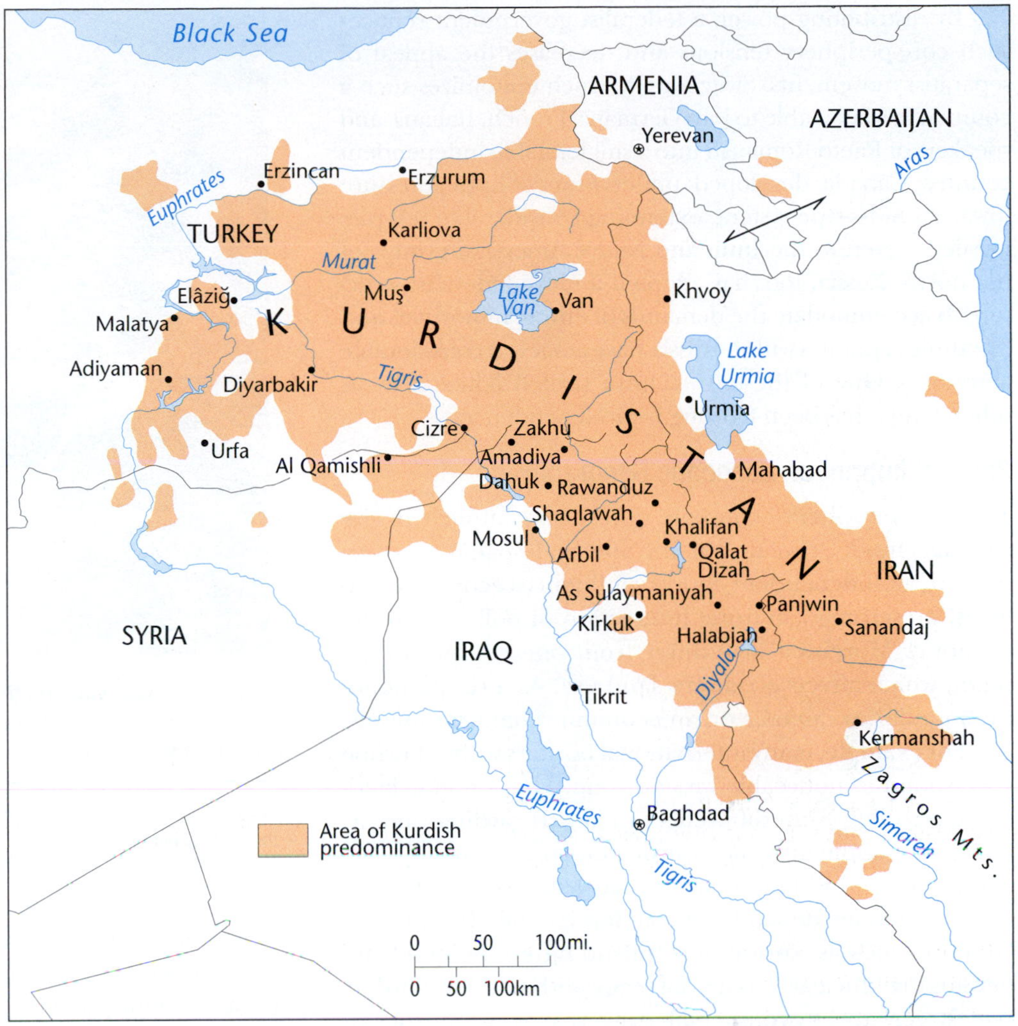

FIGURE 6.19 Kurdistan. This mountainous homeland of the Kurds now lies divided among several countries. The Kurds, numbering 30 million to 38 million, have lived in this region for millennia. They seek independence and have waged guerrilla war against Iran, Iraq, and Turkey, but so far independence has eluded them. (Source: Office of the Geographer, U.S. Department of State.)

THINKING GEOGRAPHICALLY What might cause so large and populous a nation to fail to achieve independence?

The Cleavage Model

Why do so many cultural minorities seek political autonomy or independence? The **cleavage model,** originally developed by Seymour Martin Lipset and Stein Rokkan to explain voting patterns in electoral geography, sheds light on this phenomenon. It proposes to explain persistent regional patterns in voting behavior (which, in extreme cases, can presage separatism) in terms of tensions pitting the national core area against peripheral districts, urban against rural, capitalists against workers, and the dominant culture against minority ethnic cultures. Frequently, these tensions coincide geographically: an urban core area monopolizes wealth and cultural and political power while ethnic minorities, excluded from the power structure, reside in peripheral, largely rural, and less-affluent areas.

cleavage model A political-geographic model suggesting that persistent regional patterns in voting behavior, sometimes leading to separatism, can usually be explained in terms of tensions pitting urban against rural, core against periphery, capitalists against workers, and power group against minority culture.

The great majority of ethnic separatist movements shown, particularly those that have moved beyond unrest to violence or secession, are composed of peoples living in national peripheries, away from the core area of the country. Every republic that seceded from the defunct Russian-dominated Soviet Union lay on the borders of that former country. Similarly, the Slovenes and Croats, who withdrew from the former Yugoslavia, occupied border territories peripheral to Serbia, which contained the former national capital of Belgrade. Kurdistan is made up of the peripheral areas of Iraq, Iran, Syria, and Turkey—the countries that currently rule the Kurdish lands (Figure 6.19). Slovakia, long poorer and more rural than Czechia (The Czech Republic) and remote from the center of power at Prague, became another secessionist ethnic periphery. In a few cases, the secessionist peripheries were actually more prosperous than the political core area, and the separatists resented the confiscation of their taxes to support the less affluent core. Slovenia and Croatia both occupied such a position in the former Yugoslavia.

By distributing power, a federalist government reduces such core-periphery tensions and decreases the appeal of separatist movements. Switzerland, which epitomizes such a country, has been able to join Germans, French, Italians, and speakers of Raeto-Romansh into a single, stable independent country. Canada developed under Francophone pressure toward a Swiss-type system, extending considerable self-rule privileges even to the Inuit and Native American groups of the north. Russia, too, has adopted a more federalist structure to accommodate the demands of ethnic minorities, and 31 ethnic republics within Russia have achieved considerable autonomy. One of these, Chechnya (called Ichkeria by its inhabitants), has been fighting for independence.

Political Imprint on Economic Geography

The core-periphery economic differences implicit in the cleavage model reveal that the internal spatial arrangement of an independent country influences economic patterns, presenting a cultural interaction of politics and the economy. Moreover, laws differ from one country to another, which affects economic land use. As a result, political boundaries can take on an economic character as well.

For example, many international borders within Europe correspond to noticeable variations on the economic landscape resulting from differences in laws regarding nuclear energy development. Some European states, such as France and the United Kingdom, believe nuclear power is the best way to alleviate dependence on foreign oil. In contrast, countries such as Norway and Ireland have refrained from developing nuclear energy facilities within their borders due largely to environmental concerns. Some European states that *have* developed nuclear energy facilities in the past, such as Germany and Spain, have committed themselves to phasing out nuclear power over the coming years because of these same environmental concerns. At present, however, the mosaic of nuclear energy production across European political boundaries clearly illustrates the ongoing interplay between politics and the economic landscape.

REFLECTING ON **GEOGRAPHY**

How might political decisions regarding the economics of nuclear energy production in one country impact land-use decisions in neighboring countries that have chosen *not* to develop nuclear energy production?

POLITICAL **LANDSCAPES**

How does politics influence landscape? How are landscapes politicized? The world over, national politics is literally written on the landscape. State-driven initiatives for frontier settlement,

FIGURE 6.20 Can you find the U.S.—Mexico border in this picture? The scene is near Mexicali, the capital of Baja California Norte. (Courtesy of Terry G. Jordan-Bychkov.)

THINKING GEOGRAPHICALLY Why does the cultural landscape so vividly reveal the political border?

economic development, and territorial control have profound effects on the landscape. Conversely, political writers and politicians look to landscape as a source of imagery to support or discredit political ideologies.

Imprint of the Legal Code

Many laws affect the cultural landscape. Among the most noticeable are those that regulate the land-survey system because they often require that land be divided into specific geometric patterns. As a result, political boundaries can become highly visible (**Figure 6.20**). In Canada, the laws of the French-speaking province of Québec encourage land survey in long, narrow parcels, but most English-speaking provinces, such as Ontario, use a rectangular system. Thus, the political border between Québec and Ontario can be spotted easily from the air.

Legal imprints can also be seen in the cultural landscape of urban areas. In Rio de Janeiro, height restrictions on buildings have been enforced for a long time. The result is a waterfront lined with buildings of uniform height (**Figure 6.21**). By contrast, most American cities have no height restrictions, allowing skyscrapers to dominate the central city. The consequence is a jagged skyline for cities such as San Francisco or New York City. Many other cities around the world lack height restrictions, such as Malaysia's Kuala Lumpur, which has the world's tallest skyscrapers.

(a)

(b)

FIGURE 6.21 **Legal height restrictions, or their absence, can greatly influence urban landscapes.** (a) Singapore, the city-state at the tip of the Malay Peninsula, lacks such restrictions, and spectacular skyscrapers punctuate its skyline. (b) In Rio de Janeiro, Brazil, by contrast, height restrictions allow the natural environment to provide the "high-rises." (a: Alamy; b: © Pixattitude/Dreamstime.com.)

THINKING GEOGRAPHICALLY Does your city have restrictions on building heights or architectural details? Explain.

Perhaps the best example of how political philosophy and the legal code are written on the landscape is the so-called *township and range system* of the United States. The system is the brainchild of Thomas Jefferson—an early U.S. president and one of the authors of the U.S. Constitution—who chaired a national committee on land surveying that resulted in the U.S. Land Ordinance of 1785. Jefferson's ideas for surveying, distributing, and settling the western frontier lands as they were cleared of Native Americans were based on a political philosophy of agrarian democracy. Jefferson believed that political democracy had to be founded on economic democracy, which in turn required a national pattern of equitable land ownership by small-scale

FIGURE 6.22 **Imposing order on the land.** This aerial view of Canyon County, Idaho, farmland reveals a landscape grid pattern. This pattern was imposed on the landscape across much of the United States following the passage of the 1785 Land Ordinance. (David Frazier/Image Works.)

THINKING GEOGRAPHICALLY Why did the United States government deem it necessary to impose a regular grid pattern on the land?

independent farmers. In order to achieve this agrarian democracy, the western lands would have to be surveyed into parcels that could then be sold at prices within reach of family farmers of modest means.

Jefferson's solution was the township and range system, which established a grid of square townships with 6-mile (9.6-kilometer) sides across the Midwest and West. Each of these was then divided into 36 sections of 1 square mile (2.6 square kilometers), which were in turn divided into quarter-sections, and so on. Sections were to be the basic landholding unit for a class of independent farmers. Townships were to provide the structure for self-governing communities responsible for public schools, policing, and tax collection. With the exception of the 13 original colonies and a few other states or portions of states, a gridlike landscape was imposed on the entire country as a result of Jefferson's political philosophy and accompanying land-survey system (**Figure 6.22**).

Physical Properties of Boundaries

Demarcated political boundaries can also be strikingly visible, forming border landscapes. Political borders are usually most visible where restrictions limit the movement of

FIGURE 6.23 National borders. Even peaceful, unpoliced international borders often appear vividly in the landscape. Sweden and Norway insist on cutting a swath through the forest to mark their common boundary. (Courtesy of Terry G. Jordan-Bychkov.)

THINKING GEOGRAPHICALLY Why would Sweden and Norway do this?

FIGURE 6.24 **The Great Wall of China.** The Great Wall is probably the most spectacular political landscape feature ever created. The wall, which is 1500 miles (2400 kilometers) long, was constructed over many centuries by the Chinese in an ultimately unsuccessful attempt to protect their northern boundary from adjacent tribes of nomadic herders. (Courtesy of Terry G. Jordan-Bychkov.)

THINKING GEOGRAPHICALLY Can you think of comparable modern structures?

people between neighboring countries. Sometimes such boundaries are even lined with cleared strips, barriers, pillboxes, tank traps, and other obvious defensive installations. At the opposite end of the spectrum are international borders, such as that between Tanzania and Kenya in East Africa, that are unfortified, thinly policed, and in many places very nearly invisible. Even so, undefended borders of this type are usually marked by regularly spaced boundary pillars or cairns, customhouses, and guardhouses at crossing points (**Figure 6.23**).

The visible aspect of international borders is surprisingly durable, sometimes persisting centuries or even millennia after the boundary becomes obsolete. Ruins of boundary defenses, some dating from ancient times, are common in certain areas. The Great Wall of China is probably the best-known reminder of past boundaries (**Figure 6.24**). Hadrian's Wall in England, which marks the northern border during one stage of Roman occupation and parallels the modern border between England and Scotland, is a similar reminder.

The Impress of Central Authority

Attempts to impose centralized government appear in many facets of the landscape. Railroad and highway patterns focused on the national core area, and radiating outward like the spokes of a wheel to reach the hinterlands of the country, provide good indicators of central authority. In Germany, the rail network developed largely before unification of the country in 1871. As a result, no focal point stands out. However, the superhighway system of autobahns, encouraged by Hitler as a symbol of national unity and power, tied the various parts of the Reich to such focal points as Berlin or the Ruhr industrial district.

The visibility of provincial borders within a country also reflects the central government's strength and stability. Stable, secure countries, such as the United States, often permit considerable display of provincial borders. Displays aside, such borders are easily crossed. Most state boundaries within the United States are marked with signboards or other features announcing the crossing. By contrast, unstable countries, where separatism threatens national unity, often suppress such visible signs of provincial borders. Also in contrast, such "invisible" borders may be exceedingly difficult to cross when a separatist effort involves armed conflict.

REFLECTING ON **GEOGRAPHY**

What visible imprints of the Washington, D.C.–based central government can be seen in the political landscape of the United States?

National Iconography on the Landscape

The cultural landscape is rich in symbolism and visual metaphor, and political messages are often conveyed through such means. Statues of national heroes or heroines and of symbolic figures such as the goddess Liberty or Mother Russia form important parts of the political landscape, as do assorted monuments. The elaborate use of national colors can be visually powerful as well. **Figure 6.25** illustrates this form of national iconography in the Netherlands during a Queen's Day celebration, on which many

FIGURE 6.25 Queen's Day in the Netherlands. Queen's Day is the annual Dutch national holiday in honor of the late Queen Juliana's birthday. In celebrations throughout the Netherlands, citizens dress in bright orange (the national color) and attend parades, concerts, and other public events to celebrate Dutch nationalism. (Superstock/agefotostock.)

THINKING GEOGRAPHICALLY Why do you think that citizens of many countries choose to partake in nationalistic celebrations that make a visible imprint on the cultural landscape?

citizens dress in orange, the Dutch national color. Landscape symbols such as the Rising Sun flag of Japan, the Statue of Liberty in New York harbor, and the Latvian independence pillar in Riga (which stood untouched throughout half a century of Russian-Soviet rule) evoke deep patriotic emotions. The sites of heroic (if often futile) resistance against invaders, as at Masada in Israel, prompt similar feelings of nationalism. A memorial to the victims of the Japanese occupation of Nanjing, China, provides another example of this type of iconographic imprint of the cultural landscape (**Figure 6.26**).

FIGURE 6.26 A memorial to victims of the Nanjing Massacre in China. Beginning in December, 1937, Japanese troops occupying the (then) capital city of China killed an estimated 300,000 civilians and unarmed soldiers in a brutal spree. To this day, this event lives on in the Chinese national memory. (Han Yuqing/Xinhua/ZUMAPRESS.com.)

THINKING GEOGRAPHICALLY Can you think of some examples of iconographic memorials on the U.S. landscape that mark national tragedies?

Some geographers theorize that the political iconography of landscape derives from an elite, dominant group in a country's population and that its purpose is to legitimize or justify its power and control over an area. The dominant group seeks both to rally emotional support and to arouse fear in potential or real enemies. As a result, the iconographic political landscape is often controversial or contested, representing only one side of an issue. Consider Mount Rushmore in the Black Hills of South Dakota, which features the carved heads of four U.S. presidents. The Black Hills are sacred to the Native Americans who controlled the land before whites seized it. How might these Native Americans, the Lakota Sioux, perceive this monument? Are any other political biases contained in it? Cultural landscapes are always complicated and subject to differing interpretations and meanings, and political landscapes are no exception.

Key Terms

buffer state, 152
centrifugal force, 155
centripetal force, 155
cleavage model, 171
colonialism, 152
core area, 161
electoral geography, 158
enclave, 152
ethnographic boundary, 153
European Union (EU), 155

exclave, 152
federal state, 154
folk fortress, 166
geometric boundary, 153
geopolitics, 166
gerrymandering, 159
heartland, 166
heartland theory, 166
intergovernmental
   organization, 156

Political Geography on the Internet

You can learn more about political geography on the Internet at the following web sites:

European Union

http://europa.eu

Here you can find information about the 27-member supranational organization that is increasingly reshaping the internal political geography of Europe.

Forced Migration Online

http://www.forcedmigration.org

Explore a growing collection of resources relating to refugees and forced migration, assembled and managed by the Refugee Studies Centre at the University of Oxford. The collection includes searchable full-text documents, podcasts of lectures from the Refugee Studies Centre relating to forced migration, and a discussion forum for the exchange of information regarding refugees and internal displacement.

International Boundary News Archive

http://www-ibru.dur.ac.uk/resources/newsarchive.html

This database contains more than 10,000 boundary-related reports from a wide range of news sources around the world dating from 1991 to March 2001, with additional reports from 2006 onward.

Political Geography Specialty Group, Association of American Geographers

http://www.politicalgeography.org/

This site provides details about the activities and meetings of specialists in political geography and includes useful links to other sites featuring political geography and geopolitics.

United Nations

http://www.un.org

Search the worldwide organization with a membership that includes the large majority of independent countries. The site contains politically diverse information about such ventures as peacekeeping and conflict resolution.

Recommended Books on Political Geography

Agnew, John. 1998. *Geopolitics: Re-Visioning World Politics*. London: Routledge. A leading political geographer critically examines the historical European perspective on world politics and shows how that vision of world order continues to influence geopolitical thinking.

Agnew, John. 2002. *Making Political Geography*. London: Arnold. This book provides an excellent overview of the field of political geography, highlighting the contributions of key thinkers from the nineteenth century to the present.

Brauer, Jurgen. 2009. *War and Nature: The Environmental Consequences of War in a Globalized World*. Lanham, MD: AltaMira Press. This volume explores the hidden costs of violent conflict on the environment by drawing on examples from Vietnam, the Persian Gulf, Central Africa, and Afghanistan.

Cohen, Saul Bernard. 2003. *Geopolitics of the World System*. Lanham, MD: Rowman & Littlefield. A leading political geographer of the twentieth century looks at the post–cold war world system of the early twenty-first century to explain its political geography.

Herb, Guntram H., and David H. Kaplan (eds.). 1999. *Nested Identities: Nationalism, Territory, and Scale*. Lanham, MD.: Rowman & Littlefield. This collection of essays by 14 leading political geographers focuses on the geographical issue of territoriality using case studies of troubled countries and regions at different scales.

O'Loughlin, John (ed.). 1994. *Dictionary of Geopolitics*. Westport, CT.: Greenwood Press. This is a basic reference book on political geography.

Olwig, Kenneth. 2002. *Landscape, Nature, and the Body Politic: From Britain's Renaissance to America's New World*. Madison: University of Wisconsin Press. This is an impressively researched historical study of the importance of landscape in shaping the ideas of nation and national identity in England and the United States.

Wallerstein, Immanuel. 1991. *Geopolitics and Geoculture: Essays on the Changing World-System*. Cambridge: Cambridge University Press. This collection of Wallerstein's essays links the collapse of the Soviet Union to the end of U.S. hegemony around the world.

Political-Geographical Journals

Geopolitics. This journal explores contemporary geopolitics and geopolitical change with particular reference to territorial problems and issues of state sovereignty. Published by Frank Cass. Volume 1 appeared in 1996.

Political Geography. This is a journal devoted exclusively to political geography. Formerly titled *Political Geography Quarterly*, the journal changed its name in 1992. Published by Elsevier. Volume 1 appeared in 1982.

Space and Polity. This journal is dedicated to understanding the changing relationships between the state and regional/local forms of governance. It highlights the work of scholars whose research interests lie in studying the relationships among space, place, and politics. Published by Carfax Publishing. Volume 1 appeared in 1997.

Answers to Thinking Geographically Questions

Figure 6.1: Some possible suggestions include independence for the Palestinians, Scotland, and Chechnya, although such developments are very speculative. The status of Taiwan also remains uncertain.

Figure 6.2: Elongation and fragmentation (including exclaves) make communication among the various parts of the country difficult. The different parts of the country may be physically, ethnically, culturally, and economically very different. These factors make unification difficult. Enclaves introduce a foreign force into the middle of a country, leading to potential conflict.

Figure 6.3: Although the future is impossible to predict with certainty, the demise of the Berlin Wall shows how impermanent such barriers can be.

Figure 6.4: More and more countries are joining together in various ways and in various degrees for common purposes. The idea of doing so is spreading to more parts of the world.

Figure 6.5: The (former) Soviet Union (USSR), which controlled Europe's Eastern Bloc countries, disbanded in 1991. In the years following this dissolution, many of the newly independent states of Eastern Europe sought to join the EU as a means of strengthening economic ties and trade relationships with more powerful and affluent western European states.

Figure 6.6: Many formerly communist areas continue to vote socialist or leftist (note especially Germany). Major exceptions appear in eastern Poland and western Ukraine.

Figure 6.7: Districts may be gerrymandered to give advantage to a particular party (as was the case in the original gerrymander in Massachusetts) or to give an incumbent an advantage for reelection.

Figure 6.8: "Red" states tended to be in the central part of the country (except for the southern Rocky Mountains) and in the Southeast (except for Virginia, North Carolina, and Florida). "Blue" states were along the Pacific Coast, the southern Rocky Mountains, the Northeast with a southward extension along the Atlantic Coast, and Florida.

Figure 6.9: Very few areas voted 100 percent for either party. Variations within states show up on this map, for example, the strongly Democratic votes in the lower Mississippi Delta of Mississippi and Arkansas, the "black belt" of Alabama and Mississippi, and the Indian reservations of the Southwest. Nebraska and Kansas show more strongly Republican voting in their western parts than in their eastern part.

Figure 6.10: To the west were stronger powers to resist Russian expansion; the east was largely uninhabited, and peoples who lived there were small in number and had little or no sophisticated military technology. Goals included acquisition of a warm-water seaport and sources of furs, a valuable resource for a country in a cold climate.

Figure 6.11: Namibia was under the rule of South Africa, where a powerful, entrenched white minority, who controlled most of the economy as well as the government, long resisted black majority rule for Namibia or South Africa. The dictatorial government of Portugal was also loath to allow its colonies of Angola and Mozambique to gain independence. Zimbabwe also had a powerful white population that did not want to allow black majority rule.

Figure 6.12: There continues to be a more traditional attitude, some of it derived from religious beliefs, about appropriate roles for men and women in the Deep South.

Figure 6.13: Over the course of the Arab Spring movement, both hierarchical and contagious diffusion occurred. The geographic proximity of the countries involved indicates that contagious forces were at work in spreading the revolutionary political ideas of the movement. However, the importance of the Internet and social media in spreading these ideas through virtual space also highlights the operation of hierarchical diffusion.

Figure 6.14: Military aircraft have diminished the effectiveness of landforms in forming a folk fortress, as the French learned in

World War I and World War II. Nevertheless, even if an initial assault is made by air, ground troops must still secure the conquest.

Figure 6.15: Americans today are concerned with terrorists, whose center is perceived to be in Southwest Asia. However, terrorist operations are more widely scattered and do not display the kind of expansion from a heartland that Mackinder's model describes.

Figure 6.16: Answers will vary, but it is likely that habitat-preservation organizations would encounter resistance from governments attempting to provide assistance to refugees and other war-ravaged populations. These individuals urgently need land for housing and food production in addition to medical care, educational services, and other assistance. In the minds of government officials, these needs would likely take precedence over the protection of wildlife and the provision of scarce lands for their survival.

Figure 6.17: Among the possibilities are the status of Canada because of the separatism of Québec and the status of North Korea, which is ethnically homogeneous.

Figure 6.18: Communication, and therefore organization to fight apartheid, among various African tribal groups would be hindered by their different languages.

Figure 6.19: The Kurds are a minority in each of the countries where they live, and none of the countries wants to lose its territory, which also includes the location of valuable resources: the sources of the Tigris and Euphrates Rivers oil deposits.

Figure 6.20: The cultural element here is the land survey system, which produces the dominant pattern of square fields and ownership parcels on the United States side.

Figure 6.21: Depending on where you live, there may be restrictions on heights or on architectural styles that are deemed incompatible with the history and traditions of the place.

Figure 6.22: Surveying the land in regular north-south and east-west lines, forming squares, made it easy to describe and thus to sell land to settlers. Settlement of the frontier was an important way to tame it and to bring the land firmly within the control of the United States.

Figure 6.23: Even neighbors at peace need to know exactly where their territory begins and ends because modern governments have great responsibilities throughout their territories.

Figure 6.24: Examples include the wall between the United States and Mexico and between Israel and the occupied Palestinian territories of the West Bank. The Berlin Wall and other walls between East and West Germany, now demolished, are other examples.

Figure 6.25: Answers will vary, but you may consider that the political and cultural ideas of a nation are symbolically reinforced when people interact through shared celebrations that involve dressing in national colors or costumes, displaying national flags or other symbols, or otherwise exhibiting membership in a particular nation. National identity is strengthened and perpetuated and a sense of belonging is created.

Figure 6.26: Answers will vary, but you may think of the Pearl Harbor memorial in Hawaii, the memorial to victims of the Oklahoma City bombing, and the 9/11 memorial at Ground Zero in New York City.

Parking lot shrine to the Virgin of Guadalupe. Self-Help Graphics and Art, East Los Angeles. (Courtesy of Patricia L. Price.)

The Geography of Religion

7

Religion is a core component of culture, lending vivid hues to the human mosaic. For many, religion is the most profoundly felt dimension of their identity. For this reason it is important to clearly state what is meant by the term and provide a sense of the many ways that religion can be manifest in people's lives. **Religion** can be defined as a relatively structured set of beliefs and practices through which people seek mental and physical harmony with the powers of the universe. The rituals of religion provide milestones along the course of our lives—birth, puberty, marriage, having children, and death—that are observed and celebrated. Religions often attempt not only to accommodate but also to influence the awesome forces of nature, life, and death. Religions help people make sense of their place in the world.

> **religion** A relatively structured set of beliefs and practices through which people achieve mental and physical harmony with the universe and attempt to accommodate or influence the forces of nature, life, and death.

179

Religion goes beyond a merely pragmatic set of rules for dealing with life's joys and sorrows. Most religions incorporate a sense of the supernatural that can be manifest in the concept of a God or gods who play a role in shaping human existence, in the notion of an afterlife that may involve a place of rest (or torment) for those who have died, or ideas of a soul that exists apart from our physical body and which may be released, or even born again, once we have died. This sense of the otherworldly is often spatially demarcated through the designation of sacred spaces, such as cemeteries, religious buildings, and sites of encounters with the supernatural.

Each of the world's major religions is organized according to more or less standardized practices and beliefs, and each is practiced in a similar fashion by millions, even billions, of adherents worldwide. Yet many people also express their religious faith in individual ways. Rituals and prayers can be adapted to fit particular circumstances or performed at home alone. Some religions, including the Taoic religions of East Asia, as well as Hinduism and Buddhism, are largely individual or family-oriented practices. Some people do not observe a widely recognized religion at all. They may be secular, holding no religious beliefs, or express skepticism—even hostility—toward organized religion. They may consider themselves to be faithful but not follow an organized expression of their beliefs. Or they may practice an unconventional belief system, or cult. The term *cult* is often used in a pejorative sense because it conjures images of mind control, mass suicide, and extreme veneration of a human leader. It is important to keep in mind that the practitioners of belief systems falling outside the mainstream—for example, Mormonism, Scientology, and even Alcoholics Anonymous—strongly object to being labeled cult members.

CLASSIFYING RELIGIONS

Different types of religion exist in the world. One way to classify them is to distinguish between proselytic and ethnic faiths. **Proselytic religions,** such as Christianity and Islam, actively seek new members and aim to convert all humankind. For this reason, they are sometimes also referred to as **universalizing religions.** They instruct their faithful to spread the Word to all the Earth, using persuasion and sometimes violence to convert the "heathen." The colonization of peoples and their lands is sometimes a result of the desire to convert them to the conqueror's reli-

proselytic religion A religion that actively seeks new members and aims to convert all humankind.

universalizing religion Also called proselytic religion, it expands through active conversion of new members and aim to encompass all of humanity.

gion. By contrast, each **ethnic religion** is identified with a particular ethnic or tribal group and does not seek converts. Judaism provides an example. At its most basic, a Jew is anyone born of a Jewish mother. Though a person can convert to Judaism, it is a complex process that has traditionally been discouraged. Proselytic religions sometimes grow out of ethnic religions—the evolution of Christianity from its parent Judaism is a good example.

ethnic religion A religion identified with a particular ethnic or tribal group; it does not seek converts.

Another distinction among religions is the number of gods worshipped. **Monotheistic religions,** such as Islam, Christianity, and Judaism, believe in only one God and may expressly forbid the worship of other gods. **Polytheistic religions** believe there are many gods or spirits. For example, Vodun (also spelled Voudou in Haiti or Voodoo in the southern United States) is a West African religious tradition with adaptations in the Americas wherever the enslavement of Africans was once practiced. Though, as with most major religions, there is one supreme God, it is the hundreds of spirits, or *iwa*, that Vodun adherents turn to in times of need.

monotheism The worship of only one god.

polytheism The worship of many gods.

Finally, the distinction between syncretism and orthodoxy is important. **Syncretic religions** combine elements of two or more different belief systems. Umbanda, a religion practiced in parts of Brazil, blends elements of Catholicism with a reverence for the souls of Indians, wise men, and historical Brazilian figures, along with a dash of nineteenth-century European *spiritism*, which is a set of beliefs about contacting spirits through mediums. Caribbean and Latin American religious practices often combine elements of European, African, and indigenous American religions. Sometimes, in order to continue practicing their religions, people in this region would hide statues of Afrocentric deities within images of Catholic saints. Or they would determine which Catholic figures were most like their own deities. Note in **Figure 7.1** the equation of Danbala, the snake-god of Haitian Voudou, with Saint Patrick, who is also associated with snakes.

syncretic religion A religion or strands within a religion, that combines elements of two or more belief systems.

Orthodox religions, by contrast, emphasize purity of faith and are generally not open to blending with elements of other belief systems. The word *orthodoxy* comes from Greek and literally means "right" (*ortho*) "teaching"

orthodox religion Strand within most major religions that emphasizes purity of faith and is not open to blending with elements of other belief systems.

(a)   (b)

FIGURE 7.1 **(a) Danbala and (b) Saint Patrick.** (a) Danbala in Voudou is parallel to (b) the Catholic Saint Patrick; both are associated with snakes. (a: Louise Batalla Duran/Alamy; b: Mary Evans Picture Library.)

THINKING GEOGRAPHICALLY Why do snakes figure in so many religious traditions?

(*doxy*). Many religions, including Christianity, Judaism, Hinduism, and Islam, have orthodox strains. So, for instance, although some orthodox Jews closely follow a strict interpretation of the Oral Torah (a specific version of the Jewish holy book), moderate but observant Jews may observe only some or perhaps none of the dietary, marriage, and worship proscriptions observed by orthodox Jews. Intolerance of other

> **fundamentalism** A movement to return to the founding principles of a religion, which can include literal interpretation of sacred texts or the attempt to follow the ways of a religious founder as closely as possible.

religions, or of those fellow believers not seeming to follow the "proper" ways, is associated with fundamentalism rather than orthodoxy. Many who consider themselves orthodox are in fact quite tolerant of other beliefs. **Fundamentalism** is a movement to return to the founding principles of a religion,

which can include literal interpretation of sacred texts or the attempt to follow the ways of a religious founder as closely as possible.

RELIGIOUS CULTURE **REGIONS**

What is the spatial patterning of religious faiths? Because religion, like all of culture, has a strong territorial tie, religious culture regions abound. The most basic kind of formal religious culture region depicts the spatial distribution of organized religions (**Figure 7.2** on the next page). Some parts of the world exhibit an exceedingly complicated pattern of religious adherence, and the boundaries of formal religious culture regions, like most

MAJOR RELIGIONS

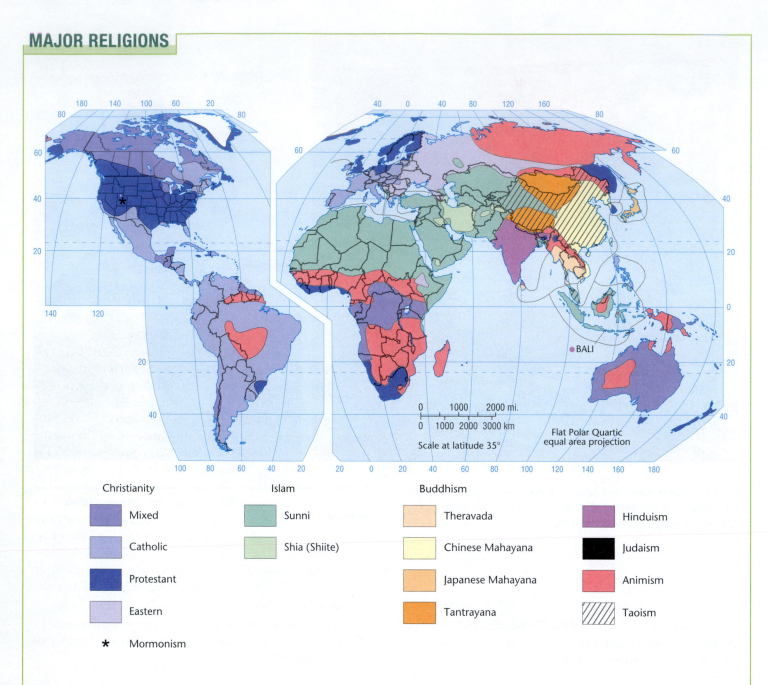

Christianity

- Mixed
- Catholic
- Protestant
- Eastern
- ✱ Mormonism

Islam

- Sunni
- Shia (Shiite)

Buddhism

- Theravada
- Chinese Mahayana
- Japanese Mahayana
- Tantrayana
- Hinduism
- Judaism
- Animism
- Taoism

FIGURE 7.2 **The world distribution of major religions.** Much overlap exists that cannot be shown on a map of this scale. The attempt is to show which faith is dominant. Animism includes a wide array of diverse belief systems. Taoistic areas are cross-hatched to show the overlap with Buddhism in these areas. Mixed Christianity means that none of the three major branches of that faith has a majority.

THINKING GEOGRAPHICALLY What kinds of areas are dominantly animist? How does this relate to patterns of language and the distribution of indigenous peoples?

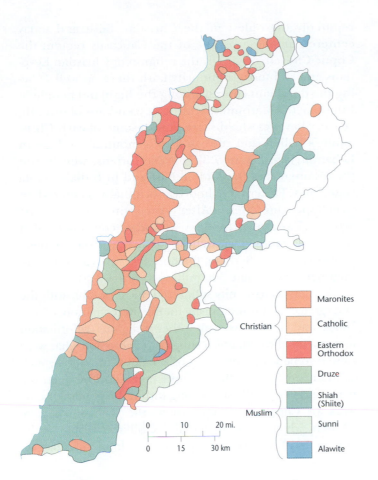

FIGURE 7.3 **Distribution of religious groups in Lebanon.** A land torn by sectarian warfare in recent times, Lebanon is one of the most religiously diverse parts of the world. Overall, Lebanon is today 34 percent Christian, 27 percent Shiite Muslim, 27 percent Sunni Muslim, 5 percent Druze, and 7 percent "irreligious." Unshaded areas are largely uninhabited. (Source: Derived, with changes, from Klaer, 1996.)

THINKING GEOGRAPHICALLY What is the impact of the concentration of Muslims in the southern part of Lebanon, along the border with Israel?

cultural borders, are rarely sharp (**Figure 7.3**). Persons of different faiths often live in the same province or town.

Judaism

Judaism is a 4000-year-old religion and the first major monotheistic faith to arise in southwestern Asia. It is the parent religion of Christianity and is also closely related to Islam. Jews believe in one God who created humankind for the purpose of bestowing kindness on them. As with Islam and Christianity, people are rewarded for their faith, are punished for violating God's commandments, and can atone for their sins. The Jewish holy book, or *Torah*, comprises the first five books of the Hebrew Bible. In contrast to the other monotheistic faiths, Judaism is not proselytic and has remained an ethnic religion through most of its existence.

Judaism has split into a variety of subgroups, partly as a result of the *Diaspora*, a term that refers to the forced dispersal of the Jews from Palestine in Roman times and the subsequent loss of contact among the various colonies. Jews, scattered to many parts of the Roman Empire, became a minority group wherever they went. In later times, they spread throughout much of Europe, North Africa,

and Arabia. Those Jews who lived in Germany and France before migrating to central and eastern Europe are known as the *Ashkenazim*; those who never left the Middle East and North Africa are called *Mizrachim; and* those from Spain and Portugal are known as *Sephardim.* Spain expelled its Sephardic Jews in 1492, the same year that Christopher Columbus set sail for the New World. It was not until the quincentennial of both events, in 1992, that the Spanish government issued an official apology for the expulsion.

The late nineteenth and early twentieth centuries witnessed large-scale Ashkenazic migration from Europe to America. The Holocaust that befell European Judaism during the Nazi years involved the systematic murder of perhaps a third of the world's Jewish population, mainly Ashkenazim. Europe ceased to be the primary homeland of Judaism, and many of the survivors fled overseas, mainly to the Americas and later to the newly created state of Israel. Today, Judaism has about 13 million adherents throughout the world. At present, roughly 6.5 million, or about half, live in North America, and slightly more than 5 million live in Israel.

Christianity

Christianity, a proselytic faith, is the world's largest religion, both in area covered and in number of adherents, claiming about a third of the global population (**Figure 7.4** on page 184). Christians are monotheistic, believing that God is a Trinity consisting of three persons: the Father, the Son, and the Holy Spirit. Jesus Christ is believed to be the incarnate Son of God who was given to humankind for the sake of human redemption some 2000 years ago. Through Jesus' death and resurrection, all of humanity is offered redemption from sin and provided eternal life in heaven and a relationship with God.

Christianity, Islam, and Judaism are the world's three great monotheisms and share a common culture hearth in southwestern Asia (see Figure 7.12, page 190). All three

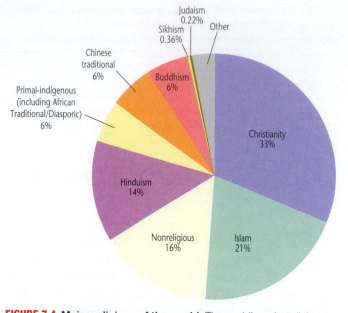

FIGURE 7.4 **Major religions of the world.** The world's major religions, by numbers of adherents expressed as a percentage of world population. Note that although Judaism is a prominent religion in the United States, it does not rank among the large religions of the world in terms of total number of adherents. Also note that because some religions overlap and because of rounding, percentages do not add up to 100. (Source: Derived from 2000 *Britannica Book of the Year; World Almanac and Book of Facts 2001.*)

THINKING GEOGRAPHICALLY Why does Hinduism, an ethnic religion, have such a large number of adherents?

religions venerate the patriarch Abraham and thus can be called Abrahamic religions. Because Judaism is the parent religion of Christianity, the two faiths share many elements, including the Torah, the first five books of what Christians call the Old Testament (which forms part of the Christian Bible or holy book); prayer; and a clergy. This is the basis for the term *Judeo-Christian*, which is used to describe beliefs and practices shared by the two faiths. Because Jesus was born into the Jewish faith, many Christians still accord Jews a special status as a chosen people and see Christianity as the natural continuation or fulfillment of Judaism.

Christianity has long been fragmented into separate branches (see Figure 7.2). The major division is threefold, made up of Roman Catholics, Protestants, and Eastern Christians. Western Christianity, which now includes Catholics and Protestants, was initially identified with Rome and the Latin-speaking areas that are largely congruent with modern western Europe, whereas the Eastern Church dominated the Greek world from Constantinople (now the city of Istanbul, Turkey). Belonging to the Eastern group are the Armenian Church,

reputedly the oldest in the Christian faith and today centered among a people of the Caucasus region; the Coptic Church, originally the religion of Christian Egyptians and still today a minority faith there, as well as being the dominant church among the highland people of Ethiopia; the Maronites, Semitic descendants of seventh-century heretics who disagreed with some of early Christianity's beliefs and retreated to a mountain refuge in Lebanon (see Figure 7.3); the Nestorians, who live in the mountains of the Middle East and in India's Kerala state; and Eastern Orthodoxy, originally centered in Greek-speaking areas. After converting many Slavic groups, Eastern Orthodoxy is today composed of a variety of national churches, such as Russian, Greek, Ukrainian, and Serbian Orthodoxy, with a collective membership of some 300 million (**Figure 7.5**).

Western Christianity splintered, most notably with the emergence of Protestantism in the Reformation of the 1400s and 1500s. As with other religious reformation movements, Protestantism sought to overcome what were viewed to be wrong practices of Roman Catholicism, such as the need for priestly or saintly mediation between humans and God, while still staying within the general framework of Christianity. Since then, the Roman Catholic Church, which alone included 1.2 billion people at the end of 2010, or about 15 percent of humanity, has re-

FIGURE 7.5 **Greek Orthodox monastery.** The bells of St. John's Monastery in Patmos, Greece. (Courtesy of Patricia L. Price.)

THINKING GEOGRAPHICALLY How do the height and durability of this structure demonstrate its importance?

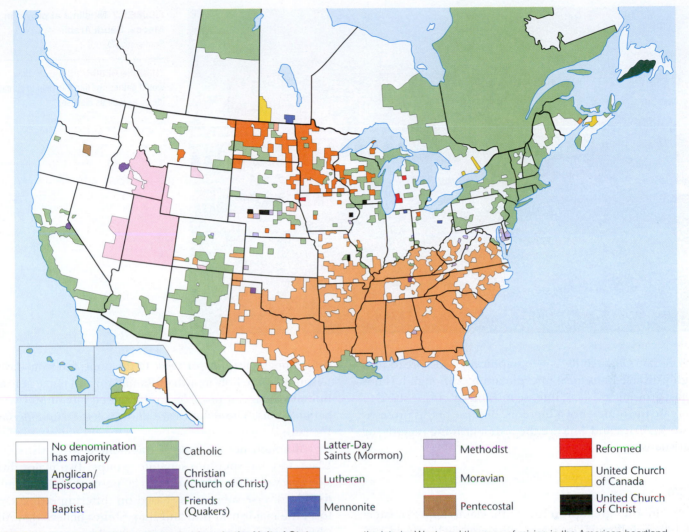

No denomination has majority | Catholic | Latter-Day Saints (Mormon) | Methodist | Reformed

Anglican/ Episcopal | Christian (Church of Christ) | Lutheran | Moravian | United Church of Canada

Baptist | Friends (Quakers) | Mennonite | Pentecostal | United Church of Christ

FIGURE 7.6 **Leading Christian denominations in the United States and Canada, shown by counties (for the United States) or census districts (for Canada).** In the shaded areas, the indicated church or denomination claims 50 percent or more of the total church membership. The most striking features of the map are the Baptist dominance through the South, a Lutheran zone in the upper Midwest, Mormon dominance in the interior West, and the zone of mixing in the American heartland.
(Sources: Simplified from Bradley et al., 1992; Government of Canada, 1974.)

THINKING GEOGRAPHICALLY How does the pattern in the Midwest and Great Plains relate to the presence of ethnic islands? (See Figure 5.7.)

mained unified; but Protestantism, from its beginnings, tended to divide into a rich array of sects, which together have a total membership today of about 593 million worldwide.

In the United States and Canada, the denominational map vividly reflects the fragmentation of Western Christianity and the resulting complex pattern of religious culture regions (**Figure 7.6**). Numerous denominations imported from Europe were later augmented by Christian denominations developed in America. The American frontier was a breeding ground for new religious groups, as individualistic pioneer sentiment found expression in splinter Protestant denominations. Today, in many parts of the country,

even a relatively small community may contain the churches of half a dozen religious groups. As a result, we find today about 2000 different religious denominations and cults in the United States alone. In a broad Bible Belt across the South, Baptist and other conservative fundamentalist denominations dominate, and Utah is at the core of the Mormon realm. A Lutheran belt stretches from Wisconsin westward through Minnesota and the Dakotas, and Roman Catholicism dominates southern Louisiana, the southwestern borderland, and the heavily industrialized areas of the Northeast. The Midwest is a thoroughly mixed zone, although Methodism is the largest single denomination.

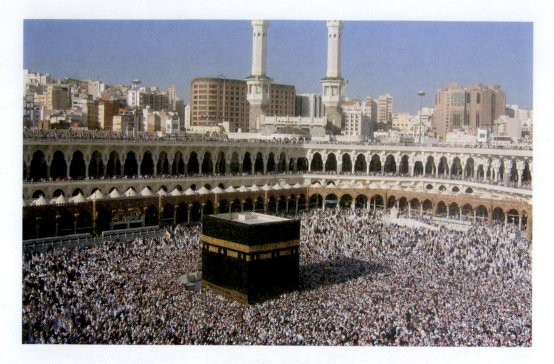

FIGURE 7.7 Muslims at prayer in Mecca, Saudi Arabia (Ayazad/Shutterstock.)

THINKING GEOGRAPHICALLY How would gatherings like this help to spread ideas, as well as diseases?

Today, Christianity is geographically widespread and highly diverse in its local interpretation. Although it is not as fast-growing as Islam, intense missionary efforts strive to increase the number of adherents. Because of this proselytic work, Africa and Asia are the fastest-growing regions for Christianity.

Islam

Islam, another great proselytic faith, claims approximately 1.6 billion followers, largely in the desert belt of Asia and northern Africa and in the humid tropics as far east as Indonesia and the southern Philippines (see Figures 7.2 and 7.4). Adherents of Islam, known as Muslims—literally, "those who submit to the will of God"—are monotheists and worship one absolute God known as Allah. Islam was founded by Muhammad, considered to be the last and most important in a long line of prophets. The word of Allah is believed to have been revealed to Muhammad by the angel Gabriel (Jibrail) beginning in A.D. 610 of the Christian calendar in the Arabian city of Mecca (**Figure 7.7**). The Qur'an (Koran), Islam's holy book, is the text of these revelations and also serves as the basis of Islamic law, or *sharia*. Most Muslims consider both Jews and Christians to be, like themselves, "people of the Book," and all three religions share beliefs in heaven, hell, and the resurrection of the dead. Many biblical figures familiar to Jews and Christians, such as Moses, Abraham, Mary, and Jesus, are also venerated as prophets in Islam. Adherents to Islam are expected to profess belief in Allah, the one God whose prophet was Muhammad; pray five times daily at estab-

lished times; give alms, or *zakat*, to the poor; fast from dawn to sunset in the holy month of Ramadan; and make at least one pilgrimage, if possible, to the sacred city of Mecca in Saudi Arabia. These duties are known as the *Five Pillars* of the faith.

Although not as severely fragmented as Christianity, Islam, too, has split into separate groups. Two major divisions prevail. Shiite Muslims, 5 to 20 percent of the Islamic total in diverse subgroups (based on differing estimates), form the majority in Iran and Iraq. Shiites believe that Ali, who was Muhammad's son-in-law, should have succeeded Muhammad. Sunni Muslims, who represent the Islamic orthodoxy (the word *Sunni* comes from *sunnah*, meaning tradition), form the large majority worldwide (see Figure 7.2). Islam has not undergone a reformation parallel to that undergone by Christianity with Protestantism. However, liberal movements within Sunni and Shiite Islam do have reformation as their goal.

Islam's strength is greatest in the Arabic-speaking lands in Southwest Asia and North Africa, but the world's largest Islamic population is found in Indonesia. Other large clusters live in Pakistan, India, Bangladesh, and western China. Because of successful conversion efforts in non-Muslim areas and high birth rates in predominantly Muslim areas, Islam is the fastest-growing world religion.

Hinduism

Hinduism, a religion closely tied to India and its ancient culture, claims about 900 million to 1 billion adherents (see Figures 7.2 and 7.4). Hinduism is a decidedly polytheistic

(Vaishyas), and workers (Shudras). Castes are related to dharma inasmuch as dharma implies a set of rules for each varna that regulate their behaviors with regard to eating, marriage, and use of space. All Hindus also share a belief in *reincarnation*, the idea that although the physical body may die, the soul lives on and is reborn in another body. Related to this is the notion of *karma*. Karma can be viewed almost as a causal law, which holds that what an individual experiences in this life is a direct result of that individual's thoughts and deeds in a past life. Likewise, all thoughts and deeds, both good and bad, affect an individual's future lives. Ultimately *moksha*, or liberation of the soul from the cycle of death and rebirth, will occur when the worldly bonds of the material self fall away and one's pure essence is freed. *Ahimsa*, or the principle of nonviolence, involves veneration of all forms of life. This implies a principle of noninjury to all sentient creatures, which is why many Hindus are vegetarians.

Hinduism has splintered into diverse groups, some of which are so distinctive as to be regarded as separate religions. Jainism, for instance, is an ancient outgrowth of Hinduism, claiming perhaps 6 million adherents, almost all of whom live in India, and traces its roots back more than 25 centuries. Although they reject Hindu scriptures, rituals, and priesthood, the Jains share the Hindu belief in ahimsa and reincarnation. Jains adhere to a strict asceticism, a practice involving self-denial and austerity. For example, they practice veganism, a form of vegetarianism that prohibits the consumption of all animal-based products, including milk and eggs. Sikhism, by contrast, arose much later, in the 1500s, in an attempt to unify Hinduism and Islam. Centered in the Punjab state of northwestern India, where the Golden Temple at Amritsar serves as the principal shrine, Sikhism has an estimated 30 million followers. Sikhs are monotheistic and have their own holy book, the Adi Granth.

No standard set of beliefs prevails in Hinduism, and the faith takes many local forms. Hinduism includes very diverse peoples. This is partly a result of its former status as a proselytic religion. A Hindu majority on the Indonesian island of Bali suggests the religion's former missionary activity (**Figure 7.9**, page 188). Today, most Hindus consider their religion an ethnic one, whereby one is acculturated into the Hindu community by birth; however, conversion to Hinduism is also allowed.

Buddhism

Hinduism is the parent religion of Buddhism, which began 25 centuries ago as a reform movement based on the teachings of Prince Siddhartha Gautama, "the awakened one" (**Figure 7.10**, page 188). He promoted the four "noble truths": life is full of suffering, desire is the cause of this suffering,

FIGURE 7.8 Ganesha. The elephant-headed god of wisdom, intelligence, and education is revered by many Hindu university students. (Punit Puranjpe/Reuters/Landov.)

THINKING GEOGRAPHICALLY What other examples can you cite of students' regard for a figure, whether religious or other, in their quest for success?

religion. Although Hindus recognize one supreme god, Brahman, it is his many manifestations that are worshipped directly. Some principal Hindu deities include Vishnu, Shiva, and the mother goddess Devi. Ganesha, the elephant-headed god depicted in **Figure 7.8**, is often revered by Hindu university students because Ganesha is the god of wisdom, intelligence, and education. Believing that no one faith has a monopoly on the truth, most Hindus are notably tolerant of other religions.

Hindus strive to locate the harmonious and eternal truth, *dharma*, that is within each human being. Social divisions, or *castes*, separate Hindu society into four major categories, or *varna*, based on occupational categories: priests (Brahmins), warriors (Kshatriyas), merchants and artisans

FIGURE 7.9 **Hindu temple in Bali, Indonesia.** Basakih, known as the "Mother Temple of Bali," is the largest Hindu temple complex on the island, occupying the slopes of Mount Agung. (Courtesy of Patricia L. Price.)

THINKING GEOGRAPHICALLY
What other examples can you cite of a religion that formerly had a widespread following in an area but now exists only in pockets?

cessation of suffering comes with the quelling of desire, and an "Eight-Fold Path" of proper personal conduct and meditation permits the individual to overcome desire. The resultant state of enlightenment is known as *nirvana*. Those few individuals who achieve nirvana are known as Buddhas. The status of Buddha is open to anyone regardless of social status, gender, or age. Because Buddhism derives from Hinduism, the two religions share many beliefs, such as dharma, reincarnation, and ahimsa. Buddhism and its parent religion, Hinduism, as well as the related faiths Jainism and Sikhism, are known as Dharmic religions.

Today, Buddhism is the most widespread religion in South and East Asia, dominating a culture region stretching from Sri Lanka to Japan and from Mongolia to Vietnam. In the process of its proselytic spread, particularly in China and Japan, Buddhism fused with native ethnic religions such as Confucianism, Taoism, and Shinto (sometimes called Shintoism) to form syncretic faiths that fall into the Mahayana division of Buddhism. Southern, or Theravada, Buddhism, dominant in Sri Lanka and mainland Southeast Asia, retains the greatest similarity to the religion's original form, whereas a variation known as Tantrayana, or Lamaism, prevails in Tibet and Mongolia (see Figure 7.2). Buddhism's tendency to merge with native religions, particularly in China, makes it difficult to determine the number of its adherents. Many estimates put the number of Buddhists at about 400 million. Although Buddhism in China has become mingled with local faiths to become part of a composite ethnic religion, elsewhere it remains one of the three great proselytic religions in the world, along with Christianity and Islam.

FIGURE 7.10 **Buddhism is one of the religious faiths of South Korea.**
Here an image of the Buddha is carved from a rocky bluff to create a sacred space and a local pilgrimage shrine. (Courtesy of Terry G. Jordan-Bychkov.)

THINKING GEOGRAPHICALLY From where would Buddhism have diffused to Korea?

Taoic Religions

Confucianism, Shinto, and Taoism together make up the faiths that center on Tao, or the force that balances and orders the universe. Derived from the teachings of philosopher Kong Fu-tzu (551–479 B.C.), Confucianism was later promoted by China's Han dynasty (206 B.C.–A.D. 220) as the official state philosophy. Thus, it has bureaucratic, ethical, and hierarchical overtones that have led some to question whether it is more a way of life than a religion in the proper sense. In China, Confucianism's formalism is balanced by Taoism's romanticism. Taoism, both an established religion and a philosophy, emphasizes the dynamic balance depicted by the Chinese yin/yang symbol. The "three jewels of Tao" are humility, compassion, and moderation. Shinto was once the state religion of Japan. Shinto has long been blended with Buddhism, as well as infused with a Confucian-derived legal system, but at its core Shinto is an animistic religion (see next section). Because Tao is found in nature, there is some overlap with the animistic faiths discussed next. In addition, because of their fluid nature, Taoic religions tend toward syncretism and have blended with other faiths, particularly Buddhism. People may simultaneously practice elements of all of these faiths. For these reasons, it is difficult to provide an exact number of adherents, although estimates hold there to be about 400 billion to 500 billion. Taoic religions center on East Asia (see Figure 7.2).

animist An adherent of animism, the idea that souls or spirits exist not only in humans but also in animals, plants, rocks, natural phenomena such as thunder, geographic features such as mountains or rivers, or other entities of the natural environment.

Animism/Shamanism

Peoples in diverse parts of the world often retain indigenous religions and are usually referred to collectively as **animists** (**Figure 7.11**).

For the most part animists do not form organized or recognized religious groups but instead practice as ethnic religions common to clans or tribes. In addition, most animists follow oral, rather than written, traditions and thus do not have holy books. A tribal religious figure, called by some a *shaman*, usually serves as an intermediary between the people and the spirits.

Currently numbering perhaps 230 million to 240 million, animists believe that nonhuman beings and inanimate objects possess spirits or souls. These spirits are believed to live in rocks and rivers, on mountain peaks and celestial bodies, in forests and swamps, and even in everyday objects. As such, objects are considered to be alive, and depending on the local beliefs, objects and animals may assume human forms or engage in human activities such as speech. For some animists, the objects in question do not actually possess spirits but rather are valued because they have a particular potency to serve as a link between a person and the omnipresent god. Followers of Wicca, a contemporary neopagan religion derived from pre-Christian European practices of reverence for the mother goddess and the horned god, worship the god and goddess who are thought to inhabit everything. Ritual tools used by Wiccans include the *athame* (dagger), *boline* (knife), and *besom* (broom), which are used in ceremonies to direct energy as practitioners communicate with the god and goddess.

Animistic elements can also pervade established religions. Japanese Shinto adherents worship *kami*, or spirits, that inhabit natural objects such as waterfalls and mountains. We should not classify such systems of belief as primitive or simple, because they can be extraordinarily complex. Even in places such as the United States, where the majority of the inhabitants would not see themselves as animists, such beliefs are pervasive. For example, many people in the United States believe that their pets possess souls that ascend to heaven on death.

FIGURE 7.11 Druids from the Mistletoe Foundation bless mistletoe in Worcestershire, England. Celtic Druids have their roots in ancient, pre-Christian reverence for natural elements such as streams, hills, and plants. Mistletoe is sacred to the Druidic faith. (Andrew Fox/Corbis.)

THINKING GEOGRAPHICALLY What appeal do animistic religions have for modern people?

RELIGIOUS **DIFFUSION**

How did the geographical distribution of religions and denominations, of regions and places, come about? What roles did hierarchical and contagious diffusion play? The spatial patterning of religions, denominations, and secularism is the product of innovation and cultural diffusion. To a remarkable degree, the origin of the major religions was concentrated spatially in three principal culture hearth areas (**Figure 7.12**).

A **culture hearth** is a focused geographic area where important innovations are born and from which they spread. Many religions mandate periodic return of the faithful to these culture hearths in order to confirm or renew their faith.

culture hearth A focused geographic area where important innovations are born and from which they spread.

The Semitic Religious Hearth

All three of the great monotheistic faiths—Judaism, Christianity, and Islam—arose among Semitic peoples who lived in or on the margins of the deserts of southwestern Asia, in the Middle East (see Figure 7.12). Judaism, the oldest of the three, originated some 4000 years ago. Only gradually did its followers acquire dominion over the lands between the Mediterranean and the Jordan River—the territorial core of modern Israel. Christianity, child of Judaism, originated there about 2000 years ago. Seven centuries later, the Semitic culture hearth once again gave birth to a major faith when Islam arose in western Arabia, partly from Jewish and Christian roots.

Religions spread by both relocation and expansion diffusion. Recall from Figure 1.10 (page 10) that expansion diffusion can be divided into hierarchical and contagious subtypes. In hierarchical diffusion, ideas become implanted

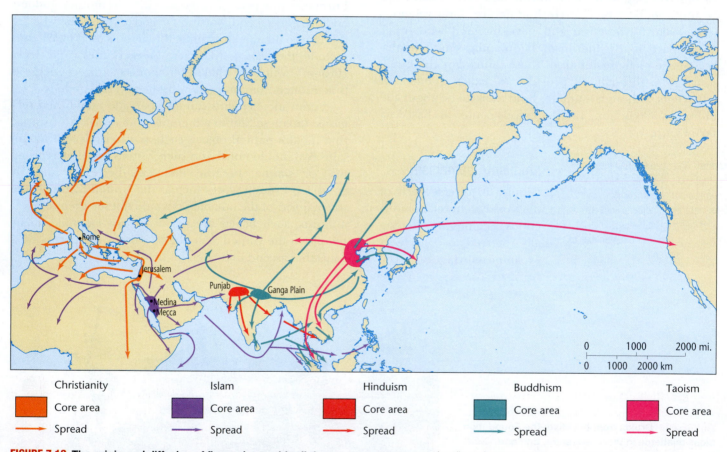

Christianity	Islam	Hinduism	Buddhism	Taoism
■ Core area	■ Core area	■ Core area	■ Core area	■ Core area
→ Spread	→ Spread	→ Spread	→ Spread	→ Spread

FIGURE 7.12 The origin and diffusion of five major world religions. Christianity and Islam, the two great proselytic monotheistic faiths, arose in Semitic southwestern Asia and spread widely through the Old World. Hinduism and Buddhism both originated in the northern reaches of the Indian subcontinent and spread throughout southeastern Eurasia. Taoic religions originated in East Asia, relocated with Chinese and Japanese presence regionally and more recently diffused to North and South America.

THINKING GEOGRAPHICALLY Why did religions developed by Native Americans not become world religions?

at the top of a society, leapfrogging across the map to take root in cities and bypassing smaller villages and rural areas. Because their main objective is to convert nonbelievers, proselytic faiths are more likely to diffuse than ethnic religions, and it is not surprising that the spread of monotheism was accomplished largely by Christianity and Islam, rather than Judaism. From Semitic southwestern Asia, both of the proselytic monotheistic faiths diffused widely.

Christians, observing the admonition of Jesus in the Gospel of Matthew—"Go ye therefore and teach all nations, baptizing them in the name of the Father, and of the Son, and of the Holy Ghost, teaching them to observe all things whatsoever I have commanded you"—initially spread through the Roman Empire, using the splendid system of imperial roads to extend the faith. In its early centuries of expansion, Christianity displayed a spatial distribution that clearly reflected hierarchical diffusion (**Figure 7.13** on page 192). The early congregations were established in cities and towns, temporarily producing a pattern of Christianized urban centers and pagan rural areas. Indeed, traces of this process remain in our language. The Latin word *pagus*, "countryside," is the root of both *pagan* and *peasant*, suggesting the ancient connection between non-Christians and the countryside.

The scattered urban clusters of early Christianity were created by such missionaries as the apostle Paul, one of Jesus' disciples who moved from town to town bearing the news of the emerging faith. In later centuries, Christian missionaries often used the strategy of converting kings or tribal leaders, setting in motion additional hierarchical diffusion. The Russians and Poles were converted in this manner. Some Christian expansion was militaristic, as in the reconquest of Iberia from the Muslims and the invasion of Latin America. Once implanted in this manner, Christianity spread farther by means of contagious diffusion. When applied to religion, this method of spread is called **contact conversion** and is the result of everyday associations between believers and nonbelievers.

contact conversion The spread of religious beliefs by personal contact.

The Islamic faith spread from its Semitic hearth area in a predominately militaristic manner. Obeying the command in the Qur'an that they "do battle against them until there be no more seduction from the truth and the only worship be that of Allah," the Arabs expanded westward across North Africa in a wave of religious conquest. The Turks, once converted by the Arabs, carried out similar Islamic conquests. In a different sort of diffusion, Muslim missionaries followed trade routes eastward to implant Islam hierarchically in the Philippines, Indonesia, and the interior of China. Tropical Africa is the current major scene of Islamic expansion,

an effort that has produced competition with Christians for the conversion of animists. As a result of missionary successes in sub-Saharan Africa and high birth rates in its older sphere of dominance, Islam has become the world's fastest-growing religion in terms of the number of new adherents.

The Indus-Ganges Hearth

The second great religious hearth area lies in the plains fringing the northern edge of the Indian subcontinent. This lowland, drained by the Ganges and Indus rivers, gave birth to Hinduism and Buddhism. Hinduism, which is at least 4000 years old, was the earliest faith to arise in this hearth. Its origin apparently lay in Punjab, from which it diffused to dominate the subcontinent, although some historians believe that the earliest form of Hinduism was introduced from Iran by emigrating Indo-European tribes about 1500 B.C. Missionaries later carried the faith, in its proselytic phase, to overseas areas, but most of these converted regions were subsequently lost to other religions.

Branching off from Hinduism, Buddhism began in the foothills bordering the Ganges Plain about 500 B.C. (see Figure 7.12). For several centuries it remained confined to the Indian subcontinent, but missionaries later carried the religion to China (100 B.C. to A.D. 200), Korea and Japan (A.D. 300 to 500), Southeast Asia (A.D. 400 to 600), Tibet (A.D. 700), and Mongolia (A.D. 1500). Buddhism developed many regional forms throughout Asia, except in India, where it was ultimately accommodated within Hinduism.

The diffusion of Buddhism, like that of Christianity and Islam, continues to the present day. It is estimated that some 1.2 million Buddhists live in the United States today, many in Southern California. Mostly, their presence is the result of relocation diffusion by Asian immigrants to the United States, where immigrant Buddhists outnumber Buddhist converts by three to one.

The East Asian Religious Hearth

Kong Fu-tzu and Lao Tzu, the respective founders of Confucianism and Taoism, were contemporaries who reputedly once met with each other. Both religions were adopted widely throughout China only when the ruling elite promoted them. In the case of Confucianism, this was several centuries after the master's death. During his life, Kong Fu-tzu wandered about with a small band of disciples trying to persuade rulers to put his ideas on good governance into practice. He was shunned even by lowly peasants, who criticized him as "a man who knows he cannot succeed but keeps trying." Thus, early attempts at

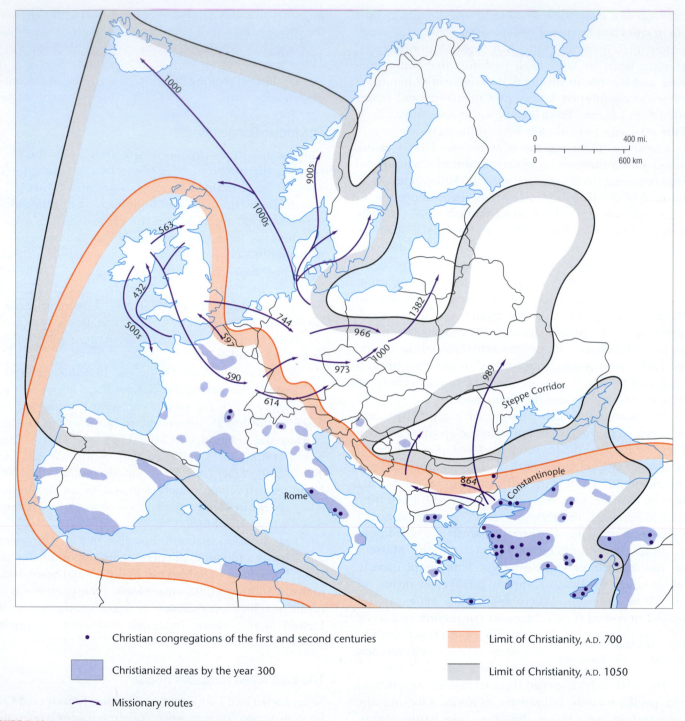

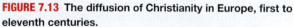

- • Christian congregations of the first and second centuries
- ▨ Christianized areas by the year 300
- ⌒ Missionary routes

- Limit of Christianity, A.D. 700
- Limit of Christianity, A.D. 1050

FIGURE 7.13 **The diffusion of Christianity in Europe, first to eleventh centuries.**

THINKING GEOGRAPHICALLY How does the limit of Christianity in A.D. 700 relate to the farthest extent of the Roman Empire?

contagious diffusion were unsuccessful, whereas hierarchical diffusion from politicians and schools spread Confucianism from the top down. Taoism, as well, did not gain wide acceptance until it was promoted by the ruling Chinese elite.

After 1949, China's communist government officially repressed organized religious expression, dismissing it as a relic of the past. In other words, the government attempted to erect an absorbing barrier that would not only halt the spread of religion but that also would erase it from

Chinese public and private life. As noted in Chapter 1, absorbing barriers are rarely completely successful, and this example is no exception. Although temples were converted to secular uses and even looted and burned, religion was driven underground rather than eradicated. After the end of the Cultural Revolution (1966–1976), which aimed to purify Chinese society of bourgeois excesses such as religion, tolerance for religious expression grew. However, the Chinese Communist Party's official stance still holds that religious and party affiliations are incompatible; thus, some party officials are reluctant to divulge their religious status. This, combined with the fact that many Chinese practice elements of several Taoic faiths simultaneously, makes the precise enumeration of adherents difficult.

Taoism and Confucianism have spread with the Chinese people through trade and military conquest. Thus, people in Taiwan, Malaysia, Singapore, Korea, Japan, and Vietnam, along with mainland China, practice these beliefs or at the least have been influenced by them. Today, Chinese and Japanese migrants alike have relocated their belief systems across the globe, as the image of the Tsubaki Shinto Shrine in Washington State in **Figure 7.14** confirms.

> ### REFLECTING ON **GEOGRAPHY**
> Why, in the entire world, did only three hearths produce so many major religious faiths?

Barriers and Time-Distance Decay

Religious ideas move in the manner of all innovation waves. They weaken with increasing distance from their places of origin and with the passage of time. Barriers are usually of the permeable type, allowing part of the innovation to diffuse through it but weakening it and retarding its spread. An example is the partial acceptance of Christianity by various Indian groups in Latin America and the western United States, which serves in some instances as a camouflage behind which many aspects of the tribal ethnic religions survive. In fact, permeable barriers are *normally* present in the expansion diffusion of religious faiths. Most religions are modified by older, local beliefs as they diffuse spatially.

FIGURE 7.14 Shinto shrine in the United States. With the immigration of Japanese people in the nineteenth and twentieth centuries, particularly to the West Coast of the United States, Shinto traditions and religious structures followed. This image depicts the Tsubaki Shrine in Granite Falls, Washington. (Copyright Alexander Matten Zhang.)

THINKING GEOGRAPHICALLY What other religious houses of worship reveal immigration from Asia?

Rarely do new ideas, whether religious or not, gain unqualified acceptance in a region.

Absorbing barriers can also exist in religious diffusion. The attempt to introduce Christianity into China provides a good example. When Catholic and Protestant missionaries reached China from Europe and the United States, they expected to find millions of people ready to receive the Word of God. However, they had crossed the boundaries of a culture region thousands of years old, in which some basic social ideas left little opening for Christianity and its doctrine of original sin. The Christian image of humankind as flawed, of a gap between creator and created, of the Fall and the impossibility of returning to godhood, was culturally incomprehensible to the Chinese. Only in the early twentieth century, as China's social structure crumbled under Western assault, did a significant, though small, number of Chinese convert to Christianity. Many of these were "rice Christians," poor Chinese willing to become Christians in exchange for the food that missionaries gave them.

In addition, religion itself can act as a barrier to the spread of nonreligious innovations. Religious taboos can even function as absorbing barriers, preventing diffusion of foods, drinks, and practices that violate the taboo. Mormons, who are forbidden to consume products that contain caffeine, have not taken part in the American fascination with coffee. Sometimes these barriers are permeable. Certain Pennsylvania Dutch churches, for example, prohibit cigarette smoking but do not object to member farmers raising tobacco for sale in the commercial market.

RELIGIOUS **ECOLOGY**

What is the relationship between religion and the natural habitat? How does religion influence our modification of the environment and shape our perception of nature? Does the habitat influence religion? All these questions, and more, fit into the theme of cultural ecology.

Appeasing the Forces of Nature

One of the main functions of many religions is the maintenance of harmony between a people and their physical environment. Thus, religion is perceived by its adherents to be part of the *adaptive strategy* (one of the cultural tools needed to survive in a given environment), and for that reason, physical environmental factors, particularly natural hazards and disasters, exert a powerful influence on the development of religions.

Environmental influence is most readily apparent in animistic faiths. In fact, an animistic religion's principal goal is to mediate between its people and the spirit-filled forces of nature. Animistic ceremonies are often intended to bring rain, quiet earthquakes, end plagues, or in some other way manipulate environmental forces by placating the spirits believed responsible for these events. Sometimes the link between religion and natural hazard is visual. The great pre-Columbian temple pyramid at Cholula, near Puebla in central Mexico, strikingly mimics the shape of the awesome nearby volcano Popocatépetl, which towers to the menacing height of nearly 18,000 feet (5500 meters).

Rivers, mountains, trees, forests, and rocks often achieve the status of sacred space, even in the great religions. The Ganges River and certain lesser streams such as the Bagmati in Nepal are holy to the Hindus, and the Jordan River has special meaning for Christians, who often transport its waters in containers to other continents for use in baptism (**Figure 7.15**). Most holy rivers are believed to possess soul-cleansing abilities. Hindu geographer Rana Singh speaks of the "liquid divine energy" of the Ganges "nourishing the inhabitants and purifying them."

FIGURE 7.15 Holy water from the Jordan River. This water is taken from the Jordan River, at the site where Jesus was thought to have been baptized by John the Baptist. It can be purchased and used in rituals. (Courtesy of Patricia L. Price.)

THINKING GEOGRAPHICALLY How does this purchase of water from the Jordan River reflect other efforts to preserve memories of place?

FIGURE 7.16 A high place that has evolved into sacred space. Snowy Mount Shasta in California is venerated by some 30 New Age cults. (iStockphoto/Thinkstock; See Huntsinger and Fernández-Giménez, 2000.)

THINKING GEOGRAPHICALLY Why do mountains so often inspire such worship?

Mountains and other high places likewise often achieve sacred status among both animists and adherents of the great religions. Mount Fuji is sacred in Japanese Shinto, and many high places are revered in Christianity, including Mount Sinai. Some mountains tower so impressively as to inspire cults devoted exclusively to them. Mount Shasta, a massive snowcapped volcano in northern California, near the Oregon border, serves as the focus of no fewer than 30 New Age cults, the largest of which is the "I Am" religion, founded in the 1930s (**Figure 7.16**). These cults posit that the Lemurians, denizens of a lost continent, established a secret city inside Mount Shasta. Geographer Claude Curran, who studied the Shasta cults, found that, although few of the adherents live near the mountain, pilgrimages and festivals held during the summer swell the population of nearby towns and contribute to the local economy.

Animistic nature-spirits lie behind certain practices found in the great religions. Geomancy, or feng shui, which literally means "wind and water," refers to the practice of harmoniously balancing the opposing forces of nature in the built environment. A feature of Asian religions that emphasizes Tao, the dynamic balance found in nature, feng shui involves choosing environmentally auspicious sites for houses, villages, temples, and graves. The homes of the living and the resting places of the dead must be aligned with the cosmic forces of the world in order to assure good luck, health, and prosperity. Although the practice of feng shui dates back some 7000 years, contemporary people practice its principles. **Figure 7.17** on page 196 depicts a high-rise condominium in Hong Kong's Repulse Bay neighborhood. The square opening in the building's center is said to provide passage for the dragon that dwells in the hill behind, allowing the dragon to drink from the waters of the bay and return to its abode unencumbered. Some Westerners have also adopted the principles of feng shui. Office spaces as well as homes are arranged according to its basic ideas. For example, artificial plants, broken articles, and paintings depicting wars are thought to bring negative energy into living spaces and so should be avoided.

Although the physical environment's influence on the major Western religions is less pronounced, it is still evident. Some contemporary adherents to the Judeo-Christian tradition believe that God uses plagues to punish sinners, as in the biblical account of the 10 plagues inflicted on Egypt, which forced the Israelites into the desert. Modern-day droughts, earthquakes, tsunamis, and hurricanes are interpreted by some as God's punishment for wrongdoing, whereas others argue that this is not so. Environmental stress can, however, evoke a religious response not so different from that of animistic faiths. Local ministers and priests often attempt to alter unfavorable weather conditions with special services, and there are few churchgoing people in the Great Plains of the United States who have not prayed for rain in dry years.

Ecotheology

Ecotheology is the name given to a rich and abundant body of literature studying the role of religion in habitat modification. More exactly, ecotheologians ask how the teachings and worldviews of religion are related to our attitudes about modifying the physical environment. In some faiths, human power over natural forces is assumed. The Maori people of New Zealand, for example, believe that humans represent one of six aspects of creation, the others being forest/animals, crops, wild food, sea/fish, and winds/storms. In the Maori worldview, people rule over all of these except winds/storms.

The Judeo-Christian tradition also teaches that humans have dominion over nature, but it goes further, promoting the view that Earth was created especially for human beings, who are separate from and superior to the

ecotheology The study of the interaction between religious belief and habitat modification.

FIGURE 7.17 **Condominium in Hong Kong.** The square opening in this building is supposed to allow for the passage of the dragon who resides in the hill behind. (Courtesy of Ari Dorfsman.)

THINKING GEOGRAPHICALLY Feng shui calls for siting buildings in relation to the sun, facing south, with the back to the north. In the northern hemisphere, what does that accomplish in modern terms?

natural world. This doctrine, called **teleology**, is implicit in God's message to Noah after the Flood, promising that "every moving thing that lives shall be food for you, and as I gave you the green plants, I give you everything." The same theme is repeated in the Psalms: "The heavens are the Lord's heavens, but the Earth he has given to the sons of men." Humans are not part of nature but separate, forming one member of a God-nature-human trinity.

> **teleology** The belief that Earth was created especially for human beings, who are separate from and superior to the natural world.

Believing that Earth was given to humans for their use, early Christian thinkers adopted the view that humans were God's helpers in finishing the task of creation, that human modifications of the environment were God's work. Small wonder, then, that the medieval period in Europe witnessed an unprecedented expansion of agricultural acreage, involving the large-scale destruction of woodlands and the drainage of marshes. Nor is it surprising that Christian monastic orders, such as the Cistercian and Benedictine monks, supervised many of these projects, directing the clearing of forests and the establishment of new agricultural colonies.

Subsequent scientific advances permitted the Judeo-Christian West to modify the environment at an unprecedented rate and on a massive scale. Some critics argue that this marriage of technology and teleology is one cause of our modern ecological crisis. The Judeo-Christian religious heritage, they believe, has for millennia promoted a view of nature that is potentially far more damaging to the habitat than an organic view of nature in which humans and nature exist in balance. Yet there is considerable evidence to the contrary. For example, in Russia, the Orthodox Church is working to create wildlife preserves on monastery lands. The Patriarch of Constantinople, leader of Eastern Orthodox Christianity, has made the fight against pollution a church policy, declaring that damaging the natural habitat constitutes a sin against God. The Church of England has declared that abuse of nature is blasphemous, and throughout the monotheistic religions, the green teachings of long-dead saints such as Christianity's St. Francis of Assisi, who treasured birds and other wildlife, now receive heightened attention.

The great religions of southern and eastern Asia, and many animistic faiths, also highlight teachings and beliefs that protect nature. In Hinduism, for example, geographer Deryck Lodrick found that the doctrine of ahimsa had resulted in the establishment of many animal homes, refuges, and hospitals, particularly in the northwestern part of India. The hospitals, or *pinjrapoles*, are maintained by the Jains.

FIGURE 7.18 **Wood gathered for Hindu cremations at Pashupatinath, on the sacred river Bagmati in Nepal.** These cremations contribute significantly to the ongoing deforestation of Nepal and reveal the underlying internal contradiction in Hinduism between conservation as reflected in the doctrine of ahimsa and sanctioned ecologically destructive practices. (Courtesy of Terry G. Jordan-Bychkov.)

THINKING GEOGRAPHICALLY What kind of diffusion is attempting to cut down on the amount of wood consumed in Hindu cremations?

However, real-world practices do not always reflect the stated ideals of religions. Buddhism, like Hinduism, protects temple trees but demands huge quantities of wood for cremations (**Figure 7.18**). Traditional Hindu cremations, for example, place the corpse on a pile of wood, cover it with more wood, and burn it during an open-air rite that can last up to six hours. The construction of funeral pyres is estimated to strip some 50 million trees from India's countryside annually. In addition, the ashes are later swept into rivers, and the burning itself releases carbon dioxide into the atmosphere, contributing to the pollution of waterways and the atmosphere. A "green cremation system" is currently under development by a New Delhi–based nonprofit organization. Placing the first layer of wood on a grate and placing a chimney over the pyre can reduce wood use by 75 percent.

CULTURAL **INTERACTION** IN RELIGION

How does religion influence other aspects of culture? Can nonreligious elements within a culture help shape a faith? Just as the interaction between religious belief and the environment can shape both religions and the land, religious faith is similarly intertwined with other aspects of culture.

Religion and Economy

In the economic sphere, religion can guide commerce, determine which crops and livestock are raised by farmers and what foods and beverages people consume, and even help decide a person's type of employment and the neighborhood in which he or she resides. Every known religion expresses itself in food choices to one degree or another. In some faiths, certain plants and livestock, as well as the products derived from them, are in great demand because of their roles in religious ceremonies and traditions. When this is the case, the plants or animals tend to spread or relocate with the faith.

For example, in some Christian denominations in Europe and the United States, celebrants drink from a cup of wine that they believe is the blood of Christ during the sacrament of Holy Communion. The demand for wine created by this ritual aided the diffusion of grape growing from the sunny lands of the Mediterranean to newly Christianized districts beyond the Alps in late Roman and early medieval times. The vineyards of the German Rhine were the creation of monks who arrived from the south between the sixth and ninth centuries. For the same reason, Catholic missionaries introduced the cultivated grape to California in an example of relocation diffusion. In fact, wine was associated with religious worship even before Christianity arose. Vineyard keeping and winemaking spread westward across the Mediterranean lands in ancient times in association with worship of the god Dionysus.

Religion can often explain the absence of individual crops or domestic animals in an area and can also explain regional variations in agricultural practices. For example, taboos associated with the consumption of pork in religions such as Islam, Judaism, and Seventh-Day Adventism correspond closely to certain regions' lack of pork production.

Another example of the influence of religious beliefs on agricultural practices can be found when comparing India, a predominantly Hindu state, with other regions that are not as heavily influenced by Hinduism.

The Hindu practice of avoiding beef consumption produces a recognizable pattern on the economic landscape. Figure 7.19 illustrates that large swaths of North America, South America, and Europe produce grain crops primarily for the feeding of livestock such as cows and pigs. Though inefficient from an energy standpoint, this use of grain supports the production of highly desired meat for human consumption in these regions. In India, however, a strikingly different agricultural practice is evident on the landscape. Due in large part to the Hindu prohibition against the consumption of beef, most grain crops produced within the country go directly toward feeding the massive human population, rather than feeding a cattle population destined for human consumption. In many other developing regions of the world with rapidly increasing populations, crops are similarly directed toward human consumption.

The case of India's sacred cows provides an intriguing example of how religious beliefs shape the role of animals in society, in this instance avoiding the use of cows as food while emphasizing the animal's sacred as well as its practical functions. Although only one in three of India's Hindus practices vegetarianism, almost none of India's Hindu population will eat beef, and many will not use leather. Why, in a populous country such as India, where many people are poor and where food shortages have plagued regions of the country in previous decades, do people refuse to consume beef? There are several quite legitimate reasons. First, the dairy products provided by cow's milk and its by-products, such as yogurt and *ghee* (clarified butter), are central to many regional Indian cuisines. If the cow is slaughtered, it will no longer be able to provide milk. Second, in areas of the world that rely heavily on local agriculture for food production, such as India, cows provide valuable agricultural labor in tilling fields. Cows also provide fertilizer in the form of dung. Finally, the value of the cow has been incorporated into Indian Hindu beliefs and practices over many centuries and has become a part of culture. Krishna, an incarnation of the major Hindu deity Vishnu, is said to be both the herder and the protector of cows. Nandi, who is the deity Shiva's attendant, is represented as a bull.

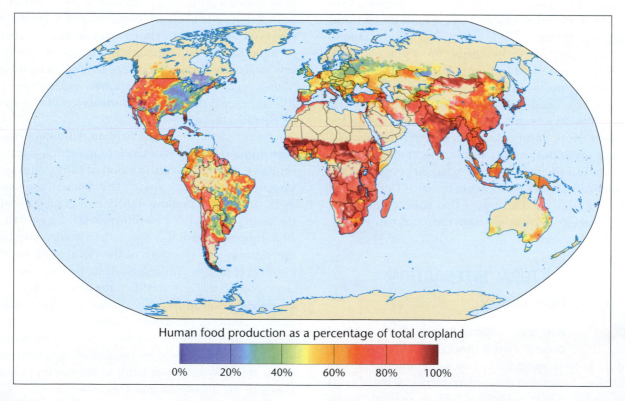

Human food production as a percentage of total cropland

0%   20%   40%   60%   80%   100%

FIGURE 7.19 Agricultural practices are influenced by many factors. Feeding crops to animals produced for meat predominates in affluent world regions where meat is a popular food choice. In India, religious prohibitions influence the prevailing agricultural choice to grow crops for direct consumption by humans rather than cattle.

(Source: Foley, Jonathan A., et al., "Solutions for a Cultivated Planet." *Nature* 478[7369]: 337.)

THINKING GEOGRAPHICALLY How might the pattern shown on the map change in a developing region that becomes more economically stable and affluent over time?

Religious Pilgrimage

For many religious groups, journeys to sacred places, or **pilgrimages,** are an important aspect of their faith (**Figure 7.20**). Pilgrimages are typical of both ethnic

> **pilgrimage** A journey to a place of religious importance.

and proselytic religions. They are particularly significant for followers of Islam, Hinduism, Shinto, and Roman Catholicism. Along with missionaries, pilgrims constitute one of the largest groups of people voluntarily on the move for religious reasons. Pilgrims do not aim to convert people to

their faith through their journeys. Rather, they enact in their travels a connection with the sacred spaces of their faith. Indeed, some religions mandate pilgrimage.

The sacred places visited by pilgrims vary in character. Some have been the setting for miracles; a few are the regions where religions originated or areas where the founders of the faith lived and worked; others contain sacred physical features such as rivers, caves, springs, and mountain peaks; and still others are believed to house gods or are religious administrative centers where leaders of the religion reside. Examples include the Arabian cities

(a)

(b)

(c)

FIGURE 7.20 Religious pilgrimage sites. Visitors number in the millions annually and can provide a holy site's main source of revenue. (a) The grotto where the Virgin Mary appeared to a girl in 1858 in Lourdes, France. (b) The Temple of the Emerald Buddha in Bangkok, Thailand. (c) The Great Mosque in Touba, Senegal. (a: Franz-Peter Tschauner/dpa/Corbis; b: Devin R. Morris/Corbis; c: Richard List/Corbis.)

THINKING GEOGRAPHICALLY What other kinds of places could be considered pilgrimage destinations?

of Mecca and Medina in Islam, which are cities where Muhammad was born and resided and thus form Islam's culture hearth; Rome, the home of the Roman Catholic Pope; the French town of Lourdes, where the Virgin Mary is said to have appeared to a girl in 1858; the Indian city of temples on the holy Ganges River, Varanasi, which is a destination for Hindu, Buddhist, and Jain pilgrims; and Ise, a shrine complex located in the culture hearth of Shinto in Japan. Places of pilgrimage might be regarded as places of spatial convergence, or nodes, of functional culture regions.

In some localities, the pilgrim trade provides the only significant source of revenue for the community. Lourdes, a town of about 16,000, attracts 4 million to 5 million pilgrims each year, many seeking miraculous cures at the famous grotto where the Virgin Mary reportedly appeared. Not surprisingly, among French cities, Lourdes ranks second only to Paris in number of hotels, although most of these are small. Mecca, a small city, annually attracts millions of Muslim pilgrims from every corner of the Islamic culture region as they perform the *hajj*, or fifth pillar of Islam, which every able-bodied Muslim who can afford to do so should do at least once in his or her lifetime. By land, by sea, and (mainly) by air, the faithful come to this hearth of Islam, a city closed to all non-Muslims (**Figure 7.21**). Such mass pilgrimages obviously have a major impact on the development of transportation routes and carriers, as well as other provisions such as inns, food, water, and sanitation facilities.

Religion and Political Geography

Cultural, economic, social, and political arenas often work together in societies around the world. An example of this interaction can be found in the Muslim practice of sharia. **Sharia** means "the way" or "the path to the water source" in Arabic, and it constitutes a religion-based method of lawmaking that guides many aspects of Muslim life, including marriage and divorce, family obligation, religious obligation, financial agreements, and criminal justice. Sharia developed from precedents and analogies found in the Qur'an and in teachings of the Prophet Mohammed, though sharia rulings are sometimes grounded in Muslim community consensus.

Due to different schools of thought within Islam, sharia practices vary widely from state to state throughout the Muslim world. Some contemporary Muslim states, including Saudi Arabia and Iran, follow a very traditional form of sharia and therefore do not have democratic constitutions or legislatures. In these states, secular rulers have limited authority to make or change laws because they are based on

sharia A legal framework, derived from the religious interpretation of Islamic texts, that governs almost all aspects of public life and some aspects of private life for Muslims living under this system of jurisprudence.

FIGURE 7.21 Religious segregation. Access to Mecca is restricted, allowing Muslims only. The Saudi Arabian government believes that allowing non-Muslim tourists to visit Mecca and Medina would disturb the sanctity of these sacred places. (Vario images GmbH & Co. KG./Alamy.)

THINKING GEOGRAPHICALLY What other examples can you cite of places that are restricted to members of a particular religion?

sharia as interpreted by religious scholars. In contrast to these traditional states, many majority-Muslim countries (including Tanzania, Indonesia, and Lebanon) have a dual system wherein the government is secular, but Muslim citizens can opt to bring family and financial disputes to special sharia courts. Other Muslim-majority states, including Azerbaijan, Chad, and Senegal, have declared their governments to be purely secular. However, in these countries, sharia may still influence some local customs. The Arab Spring of 2011 (see Chapter 6, page 164) in Libya, Tunisia, and Egypt led to the ousting of "pro-Western" autocratic governments in favor of governments led by Islamic political parties. As a result, support for certain traditional sharia practices has been renewed in these states.

Another example demonstrating how contemporary religion and politics work together can be found in Latin America and the Caribbean. In this region, Roman Catholicism has dominated the religious landscape since the time of the Iberian conquerors in the late 1400s. More Catholics live in this region than in any other on Earth, with three out of every five Catholics residing here. Brazil is the largest Catholic country in the world in terms of population. However, the Catholic Church has been on the decline in Brazil and throughout the region in recent decades. Fewer than 80 percent of Brazilians are Catholics today, whereas 50 years ago that figure was more than 95 percent. What has happened to the region's Catholics?

Briefly, more and more Latin Americans seem to believe that the Catholic Church has failed to keep in touch

FIGURE 7.22 **Worshippers at an Evangelical church.** Worshippers pray at the Crescendo na Graça church in the Sao Paulo, Brazil suburb of Itaim Paulista. (© CAETANO BARREIRA/X01990/Reuters/Corbis.)

THINKING GEOGRAPHICALLY What kind of diffusion seems to be operating here?

with the needs and concerns of contemporary urban societies. Birth control, divorce, and persistent poverty are issues that have not been addressed by Roman Catholicism to the satisfaction of many Latin Americans. As a result, more and more are turning toward evangelical Protestant faiths, such as the Seventh-Day Adventist and Pentecostal churches (**Figure 7.22**). The focus of these churches on thrift, sobriety, and resolving problems directly rather than through the mediation of priests is appealing to many. Others find their needs are best addressed by an array of African-based spiritist religions, which include Umbanda, Candomblé, and Santería.

Religion on the Internet

In the mid-fifteenth century, Gutenberg's printing press made the Bible available to a mass audience. The Internet, proponents argue, represents a similar technological revolution, one that will inevitably attract new adherents to religious faiths. Gathering together regularly in a holy place—a mosque, temple, church, or shrine—is at the heart of most organized religions. Traditionally, people have come together to receive sacraments, sing, pray, and celebrate major life events in houses of worship. With the tens of thousands of religious web sites now in existence, it is no longer necessary to physically go to a place of worship. Rather, you can sit in front of a computer, at any time of the day or night, and read a holy book, submit and read online prayers, engage in theological debate, or watch broadcasts of religious services.

For cultural geographers, the most interesting dimension of online worship concerns what it does to the role of place in a global society. Clearly, the Internet has made it possible to practice religion in a way that is not linked to a specific place of worship, but is this a good thing? Supporters of practicing religion online contend that it *is* a good thing, because people can become members of virtual communities that would not otherwise have been available to them in times of spiritual need. Now illness, invalidism, or the pressures of a busy life need not present a barrier to worship. Detractors argue that solitary worship online erodes the place-based communities that are at the heart of many religions. As people pick and choose elements from the different faiths available online, will religions converge into a watered-down, homogeneous "McFaith"?

> **REFLECTING ON GEOGRAPHY**
>
> Are virtual religious communities created through the Internet an acceptable substitute for traditional place-based religious communities?

Religion's Relevance in a Global World

In the United States, geographer Roger Stump points to a twentieth-century trend toward fragmentation and dominance of particular religions in regions of the United States. Baptists in the South, Lutherans in the upper Midwest, Catholics in the Northeast and Southwest, and Mormons in the West each dominate their respective regions more thoroughly today than at the turn of the twentieth century. Each of the four traditions is socially conservative and has a strong, long-standing infrastructure. Other experts, however, believe that American culture is becoming more religiously mixed, with weakening regional borders around religious core areas.

Newer religious influences, too, are making an entrance onto the American religious stage. More recently arrived religions to the United States tend to be regionally clustered as a result of their strong association with particular migrant groups (**Figure 7.23** on page 202). In other words, as various migrant populations arrived and settled in particular regions of the country during recent decades, they carried their religious traditions with them. Along with Christianity, these growing religions exert influence on many aspects of American culture. For example, if you have ever practiced yoga or meditation, you are engaging in Hindu and Buddhist practices. Many Americans do yoga or meditate as a means of physical or spiritual growth, for stress relief, or even to keep up with the latest trends, rather than as part of a religious practice. Yet they are among the growing number of Americans who have found a blend of Eastern and Western religious practices to be compatible with their beliefs and lifestyles.

Despite the influence of an increasing number of religions and *blends* of religions on the U.S. cultural landscape, overall, religious adherence has been on the decline in the United States in recent decades. The American Religious

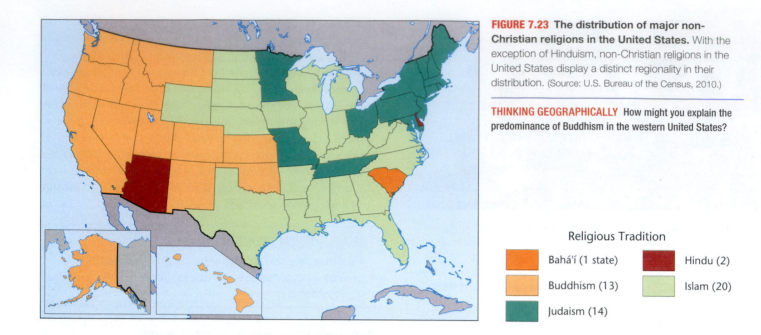

FIGURE 7.23 **The distribution of major non-Christian religions in the United States.** With the exception of Hinduism, non-Christian religions in the United States display a distinct regionality in their distribution. (Source: U.S. Bureau of the Census, 2010.)

THINKING GEOGRAPHICALLY How might you explain the predominance of Buddhism in the western United States?

Religious Tradition

Bahá'í (1 state)

Buddhism (13)

Judaism (14)

Hindu (2)

Islam (20)

Identification Survey, compiled in 2008, found 46 million nonreligious adults in the United States, or about 20.2 percent of the population. In the state of Vermont, for example, approximately 40 percent of adults are not affiliated with a church. As a result of these shifts in American religious adherence, areas of religious vitality now often lie alongside **secular** areas (that is, areas with little or no religious belief or practice) in a disorderly jumble. **Figure 7.24** illustrates this increasingly diverse cultural pattern on the American landscape.

secular Characterized by little or no religious belief or practice.

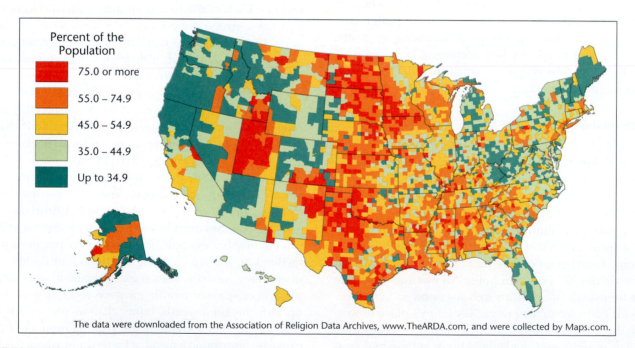

Percent of the Population

75.0 or more

55.0 – 74.9

45.0 – 54.9

35.0 – 44.9

Up to 34.9

The data were downloaded from the Association of Religion Data Archives, www.TheARDA.com, and were collected by Maps.com.

FIGURE 7.24 **Religious adherence in the United States.** Increasing secularization in the United States has led to a complex mosaic on the American religious landscape. Secularization is most prevalent in the western United States, while religious adherence remains quite strong in the midsection of the country. (Source: Clifford Grammich, Kirk Hadaway, Richard Houseal, Dale E. Jones, Alexei Krindatch, Richie Stanley, and Richard H. Taylor. 2012. 2010 *U.S. Religion Census: Religious Congregations & Membership Study.* Association of Statisticians of American Religious Bodies.)

THINKING GEOGRAPHICALLY What reason can you give for the higher rate of religious adherence in the highly urbanized areas of southern and central California as compared to much of the rest of the American West?

Outside of the United States, other regions have also experienced a trend toward secularization, particularly much of Europe (**Figure 7.25**). Globally, however, there has been a notable overall trend toward *increasing* religious adherence. This shift is largely attributable to population increases in the Middle East and parts of Africa, where most people strongly adhere to religious traditions such as Islam and Christianity. As a result of this global trend toward increasing religious adherence, the number of secular persons in the world decreased from 913 million in 2000 to approximately 640 million in 2010. However, there have been some national and regional instances of retreat from organized religion resulting from governments' active hostility toward a particular faith or toward religion in general. For example, the French government has long discouraged public displays of religious faith. France's long-standing secularism is evident in Figure 7.25. This secularization extends not only to France's traditional Roman Catholicism, but to Jews and the rising Muslim population as well. In 2004, French law banned the wearing of skullcaps, headscarves, or large crosses in public schools. Such patterns once again reveal the inherent spatial variety of humankind and invite analysis by the cultural geographer.

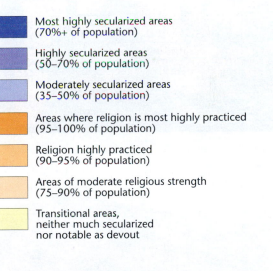

RELIGIOUS **LANDSCAPES**

In what forms does religion appear in the cultural landscape? How does the visibility of religion differ from one faith or denomination to another? Religion is a vital aspect of culture; its visible presence can be striking, reflecting the role religious motives play in the human transformation of the landscape. In some regions, the religious aspect is the dominant visible evidence of culture, producing sacred landscapes. At the opposite extreme are areas almost purely secular in appearance. Religions differ greatly in visibility, but even those least apparent to the eye usually leave some subtle mark on the countryside.

Religious Structures

The most obvious religious contributions to the landscape are the buildings erected to house divinities or to shelter worshippers. These structures vary greatly in size, function, architectural style, construction material, and degree of ornateness (**Figure 7.26** on the next page). To Roman Catholics, for example, the church building is literally the house of God, and the altar is the focus of

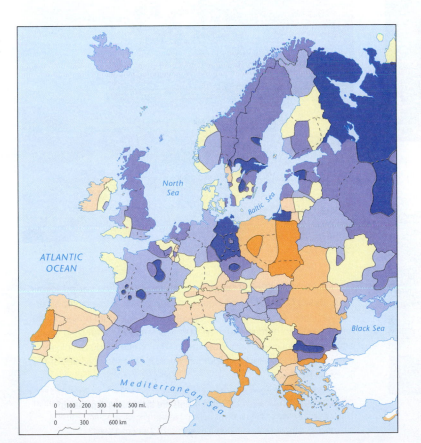

Most highly secularized areas
(70%+ of population)

Highly secularized areas
(50–70% of population)

Moderately secularized areas
(35–50% of population)

Areas where religion is most highly practiced
(95–100% of population)

Religion highly practiced
(90–95% of population)

Areas of moderate religious strength
(75–90% of population)

Transitional areas,
neither much secularized
nor notable as devout

FIGURE 7.25 Secularized areas in Europe. These areas, in which Christianity has ceased to be of much importance, occur in a complicated pattern. In all of Europe, some 190 million people report no religious faith, amounting to 27 percent of the population. (Source: Jordan-Bychkov and Jordan, 2002: 104.)

THINKING GEOGRAPHICALLY What causal forces might be at work here?

key rituals. Partly for these reasons, Catholic churches are typically large, elaborately decorated, and visually imposing. In many towns and villages, the Catholic house of worship is the focal point of the settlement, exceeding all other structures in size and grandeur. In medieval European towns, Christian cathedrals were the tallest buildings, representing the supremacy of religion over all other aspects of life.

For many Protestants—particularly the traditional Calvinistic chapel-goers of British background, including Presbyterians and Baptists—the church building is, by contrast, simply a place to assemble for worship. The result is an unsanctified, small, simple structure that appeals less to

the senses and more to personal faith. For this reason, traditional Protestant houses of worship typically are not designed for comfort, beauty, or high visibility but instead appear deliberately humble (see Figure 7.26b). Similarly, the religious landscape of the Amish and Mennonites in rural North America, the "plain folk," is very subdued because they reject ostentation in any form. Some of their adherents meet in houses or barns, and the churches that do exist are very modest in appearance, much like those of the southern Calvinists.

Islamic mosques are usually the most imposing items in the landscape, whereas the visibility of Jewish synagogues varies greatly. Hinduism has produced large numbers of

(a)

(b)

(c)

FIGURE 7.26 Traditional religious architecture takes varied forms.
(a) St. Basil's Church on Red Square in Moscow reflects a highly ornate Russian landscape presence, whereas (b) the plain board chapel in the American South demonstrates an opposite tendency of favoring visual simplicity by British-derived Protestants. The ornate Hindu temple in Varanasi, India (c), offers still another sacred landscape. (Courtesy of Terry G. Jordan-Bychkov.)

THINKING GEOGRAPHICALLY What beliefs are these icons expressing?

THINKING GEOGRAPHICALLY What parallels to these shrines can be found in your community?

visually striking temples for its multiplicity of gods, but much worship is practiced in private households with their own personal shrines (**Figure 7.27**).

Nature religions such as animism generally place only a subtle mark on the landscape. Nature itself is sacred, and imposing features already present in the landscape—mountains, waterfalls, and grottos, for example—mean that few human-made shrines are needed. There are, not surprisingly, exceptions. In Korea, for example, where animism merged with Buddhism and the Chinese composite religion, animistic shrines survive in the landscape (**Figure 7.28**).

Houses of worship can also reveal subtler content and messages. In Polynesian Maori communities of New Zealand, for example, the *marae*, a structure linked to the pagan gods of the past, generally stands alongside the Christian chapel, which reflects the Maoris' conversion (**Figure 7.29** on the next page). The landscape thus reflects the blending of two faiths.

Paralleling this contrast in church styles are attitudes toward roadside shrines and similar manifestations of faith. Catholic culture regions typically abound with shrines, crucifixes, and other visual reminders of religion, as do some Eastern Orthodox Christian areas. Protestant

FIGURE 7.28 **Stacked stones at a pilgrimage site in South Korea perpetuate an ancient animistic/ shamanistic practice.** These stones had meaning in ancient times, meaning that has since been lost.
(Courtesy of Terry G. Jordan-Bychkov.)

THINKING GEOGRAPHICALLY What do you speculate these stones might mean? What do they remind you of?

FIGURE 7.29 **Tikitiki, New Zealand.** In the Maori community of Tikitiki, on North Island, New Zealand, a traditional Polynesian *marae*, or shrine (on the left), stands beside a Christian chapel (on the right). (Courtesy of Terry G. Jordan-Bychkov.)

THINKING GEOGRAPHICALLY What does this juxtaposition tell us about the faith of the modern Maori?

areas, by contrast, are bare of such symbolism. Their landscapes, instead, display such features as signboards advising the traveler to "Get Right with God," a common sight in the southern United States (**Figure 7.30**).

Faithful Details

Not only buildings but also other architectural choices, such as color and landscaping elements, can convey spiritual meaning through the landscape. Consider the color green. The image of a Muslim mosque in northern Nigeria in **Figure 7.31** depicts how green—sacred in Islam—appears often on mosque domes and *minarets*, the tall

circular towers from which Muslims are called to prayer. The prophet Muhammad declared his favorite color to be green, and his cloak and turban were green. In Christianity, green also has positive connotations, symbolizing fertility, freedom, hope, and renewal. It is used in decorations, such as banners, and for clergy vestments, especially those associated with the Trinity and Epiphany seasons.

Water, often used as a decorative element in fountains, baths, and pools, has religious meaning. In fact, all three of the great desert monotheisms—Islam, Christianity, and Judaism—consider water to have a special status in religious rituals. In general, followers of all three religions believe water has the ability to purify and cleanse the body as well as to purify and cleanse the soul of sins. Muslims use water in ablutions, or washing, of the hands, face, or ears before daily prayers and other rituals. Footbaths, or ablution fountains, are important features located outside

FIGURE 7.31 **Muslim mosque in northern Nigeria.** Areas with Muslim populations enjoy a landscape where mosque minarets and domes are frequently green. (Diane Rawson/Photo Researchers.)

THINKING GEOGRAPHICALLY Besides being Muhammad's favorite color, why might green be a sacred color to Muslims?

FIGURE 7.30 **Billboard in Three Forks, Montana.** Protestant billboards often use American-style advertising slogans to promote church attendance. (Nancy H. Belcher.)

THINKING GEOGRAPHICALLY Why do American churches see the need to market themselves?

FIGURE 7.32 **Landscape of the dead, landscape of the living.** The Beni Hassan Islamic necropolis in central Egypt lies in the desert, just beyond the irrigable land; nearby, the living make intensive use of every parcel of land watered by the Nile River. The mud brick structures are all tombs. Thus, the doubly dead landscape is sacred, whereas the realm of living plants and people is profane. (Courtesy of Terry G. Jordan-Bychkov.)

THINKING GEOGRAPHICALLY Why is this a good place to build a cemetery?

mosques, and decorating with water, particularly pools and fountains, is a common practice in Islamic architecture. Central to Christianity is baptism, during which the newborn infant or adult convert is sprinkled with or fully immersed in holy water, symbolizing the cleansing of original sin for infants and repentance or cleansing from sin for adults. Jews may engage in a ritual bath, known as a *mikveh*, before important events, such as weddings (for women), or on Fridays, the evening of the Jewish Sabbath (for men). The story of the Great Flood, found in the Book of Genesis in the Old Testament (common to both Jews and Christians), depicts the washing away of the sins of the world so that it could be born anew.

Landscapes of the Dead

Religions differ greatly in the type of tribute each awards to its dead. These variations appear in the cultural landscape. The few remaining Zoroastrians, called Parsees, who preserve a once-widespread Middle Eastern faith now confined to parts of India, have traditionally left their dead exposed to be devoured by vultures. Thus, the Parsee dead leave no permanent mark on the landscape. Hindus cremate their dead. Having no cemeteries, their dead, as with the Parsees, leave no obvious mark on the land. Yet on the island of Bali in Indonesia, Hinduism has blended with animism in such a way that temples to family ancestors occupy prominent places outside the houses of the Balinese, a landscape feature that does not exist in India itself (see Figure 7.27).

In Egypt, spectacular pyramids and other tombs were built to house dead leaders. These monuments, as well as the modern graves and tombs of the rural Islamic folk of Egypt, lie on desert land not suitable for farming (**Figure 7.32**). Muslim cemeteries are usually modest in appearance, but spectacular tombs are sometimes erected for aristocratic persons, giving us such sacred structures as the Taj Mahal in India (**Figure 7.33**), one of the architectural wonders of the world.

Taoic Chinese typically bury their dead, setting aside land for that purpose and erecting monuments to their deceased kin. In parts of pre-Communist China, cemeteries and ancestral shrines take up as much as 10 percent of the land in some districts, greatly reducing the acreage available for agriculture.

Christians also typically bury their dead in sacred places set aside for that purpose. These sacred places vary significantly from one Christian denomination to another. Some graveyards, particularly those of Mennonites and southern Protestants, are very modest in appearance, reflecting the reluctance of these groups to use any symbolism that might be construed as idolatrous. Among certain

FIGURE 7.33 **The Taj Mahal in Agra, India.** Built as a Muslim tomb, the Taj Mahal is perhaps the most impressive religious structure in the world. (Pallava Bagda/Corbis.)

THINKING GEOGRAPHICALLY Compare this structure with the mosque in Figure 7.31. How is it similar? How is it different?

(a)

(b)

FIGURE 7.34 Two different Christian landscapes of the dead. (a) In the Yucatán Peninsula of Mexico, the dead rest in colorful aboveground crypts. (b) In Amana, Iowa, communalistic Germans prefer tidiness, order, and equality. (a: Macduff Everton/Corbis; b: Courtesy of Terry G. Jordan-Bychkov.)

THINKING GEOGRAPHICALLY What do cemeteries in your community look like?

other Christian groups, such as Roman Catholics and the Orthodox Churches, cemeteries are places of color and elaborate decoration (**Figure 7.34**).

Cemeteries often preserve truly ancient cultural traits, because people as a rule are reluctant to change their practices relating to the dead. The traditional rural cemetery of the southern United States provides a case in point. Freshwater mussel shells are placed atop many of the elongated grave mounds, and rose bushes and cedars are planted throughout the cemetery. Research suggests that the use of roses may derive from the worship of an ancient pre-Christian mother goddess of the Mediterranean lands. The rose was a symbol of this great goddess, who could restore life to the dead. Similarly, the cedar evergreen is an age-old pagan symbol of death and eternal life, and the use of shell decoration derives from an animistic custom in West Africa, the geographic origin of slaves in the American South. Although the present Christian population of the South is unaware of the origins of their cemetery symbolism, it seems likely that their landscape of the dead contains animistic elements thousands of years old, revealing truly ancient beliefs and cultural diffusions.

REFLECTING ON **GEOGRAPHY**

What is the most visible element of the religious landscape where you live?

Sacred Space

Sacred spaces consist of natural and/or human-made sites that possess special religious meaning that is recognized as worthy of devotion, loyalty, fear, or esteem. By virtue of their sacredness, these special places might be avoided by the faithful, sought out by pilgrims, barred to members of other religions, or removed from economic use (see Figure 7.21). Often, sacred space includes the site of supposed supernatural events or is viewed as the abode of gods. Cemeteries are generally regarded as a type of sacred space. So is Mount Sinai, described by geographer Joseph Hobbs as endowed "with special grace," where God instructed the Hebrews to "mark out the limits of the mountain and declare it sacred." Conflict can result if two religions venerate the same space. In Jerusalem, for example, the Muslim Dome of the Rock, on the site where Muhammad is believed to have ascended to heaven, stands above the Western Wall, the remnant of the ancient Jewish temple (**Figure 7.35**). These two religions literally claim the same space as their holy site, which has led to conflict.

Sacred space is receiving increased attention in the world. In the mid-1990s, the internationally funded Sacred Land Project began to identify and protect such sites, 5000 of which have been cataloged in the United Kingdom

> **sacred space** An area recognized by one or more religious groups as worthy of devotion, loyalty, esteem, or fear to the extent that they become sought out, avoided, inaccessible to the nonbeliever, and/or removed from economic use.

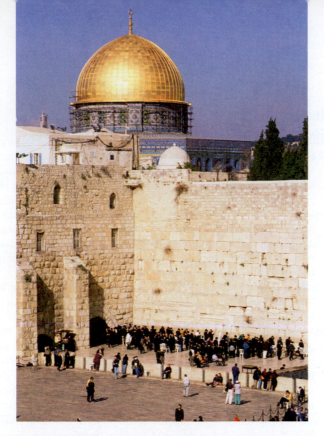

FIGURE 7.35 **Conflict over sacred space.** Jews pray at the Western Wall, the remnant of their great ancient temple in Jerusalem, Israel. Standing above the wall, on the site of the vanished Jewish temple, is one of the holiest sites for Muslims—the golden-capped Dome of the Rock, covering the place from which the Prophet Muhammad is believed to have ascended to heaven. Perhaps no other place on Earth is so heavily charged with religious meaning and conflict. (Gary Cralle/Gettyone.)

THINKING GEOGRAPHICALLY Why is the location of the Dome of the Rock so likely to lead to a heightened degree of friction?

alone. Included are such places as ancient stone circles, pilgrim routes, holy springs, and sites that convey mystery or great natural beauty. This last type falls into the category of *mystical places*: locations unconnected with established religion where, for whatever reason, some people believe that "extraordinary, supernatural things can happen." Sometimes the sacred space of vanished ancient religions never loses—or later regains—the functional status of mystical place, which is what has happened at Stonehenge in England.

In his 1957 book *The Sacred and the Profane*, the renowned religious historian Mircea Eliade suggested that all societies, past and present, have sacred spaces. According to Eliade, sacred spaces are so important because they establish a geographic center on which society can be anchored. Through what he termed *hierophany*, a sacred space emerges from the profane, ordinary spaces surrounding it.

Religious Names on the Land

"St.-Jean," "St.-Aubert," "St.-Pamphile," "St.-Adalbert"—so read the placards of town names as one drives from the St. Lawrence River south in Québec, paralleling the Maine border. All the saintliness is merely part of the French-Canadian religious landscape, as **Figure 7.36** shows. The point is that religion often inspires the names that people place on the land. Within Christianity, the use of saints' names for settlements is very common, especially in Roman Catholic and Greek Orthodox areas,

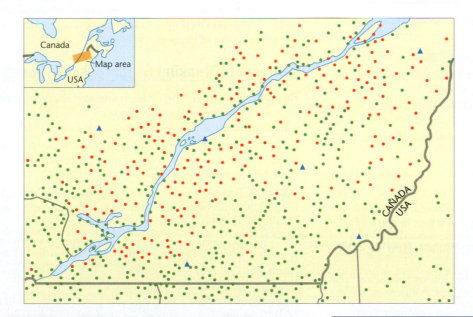

▲ Names beginning with Notre Dame

• Names beginning with Saint or Sainte

• Other names

FIGURE 7.36 **Religious place names dot the map of French Canada.** In the French-Canadian province of Québec, the dominant Roman Catholic religion finds an expression in the names given to towns and villages. Saints' names are dominant in the area of purest French settlement. Nearer the U.S.–Canadian border, in townships settled by English-speaking people, religious place names are rare.

THINKING GEOGRAPHICALLY On this basis, where exactly would you draw the French Catholic–English Protestant border at the time of initial settlement? Is that border still in the same location today?

such as Latin America and French Canada. In areas of the Old World that were settled long before the advent of Christianity, saints' names were often grafted onto pre-Christian names, as in Alcazar de San Juan in Spain, which combines Arabic and Christian elements.

Toponyms in Protestant religions display less religious influence, but some imprint can usually be found. In the southern United States, for example, the word *chapel* as a prefix or suffix, as in Chapel Hill or Ward's Chapel, is very common in the names of rural hamlets. Such names accurately convey the image of the humble, rural Protestant churches that are so common in the South.

Key Terms

animist, 189
contact conversion, 191
culture hearth, 190
ecotheology, 195
ethnic religion, 180
fundamentalism, 181
monotheism, 180
orthodox religion, 180
pilgrimage, 199

polytheism, 180
proselytic religion, 180
religion, 179
sacred space, 208
secular, 202
sharia, 200
syncretic religion, 180
teleology, 196
universalizing religion, 180

Geography of Religion on the Internet

You can learn more about the geography of religion on the Internet at the following web sites:

American Religious Identification Survey, 2008
http://b27.cc.trincoll.edu/weblogs/AmericanReligionSurvey-ARIS/reports/ARIS_Report_2008.pdf
This work by Barry A. Kosmin, Egon Mayer, and Ariela Keysar of the Graduate Center of the City University of New York contains the latest data on religious affiliations in the United States.

Glenmary Research Center
http://www.glenmary.org
Among other activities, the center conducts a census of religious denominations in the United States every 10 years; the 2000 census, listing membership by county, appeared late in 2001.

Material History of American Religion Project, New York, NY
http://www.materialreligion.org
Based at Columbia University, this site has information about religious material objects, including religious landscapes, in the United States.

World Council of Churches
http://www.oikoumene.org
An interdenominational group headquartered in Geneva, Switzerland, the World Council of Churches works for greater cooperation and understanding among different faiths.

Recommended Books on the Geography of Religion

Al-Faruqi, Isma'il R., and David E. Sopher. 1974. *Historical Atlas of the Religions of the World*. New York: Macmillan. Pretty much what its title promises, this informative atlas allows you to see cultural diffusion in action over the centuries and millennia.

Esposito, John, Susan Tyler Hitchcock, Desmond Tutu, and Mpho Tutu. 2004. *Geography of Religion: Where God Lives, Where Pilgrims Walk*. Washington, D.C.: National Geographic. In the tradition of the *National Geographic*, this comprehensive reference book is beautifully illustrated.

Gottlieb, Roger S. (ed.). 1995. *This Sacred Earth: Religion, Nature and Environment*. London: Routledge. A good introduction to ecotheology.

Halvorson, Peter L., and William M. Newman. 1994. *Atlas of Religious Change in America, 1952–1990*. Atlanta: Glenmary Research Center. This atlas shows the changing pattern of denominational membership in the United States in the second half of the twentieth century, revealing the strengthening of some religions and the weakening of others.

Pui-lan, Kwok (ed.). 1994. *Ecotheology: Voices from South and North*. New York: World Council of Churches Publications. A very readable sampling of recent ecotheological thought.

Stoddard, Robert H., and Alan Morinis (eds.). 1997. *Sacred Places, Sacred Spaces: The Geography of Pilgrimages*. Baton Rouge: Geoscience Publications. Published by the Department of Geography and Anthropology at Louisiana State University, these 14 essays, each by a different author, draw attention to Christian, Muslim, Hindu, and lesser Asian pilgrimage traditions, as seen from the perspective of cultural geography.

Stump, Roger W. 2008. *The Geography of Religion: Faith, Place, and Space*. Lanham, MD: Rowman & Littlefield. This comprehensive study of all aspects of the geography of religion includes discussion and examples of world religions, modern religions, and religions of the past.

Answers to Thinking Geographically Questions

Figure 7.1: Snakes are often dangerous, and they can seem to appear out of nowhere. They are also often found on the ground, in other words, the opposite of high.

Figure 7.2: Animist religions, like indigenous peoples and their languages, tend to be located in remote places that have not been settled by numerous members of world religions.

Figure 7.3: There has been a series of conflicts in this region, and for a while Israel occupied southern Lebanon. Israeli troops were withdrawn in 2000.

Figure 7.4: Hinduism is the faith of most people of India, the country with the second largest population in the world.

Figure 7.5: The bell panel extends well above the surrounding buildings, and the structure is built of stone.

Figure 7.6: One of the most frequently preserved institutions of an immigrant ethnic group is its religion. An island shows up as a distinctive religious affiliation if it is different from that of the surrounding area.

Figure 7.7: Pilgrims come from all over the world and meet and mingle. They then return to their homes, taking these ideas (and sometimes disease germs) with them.

Figure 7.8: Among the examples are the statue of John Harvard on Harvard University's campus, where students rub the foot for good luck. At Eastern Kentucky University, students rub the foot of a statue of Daniel Boone.

Figure 7.9: Southwest Asia was Christian for several hundred years after the time of Christ but was then largely converted to Islam. Small pockets of Christianity, however, continue to exist.

Figure 7.10: The immediate source was China; Buddhism spread through Korea to Japan.

Figure 7.11: Among the appeals are a longing for a simpler time, dissatisfaction with organized religions, and a reverence for nature, related to the ecology movement.

Figure 7.12: Native Americans were conquered by Europeans rather than the other way around, and they have not settled in other parts of the world in large numbers.

Figure 7.13: The limit is very closely correlated with the farthest extent of the Roman Empire, even though by A.D. 700, the Roman Empire had fallen. After A.D. 700 until A.D. 1050, Christianity spread to areas well beyond the old Roman lands.

Figure 7.14: In more and more communities, there are mosques and houses of worship for Sikhs, Buddhists, and other Asian groups.

Figure 7.15: Something with strong religious significance becomes even more important than a simple souvenir in the effort to bring a little bit of one's journey home.

Figure 7.16: By definition, mountains are high, and high things (and places) are important. They are also awe-inspiring in their size and frequently their beauty.

Figure 7.17: A building so oriented would get maximum benefit from solar heating.

Figure 7.18: This is a form of hierarchical diffusion, directed culture change, being introduced from educated, urban advocates.

Figure 7.19: If a developing state or region were to achieve greater economic stability with a growing affluent segment of the population, the production of crops for feeding livestock might increase, as more of the population can afford this less-efficient use of plant energy.

Figure 7.20: "Civic religion," the beliefs that unite a country, may also have places that are important to its history and belief system. In the United States, these include Independence Hall in Philadelphia, Old North Church in Boston, and the battlefield at Gettysburg, Pennsylvania.

Figure 7.21: Mormon temples are restricted to members in good standing of the Church of Jesus Christ of Latter-Day Saints. In similar fashion, the inner sanctums of ancient Jewish temples were restricted to priests.

Figure 7.22: This would be contagious diffusion. This church is not part of the establishment.

Figure 7.23: Buddhism has diffused to the United States with the migration of peoples from Asia, largely in the twentieth and twenty-first centuries. These migrants tended to settle in larger numbers on the West Coast than in the East, where other migrant groups had already settled in larger numbers during earlier migration waves.

Figure 7.24: The urban areas in central and southern California are heavily populated by people of Hispanic or Latino origin. This population as a group tends to practice a quite traditional form of Roman Catholicism. The Pacific Northwest and the intermountain West have smaller Hispanic and Latino populations in comparison. Further, these regions are populated by many American Indian groups adhering to a variety of traditional belief systems that may not be included in the "participating religious groups" shown on the map.

Figure 7.25: Besides the long-standing antireligious attitude of the French government, many other Europeans, especially Protestants, have found the source of answers to be in science, economic advancement, and other nonreligious ideologies.

Figure 7.26: The Orthodox church displays a belief that the house of God should be suitably luxurious and ornate, as does the Hindu temple. The Protestant church is simply a meeting place; the church is the people.

Figure 7.27: Among the possibilities are cemeteries, especially elaborate grave markers, and displays of photographs in the home. Families and friends of persons killed in motor vehicle accidents often erect shrines at the site along highways.

Figure 7.28: The top of each pile of stones points upward, similar to steeples and to sacred mountains.

Figure 7.29: The two buildings suggest a syncretism of ancient Maori beliefs and Christianity.

Figure 7.30: In a country and a society that practices freedom of religion, each group is free to try to gain as many members as possible, and many have turned to the same tactics as marketers of secular goods in this "marketplace of ideas."

Figure 7.31: In a desert, such as that where Islam originated, green indicates water, vegetation, and an oasis.

Figure 7.32: In a crowded country where every piece of farmland counts, putting the cemetery on land that cannot be farmed avoids wasting farmland for nonfarming purposes.

Figure 7.33: The Taj Mahal, like the mosque, features a central building with a large dome and four surrounding towers, like the minarets. Its exterior, however, is much more elaborately decorated.

Figure 7.34: Most community cemeteries in the United States and Canada feature a wide range of grave markers and designs, depending mainly on what the family could afford and chose to spend. Some cemeteries have regulations about what kinds of markers can be used, and a few modern ones even insist that all markers are flush with the surface of the ground to make maintenance easier. Some religious groups (for example, the Quakers, the Amish, and the Moravians) traditionally require markers to be small and uniform.

Figure 7.35: To Jews the site is also sacred as the site of their temple. The Western Wall is only a remnant of the supporting wall of the temple, and the temple itself was destroyed in A.D. 70.

Figure 7.36: The border would be north of the present international boundary, south of and parallel to the St. Lawrence River. The international boundary was determined and surveyed by treaty without much regard to the population. The Eastern Townships of Québec Province, between the St. Lawrence Valley and the U.S. border, are still English-speaking.

Two types of contemporary agricultural landscapes. (Left: Michael Busselle/Corbis; Right: Jim Wark/AirPhoto)

Agriculture

8

Every one of us depends, either directly or indirectly, on agriculture for our survival. It is easy to forget that urban-industrial society relies (none too securely) on the food surplus generated by farmers and herders and that without agriculture there would be no cities, universities, factories, or offices.

Agriculture, the tilling of crops and raising of domesticated animals to produce food, feed, drink, and fiber, has been the principal enterprise of humankind throughout recorded history. Even today, agriculture remains by far the most important economic activity in the world, using more land than any other activity and employing about 40 percent of the working population. In some parts of Asia and Africa, more than 75 percent of the labor force is devoted to agriculture. North Americans, on the other hand, live in an urban society in which less than 2 percent of the population work as agriculturists. Europe's labor force is as thoroughly nonagricultural as North America's. Nearly half of the world's population, however, continues to live in farm villages.

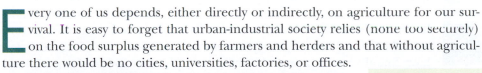

agriculture The tilling of crops and the raising of domesticated animals to produce food, feed, drink, and fiber.

AGRICULTURAL REGIONS

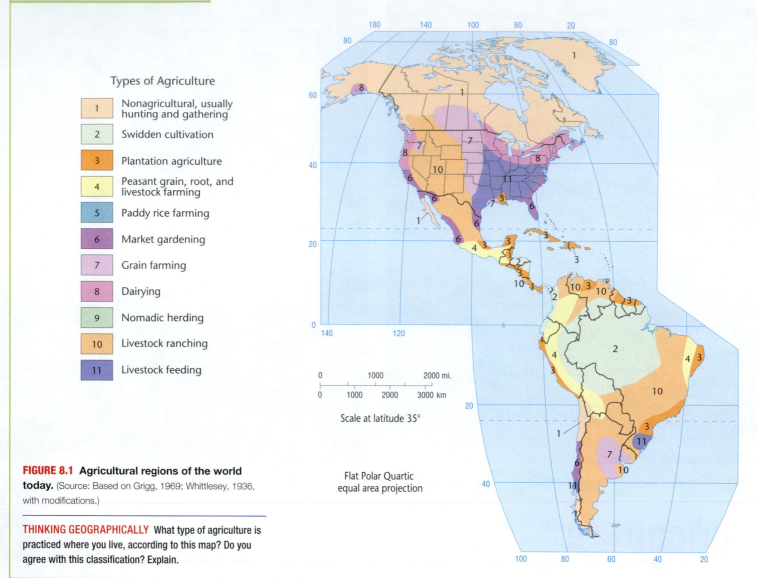

Types of Agriculture

1	Nonagricultural, usually hunting and gathering
2	Swidden cultivation
3	Plantation agriculture
4	Peasant grain, root, and livestock farming
5	Paddy rice farming
6	Market gardening
7	Grain farming
8	Dairying
9	Nomadic herding
10	Livestock ranching
11	Livestock feeding

Scale at latitude 35°

0   1000   2000 mi.
0   1000   2000   3000 km

Flat Polar Quartic
equal area projection

FIGURE 8.1 Agricultural regions of the world today. (Source: Based on Grigg, 1969; Whittlesey, 1936, with modifications.)

THINKING GEOGRAPHICALLY What type of agriculture is practiced where you live, according to this map? Do you agree with this classification? Explain.

AGRICULTURAL **REGIONS**

How is the theme of culture region relevant to agriculture? For thousands of years, farmers have found ways to cope with a range of environmental conditions, creating in the process an array of different types of food-producing systems. Collectively, these farming practices have constructed formal **agricultural regions** (**Figure 8.1**). During the past 500 years, colonialism, industrialization, and globalization have greatly altered existing practices and created new types of agricultural regions, such as plantations. Increasingly, large-scale capitalist agriculture

agricultural region A geographic region defined by a distinctive combination of physical environmental conditions; crop type; settlement patterns; and labor, cultivation, and harvesting practices.

overcomes local environmental constraints through irrigation, land reclamation, and the use of synthetic fertilizers, chemical pesticides, herbicides, and genetic engineering. This form of large-scale farming using mechanization and other forms of capital, known as **intensive agriculture,** produces large crop yields per unit of land and therefore requires less land to produce profits than less intensive forms of farming. However, farmers engaging in intensive agriculture in large countries like the United States and Canada, where much agricultural land is available, *choose* to farm larger tracts of land in order to maximize investments in machinery and other capital. In parts of the world where agricultural land is scarcer, intensive agriculture is often

intensive agriculture A form of agriculture using mechanization, labor, and other capital to produce large crop yields relative to the amount of land being farmed.

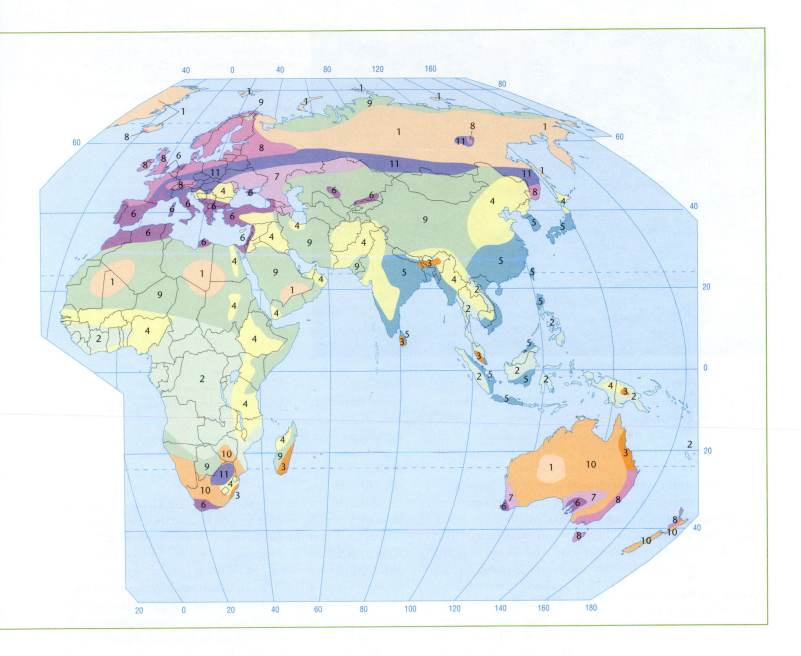

practiced on smaller tracts of land located in high-value areas close to markets.

In contrast to intensive agriculture is a less mechanized, less capital-intensive form of farming known as **extensive agriculture.** In this type of farming, productivity is heavily affected by environmental characteristics in a local area such as soil fertility, terrain, water supply, and climate.

> **extensive agriculture** A less capital-intensive form of agriculture that is highly influenced by local environmental characteristics and often produces low crop yields per unit of land being farmed.

Subsequently, in extensive agriculture, yields per unit of land are highly variable, making it necessary to farm larger tracts of land in order to ensure a profit. For this reason, extensive agriculture is normally practiced in areas where land values are low, such as in low-population-density areas

or in areas at some distance from markets. Understanding the differing influences of environmental and economic forces on these different forms of agriculture, in addition to the influence of other cultural and political factors, is important for our understanding of agricultural regions.

Swidden (Shifting) Cultivation

Many of the peoples of tropical lowlands and hills in the Americas, Africa, and Southeast Asia practice a land-rotation agricultural system known as **swidden cultivation.** The term *swidden* is derived from an old English term meaning "burned clearing." Using machetes,

> **swidden cultivation** A type of agriculture characterized by land rotation in which temporary clearings are used for several years and then abandoned to be replaced by new clearings; also known as *slash-and-burn agriculture*.

axes, and chainsaws, swidden cultivators chop away the undergrowth from small patches of land and kill the trees by removing a strip of bark completely around the trunk. After the dead vegetation dries out, the farmers set it on fire to clear the land. Because of these clearing techniques, swidden cultivation is also called *slash-and-burn agriculture*. Working with digging sticks or hoes, the farmers then plant a variety of crops in the ash-covered clearings, varying from the maize (corn), beans, bananas, and manioc of Native Americans to the yams and nonirrigated rice grown by hill tribes in Southeast Asia (**Figure 8.2**). Different crops typically share the same clearing, a practice called **intercropping** or **intertillage.** This technique allows taller, stronger crops to shelter lower, more fragile ones; reduces the chance of total crop losses from disease or pests; and provides the farmer with a varied diet. The complexity of many intercropping systems reveals the depth of knowledge acquired by swidden cultivators over many centuries. Relatively little tending of the plants is necessary until harvest time, and no fertilizer is applied to the fields because the ashes from the fire are a sufficient source of nutrients.

> **intercropping (intertillage)** The practice of growing two or more different types of crops in the same field at the same time.

The planting and harvesting cycle is repeated in the same clearings for perhaps three to five years, until soil fertility begins to decline as nutrients are taken up by crops and not replaced. Subsequently, crop yields decline. These fields then are abandoned, and new clearings are prepared to replace them. Because the farmers periodically shift their cultivation plots, another commonly used term for the system is *shifting cultivation*. The abandoned cropland lies fallow for 10 to 20 years before farmers return to clear it and start the cycle again. Swidden cultivation represents one form of **subsistence agriculture**—food production mainly for the family and local community rather than for market.

> **subsistence agriculture** Farming to supply the minimum food and materials necessary to survive.

Swidden cultivation in the tropics often has been viewed by outsiders as destructive and inefficient, because the techniques used and the landscapes that result are so different from those common in temperate-zone agriculture. However, although the technology of swidden may be simple, it has proved to be an efficient and adaptive strategy. Swidden farming, unlike some modern systems, is ecologically sustainable and has endured for millennia. Indeed, contemporary studies suggest that some swidden systems have actually enhanced biodiversity. Furthermore, swidden returns more calories of food for the calories spent on cultivation than does modern mechanized agriculture. In many tropical forest regions, swidden cultivation, unlike Western plantations, has left most of the forest intact over centuries of continuous use.

Nonetheless, swidden cultivation can be environmentally destructive under certain conditions. In poor countries with large landless populations, one often finds a front of pioneer swidden farmers advancing on the forests. In such situations, a range of institutional, economic, political, and demographic factors restrict poor farmers' abilities to employ the methods of swidden agriculture in a sustainable way. In many tropical countries, for example, a small proportion of the population owns most of the best agricultural land, forcing the majority of farmers to clear forests to gain access to land. Another condition that may diminish the sustainability of swidden cultivation occurs

FIGURE 8.3 Cultivation of rice on the island of **Bali, Indonesia.** Paddy rice farming traditionally entails enormous amounts of human labor and very high yields per unit of land. (Denis Waugh/Tony Stone Images.)

THINKING GEOGRAPHICALLY What are the disadvantages of such a system?

when a population experiences a sudden increase in its rate of growth and political or social conditions restrict its mobility. Population growth may encourage farmers to shorten the period during which the land is recuperating, a departure from past practices that can lead to environmental deterioration.

Paddy Rice Farming

Peasant farmers in the humid tropical and subtropical parts of Asia practice a highly distinctive type of subsistence agriculture called **paddy rice farming.** A paddy is a small flooded field enclosed by mud dikes, usually found in the humid areas of the Far East. Rice, the dominant paddy crop, forms the

paddy rice farming The cultivation of rice on a paddy, or small flooded field enclosed by mud dikes, practiced in the humid areas of the Far East.

basis of civilizations in which almost all the caloric intake is of plant origin. From the monsoon coasts of India through the hills of southeastern China and on to the warmer parts of Korea and Japan stretches a broad region of diked, flooded rice fields or paddies, many of which are perched on terraced hillsides (**Figure 8.3**). The terraced paddy fields form a striking cultural landscape (see Figure 1.18, p. 22).

A paddy rice farm of only 3 acres (1 hectare = 2.47 acres) usually is adequate to support a family, because irrigated rice provides a very large output of food per unit of land. Still, paddy farmers must till their small patches intensively to harvest enough food. A system of irrigation that can deliver water when and where it is needed is key to success. In addition, large amounts of fertilizer must be applied to the land. Paddy farmers often plant and harvest

the same parcel of land twice per year, a practice known as **double-cropping.**

double-cropping Harvesting twice a year from the same parcel of land.

These systems are extremely productive, yielding more food per acre than many forms of industrialized agriculture in the United States.

The modern era has witnessed a restructuring of paddy rice farming in more developed countries, such as Japan, Korea, and Taiwan. In some cases, the terrace structure has been reengineered to produce larger fields that can be worked with machines. In addition, dams, electric pumps, and reservoirs now provide a more reliable water supply, and high-yielding seeds, pesticides, and synthetic fertilizers boost production further. Most paddy rice farmers now produce mainly for urban markets.

This production of paddy rice crops for sale in external markets, rather than solely for consumption by one's own family, is one type of **commercial agriculture.** There are other types of commercial agriculture as well, including plantation agriculture, that produce food for sale to wholesale and retail markets. The production of nonfood crops like cotton and tobacco is considered commercial agriculture because these commodities cannot be used directly to feed farming families.

commercial agriculture The growing of crops, including nonfood crops, for sale rather than strictly for consumption by one's own family.

Peasant Grain, Root, and Livestock Farming

In colder, drier Asian farming regions that are climatically unsuited to paddy rice farming—as well as in the river

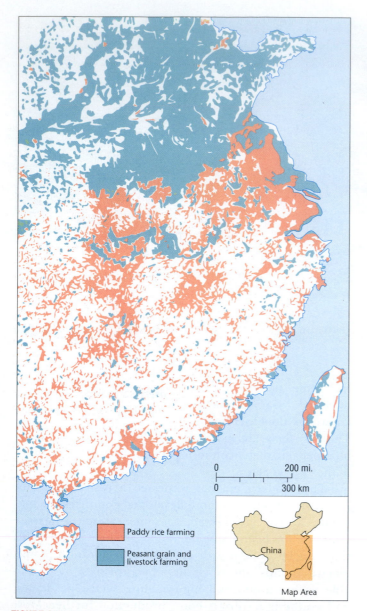

FIGURE 8.4 **Two agricultural regions in China.** The intricacies of culture region boundaries are suggested by the distribution of two types of agriculture in Taiwan and the eastern part of China. All such cultural-geographical borders are difficult to draw. (Source: Chuan-jun, 1979.)

Legend:
- Paddy rice farming
- Peasant grain and livestock farming

China — Map Area

THINKING GEOGRAPHICALLY What might account for the more fragmented distribution of paddy rice farming, as contrasted with peasant grain and livestock farming? Where would you draw the cultural boundary between the two types of agriculture?

valleys of the Middle East, in parts of Europe, in Africa, and in the mountain highlands of Latin America and New Guinea—farmers practice a diverse system of agriculture based on bread grains, root crops, and herd livestock (**Figures 8.4** and **8.5**). Many geographers refer to these farmers as **peasants**, recognizing that they often

peasant A farmer belonging to a folk culture and practicing a traditional system of agriculture.

FIGURE 8.5 **Peasant grain, root, and livestock agriculture in highland New Guinea.** Distinctive raised fields with sweet potato mounds are found here among the farms of highland New Guinea. These people raise diverse crops but give the greatest importance to sweet potatoes, together with pigs. (Courtesy of Terry G. Jordan-Bychkov.)

THINKING GEOGRAPHICALLY Why might the people go to the trouble of creating these small mounds?

represent a distinctive **folk culture** strongly rooted in the land. Peasants generally are small-scale farmers who own their fields and produce both for their own subsistence and for sale in the market. The dominant grain crops in these regions are wheat, barley, sorghum, millet, oats, and maize. Common cash crops—some of them raised for export—include cotton, flax, hemp, coffee, and tobacco.

folk culture Small, cohesive, stable, isolated, nearly self-sufficient group that is homogeneous in custom and race; characterized by a strong family or clan structure, order maintained through sanctions based in the religion or family, little division of labor other than that between the sexes, frequent and strong interpersonal relationships, and material cultures consisting mainly of handmade goods.

These farmers also raise herds of cattle, pigs, sheep, and, in South America, llamas and alpacas. The livestock pull the plow; provide milk, meat, and wool; serve as beasts of burden; and produce manure for the fields. They also consume a portion of the grain harvest. In some areas, such as the Middle Eastern river valleys, the use of irrigation helps support this peasant system. In general, however, modern agricultural technologies are beyond the financial reach of most peasants.

Plantation Agriculture

In certain tropical and subtropical areas, Europeans and Americans introduced a commercial agricultural system

FIGURE 8.6 **Plantation agriculture.** This sign was erected by the management at the entrance to a banana plantation in Costa Rica. "Welcome to Freehold Plantation, a workplace where labor harmony reigns; in mutual respect and understanding, we united workers produce and export quality goods in peace and harmony." (Courtesy of Terry G. Jordan-Bychkov.)

THINKING GEOGRAPHICALLY How does this message suggest that in fact not all is harmonious here and that the tension of the two-class plantation system simmers below the surface?

called **plantation agriculture**. A **plantation** is a landholding devoted to capital-intensive, large-scale, specialized production of one tropical or subtropical crop for the global market. Each plantation district in the tropical and subtropical zones tends to specialize in one crop. Plantation agriculture has long relied on large amounts of manual labor, initially in the form of slave labor and later as wage labor. The plantation system originated in the 1400s on Portuguese-owned sugarcane-producing islands off the coast of tropical West Africa—São Tomé and Príncipe—but the greatest concentrations are now in the American tropics, Southeast Asia, and tropical South Asia. Most plantations lie near the seacoast, close to the shipping lanes that carry their produce to nontropical lands such as Europe, the United States, and Japan.

Workers usually live right on the plantation, where a rigid social and economic segregation of labor and management produces a two-class society of the wealthy and the poor. As a result of the concentration of ownership and production, a handful of multinational corporations, such as Chiquita and Dole, control the largest share of plantations globally. Tension between labor and management is not uncommon, and the societal ills of the plantation system remain far from cured (**Figure 8.6**).

Plantations provided the base for European and American economic expansion into tropical Asia, Africa, and Latin America. They maximize the production of luxury crops for Europeans and Americans, such as sugarcane, bananas, coffee, coconuts, spices, tea, cacao, pineapples, rubber, and tobacco (**Figure 8.7**). Similarly, textile factories require cotton, sisal, jute, hemp, and other fiber crops from the plantation areas. Much of the profit from these plantations is exported, along with the crops themselves, to Europe and North America, another source of political friction between countries of the global North and South.

> **plantation agriculture** A system of monoculture for producing export crops requiring relatively large amounts of land and capital; originally dependent on slave labor.
>
> **plantation** A large landholding devoted to specialized production of a single tropical cash crop.

FIGURE 8.7 **Tea plantation in the highlands of Papua New Guinea.** Although profitable for the owners and providing employment for a small labor force, the plantation recently displaced a much larger population of peasant grain, root, and livestock farmers. This is one result of globalization. (Courtesy of Terry G. Jordan-Bychkov.)

THINKING GEOGRAPHICALLY Should the government have prevented such a displacement?

Market Gardening

The growth of urban markets in the last few centuries also gave rise to other commercial forms of agriculture, including **market gardening,** also known as *truck farming.* Unlike

> **market gardening** Farming devoted to specialized nontropical fruit, vegetable, or vine crops for sale rather than consumption. Also known as *truck farming.*

plantations, truck farms are located in developed countries and specialize in intensively cultivated nontropical fruits, vegetables, and vines. They raise no livestock. Many districts concentrate on a single product, such as wine, table grapes, raisins, olives, oranges, apples, lettuce, or potatoes, and the entire farm output is raised for sale rather than for consumption on the farm. Many truck farmers participate in cooperative marketing arrangements and depend on migratory seasonal farm laborers to harvest their crops. Market garden districts appear in most industrialized countries. In the United States, a broken belt of market gardens extends from California eastward through the Gulf and Atlantic coast states, with scattered districts in other parts of the country. (**Figure 5.8** provides an overview of U.S. agricultural patterns.) The lands around the Mediterranean Sea are dominated by market gardens. In regions of mild climates, such as California, Florida, and the Mediterranean region, winter vegetables are a common market crop raised for sale in colder regions of the higher latitudes.

Livestock Feeding

In **livestock feeding,** farmers raise and fatten cattle and

> **livestock feeding** A commercial type of agriculture that produces cattle and hogs for meat.

hogs for slaughter. One of the most highly developed meat industry areas is the famous Corn Belt of the

U.S. Midwest, where farmers raise corn and soybeans to feed cattle and hogs. Typically, slaughterhouses are located close to feedlots, creating a new meat-producing region, which is often dependent on mobile populations of cheap immigrant labor. A similar system prevails over much of western and central Europe, though the feed crops there more commonly are oats and potatoes. Other zones of commercial livestock feeding appear in overseas European settlement zones such as southern Brazil and South Africa.

One of the central traditional characteristics of livestock feeding is the combination of crops and animal husbandry. Farmers breed many of the animals they feed out, especially hogs. In the last half of the twentieth century, livestock feeders began to specialize their activities; some concentrated on breeding animals, others on preparing them for market. In the factory-like **feedlot,** farmers raise imported cattle and hogs on purchased feed (**Figure 8.9**). Increases in the amount of land and

> **feedlot** A factorylike farm devoted to livestock feeding or dairying; typically all feed is imported and no crops are grown on the farm.

crop harvest dedicated to beef production have accompanied the growth in feedlot size and number. In the United States, across Europe, and in European settlement zones around the globe, 51 to 75 percent of all grain raised goes to livestock feeding.

The livestock feeding and slaughtering industry has become increasingly concentrated, on both the national and global scales. In 1980 in the United States, the top four companies accounted for 41 percent of all slaughtered cattle. By 2000, the top four companies were slaughtering 81 percent of all feedlot cattle.

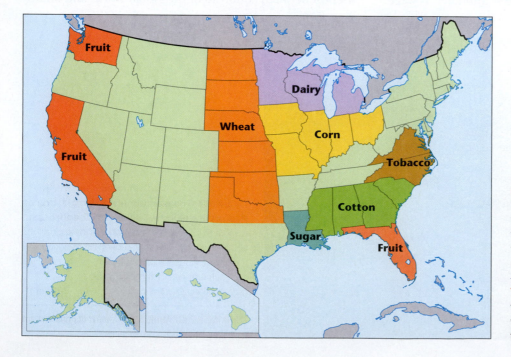

FIGURE 8.8 Agricultural patterns in the United States. Climatic conditions, soil types, terrain, and distances to markets are some of the many factors that have contributed to the formation of notable patterns in agricultural production on the U.S. landscape.

THINKING GEOGRAPHICALLY How do you explain the absence of major crop-producing regions in the western Great Plains and intermountain region of the United States?

FIGURE 8.9 **Cattle feedlot for beef production.** This feedlot is located in Saskatchewan, Canada. (Dave Reede/AgStock Images/Corbis.)

THINKING GEOGRAPHICALLY What ecological problems might such an enterprise cause?

Grain Farming

Grain farming is a type of specialized agriculture in which farmers grow primarily wheat, rice, or corn for commercial markets. The United States is the world's leading wheat and corn exporter. (Figure 8.8 highlights the dominant corn-producing area in the Midwest.) Climate and soil types in this region are ideally suited to growing corn, while proximity to the Great Lakes shipping corridor allows for convenient transport of corn crops to major markets on the East Coast and to markets abroad.

The United States, Canada, Australia, the European Union (EU), and Argentina together account for more than 85 percent of all wheat exports, while the United States alone accounts for about 70 percent of world corn exports. Wheat belts stretch through Australia, the Great Plains of interior North America, the steppes of Russia and Ukraine, and the pampas of Argentina. Farms in these areas generally are very large, ranging from family-run wheat farms to giant corporate operations (**Figure 8.10**).

Widespread use of machinery, synthetic fertilizers, pesticides, and genetically engineered seed varieties enables grain farmers to operate on this large scale. The planting and harvesting of grain are more completely mechanized than any other form of agriculture. Commercial rice farmers employ such techniques as sowing grain from airplanes. Harvesting is usually done by hired migratory crews using corporation-owned machines (**Figure 8.11** on the next page). Perhaps grain farming's ultimate development is the **suitcase farm,** which is found in the Wheat Belt of the northern Great Plains of the United States. The people who own and operate these farms do not live on the land. Most of them own several suitcase farms, lined up in a south-to-north row through the Plains states. They

suitcase farm In U.S. commercial grain agriculture, a farm on which no one lives; planting and harvesting are done by hired migratory crews.

FIGURE 8.10 **A wheat landscape in the Palouse, a grain-farming region on the borders of Washington and Oregon.** Grain elevators are a typical part of such agricultural landscapes. The raising of one crop, such as wheat, across entire regions is called monoculture. (Courtesy of Terry G. Jordan-Bychkov.)

THINKING GEOGRAPHICALLY What problems might be linked to monoculture?

FIGURE 8.11 **Mechanized wheat harvest in the United States.**
Montana grain farmers operate in a capital-intensive manner, investing in machines, chemical fertilizers, and pesticides. (Timothy Hearsum/AgStock Images/Corbis.)

THINKING GEOGRAPHICALLY What long-term problems might such methods cause? What benefits are realized in such a system?

keep fleets of farm machinery, which they send north with crews of laborers along the string of suitcase farms to plant, fertilize, and harvest the wheat. The progressively later ripening of the grain as one moves north allows these farmers to maintain and harvest crops on all their farms with the same crew and the same machinery. Except for visits by migratory crews, the suitcase farms are uninhabited.

Such highly mechanized, absentee-owned, large-scale operations, or **agribusinesses,** have mostly replaced the small, husband- and wife-operated American family farm, an important part of the U.S. rural heritage. Geographer Ingolf Vogeler documented the decline of the small family farms in the American countryside and argued that U.S. governmental policies, prompted by the forces of globalization, have consistently favored the interests of agribusiness, thereby hastening the decline. Family-owned farms continue to play an important role, but they now operate mostly as large agribusinesses that own or lease many far-flung grain fields.

> **agribusiness** Highly mechanized large-scale farming, usually under corporate ownership.

Dairying

In many ways, the specialized production of dairy goods closely resembles livestock farming. In the large dairy belts of the northern United States from New England to the upper Midwest, western and northern Europe, southeastern Australia, and northern New Zealand, the keeping of dairy cows depends on the large-scale use of pastures. In colder areas, some acreage must be devoted to winter feed crops, especially hay. Dairy products vary from region to region, depending in part on how close the farmers are to

their markets. Dairy belts near large urban centers usually produce milk, which is more perishable, while those farther away specialize in butter, cheese, or processed milk. An extreme case is New Zealand, which, because of its remote location from world markets, produces much butter.

> **REFLECTING ON GEOGRAPHY**
>
> Why is dairying confined to northern Europe and the overseas lands settled by northern Europeans? (See Figure 8.1, p. 214.)

As with livestock feeding, in recent decades a rapidly increasing number of dairy farmers have adopted the feed-lot system and now raise their cattle on feed purchased from other sources. Often situated on the suburban fringes of large cities for quick access to market, the dairy feedlots operate like factories. Like industrial factory owners, feed-lot dairy owners rely on hired laborers to help maintain their herds. Dairy feedlots are another indicator of the rise of globalization-induced agribusiness and the decline of the family farm. By easing trade barriers, globalization compels U.S. dairy farmers to compete with producers in other parts of the world. Huge feedlots, a factory-style organization of production, automation, the concentration of ownership, and the increasing size of dairy farms are responses to this intense competition.

Nomadic Herding

In the dry or cold lands of the Eastern Hemisphere, particularly in the deserts, prairies, and savannas of Africa, the Arabian Peninsula, and the interior of Eurasia,

nomadic livestock herders graze cattle, sheep, goats, and camels. North of the tree line in Eurasia, the cold tundra forms a zone of nomadic herders who raise reindeer. The common characteristic of all nomadic herding is mobility. Herders move with their livestock in search of forage for the animals as seasons and range conditions change. Some nomads migrate from lowlands in winter to mountains in summer; others shift from desert areas during the rainy season to adjacent semiarid plains in the dry season or from tundra in summer to nearby forests in winter. Some nomads herd while mounted on horses, such as the Mongols of East Asia, or on camels, such as the Bedouin of the Arabian Peninsula. Others, such as the Rendile of East Africa, herd cattle, goats, and sheep on foot.

> **nomadic livestock herder** A member of a group that continually moves with its livestock in search of forage for its animals.

Their need for mobility dictates that the nomads' few material possessions be portable, including the tents or huts used for housing (**Figure 8.12**). Their mobile lifestyle also affects how wealth is measured. Typically, in nomadic cultures wealth is based on the size of livestock holdings rather than on the accumulation of property and personal possessions. Usually, the nomads obtain nearly all of life's necessities from livestock products or by bartering with the farmers of adjacent river valleys and oases.

For a number of different reasons, nomadic herding has been in decline since the early twentieth century. Some national governments established policies encouraging nomads to practice **sedentary cultivation** of the land, meaning farming in fixed and permanent fields. This practice was begun in the

> **sedentary cultivation** Farming in fixed and permanent fields.

nineteenth century by British and French colonial administrators in North Africa, because it allowed greater control of the people by the central governments. Today, many nomads are voluntarily abandoning their traditional life to seek jobs in urban areas or in the Middle Eastern oil fields. Severe drought in sub-Saharan Africa's Sahel region, which decimated nomadic livestock herds, was a further impetus to abandon nomadic life.

In recent decades, research conducted by geographers and anthropologists in Africa's semiarid environments has revealed the sound logic of nomadic herding practices. These studies demonstrate that nomadic cultures' pasture and livestock management strategies are rational responses to an erratic and unpredictable environment. Rainfall is highly irregular in time and space, and herding practices must adjust. The most important nomadic strategy is mobility, which allows herders to take fullest advantage of the resulting variations in range productivity. These findings have led to a new appreciation of nomadic herding cultures and may cause governments to reconsider sedentarization programs, thus postponing the demise of herding cultures.

Livestock Ranching

Superficially, **ranching** might seem similar to nomadic herding. It is, however, a fundamentally different livestock-raising system. Although both nomadic herders and livestock ranchers specialize in animal husbandry and both live in arid or semiarid regions, livestock ranchers have fixed places of residence and operate as individuals rather than within a communal or tribal organization. In

> **ranching** The commercial raising of herd livestock on a large landholding.

FIGURE 8.12 Nomadic pastoralists in the Kashmir Valley, India. Mobility is key to nomadic pastoralists' successful use of variable and unpredictable environment. (Getty Images/Flickr RF.)

THINKING GEOGRAPHICALLY What are the hazards of this agricultural system?

addition, ranchers raise livestock on a large scale for market, not for their own subsistence.

Livestock ranchers are found worldwide in areas with environmental conditions that are too harsh for crop production. They raise only two kinds of animals in large numbers: cattle and sheep. Ranchers in the United States, Canada, tropical and subtropical Latin America, and the warmer parts of Australia specialize in cattle raising. Mid-latitude ranchers in cooler and wetter climates specialize in sheep. Sheep production is geographically concentrated to such a degree that only three countries (Australia, China, and New Zealand) account for 56 percent of the world's export wool.

Urban Agriculture

The United Nations (UN) calculates that in 2008 the human species passed a milestone. For the first time in history, more people are living in cities than in the countryside. As this global-scale rural-to-urban migration gained momentum, a distinct form of agriculture rose in significance. We might best call this **urban agriculture.** Millions of city dwellers, especially in developing countries, now produce enough vegetables, fruit, meat, and milk from tiny urban or suburban plots to provide most of their food, often with a surplus to sell. In China, urban agriculture now provides 90 percent or more of all the vegetables consumed in the cities. In the African metropolises of Kampala and Dar es Salaam, 70 percent of the poultry and 90 percent of the leafy vegetables consumed in the cities, respectively, come from urban lands. Even a developed country such as Russia derives nearly half of its food from such operations. Similarly, neighborhood gardens increasingly are found in inner-city areas of North America.

> **urban agriculture** The raising of food, including fruit, vegetables, meat, and milk, inside cities, especially common in developing countries.

Geographer Susanne Freidberg has conducted research demonstrating the importance of urban agriculture to family income and food security in West Africa. Focusing on the city of Bobo-Dioulasso in Burkina Faso, Freidberg showed that though plots were small, urban agriculture offered residents "a culturally meaningful way to fulfill their roles as food producers and family providers." In its heyday in the 1970s and 1980s, urban farming provided substantial incomes from vegetable sales in both the domestic and export markets. Since then, collapsing demand and the deterioration of environmental conditions have threatened the enterprise and undermined cooperation and trust within Bobo-Dioulasso's urban agricultural communities.

REFLECTING ON **GEOGRAPHY**

What types of agriculture occur in the three main densely populated areas of the world? (Compare Figures 3.1 and 8.1.) Are these two characteristics—type of agriculture and population density—linked?

Farming the Waters

Most of us don't think of the ocean when discussing agriculture. In fact, every year more and more of our animal protein is produced through **aquaculture:** the cultivation and harvesting of aquatic organisms under controlled conditions. Aquaculture includes **mariculture,** which often involves the transformation of coastal environments and the production of distinctive landscapes. Modern aquaculture includes shrimp farming, oyster farming, fish farming, pearl cultivation, and more. In the heart of Brooklyn, New York, urban aquaculture thrives in a laboratory, where fish bound for New York City restaurants are harvested from indoor tanks. Aquaculture has its own distinctive cultural landscapes of containment ponds, rafts, nets, tanks, and buoys (**Figure 8.13**). Like terrestrial agriculture, there is a marked regional character to aquacultural production.

> **aquaculture** The cultivation and harvesting of aquatic organisms under controlled conditions.
>
> **mariculture** A branch of aquaculture specific to the cultivation of marine organisms, often involving the transformation of coastal environments and the production of distinctive new landscapes.

Aquaculture is an ancient practice, dating back at least 4500 years. Older local practices, sometimes referred to as traditional aquaculture, involve simple techniques such as constructing retention ponds to trap fish or seeding flooded rice fields with shrimp. Commercial aquaculture, the industrialized, large-scale protein factories that are driving today's production growth, is a contemporary phenomenon. The greatest leaps in technology and production have occurred mostly since the 1970s.

FIGURE 8.13 Emerging mariculture landscape. As mariculture expands, coastal landscapes and ecosystems are transformed, as in the case of marine fish farming off Langkawi Island, Malaysia. (Copyright © age fotostock/SuperStock.)

THINKING GEOGRAPHICALLY Is mariculture a better solution to the need for protein in a growing population than land-based livestock farming? Explain.

In the twenty-first century, aquaculture is experiencing phenomenal growth worldwide, growing nearly four times faster than all terrestrial animal food-producing sectors combined. Aquaculture is a key reason that per capita protein availability has mostly kept pace with population growth in recent decades. Aquaculture's share of production is projected to continue growing as demand for seafood increases and wild fish stocks decline.

The extraordinary expansion of food production by aquaculture has come with high costs to the environment and human health. As with the rest of industrialized agriculture, most commercial aquaculture relies on high energy and chemical inputs, including antibiotics and artificial feeds made from the wastes of poultry and hog

processing. Such production practices tend to concentrate toxins in farmed fish, creating a potential health threat to consumers. The discharge from fish farms, which can be equivalent to the sewage from a small city, can pollute nearby natural aquatic ecosystems.

Although aquaculture can take place just about anywhere that water is found, strong regional patterns do exist. Mariculture is prevalent along tropical coasts, particularly in the mangrove forest zone. Marine coastal zones in general, especially in protected gulfs and estuaries, have high concentrations of mariculture. On a global scale, China dwarfs all other regions, producing over two-thirds of the world's farmed seafood (**Figure 8.14**). Asia and Pacific regions combined produce over 90 percent of the world's total.

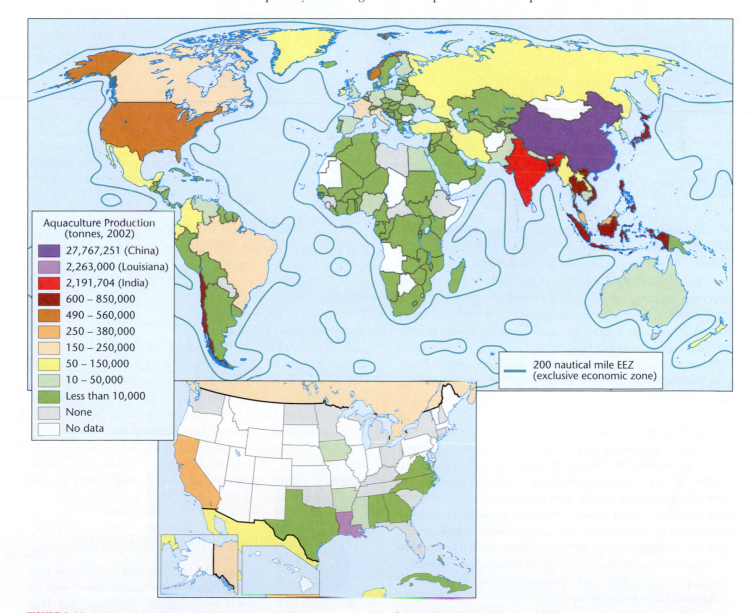

FIGURE 8.14 Global aquaculture and fisheries. Aquaculture is expanding rapidly around the globe, led by China, the world's top producer. Virtually all mariculture takes place within a country's 200-mile coastal territory, known as the EEZ (exclusive economic zone). The EEZ is also the site of most commercial fishing, which has greatly depleted wild fish stocks and raised the need for increased mariculture production.
(Source: The Global Education Project, Fishing and Aquaculture.)

THINKING GEOGRAPHICALLY Why would China be so interested in developing aquaculture?

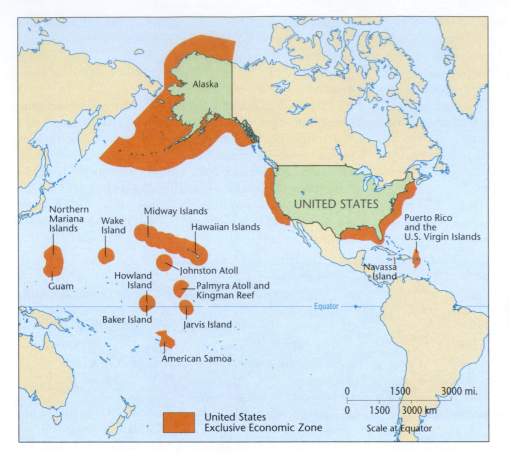

FIGURE 8.15 The U.S. exclusive economic zone (EEZ). These delimited zones are reserved exclusively for U.S. fishing, mariculture, mining, and other maritime economic activity.

THINKING GEOGRAPHICALLY What factors might cause the delimitation and management of EEZs for the world's many countries possessing coastlines to be politically problematic?

Globally, the right to engage in mariculture, along with other maritime economic activities such as fishing and undersea mining, is governed by the delimitation of *exclusive economic zones (EEZs)*. Under the United Nations Law of the Sea, states possessing coastline, or governing territories with coastline, are granted an extension of their boundaries out into the sea for the purposes of engaging in exploration and the use of marine resources. Generally, this zone is defined by a 200-nautical-mile limit from a country's coast. The United States has the largest EEZ of any country on Earth due to its extensive coastline and many island territories (**Figure 8.15**). This expansive EEZ has often been a source of political tension between the United States and other states possessing lesser opportunities for exploiting ocean resources for economic gain. In recent decades, debates between countries regarding EEZ rights have escalated in light of increasing global resource scarcities and environmental degradation. Overfishing, oil spills, and the dumping of waste are but a few potential sources of marine resource endangerment. As resource demands continue to increase, the conservation and protection of global waters within the world's EEZs will become even more vital.

Nonagricultural Areas

Areas of extreme climate, particularly deserts and subarctic forests, do not support any form of agriculture. Such lands are found predominantly in much of Canada and Siberia. Often these areas are inhabited by **hunting and gathering** groups of native peoples, such as the Inuit, who gain a livelihood by hunting game, fishing where possible, and gathering edible and medicinal wild plants. At one time all humans lived as hunter-gatherers. Today, fewer than 1 percent of humans do. Given the various inroads of the modern world, even these people rarely depend entirely on hunting and gathering. In most hunting-and-gathering societies, a division of labor by gender occurs. Males perform most of the hunting and fishing, whereas females carry out the equally important task of gathering harvests from wild plants. Hunter-gatherers generally rely on a great variety of animals and plants for their food.

> **hunting and gathering** The killing of wild game and the harvesting of wild plants to provide food in traditional cultures.

AGRICULTURAL **DIFFUSION**

How does the theme of cultural diffusion help us understand the spatial and cultural patterns of agricultural production and food consumption? Some of the variation among the agriculture regions we've discussed results from cultural diffusion. Agriculture and its many components are inventions; they arose as innovations in certain source areas and diffused to other parts of the world.

Origins and Diffusion of Plant Domestication

Agriculture probably began with the domestication of plants. A **domesticated plant** is one that is deliberately planted, protected, cared for, and used by humans. Such plants are genetically distinct from their wild ancestors because they result from selective breeding by agriculturists. Accordingly, they tend to be bigger than wild species, bearing larger and more abundant fruit or grain. For example, the original wild "Indian maize" grew on a cob only one-tenth the size of the cobs of domesticated maize.

> **domesticated plant** A plant deliberately planted, protected, cared for, and used by humans; it is genetically distinct from its wild ancestors as a result of selective breeding.

Plant domestication and improvement constituted a process, not an event. It began as the gradual culmination of hundreds, or even thousands, of years of close association between humans and the natural vegetation. The first step in domestication was perceiving that a certain plant was useful, which led initially to its protection and eventually to deliberate planting.

Cultural geographer Carl Johannessen suggests that the domestication process can still be observed. He believes that by studying current techniques used by native subsistence farmers in places such as Central America, we can gain insight into the methods of the first farmers of prehistoric antiquity. Johannessen's study of the present-day cultivation of the *pejibaye* palm tree in Costa Rica revealed that native cultivators actively engage in seed selection. All choose the seed of fresh fruit from superior trees, ones that bear particularly desirable fruit, as determined by size, flavor, texture, and color. Superior seed stocks are built up gradually over the years, with the result that elderly farmers generally have the best selections. Seeds are shared freely within family and clan groups, allowing rapid diffusion of desirable traits.

The widespread association of female deities with agriculture suggests that it was women who first worked the land. Recall the almost universal division of labor in hunting-gathering societies. Because women had day-to-day contact with wild plants and their mobility was constrained by childbearing, they probably played the larger role in early plant domestication.

Locating Centers of Domestication

When, where, and how did these processes of plant domestication develop? Most experts now believe that the process of domestication was independently invented at many different times and locations. Geographer Carl Sauer, who conducted pioneering research on the origins and dispersal of plant and animal domestication, was one of the first to propose this explanation.

Sauer believed that domestication did not develop in response to hunger. He maintained that necessity was not the mother of agricultural invention, because starving people must spend every waking hour searching for food and have no time to devote to the leisurely experimentation required to domesticate plants. Instead, he suggested this invention was accomplished by peoples who had enough food to remain settled in one place and devote considerable time to plant care. The first farmers were probably sedentary folk rather than migratory hunter-gatherers. He reasoned that domestication did not occur in grasslands or large river floodplains, because primitive cultures would have had difficulty coping with the thick sod and periodic floodwaters. Sauer also believed that the hearth areas of domestication must have been in regions of great biodiversity where many different kinds of wild plants grew, thus providing abundant vegetative raw material for experimentation and crossbreeding. Such areas typically occur in hilly districts, where climates change with differing sun exposure and elevation above sea level.

Geographers, archaeologists, and, increasingly, genetic scientists continue to investigate the geographic origins of domestication. Because the conditions conducive to domestication are relatively rare, most agree that agriculture arose independently in at most nine regions (**Figure 8.16**). All of these have

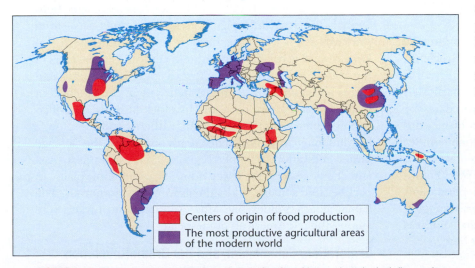

Centers of origin of food production

The most productive agricultural areas of the modern world

FIGURE 8.16 Ancient centers of plant domestication. New archaeological discoveries and new technologies such as genetic science are changing our understanding of the geography and history of domestication. This map represents a synthesis of the latest findings. (Source: Diamond, 2002.)

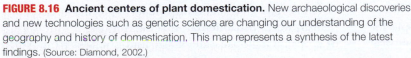

THINKING GEOGRAPHICALLY Why is Southwest Asia, one of the first sites of agricultural domestication, no longer a leading agricultural area?

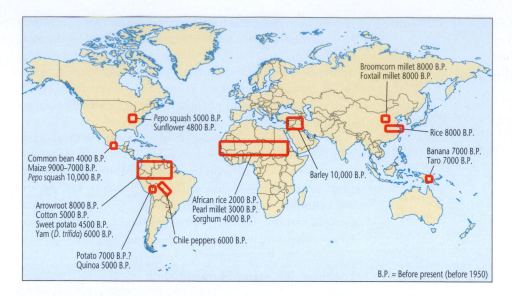

FIGURE 8.17 Ancient domestication regions for selected crops. Many of the crops still central to modern diets around the world were domesticated thousands of years ago in Asia, Africa, and Middle/South America.

THINKING GEOGRAPHICALLY Why do you think that most ancient plant domestication took place in the mid-latitudes rather than in more extreme portions of the Northern or Southern hemispheres?

made significant contributions to the modern global food system. For example, the Fertile Crescent in the Middle East is the origin of the great bread grains (wheat, barley, rye, and oats) that are so key to our modern diets. This region is also home to the first domesticated grapes, apples, and olives. China and New Guinea provided rice, bananas, and sugarcane, while the African centers gave us peanuts, yams, and coffee. Native Americans in Mesoamerica created another important center of domestication, from which came crops such as maize (corn), tomatoes, and beans. Farmers in the Andes domesticated the potato. While crop diffusions out of these nine regions have occurred over the millennia, the forces of globalization have now made even the rarest of local domesticates available around the world. **Figure 8.17** shows the likely areas of domestication for several common crops.

The dates of earliest domestication are continually being updated by new research findings. Until recently, archaeological evidence suggested that the oldest center is the Fertile Crescent, where crops were first domesticated roughly 10,000 years ago. However, domestication dates for other regions are constantly being pushed back by new discoveries. Most dramatically, in the Peruvian Andes archaeologists recently excavated domesticated seeds of squash and other crops that they dated to 9,240 years before the present. These seeds were associated with permanent dwellings, irrigation canals, and storage structures, suggesting that farming societies were established in the Americas 10,000 years ago, similar to the Fertile Crescent date.

Animal Domestication

A **domesticated animal** is one that depends on people for food and shelter and that differs from wild species in physical appearance and behavior as a result of controlled breeding and frequent contact with humans. Animal domestication apparently occurred later in prehistory than did the first planting of crops—with the probable exception of the dog, whose companionship with humans appears to be much more ancient. Typically, people value domesticated animals and take care of them for some utilitarian purpose. Certain domesticated animals, such as the pig and the dog, probably attached themselves voluntarily to human settlements to feast on garbage. At first, perhaps, humans merely tolerated these animals, later adopting them as pets or as sources of meat.

The early farmers in the Fertile Crescent deserve credit for the first great animal domestications, most notably that of herd animals. The wild ancestors of major herd animals—such as cattle, pigs, horses, sheep, and goats—lived primarily in a belt running from Syria and southeastern Turkey eastward across Iraq and Iran to central Asia. Farmers in the Middle East were also the first to combine domesticated plants and animals in an integrated system, the antecedent of the peasant grain, root, and livestock farming described earlier. These people began using cattle to pull the plow, a revolutionary invention that greatly increased the acreage under cultivation. In other regions, such as southern Asia and the Americas, far fewer domestications took place, in part because suitable wild animals were less numerous. The llama, alpaca, guinea pig, Muscovy duck, and turkey were among the few American domesticates.

Exploration, Colonialism, and the Green Revolution

Over the past 500 years, European exploration and colonialism were instrumental in redistributing a wide variety of crops on a global scale: maize and potatoes from North America to Eurasia and Africa, wheat and grapes from the Fertile Crescent to the Americas, and West African rice to the Carolinas and Brazil.

> **domesticated animal** An animal that depends on people for food and shelter and that differs from wild species in physical appearance and behavior as a result of controlled breeding and frequent contact with humans.

The diffusion of specific crops continues, extending the process begun many millennia ago. The introduction of the lemon, orange, grape, and date palm by Spanish missionaries in eighteenth-century California, where no agriculture existed in the Native American era, is a recent example of relocation diffusion. This was part of a larger process of multidirectional diffusion. Eastern Hemisphere crops were introduced to the Americas, Australia, New Zealand, and South Africa through the mass emigrations from Europe over the past 500 years. Crops from the Americas diffused in the opposite direction. For example, chili peppers and maize, carried by the Portuguese to their colonies in South Asia, became staples of diets all across that region (**Figure 8.18**).

In cultural geography, our understanding of agricultural diffusion focuses on more than just the crops; it also includes an analysis of the cultures and indigenous technical knowledge systems in which they are embedded. For example, geographer Judith Carney's study of the diffusion of African rice (*Oryza glaberrima*), which was domesticated independently in the inland delta area of West Africa's Niger River, shows the importance of indigenous knowledge. European planters and slave owners carried more than seeds across the Atlantic from Africa to cultivate in the Americas. The Africans taken into slavery, particularly women from the Gambia River region, had the knowledge and skill to cultivate rice. Slave owners actively sought slaves from specific ethnic groups and geographic locations in the West African rice-producing zone, suggesting that they knew about and needed Africans' skills and knowledge.

Not all innovations involve expansion diffusion and spread wavelike across the land; less orderly patterns are more typical. The **green revolution** in Asia provides an example. The green revolution is a product of modern agricultural science that involves the development of high-yield hybrid varieties of crops, increasingly genetically engineered, coupled with extensive use of chemical fertilizers. The high-yield crops of the green revolution tend to be less resistant to insects and diseases, necessitating the widespread use of pesticides. The green revolution, then, promises larger harvests but ties the farmer to greatly increased expenditures for seed, fertilizer, and pesticides. It enmeshes the farmer in the global corporate economy. In some countries, most notably India, the green revolution diffused rapidly in the latter half of the twentieth century. By contrast, countries such as Myanmar (Burma) resisted the revolution, favoring traditional methods. An uneven pattern of acceptance still characterizes the paddy rice areas today.

> **green revolution** The fairly recent introduction of high-yield hybrid crops and chemical fertilizers and pesticides into traditional agricultural systems, most notably paddy rice farming, with attendant increases in production and ecological damage.

(a)

(b)

FIGURE 8.18 Chili peppers in Nepal and Korea. A Tharu tribal woman of lowland Nepal prepares a condiment made of chili peppers from her garden (a), and in South Korea chili peppers dry under a plastic-roofed shed (b). This crop comes from the Indians of Mexico. (Courtesy of Terry G. Jordan-Bychkov.)

THINKING GEOGRAPHICALLY How might the chili pepper have diffused so far and become so important in Asia?

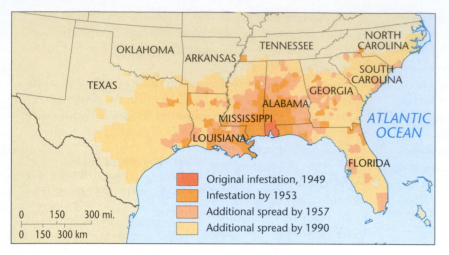

FIGURE 8.19 Relocation diffusion of the imported fire ant in the American South. Accidentally introduced from tropical Latin America with a boatload of bananas in Mobile, Alabama, the fire ant has since resisted efforts to eradicate it and has continued to spread. (Source: Lofgren and Vander Meer, 1986.)

Original infestation, 1949
Infestation by 1953
Additional spread by 1957
Additional spread by 1990

THINKING GEOGRAPHICALLY How might this relocation diffusion have proceeded, and what barriers would the fire ant encounter? How did outliers develop? Why is this cultural diffusion?

The green revolution illustrates how cultural and economic factors influence patterns of diffusion. In India, for example, new hybrid rice and wheat seeds first appeared in 1966. These crops required chemical fertilizers and protection by pesticides, but with the new hybrids India's 1970 grain production output was double its 1950 level. However, poorer farmers—the great majority of India's agriculturists—could not afford the capital expenditures for chemical fertilizer and pesticides, and the gap between rich and poor farmers widened. Many of the poor became displaced from the land and flocked to the overcrowded cities of India, aggravating urban problems. To make matters worse, the use of chemicals and poisons on the land heightened environmental damage.

The widespread adoption of hybrid seeds has created another problem: the loss of plant diversity or genetic variety. Before hybrid seeds diffused around the world, each farm developed its own distinctive seed types through the annual harvest-time practice of saving seeds from the better plants for the next season's sowing. Enormous genetic diversity vanished almost instantly when farmers began purchasing hybrids rather than saving seed from the last harvest. "Gene banks" have belatedly been set up to preserve what remains of domesticated plant variety, not just in the areas affected by the green revolution but also in the American Corn Belt and many other agricultural regions where hybrids are now dominant. In sum, the green revolution has been a mixed blessing.

Not all diffusion related to agriculture has been intentional. In fact, accidental diffusion accomplished by humans probably occurs more commonly than the purposeful type. An example is the diffusion of the tropical American fire ant, so named because of its very painful sting (**Figure 8.19**). A shipload of plantation-grown bananas from tropical America accidentally brought the fire ant to Mobile, Alabama, in 1949, and a continuing relocation diffusion has since brought the fire ant across most of the American South. These vicious ants now endanger livestock and poultry raising, because swarms of them can attack and kill young animals.

In addition to animal species, pest plant species have also diffused to regions where they now interfere with commercial agriculture. One such example is the unintentional spread of a plant commonly known as tumbleweed. Though perhaps seeming quite harmless, and maybe even picturesque when viewed rolling across rural landscapes of the American West, this Russian form of the thistle plant is actually an invasive plant species. It diffused to the United States from Asia over 100 years ago, but in recent decades it has become increasingly troublesome to farmers and livestock producers as it has spread throughout most of the western United States and continues to diffuse eastward (**Figure 8.20**). This spread has been possible because, in order to survive, the tumbleweed plant dries and detaches from the location where it grows and rolls across the landscape to disburse its seeds. In doing so, the tumbleweed often acts as a vehicle for other non-native pest plant and insect species that are picked up within it as it tumbles. As the rolling plants eventually come to rest and pile up along fences and buildings in agricultural areas, they cause extensive and expensive structural damage. Tumbleweeds have also been known to contribute to the spread of wildfires, further endangering valuable agricultural lands. The plant can be a hazard to the general public as well, as it is credited with causing numerous road accidents.

REFLECTING ON GEOGRAPHY

Agricultural diffusion continues today. You participate in it when you decide whether to eat a food you have not previously known. What goes through your mind in, say, an ethnic restaurant when you decide to try, or not to try, a new food?

AGRICULTURAL ECOLOGY

How are environment-culture relations expressed through the production and consumption of food? Agriculture has been the fundamental encounter between culture and the environment for more than 10,000 years, as human labor is mixed with

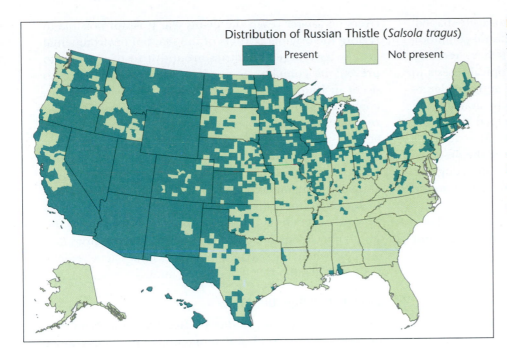

FIGURE 8.20 Modern distribution of the Russian Thistle plant (tumbleweed) in the United States. The tumbleweed was introduced to the United States through relocation diffusion from Asia over 100 years ago. It now thrives over approximately one half of the land area of the continental United States, continuing its invasive spread east and southward. (Source: Center for Invasive Species and Ecosystem Health, University of Georgia.)

THINKING GEOGRAPHICALLY What physical environmental factors do you think may have slowed the spread of the tumbleweed into the southeastern United States?

nature's bounty to produce our sustenance. What we eat and how we eat it is a basic source of cultural identity. In many ways, the map of agricultural regions (see Figure 8.1, p. 214) reflects human adaptation to environmental influences. At the same time, thousands of years of agricultural use of the land has led to massive alterations in our natural environment.

Cultural Adaptation, Nature, and Technology

Historically, climate and the physical environment have exerted the greatest influence on agriculture. People have had to adjust their subsistence strategies and techniques to the prevailing regional climate conditions. In addition,

soils have played an influential role in both agricultural practices and food provisioning. Swidden cultivation in part reflects an adaptation to poor tropical soils, which rapidly lose their fertility when farmed. Peasant agriculture, by contrast, often owes its high productivity to the long-lived fertility of local volcanic soils. Terrain has also influenced agriculture, as farmers tended to cultivate relatively level areas (**Figure 8.21**). In sum, the constraints of climate, soil, and terrain historically have limited the types of crops that could be grown and the cultivation methods that could be practiced.

In recent centuries, markets, technology, and capital investment have greatly altered the spatial patterns of

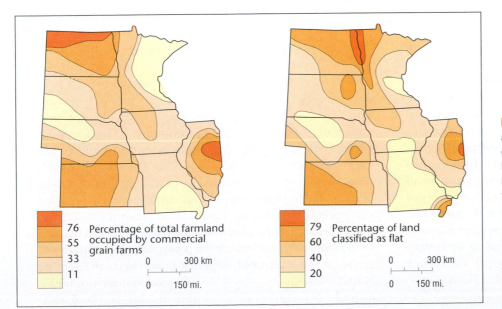

76	Percentage of total farmland
55	occupied by commercial
33	grain farms
11	

0   300 km
0   150 mi.

79	Percentage of land
60	classified as flat
40	
20	

0   300 km
0   150 mi.

FIGURE 8.21 The influence of terrain on agriculture. The spatial relationship of commercial grain farming and flat terrain appears in the American Midwest in about 1960. *Flat* terrain is defined as any land with a slope of 3° or less. Commercial grain farming is completely mechanized, and flat land permits more efficient machine operation. The result is this striking correlation between a type of agriculture and a type of terrain. (Source: After Hidore, 1963.)

THINKING GEOGRAPHICALLY What other factor might attract mechanized grain farming to level land?

agriculture that climate and soils had historically shaped. Expanding global-scale markets for agricultural commodities such as sugar, coffee, and edible oils have reduced millions of acres of biologically diverse tropical forests to monocrop plantations. Synthetic fertilizers and petroleum-based insecticides and herbicides, widely available in developed countries after World War II, helped boost agricultural productivity to unimagined levels. Massive dams and large-scale irrigation systems have caused the desert to bloom from central Asia to the Americas, converting, for example, the semiarid Central Valley of California into the world's most productive agricultural region.

The ecological price of such technological miracles is high. Drainage and land reclamation destroy wetlands and associated biodiversity. The application of synthetic fertilizers results in nutrient-rich runoff from farms that enters freshwater systems as pollution, lowering water quality, destroying aquatic habitat, and reducing biodiversity. Agrochemicals also enter the environment as runoff, as residue on food crops, and in the tissues of livestock. These chemicals ultimately reduce biodiversity, pollute water systems, and cause increases in the rates of cancer and birth defects in humans and animals.

In the context of global climate change, great concern exists about the long-term sustainability of modern agricultural practices. In many parts of the world, groundwater is being pumped to the surface faster than it can be replenished and reservoirs are approaching the end of their life spans. Well-and-pump irrigation has drastically lowered the water table in parts of the American Great Plains, particularly Texas, causing ancient springs to go dry. Climate change data from the distant past and computer models of future climate conditions suggest that the U.S. Southwest is likely entering a drier climate era. Lake Mead, the giant reservoir in Arizona and Nevada that helps supply water to California's cities and farms, was only 49 percent full in 2006 and is unlikely ever to be full again. This region, the most agriculturally productive in the world, will face increasingly difficult questions regarding the viability of current farming and land-use practices.

Another area where arid land irrigation has had severe ecological consequences lies in the Aral Sea on the borderland between Kazakhstan and Uzbekistan in central Asia. Since the latter half of the twentieth century, this sea—once the world's fourth largest—has shrunk to a fraction of its former surface area (**Figure 8.22**), forming two separate and much smaller seas, Large Aral and Small Aral. This environmental crisis is the result of excessive extraction of water from the rivers that feed the lake. The water is used to irrigate farmlands surrounding the sea, which would otherwise be infertile, as well as to sustain people and livestock in the region. As a result, the source

rivers no longer reach the dying sea. It is not possible to restore the original borders of the sea, but environmental groups have urged countries around it to take action to prevent its complete extinction, which is predicted to occur in 10 to 15 years. Nevertheless, four out of the five countries rimming the lake, including Kazakhstan, plan to *increase* extraction from source rivers to meet the needs of agricultural development for their growing populations.

The consequences of the diminishing Aral Sea extend far beyond the loss of habitat for many marine plants and animals. Contamination of the lake with agrochemicals that wash in from adjacent farm fields has resulted in tons of toxic dust that has dispersed to areas as far away as the Antarctic, Greenland, and Europe. Sandstorms, from desert areas where the sea used to be, also now plague regions as far as 310 miles (500 kilometers) away.

Sustainable Agriculture

As cultural geography studies by Zimmerer and many others have shown, local and indigenous knowledge about ecological conditions can be a foundation for sustainable agriculture. **Sustainability**—the survival of a land-use system for centuries or millennia without destruction of the environmental base—is the central ecological issue confronting agriculture today. The case of the Quichua peasants offers an optimistic

> **sustainability** The survival of a land-use system for centuries or millennia without destruction of the environmental base.

assessment of indigenous knowledge as the basis for long-term sustainability. Their response to contemporary market pressures suggests that development and conservation can be compatible. Their knowledge of complex and variable ecological conditions in the Andes has allowed them to farm highly diverse crop varieties, a practice that in some cases has been strengthened by economic development.

Another example of sustainable indigenous agriculture is the paddy rice farming that occurs near the margins of the Asian wet-rice region, where unreliable rainfall causes harvests to vary greatly from one year to the next. Farmers have developed complex cultivation strategies to avert periodic famine, including growing many varieties of rice. These farmers, including those in parts of Thailand, almost universally rejected the green revolution. The simplistic advice given to them by agricultural experts working for the Thai government was inappropriate for their marginal lands. Based on generations of experimentation, the local farmers knew that their traditional diversified adaptive strategy was superior. In West Africa, peasant grain, root, and livestock farmers have also developed adaptations to local environmental influences. They raise a multiplicity of crops on the more humid lands near the coast. Moving

1989

2012

(a)   (b)

FIGURE 8.22 **The shrinking of the Aral Sea 1989–2012.** Due to excessive water extraction from source rivers, the Aral Sea has shrunk dramatically. The result has been widespread environmental impacts through Central Asia and beyond. [a: The six scenes were collected by Landsat 4's Thematic Mapper Sensor in four passes: July 22, August 7 and 16, and September 7, 1989. Courtesy of Global Land Cover Facility (www.landcover.org); b: NASA Earth Observatory.]

THINKING GEOGRAPHICALLY How might the ecological crisis of the Aral Sea also become a political issue?

inland toward the drier interior, farmers plant fewer kinds of crops but grow more drought-resistant varieties. Having observed many cases like these in which local practices have proved effective and sustainable, most geographers now agree that agricultural experts need to consider indigenous knowledge when devising development plans.

Intensity of Land Use

Great spatial variation exists in the intensity of rural land use. As noted earlier in this chapter, intensive agriculture means that a large amount of human labor or investment capital, or both, is put into each acre or hectare of land, with the goal of obtaining the greatest output. Intensity can be calculated by measuring either energy input or level of productivity. In much of the world, especially the paddy rice areas of Asia, high intensity is achieved through prodigious use of human labor, which results in a rice output per unit of land that is the highest in the world. In Western countries, high intensity is achieved through the use of massive amounts of investment capital for machines,

fertilizers, and pesticides, resulting in the highest agricultural productivity per capita found anywhere.

Many geographers support the theory that increased land-use intensity is a common response to population growth. As demographic pressure mounts, farmers systematically discard the more geographically extensive adaptive strategies to focus on those that provide greater yield per unit of land. In this manner, the population increase is accommodated. The resultant farming system may be riskier, because it offers fewer options and possesses greater potential for environmental modification, but it does yield more food—at least in the short run. Other geographers reject this theory, arguing instead that increases in population density follow innovations, such as the introduction of new high-calorie crops, which lead to greater land-use intensity.

The Desertification Debate

Over the millennia, as dependence on agriculture grew and as population increased, humans made ever larger demands on the forests. With the rise of urban civilization

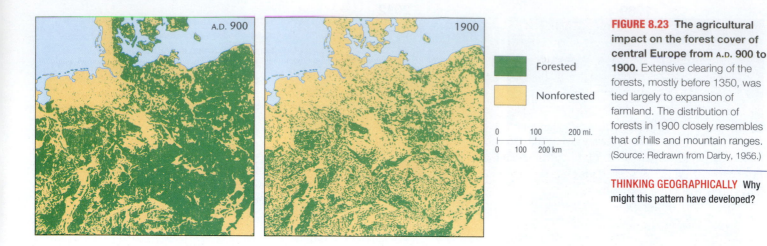

FIGURE 8.23 The agricultural impact on the forest cover of central Europe from A.D. 900 to 1900. Extensive clearing of the forests, mostly before 1350, was tied largely to expansion of farmland. The distribution of forests in 1900 closely resembles that of hills and mountain ranges. (Source: Redrawn from Darby, 1956.)

THINKING GEOGRAPHICALLY Why might this pattern have developed?

and conquering empires, the human transformation of forests to fields accelerated and expanded. In many parts of China, India, and the Mediterranean lands, forests virtually vanished. In transalpine Europe, the United States, and some other areas, they were greatly reduced (**Figure 8.23**).

Grasslands suffered similar modifications. Farmers occasionally plowed grasslands that were too dry for sustainable crop production, and herders sometimes damaged semiarid pastures through overgrazing. The result could be **desertification,** a process whereby human actions unintentionally turn productive lands into deserts through agricultural and pastoral misuse, destroying vegetation and soil to the point where they cannot regenerate. Desertification was first studied half a century ago by geographer Rhoads Murphey. He argued that farmers caused substantial parts of North Africa to be added to the margins of the Sahara Desert. He noted the catastrophic decline of countries such as Libya and Tunisia in the 1500 years since the time of Roman rule, when North Africa served as the "granary of the Empire," yielding huge wheat harvests.

More recent research on desertification has centered on the Sahel, a semiarid tropical savanna region just south of the Sahara Desert in Africa (**Figure 8.24**). A series of droughts in the 1970s and 1980s raised concern that the Sahel was becoming desertified. This was a focal point of discussion at the 1977 UN Conference on Desertification. One theory advanced at the conference was that farmers and pastoralists were overusing the land, destroying vegetation to such an extent that the plant life could not regenerate. Lands that had been covered with pastures and fields could become permanently joined with the adjacent Sahara. In sum, researchers theorized that Africans were overgrazing rangelands and using poor cultivation practices, which were causing the desert to spread southward.

> **desertification** A process whereby human actions unintentionally turn productive lands into deserts through agricultural and pastoral misuse, destroying vegetation and soil to the point where they cannot regenerate.

In the intervening years, substantial evidence, much of it obtained through satellite images, has raised questions about this theorized link between land use and the advancing Sahara Desert. Satellite imagery suggests to many researchers that the semiarid lands possess more resiliency than was once thought. Since 1960, they claim that the Sahara-Sahel boundary has not migrated steadily south but fluctuated as it always has, responding to wetter and drier years. New research by geographers and anthropologists also challenges the general claim that Africans were misusing the land. These findings suggest that African land-use practices in the region are highly adapted to the variable and unpredictable environments of the Sahel. A great deal of careful research is needed to distinguish the fluctuations caused by climate variability from permanent ecological damage caused by human misuse. For now, the debate continues over the extent and sources of desertification in the Sahel.

REFLECTING ON GEOGRAPHY

What might be some ecological consequences of expanding cropland to meet the world's rising food needs?

Environmental Perception by Agriculturists

People perceive the physical environment through the lenses of their culture. Each person's agricultural heritage can be influential in shaping these perceptions. This is not surprising, because human survival depends on how successfully people can adjust their ways of making a living to environmental conditions.

The Great Plains of the United States provide the setting for how an agricultural experience in one environment influenced farmers' environmental perceptions and subsequent behavior when they began farming in the Great Plains. The Plains farmers who came from the humid eastern United States consistently underestimated the problem of drought in their new home. In the 1960s, geographer Thomas Saarinen pointed out that almost every Great Plains farmer, even the oldest and most experienced of the farmers, underestimated the frequency of the dry periods. By contrast, culturally preadapted German-speaking immigrants from the steppes of Russia and Ukraine, an area very much like the American Great Plains, accurately perceived the new land and experienced fewer problems due to drought.

Farmers rely on climatic stability. A sudden spell of unusual weather can change agriculturists' environmental perceptions. Geographer John Cross studied Wisconsin agriculture following a series of floods, droughts, and other anomalies. He found that two-thirds of all Wisconsin dairy farmers now believe the climate is changing for the worse, and fully one-third told him that continued climatic variability threatened their operations. Perhaps they perceive the environmental hazard to be greater than it really is, but they make decisions based on their perceptions.

Environmental perceptions also operate at the level of society. Such is the case in the U.S. Southwest, where early twentieth century dam building on the Colorado River allowed for the expansion of irrigated agriculture and an urban population boom that is still going. The widespread perception that water is abundant, however, has been based on a misreading of the climate. The early 1920s, when the Colorado River flow was divided among western states, was an unusually wet period in the region's climate. Officials greatly overestimated average river flow when allocating water rights, thus creating a perception of water abundance. The region is now entering a drier phase, partly associated with global warming, which is bound to put agriculture in increasing conflict with urban demands. Because migration into the Southwest continues, where urban growth outpaces that of the rest of the country, we can conclude that perceptions have yet to catch up with the reality of water scarcity.

Organic Agriculture

Alarmed by the ecological and health hazards of chemical-dependent, industrialized agriculture, a small counterculture movement emerged in the United States and Europe in the 1960s and 1970s. Geographer Julie Guthman labels this the organic farming movement. For Guthman, the movement in the United States saw **organic agriculture** as a solution to a range of social, cultural, political, and

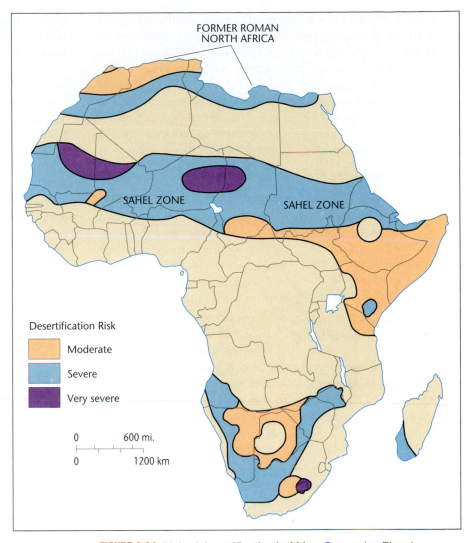

FIGURE 8.24 **Risk of desertification in Africa.** Geographer Rhoads Murphey began the debate about the relation between desertification and agricultural land use a half century ago, and it continues. Africa has been at the center of the debate. (Sources: Murphey, 1951; Thomas, 1993.)

THINKING GEOGRAPHICALLY Are the deserts of the world expanding because of agricultural land use, or do desert boundaries advance and retreat primarily in response to climate fluctuations?

environmental ills. These included the loss of small, family-owned and -operated farms, environmental pollution from industrial agriculture, corporate control of the food system, and the nutritional deficiencies of highly processed foods. Organic agriculture relies on manuring, mulching, and biological pest control and rejects the use of synthetic fertilizers, insecticides, herbicides, and genetically modified crops.

By legislating regulatory standards for organic agriculture, the U.S. state and federal governments provided

organic agriculture A form of farming that relies on manuring, mulching, and biological pest control and rejects the use of synthetic fertilizers, insecticides, herbicides, and genetically modified crops.

organic farmers the basis for differentiating their product in the marketplace. This in turn has allowed producers and retailers to charge consumers a premium. Premium pricing has encouraged many producers to switch to organic agriculture. The organic food market is now the most rapidly growing (and most profitable) agricultural sector. Affluent consumer demand for organics in developed countries, while only 1 to 3 percent of national retail sales, is growing at the phenomenal pace of 20 to 25 percent annually.

Green Fuels from Agriculture

Henry Ford fueled his first car with alcohol, and Rudolf Diesel ran his engine on fuel made from peanut oil. They soon abandoned these **biofuels,** however, for nonrenewable fossil fuels derived from "rock oil," and the rest is history. The modern global economy is dependent on fossil fuels to produce everything that we eat, wear, listen to, read, and live in. Now fossil fuels, particularly oil, have become scarce, causing prices and political uncertainties to increase. In addition, fossil-fuel combustion is a source of atmospheric greenhouse gases, which are partly responsible for global warming. Thus, an urgent search is under way to find alternative, renewable fuel supplies, and agriculture has become one of the main sources.

> **biofuel** Broadly, any form of energy derived from biological matter, increasingly used in reference to replacements for fossil fuels in internal combustion engines, industrial processes, and the heating and cooling of buildings.

How can agriculture be a source of energy for industry? By tapping the simple process of fermentation, many plant materials can be converted to a combustible alcohol, ethanol. In the United States, corn is the main source of ethanol, and new ethanol plants are springing up across the Corn Belt in the midwestern United States. To encourage the growth of these new modes of fuel production, the U.S. government instituted subsidies for America's corn ethanol producers amounting to almost $6 billion per year. These subsidies recently expired, so corn ethanol producers now compete on the U.S. energy market without the assistance of governmental supports. Experts predict, however, that rising oil prices globally will likely ensure that corn ethanol production remains competitive in years to come without the help of subsidies. In addition, other forms of biofuel from nonfood crops such as wood chips and switchgrass (grown in warm-season climates throughout North America) are still supported by federal subsidies due to higher research and development costs associated with this type of ethanol production.

While the use of food crops (mainly corn) for the production of biofuels may help ease demands on dwindling world fossil-fuel reserves, it has also elevated corn prices worldwide. Higher prices, though beneficial to corn producers, have led to less affordable corn available to countries using it primarily for human consumption. As a result, some humanitarian organizations, as well as the United Nations, have taken a position against increased biofuel production, arguing that it is jeopardizing food security for some developing countries.

In addition, the production of biofuels from tropical crops, such as oil palm and sugarcane, has led to the clear-cutting of rainforest vegetation in many regions. Many climate researchers oppose this deforestation, warning that the alteration of delicate tropical ecosystems will contribute to global warming. Debates among environmentalists, biofuel producers, and adherents will likely continue on this issue, though alternatives for biofuel crop production on abandoned agricultural lands and other nonforested lands are being sought. Ultimately, whether biofuels from agriculture will produce a sustainable, environmentally friendly alternative to fossil fuels remains to be seen.

CULTURAL **INTERACTION** IN AGRICULTURE

How does agriculture interact with other facets of culture? Let us next observe some of the ways other cultural forces influence agricultural activities. Religious taboos, politically based tariff restrictions, rural land-use zoning policies, population density, and many other human factors influence the type and distribution of agricultural activities.

Agriculture and the Economy: Local-Global Food Provisioning

As European maritime explorers brought far-flung cultures into contact with one another, a multitude of crops were diffused around the globe. The processes of exploration, colonization, and globalization created new regional cuisines (imagine Italian cuisine without tomatoes from the Americas) and at the same time simplified the global diet to a disproportionate reliance on only three grains: wheat, rice, and maize.

The expansion of European empires in the seventeenth and eighteenth centuries was inseparable from the expansion of tropical plantation agriculture. Plantations in warm climates produced what were then luxury foods for markets in the global North, which had developed seemingly insatiable appetites for sugar, tea, coffee, and other tropical crops. The expansion of plantation agriculture had profound effects on local ecology, all but obliterating, for example, the forests of the Caribbean and the tropical coasts of the Americas.

In short, we have witnessed the development of a global food system, which, for better or worse, has freed consumers in the affluent regions of the world from the constraints of local ecologies. Fresh strawberries, bananas, pears, avocados, pineapples, and many other types of temperate, subtropical, and tropical produce are available in our urban supermarkets any day, any time of the year. On the other hand, the emphasis on a relatively small number of staple crops desired by northern consumers can mean the abandonment of local crop varieties and a decline in the associated biological diversity. Imported refined wheat from the global North enters poor tropical countries by the shipload, altering local dietary cultures and undercutting the ability of local farmers to sell their crops at a profitable price.

These are the general patterns, but the globalization of food and agriculture has complex effects on culture and ecology that vary by location and spatial scale. These complexities are best illustrated by a case study from the Peruvian Andes. Geographer Karl Zimmerer conducted extensive fieldwork among Quichua peasant farmers in the Paucartambo Andes to determine the effects of economic change on indigenous agricultural practices and the genetic diversity of local crops. Zimmerer's study produced surprising findings on the complex relationships among culture, economy, and the environment. On the question of whether farmers must abandon crop diversity in order to adopt new, commercially oriented high-yielding varieties, he found there was no simple answer. In fact, it was the better-off peasants, heavily involved in commercial farming, who had the resources and land to cultivate diverse crops and "enjoy their agronomic, culinary, cultural, and ritual values." Among these values was the use of diverse noncommercial potato varieties in local bartering. The ability to use

noncommercial varieties in this way is valued in the local culture because it is a traditional way to cement interpersonal bonds. Such uses emphasize the cultural importance of crop diversity. Zimmerer discovered that the cultural relevance of crops was a strong motivation for planting by well-off farmers. At least 90 percent of the genetically diverse crops had been conserved, even as the Quichua were further integrated into commercial production for the market. In short, cultural values, and not merely a strict economic or ecological calculus, critically influenced Quichua farming decisions.

Agriculture and Transportation Costs: The von Thünen Model

Geographers and others have long tried to understand the distribution and intensity of agriculture based on transportation costs to market. Long before globalization took hold, the nineteenth-century German scholar-farmer Johann Heinrich von Thünen developed a core-periphery model to address the problem. In his model, von Thünen proposed an "isolated state" that had no trade connections with the outside world; possessed only one market, located centrally in the state; had uniform soil and climate; and had level terrain throughout. He further assumed that all farmers located the same distance from the market had equal access to it and that all farmers sought to maximize their profits and produced solely for market. von Thünen created this model to study the influence of distance from market and the concurrent transport costs on the type and intensity of agriculture.

Figure 8.25 presents a modified version of von Thünen's isolated-state model, which reflects the effects of

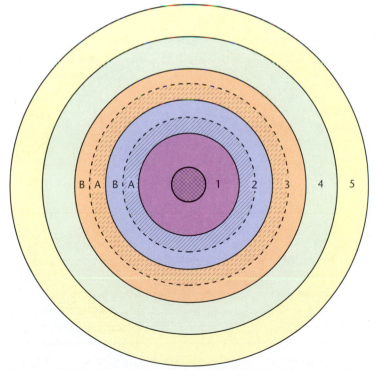

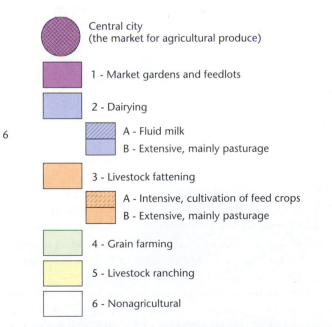

Central city (the market for agricultural produce)

1 - Market gardens and feedlots

2 - Dairying
- A - Fluid milk
- B - Extensive, mainly pasturage

3 - Livestock fattening
- A - Intensive, cultivation of feed crops
- B - Extensive, mainly pasturage

4 - Grain farming

5 - Livestock ranching

6 - Nonagricultural

FIGURE 8.25 Von Thünen's isolated-state model. The model is modified to fit the modern world better and shows the hypothetical distribution of types of commercial agriculture. Other causal factors are held constant to illustrate the effect of transportation costs and differing distances from the market. The more intensive forms of agriculture, such

as market gardening, are located nearest the market, whereas the least intensive form (livestock ranching) is the most remote.

THINKING GEOGRAPHICALLY Why does the model have the configuration of concentric circles?

improvements in transportation since the 1820s, when von Thünen proposed his theory. The model's fundamental feature is a series of concentric zones, each occupied by a different type of agriculture, located at progressively greater distances from the central market.

REFLECTING ON **GEOGRAPHY**

Why should we study spatial models, such as von Thünen's, when they do not depict reality?

For any given crop, the intensity of cultivation declines with increasing distance from the market. Farmers near the market have minimal transportation costs and can invest most of their resources in labor, equipment, and supplies to augment production. Indeed, because their land is more valuable and subject to higher taxes, they have to farm intensively to make a bigger profit. With increasing distance from the market, farmers invest progressively less in production per unit of land because they have to spend progressively more on transporting produce to market. The effect of distance means that highly perishable products such as milk, fresh fruit, and garden vegetables have to be produced near the market, whereas peripheral farmers have to produce nonperishable products or convert perishable items into a more durable form, such as cheese or dried fruit.

The concentric-zone model describes a situation in which highly capital-intensive forms of commercial agriculture, such as market gardening and feedlots, lie nearest to market. The increasingly distant, successive concentric belts are occupied by progressively less intensive types of agriculture, represented by dairying, livestock feeding, grain farming, and ranching.

How well does this modified model describe reality? As we would expect, the real world is far more complicated. For example, the emergence of **cool chains** for agricultural commodities—the refrigeration and transport technologies that bring fresh produce from fields around the globe to our dinner tables—have collapsed distance. Still, on a world scale, we can see that intensive commercial

> **cool chain** The refrigeration and transport technology that allows for the distribution of perishables.

types of agriculture tend to occur most commonly near the huge urban markets of northwestern Europe and the eastern United States (see Figure 8.1, p. 214). An even closer

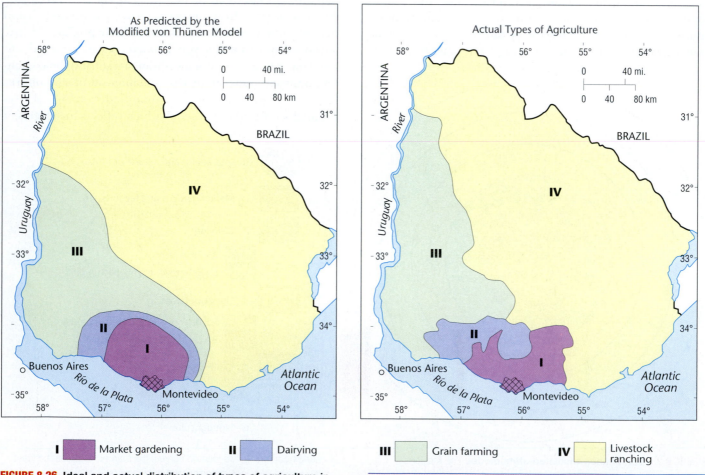

FIGURE 8.26 Ideal and actual distribution of types of agriculture in Uruguay. This South American country possesses some attributes of von Thünen's isolated state, in that it is largely a plains area dominated by one city.

Legend:
I — Market gardening
II — Dairying
III — Grain farming
IV — Livestock ranching

THINKING GEOGRAPHICALLY In what ways does the spatial pattern of Uruguayan agriculture conform to von Thünen's model? How is it different? What might cause the anomalies?

match can be observed in smaller areas, such as in the South American nation of Uruguay (**Figure 8.26**).

The value of von Thünen's model can also be seen in the underdeveloped countries of the world. Geographer Ronald Horvath made a detailed study of the African region centering on the Ethiopian capital city of Addis Ababa. Although noting disruptions caused by ethnic and environmental differences, Horvath found "remarkable parallels between von Thünen's crop theory and the agriculture around Addis Ababa." Similarly, German geographer Ursula Ewald applied the model to the farming patterns of colonial Mexico during the period of Spanish rule, concluding that even this culturally and environmentally diverse land provided "an excellent illustration of von Thünen's principles on spatial zonation in agriculture."

Can the World Be Fed?

Are famine and starvation inevitable as the world's population grows, as Thomas Malthus predicted (see Chapter 3)? Or can our agricultural systems successfully feed nearly 7 billion people? In trying to answer these questions, we face a paradox. Today, nearly 1 billion people are malnourished, some to the point of starvation. Almost every year, we read of famines occurring somewhere in the world. In 2008, there were food riots in 30 countries around the world when food prices shot beyond the reach of hundreds of thousands of the urban poor. Between 1990 and 2010 the number of hungry people in western and southern Asia and sub-Saharan Africa increased by tens of millions (**Figure 8.27**).

Yet—and this would astound Malthus—food production has grown more rapidly than the world population over the past 40 or 50 years. Per capita, more food is available today than in 1950, when only about half as many people lived on Earth. Production continues to increase. From 1996 to 2006 world food production increased at an annual rate of 2.2 percent, and hunger was reduced by 30 percent in more than 30 countries. Thus, paradoxically, on a global scale there is enough food produced to feed everyone, while famines and malnutrition prevail.

What explains this paradox? If the world food supply is sufficient to feed everyone and yet hunger afflicts one of every six or seven persons, then cultural or social factors must be responsible. Ultimately, international political economics, not global food shortages, causes hunger and starvation. International trade favors the farmers of wealthier countries through systems of government subsidies that make the prices of their agricultural exports artificially low. Farmers in developing countries find it difficult to compete. Many developing countries do not grow enough food to feed their populations, and they cannot afford to purchase enough imported food to make up the difference. As a result, famines can occur even when plenty of food is available. Millions of Irish people starved in the 1840s while

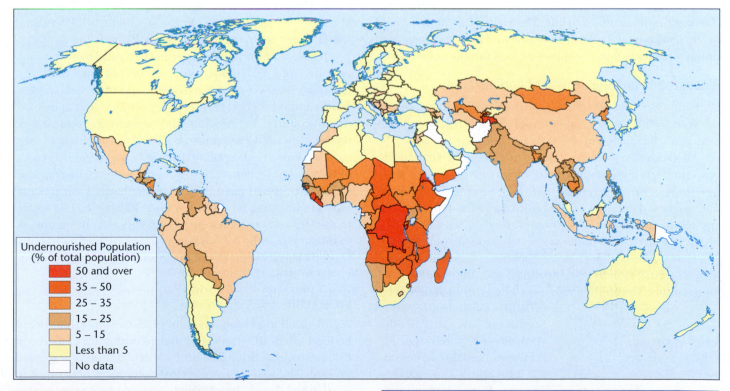

Undernourished Population
(% of total population)

- 50 and over
- 35 – 50
- 25 – 35
- 15 – 25
- 5 – 15
- Less than 5
- No data

FIGURE 8.27 Mapping hunger worldwide. While food supply has outpaced population growth on a global scale, many world regions continue to suffer from malnutrition. (Source: UN FAO, 2008.)

THINKING GEOGRAPHICALLY What geographic patterns does this map reveal? How might they be explained?

adjacent Britain possessed enough surplus food to have prevented this catastrophe. Bangladesh suffered a major famine in 1974, a year of record agricultural surpluses in the world.

Internal government policies are also an important cause of famine. The roots of the largest famine of the twentieth century lie in the agricultural policies of the Chinese government's 1958–1961 Great Leap Forward. The Chinese government required peasants to abandon their individual fields and work collectively on large, state-run farms. This policy of collectivization succeeded in boosting food production in some cases but failed in most. Thirty million rural Chinese died of starvation during the Great Leap Forward.

Misguided government policies triggered one of the first famines of the twenty-first century as well. In the early 2000s, Zimbabwe's President Robert Mugabe clung to power by demonizing white commercial farmers. In 2002 he threatened Zimbabwe's commercial farmers with imprisonment if they continued to farm. Other government policies discouraged planting and cultivation, thus producing another human-caused famine.

Even when major efforts are made to send food from wealthy countries to famine-stricken areas, the poor transportation infrastructure of developing countries often prevents effective distribution. Political instability can disrupt food shipments, and the donated food often falls into the hands of corrupt local officials. Such was the case in Somalia in the 1990s, when warring factions in the capital city of Mogadishu prevented food aid from getting to starving populations. So while the trigger for famine may be environmental, there are deep-seated political and economic problems that conspire to block famine relief.

Globalization and the Growth of Agribusiness

Globalization and its impact on agriculture have been referred to throughout this chapter. Such references have frequently been accompanied by the term *agribusiness*.

Globalization involves the restructuring of the world economy by multinational corporations thriving in an era of free-trade capitalism, rapid communications, improved transport, and computer-based information systems. When applied to agriculture, globalization tends to produce agribusiness: a modern farming system that is totally commercial, large-scale, mechanized, and dependent upon chemicals, hybrid seeds, genetic engineering, and the practice of **monoculture** (raising a single specialty crop on vast tracts of land). Furthermore, agribusinesses are frequently vertically integrated; that is, they own the land as well as the processing and marketing facilities. The green revolution is part of agricultural globalization, as are countless "rural development" projects

> **monoculture** The raising of only one crop on a huge tract of land in agribusiness.

in developing countries, usually funded by the World Bank or the International Monetary Fund. These projects typically displace peasant farmers to make way for agribusinesses. The family-run farm is one victim of agricultural globalization.

Genetically modified (GM) crops, the products of biotechnology, are seen by many as another aspect of globalization. Genetic engineering produces new organisms through gene splicing. Pieces of DNA can be recombined with the DNA of other organisms to produce new properties, such as pesticide tolerance or disease resistance. DNA can be transferred not only between species but also between plants and animals, which makes this technology truly revolutionary and unlike any other development since the beginning of domestication. Agribusinesses are often able to patent the processes and resulting genetically engineered organisms and thus claim legal ownership of new life-forms.

> **genetically modified (GM) crop** A new organism whose genetic characteristics have been programmed through gene splicing between plant and/or animal species.

Commercial production of GM crops began in the United States in 1996. The technology has now spread around the globe, but the United States still dominates, accounting for 43 percent of the world's total production (**Figure 8.28**). Two crops, soybeans and corn, account for the rapid growth of GM food production in the United States. By 2011, 94 percent of all soybeans and 88 percent of all corn produced in the United States were genetically modified. There have also been recent increases in the adoption of GM cotton, alfalfa, sugar beets, and papaya in the United States.

If you shop in a U.S. supermarket, you have undoubtedly ingested GM foods. Whether or not one finds this troubling is closely related to the strength of certain cultural norms and values that vary from region to region and country to country. In England and western Europe, where national identities are strongly linked to the countryside, agrarian culture, and regional cuisines, there has been a lot of opposition to biotechnology in agriculture. In response to public pressure, major supermarket chains, such as Sainsbury's in England, have refused since 1998 to sell GM foods. In the United States, the response has been far more muted, so much so that the expansion of GM crop planting has proceeded virtually without public debate.

Food Fears

A globalized food system is vulnerable to events that threaten food safety. Recent events, such as outbreaks of mad cow disease in Great Britain and the contamination of imported produce by *Salmonella* or the virus that causes hepatitis A in the United States, have raised anxiety levels among consumers about eating food transported from distant lands.

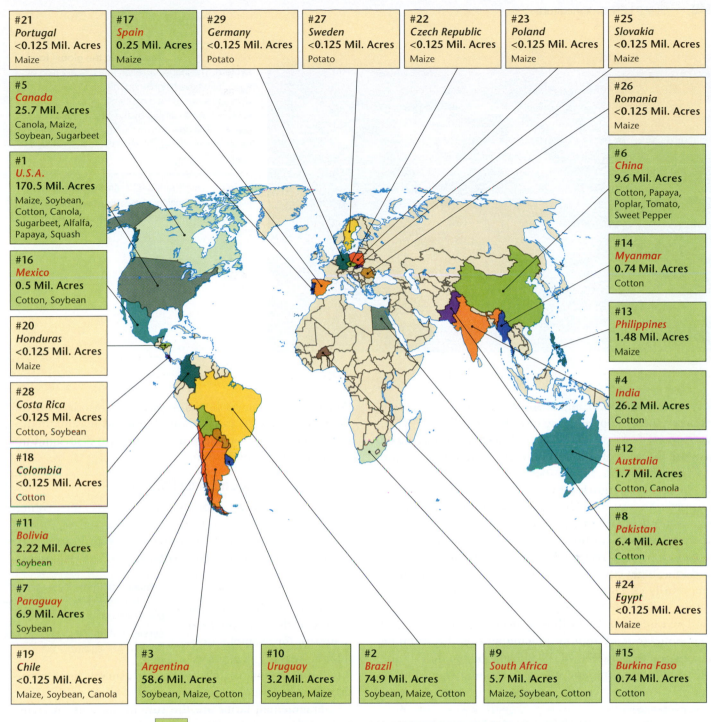

#21 *Portugal* **<0.125 Mil. Acres** Maize	**#17** *Spain* **0.25 Mil. Acres** Maize	**#29** *Germany* **<0.125 Mil. Acres** Potato	**#27** *Sweden* **<0.125 Mil. Acres** Potato	**#22** *Czech Republic* **<0.125 Mil. Acres** Maize	**#23** *Poland* **<0.125 Mil. Acres** Maize	**#25** *Slovakia* **<0.125 Mil. Acres** Maize

#5 *Canada* **25.7 Mil. Acres** Canola, Maize, Soybean, Sugarbeet

#1 *U.S.A.* **170.5 Mil. Acres** Maize, Soybean, Cotton, Canola, Sugarbeet, Alfalfa, Papaya, Squash

#16 *Mexico* **0.5 Mil. Acres** Cotton, Soybean

#20 *Honduras* **<0.125 Mil. Acres** Maize

#28 *Costa Rica* **<0.125 Mil. Acres** Cotton, Soybean

#18 *Colombia* **<0.125 Mil. Acres** Cotton

#11 *Bolivia* **2.22 Mil. Acres** Soybean

#7 *Paraguay* **6.9 Mil. Acres** Soybean

#26 *Romania* **<0.125 Mil. Acres** Maize

#6 *China* **9.6 Mil. Acres** Cotton, Papaya, Poplar, Tomato, Sweet Pepper

#14 *Myanmar* **0.74 Mil. Acres** Cotton

#13 *Philippines* **1.48 Mil. Acres** Maize

#4 *India* **26.2 Mil. Acres** Cotton

#12 *Australia* **1.7 Mil. Acres** Cotton, Canola

#8 *Pakistan* **6.4 Mil. Acres** Cotton

#24 *Egypt* **<0.125 Mil. Acres** Maize

#19 *Chile* **<0.125 Mil. Acres** Maize, Soybean, Canola	**#3** *Argentina* **58.6 Mil. Acres** Soybean, Maize, Cotton	**#10** *Uruguay* **3.2 Mil. Acres** Soybean, Maize	**#2** *Brazil* **74.9 Mil. Acres** Soybean, Maize, Cotton	**#9** *South Africa* **5.7 Mil. Acres** Maize, Soybean, Cotton	**#15** *Burkina Faso* **0.74 Mil. Acres** Cotton

17 biotech mega-countries growing 123,550 acres, or more, of biotech crops.

FIGURE 8.28 Worldwide use of genetically altered crop plants. The diffusion of GM crops has occurred despite the concerns of health scientists, environmentalists, and consumers. (Source: James, Clive. 2011. Global Status of Commercialized Biotech/GM Crops: 2011. ISAAA Brief No. 43. ISAAA: Ithaca, NY.)

THINKING GEOGRAPHICALLY What problems might arise from genetic modification?

The way that national governments and consumers respond to a particular food safety problem can fundamentally reshape geographic patterns of agricultural production on a global scale. Conversely, the idea that a culturally symbolic food may be tainted and life-threatening can shake the strongest of cultural identities.

The case of mad cow disease in Great Britain—where beef and dairy consumption has long been associated with British cultural identity—provides a good illustration. Mad cow disease is the vernacular term for bovine spongiform encephalopathy (BSE), a disease that attacks the central nervous system of cattle. The U.K. industry suffered three

FIGURE 8.29 Vulnerabilities in the global food system. The industrialization and globalization of agriculture have heightened the possibilities for and consequences of food contamination, most sensationally illustrated by the case of mad cow disease. The cow shown here has been quarantined because it was infected with BSE. (Kevin P. Casey/Corbis.)

THINKING GEOGRAPHICALLY Are there alternatives to the wholesale slaughter of animals in such cases? How might societies prepare for such outbreaks to avoid a crisis?

BSE crises (1988, 1996, and 2000) that had the net result of reducing their exports of live cows to zero (**Figure 8.29**). Consumers' fears of contaminated beef initially reduced demand for cattle worldwide, though the downward trend has been reversed, mostly because of new demand in middle-income countries. In the United Kingdom and the European Union, the BSE scares have reinforced a long-term pattern of declining demand for beef. Similarly, in Japan consumers continue to be wary of imported beef; demand has yet to recover. In 2003, when BSE was discovered in a single cow in the United States, all exports were temporarily curtailed and the industry's global competitiveness suffered long-term damage.

The case of mad cow disease demonstrates the global interconnectivity of food-producing and food-consuming regions, the vulnerability of the global food system, and the role of cultural norms and values surrounding questions of food safety.

AGRICULTURAL LANDSCAPES

What is the agricultural component of the cultural landscape? What might we learn about agriculture by examining its unique landscape?

According to the UN Food and Agriculture Organization, 37.3 percent of the world's land area is cultivated or pastured. In this huge area, the visible imprint of humankind might best be called the **agricultural landscape.** The agricultural landscape often varies even over short distances, telling us much about local cultures and subcultures.

agricultural landscape The cultural landscape of agricultural areas.

Moreover, it remains in many respects a window on the past, and archaic features abound. For this reason, the traditional rural landscape can teach us a great deal about the cultural heritage of its occupants.

Survey, Cadastral, and Field Patterns

A **cadastral pattern** is one that describes property ownership lines, whereas a *field pattern* reflects the way that a farmer subdivides land for agricultural use. Both can be greatly influenced by **survey patterns,** the lines laid out by surveyors prior to the settlement of an area. Major regional contrasts exist in survey, cadastral, and field patterns, for example, unit-block versus fragmented landholding and regular, geometric survey lines versus irregular or unsurveyed property lines.

cadastral pattern The shapes formed by property borders; the pattern of land ownership.
survey pattern A pattern of original land survey in an area.

Fragmented farms are the rule in the Eastern Hemisphere. Under this system, farmers live in farm villages or smaller **hamlets.** Their landholdings lie splintered into many separate fields situated at varying distances and lying in various directions from the settlement. One farm can consist of 100 or more separate, tiny parcels of land (**Figure 8.30**). The individual plots may be roughly rectangular, as in Asia and southern Europe, or they may lie in narrow strips. The latter pattern is most common in Europe, where farmers traditionally worked with a bulky plow that was difficult to turn. The origins of the fragmented farm system go back to an early period of peasant communalism. One of its initial

hamlet A small rural settlement, smaller than a farm village.

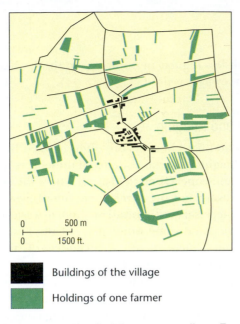

Buildings of the village

Holdings of one farmer

FIGURE 8.30 **Fragmented landholdings surrounding a French farm village.** The numerous fields and plots belonging to one individual farmer are shaded. Such fragmented farms remain common in many parts of Europe and Asia. (Source: After Demangeon, 1946.)

THINKING GEOGRAPHICALLY What are the advantages and disadvantages of this system?

justifications was a desire for peasant equality. Each farmer in the village needed land of varying soil composition and terrain. Travel distance from the village was to be equalized. From the rice paddies of Japan and India to the fields of western Europe, the fragmented holding remains a prominent feature of the cultural landscape.

Unit-block farms, by contrast, are those in which all of the farmer's property is contained in a single, contiguous

FIGURE 8.31 **American township and range survey creates a checkerboard.** The checkerboard is illustrated well by irrigated agriculture in the desert of California's Imperial Valley. (Glowimages/Getty Images.)

THINKING GEOGRAPHICALLY What are the advantages and disadvantages of such a geometric land-division system?

piece of land. Such forms are found mainly in the overseas area of European settlement, particularly the Americas, Australia, New Zealand, and South Africa. Most often, they reveal a regular, geometric land survey. The checkerboard of farm fields in the rectangular survey areas of the United States provides a good example of this cadastral pattern (**Figure 8.31**).

The American township and range system, discussed in Chapter 6, first appeared after the Revolutionary War as an orderly method for parceling out federally owned land for sale to pioneers. It imposed a rigid, square graph-paper pattern on much of the American countryside; geometry triumphed over physical geography (**Figure 8.32**). Similarly, roads follow section and township

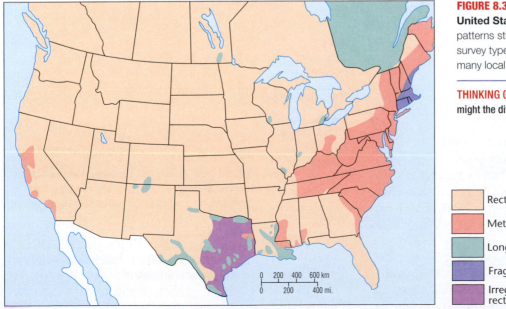

FIGURE 8.32 **Original land-survey patterns in the United States and southern Canada.** The cadastral patterns still retain the imprint of the various original survey types. The map is necessarily generalized, and many local exceptions occur.

THINKING GEOGRAPHICALLY What impact on rural life might the different patterns have?

Rectangular survey

Metes and bounds survey

Long-lot survey

Fragmented landholdings with farm villages

Irregular rectangular survey (mixture of rectangular and metes and bounds)

lines, adding to the checkerboard character of the American agricultural landscape. Canada adopted an almost identical survey system, which is particularly evident in the Prairie Provinces.

Equally striking in appearance are long-lot farms, where the landholding consists of a long, narrow unit-block stretching back from a road, river, or canal (**Figure 8.33**). Long-lots lie grouped in rows, allowing this cadastral survey pattern to dominate entire districts. Long-lots occur widely in the hills and marshes of central and western Europe, in parts of Brazil and Argentina, along the rivers of French-settled Québec and southern Louisiana, and in parts of Texas and northern New Mexico. These unit-block farms are elongated because such a layout provides each farmer with fertile valley land, water, and access to transportation facilities, either roads or rivers. In French America, long-lots appear in rows along streams, because waterways provided the chief means of transport in colonial times. In the hill lands of central Europe, a road along the valley floor provides the focus, and long-lots reach back from the road to the adjacent ridge crests.

Some unit-block farms have irregular shapes rather than the rectangular or long-lot patterns. Most of these result from *metes and bounds surveying*, which makes much use of natural features such as trees, boulders, and streams. Parts of the eastern United States were surveyed under the metes and bounds system, with the result that farms there are much less regular in outline than those where rectangular surveying was used (**Figure 8.34**).

Fencing and Hedging

Property and field borders are often marked by fences or hedges, heightening the visibility of these lines in the agricultural landscape. Open-field areas, where the dominance of crop raising and the careful tending of livestock make fences unnecessary, still prevail in much of western Europe, India, Japan, and some other parts of the Eastern Hemisphere, but much of the remainder of the world's agricultural land is enclosed.

Fences and hedges add a distinctive touch to the cultural landscape (**Figure 8.35**). Different cultures have their own methods and ways of enclosing land, so types of fences and hedges can be linked to particular groups. Fences in different parts of the world are made of substances as diverse as steel wire, logs, poles, split rails, brush, rock, and earth. Those who visit rural New England, western Ireland, or the Yucatán Peninsula will see mile upon mile of stone fence that typifies those landscapes. Barbed-wire fences swept across the American countryside a century ago, but remnants of older styles can still be seen. In Appalachia, the traditional split-rail zigzag fence of pioneer times survives here and there. As do most visible features of culture, fence types can serve as indicators of cultural diffusion.

The hedge is a living fence. In Brittany and Normandy in France and in large areas of Great Britain and Ireland, hedgerows are a major aspect of the rural landscape. To walk or drive the roads of hedgerow country is to experience a unique feeling of confinement quite different from the openness of barbed wire or unenclosed landscapes.

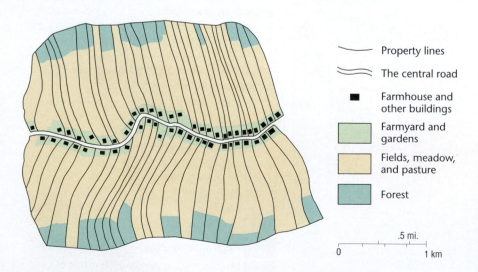

Property lines

The central road

■ Farmhouse and other buildings

Farmyard and gardens

Fields, meadow, and pasture

Forest

.5 mi.

0   1 km

FIGURE 8.33 **A long-lot settlement in the hills of central Germany.** Each property consists of an elongated unit-block of land stretching back from the road in the valley to an adjacent ridgecrest, part of which remains wooded.

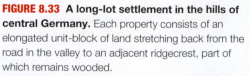

THINKING GEOGRAPHICALLY What advantages does this arrangement offer to farmers?

Original Survey Lines

Property Lines, About 1955
(Those that follow original survey
lines are shown by thicker lines.)

Field and Woodlot Borders,
About 1955

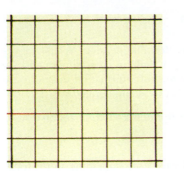  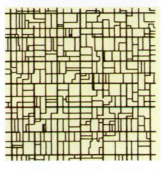

U.S. RECTANGULAR SURVEY, HANCOCK AND HARDIN COUNTIES, OHIO

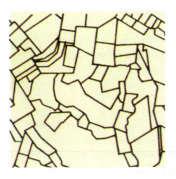 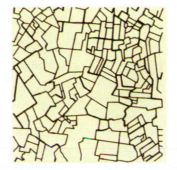 

METES AND BOUNDS SURVEY, UNION AND MADISON COUNTIES, OHIO

FIGURE 8.34 Two contrasting land-survey patterns, rectangular and metes and bounds. Both types were used in an area of west-central Ohio. Note the impact these survey patterns had on modern cadastral and field patterns. (Source: After Thrower, 1966.)

THINKING GEOGRAPHICALLY What other features of the cultural landscape might be influenced by these patterns?

FIGURE 8.35 Traditional fence in the mountains of Papua New Guinea. The fence is designed to keep pigs out of sweet potato gardens. The modern age has had an impact, as revealed in the use of tin cans to decorate and stabilize the fence. Each culture has its own fence types, adding another distinctive element to the agricultural landscape. (Courtesy of Terry G. Jordan-Bychkov.)

THINKING GEOGRAPHICALLY What kinds of fences do you find in your area? What is their purpose?

Key Terms

agribusiness, 222
agricultural landscape, 242
agricultural region, 214
agriculture, 213
aquaculture, 224
biofuel, 236
cadastral pattern, 242
commercial agriculture, 217
cool chain, 238
desertification, 234
domesticated animal, 228
domesticated plant, 227
double-cropping, 217
extensive agriculture, 215
feedlot, 220
folk culture, 218
genetically modified (GM) crop, 240
green revolution, 229
hamlet, 242
hunting and gathering, 226

intensive agriculture, 214
intercropping (intertillage), 216
livestock feeding, 220
mariculture, 224
market gardening, 220
monoculture, 240
nomadic livestock herder, 223
organic agriculture, 235
paddy rice farming, 217
peasant, 218
plantation, 219
plantation agriculture, 219
ranching, 223
sedentary cultivation, 223
subsistence agriculture, 216
suitcase farm, 221
survey pattern, 242
sustainability, 232
swidden cultivation, 215
urban agriculture, 224

Agricultural Geography on the Internet

You can learn more about agricultural geography on the Internet at the following web sites:

Agriculture, Food, and Human Values (AFHVS)

http://www.afhvs.org
Founded in 1987, AFHVS promotes interdisciplinary research and scholarship in the broad areas of agriculture and rural studies. The organization sponsors an annual meeting and publishes a journal by the same name.

The Early Detection and Distribution Mapping System (EDDMapS)

http://www.eddmaps.org/distribution/
Developed by the University of Georgia's Center for Invasive Species and Ecosystem Health, this extensive collection of distribution maps for invasive plant species is searchable at the state and county level for the entire United States. Data can also be downloaded in a variety of formats.

Food First

http://www.foodfirst.org
Founded in 1975 by author-activist Francis Moore Lappé, Food First is a nonprofit "people's" think tank and clearinghouse for information and political action. The organization highlights root causes and value-based solutions to hunger and poverty around the world, with a commitment to establishing food as a fundamental human right.

International Food Policy Research Institute, Washington, D.C.

http://www.ifpri.cgiar.org
Learn about strategies for more efficient planning for world food supplies and enhanced food production from a group concerned with hunger and malnutrition. Part of this site deals with domesticated plant biodiversity.

Resources for the Future

http://www.rff.org
This well-respected center for independent social science research was the first U.S. think tank on environment and natural resources. There is a great deal of information related to the environmental aspects of global food and agriculture.

United Nations Food and Agriculture Organization (FAO), Rome, Italy

http://www.fao.org
Discover an agency that focuses on expanding world food production and spreading new techniques for improving agriculture as it strives to predict, avert, or minimize famines.

United States Department of Agriculture, Washington, D.C.

http://www.usda.gov
Look up a wealth of statistics about American farming from the principal federal regulatory and planning agency dealing with agriculture.

Urban Agriculture Notes

http://www.cityfarmer.org
This is the site of Canada's Office of Urban Agriculture. It concerns itself with all manner of subjects, from rooftop gardens to composting toilets to air pollution and community development. It encompasses mental and physical health, entertainment, building codes, rats, fruit trees, herbs, recipes, and much more.

Worldwatch Institute

http://www.worldwatch.org
Learn about a privately financed organization focused on long-range trends, particularly food supply, population growth, and ecological deterioration.

Recommended Books on Agricultural Geography

Clay, Jason. 2004. *World Agriculture and the Environment: A Commodity-by-Commodity Guide to Impacts and Practices.* Washington, D.C., and Covelo, CA: Island Press. The book describes the environmental effects resulting from the production of 22 major crops; it is global in scope and encyclopedic in detail.

Grigg, David B. 1995. *An Introduction to Agricultural Geography,* 2nd ed. London: Routledge. A comprehensive introduction to the human and environmental factors that influence how agriculture and agricultural practices differ from place to place.

Middleton, Nick, and David S. G. Thomas (eds.). 1997. *World Atlas of Desertification,* 2nd ed. London: Arnold. Look at the cartographic evidence and decide for yourself whether the deserts of the world are enlarging at the expense of agricultural lands.

Millstone, Erik, and Tim Lang. 2003. *The Penguin Atlas of Food.* New York: Penguin Books. This book is packed with information on global agriculture on a wide range of topics, including genetically modified crops, fast food, organic farming, and more, all presented in brilliantly detailed maps.

Sachs, Carolyn E. 1996. *Gendered Fields: Rural Women, Agriculture, and Environment.* Boulder, CO: Westview Press. An exploration of the commonalities and differences in rural women's experiences and their strategies for dealing with the challenges and opportunities of rural living.

Sauer, Carl O. 1969. *Seeds, Spades, Hearths, and Herds.* Cambridge, MA: MIT Press. The renowned American cultural geographer presents his theories on the origins of plant and animal domestication—the beginnings of agriculture.

Journals in Agricultural Geography

Agriculture and Human Values. An interdisciplinary journal dedicated to the study of ethical questions surrounding agricultural practices and food. Published by Kluwer. Volume 1 appeared in 1984. Visit the homepage of the journal at http://www.kluweronline.com/issn/0889–048X/current.

Journal of Agrarian Change. A journal focusing on agrarian political economy, featuring both historical and contemporary studies of the dynamics of production, property, and power. Published by Blackwell. Volume 1 appeared in 2000. Formerly titled the *Journal of Peasant Studies,* which began publication in 1973. Visit the homepage of the journal at http://www.blackwell-publishing.com/journal. asp?ref=1471-0358&site=1.

Journal of Rural Studies. An international interdisciplinary journal ranked as the best of its kind. Published by Pergamon, an imprint of Elsevier Science, Amsterdam, the Netherlands. Volume 1 appeared in 1985. Visit the homepage of the journal at http://www.elsevier.nl/locate/jrurstud.

Answers to Thinking Geographically Questions

Figure 8.1: Answers will vary according to location.

Figure 8.2: Low-technology methods may be better for soil conservation and productivity per unit of land than large-scale machinery.

Figure 8.3: The work is hard to mechanize, especially in terraced fields on steep slopes like these. Mechanization would allow farmers to be freed for other occupations.

Figure 8.4: Rice is dominant in southern China, which is hillier than northern China, restricting the land available for fields. The common placement of the line would be between the large areas of orange and the large areas of blue, although the two crop systems are intermixed in some areas. There are also outliers of rice, not shown here, in northeastern China.

Figure 8.5: The mounds help to provide drainage around the plants' roots and avoid rotting of the developing sweet potatoes.

Figure 8.6: If the relations were really so harmonious, there would be no need for a sign to say so.

Figure 8.7: Whether the government should or should not have prevented this displacement, the government stood to collect taxes on the tea raised on the plantation, whereas the grain and livestock farmers were subsistence, with little of their production or income entering the formal marketplace.

Figure 8.8: The huge numbers of cattle produce huge amounts of waste, which must be disposed of. Raising cattle in such confined quarters also encourages the spread of disease. Food must be brought in, leading to the "carbon footprint" of the transportation system.

Figure 8.9: The western Great Plains and the intermountain region of the United States are arid regions with climates unsuited to the large-scale production of crops for commercial markets.

Figure 8.10: Monoculture produces ideal conditions for the spread of plant diseases. It also obliterates the natural biodiversity. Economically, if the one crop fails, there is no other product to provide an income.

Figure 8.11: Long-term problems include the exhaustion of the soil from continuous heavy cropping, the disadvantages of monoculture, the huge amount of fuel consumption in the machinery and the fertilizer, and the vulnerability to soil erosion. The system is also extremely expensive, leaving the farmers vulnerable to market fluctuations in price and to heavy debt. On the other hand, it has been enormously productive with a small labor force, enabling the United States to feed not only itself but much of the rest of the world while freeing many to pursue other occupations.

Figure 8.12: If population grows too much, there may be insufficient pasture, and herds may return to the same pasture before it has a chance to regrow. Drought may also dry up pastures so that they are insufficient.

Figure 8.13: In areas with fresh or salt water, mariculture produces cheaper protein, assuming that it is turned to food for humans rather than animals. However, it has ecological problems of its own and may not be suitable for dry areas.

Figure 8.14: China has a huge population, which until recently has had insufficient protein in its diet because there is little land to devote to pasture and meat is expensive. China also has a comparatively short coastline, and many of its rivers are badly polluted. Chinese people like fish, but transporting it inland has been difficult with poor transportation systems.

Figure 8.15: Answers will vary, but you may suggest that EEZ limits can overlap where countries or island territories are close to one another. Additionally, viable shipping lanes may cross EEZ territories of states that are not politically cooperative. Further, historical claims by indigenous groups to maritime rights in certain regions may conflict with contemporary EEZ claims for these same areas.

Figure 8.16: Southwest Asia became overpopulated and overused, and soil was not conserved. Salinization (deposition of salts on the surface from improper irrigation techniques) brought down some of the early kingdoms of Mesopotamia, for example.

Figure 8.17: Climatic conditions, soil types, and terrain in many equatorial and mid-latitude regions of the globe led to greater diversity in plant life in these areas in ancient times. Climatic factors also contributed to greater ease in cultivating crops during longer growing seasons in these regions.

Figure 8.18: The chili pepper was one of the food crops to be brought back to Europe in the Columbian Exchange, the great movement of food crops that resulted from the voyages of Christopher Columbus and other European explorers beginning just before 1500. Besides adding flavor to rather bland foods like tofu and rice, it contains important vitamins, especially Vitamin C.

Figure 8.19: In the immediate area, the ants probably just spread out into areas of compatible environments. Since the ants are a tropical species, a barrier to diffusion could be climatic: cold weather. They probably also hitched rides on truckloads of suitable products like nursery stock headed inland, creating the outliers as they established themselves in suitable environmental pockets. This is cultural diffusion because people created the conditions suitable to fire ants and moved products in which they could hide.

Figure 8.20: The climate of the southeastern United States is much less arid than that of the American West and Midwest. Therefore, the tumbleweed's ability to detach, dry, and roll easily across the southeastern landscape may be inhibited by higher moisture levels in the soil and groundcover vegetation.

Figure 8.21: Level land is less likely to suffer from severe gully erosion.

Figure 8.22: Countries extracting water from source rivers feeding the remains of the Aral Sea may begin to clash politically over rights to ever-shrinking water supplies for agriculture. Debates over responsibility for resulting environmental impacts outside these countries' borders will also likely result.

Figure 8.23: Steep slopes would be less attractive to farmers and would be permitted to remain forested.

Figure 8.24: With rising populations, people expand cropland and pasture more animals on marginal land. Plant life is killed off, leaving the land barren. Recurrent droughts also kill plant life, creating more desertlike conditions.

Figure 8.25: Concentric circles illustrate best the comparative distance from the market in the center.

Figure 8.26: The more intensive agriculture is closest to Montevideo, and the least intensive agriculture is farther away. Anomalies might be caused by soil quality and transportation lines.

Figure 8.27: The areas with the highest proportion of malnourished people are Africa south of the Sahara and secondarily Asia and South America. These are areas of rapidly growing populations, and especially in the case of Africa, considerable corruption, mismanagement, and warfare.

Figure 8.28: Scientists are unsure about the effects of genetic modification on human and animal health. There is also a risk of losing indigenous strains of plants, which would be needed to replenish agriculture if the genetically modified varieties were wiped out by some disaster.

Figure 8.29: Depending on the situation, quarantine might have prevented the spread of the disease. Societies can set up procedures to handle such outbreaks so that they can be controlled without causing a major crisis. Raising animals in less crowded conditions can help control the spread of diseases.

Figure 8.30: Each farmer can farm fields on a variety of soils. Living close together in a village enhances mutual aid. However, fragmentation limits mechanization, and farmers spend a lot of time commuting to their fields.

Figure 8.31: Land parcels are easy to describe for purposes of ownership, reducing the likelihood of conflicts and empty spaces. However, they take no account of environmental conditions or topography, encouraging farmers to ignore these environmental factors, too.

Figure 8.32: Although interstate highways have modified life to some extent, country roads and lines of communication tend to be in compass directions. Municipalities and counties often follow these lines, regardless of local terrain or natural corridors like rivers.

Figure 8.33: Each farmer gets some of each kind of land as well as road access.

Figure 8.34: Boundaries of municipalities and counties often follow property lines, as do roads.

Figure 8.35: In rural areas, there are fences to keep livestock and cropland separate and to keep livestock from wandering away. There are also fences in urban areas: for privacy, for keeping out trespassers, and for keeping pets and children confined. Styles and construction vary greatly.

Morning exercise at a sport shoes factory, Gao Bu Chen, Dongguan, China. (Francis Li/Alamy.)

Economic Geography: Industries, Services, and Development

9

Almost every facet of our lives is affected in some way by economic activity. On a Friday night out, you might drive in a car to a single outlet in a nationwide chain of restaurants, where you order chicken raised indoors several states away on special enriched grain, brought by refrigerated truck to a deep freeze, and cooked in an electric deep fryer. The car you drive most likely was manufactured in several different locations around the world, assembled somewhere else, brought by container ship or truck to your local automobile dealership, and purchased with financing provided by a bank or other financial service provider. Later, at a movie, you buy a candy bar manufactured halfway across the country. You then enjoy a series of machine-produced pictures that flash in front of your eyes so rapidly that they seem to be moving. Just about every object and event in your life is affected, if not actually created, by economics. What we mean by *economics* is how goods (from the food we eat to the Internet sites we visit) and services are produced, distributed, financed, sold, and consumed. These diverse activities greatly shape culture and are in turn shaped by cultural preferences, ideas, and beliefs. Economies function differently in different parts of the world. In this chapter, we examine differences and similarities of geographies of economic activities from a variety of vantage points that range from our own neighborhoods to the entire globe.

WHAT IS ECONOMIC DEVELOPMENT?

How can we understand why certain regions of the world are relatively wealthy while other regions of the world are poor? Many of you have had the experience of traveling to countries or regions where the everyday conditions of living seem so much more difficult than the conditions you are used to at home. If you haven't traveled to these places, you certainly have seen enough television shows, movies, and Internet features to understand that the everyday lives of people in different parts of the world, including their access to housing, food, and health care, may differ greatly from yours. You may also notice that there are economic differences within your own neighborhood, town, or city. On a global scale, the differences between those parts of the world that are considered wealthy and those that are less wealthy often are explained with reference to stages of economic development. In general, **economic development** refers to

> **economic development** The process by which an agricultural society moves toward industrialization and (usually) higher patterns of income.

the process by which an agricultural society moves toward industrialization and (usually) higher patterns of income.

We've discussed the idea of development throughout the book, particularly in Chapter 3, where we examined how development is often correlated with a region's demographic structure. In this chapter we focus on the economic aspects of development to help explain how and why people living in different parts of the world have differing access to resources such as food, health care, and housing. First, however, it is important to understand exactly what is meant by economic development.

The term *economic development* emerged in the post–World War II era, when scholars and government and policy officials, particularly those in the United States, became concerned that the standard of living in many regions of the world was quite low. By *standard of living*, they were referring to such things as literacy rates, infant mortality, life expectancy, and poverty levels. They realized that in general a low standard of living was correlated with an economy that was based primarily on subsistence agriculture. These officials and scholars believed that introducing new technologies and

skills would enable these regions to develop more productive forms of agriculture and that **industrialization**—the transformation of raw materials into commodities—would follow. These new economic initiatives then would lead to a higher standard of living.

> **industrialization** The transformation of raw materials into commodities or the process by which a society moves from subsistence agriculture toward mass production based on machinery and industry.

This viewpoint—categorizing regions of the world using the criterion of economic structure and then assuming that transforming economic structure could raise the standard of living in certain regions—is what came to be known as economic development.

Today, economic development as a concept refers to two related but somewhat different ideas. The first is a way of categorizing the regions of the world. Regions are categorized according to particular measures of economic growth, such as **GDP (gross domestic product).** GDP is a measurement of the total value of all goods and ser-

> **gross domestic product (GDP)** A measurement of the total value of all goods and services produced within a particular region or country within a set time period.

vices produced within a particular region or country within a set time period. Those regions with relatively high GDPs are considered *developed regions;* those with relatively low GDPs are considered *developing regions* (**Figure 9.1**). The second meaning refers to the actual process of economic growth. It can refer to the shift of a region's economy from one based on subsistence agriculture to one that relies on industrialization. Thus, economic development refers both to a process and to a way of categorizing the different regions of the world.

As we saw in Chapter 1, a measure that geographers commonly use to describe economic development is the Human Development Index (HDI). The HDI measures well-being by way of three categories: health, education, and living standards (**Figure 9.2** on page 252). Each of these dimensions incorporates several indicators, including life expectancy, mean (average) years of schooling, and per capita income. Geographers consider these measures to be important in determining overall quality of life. Unlike GDP, the HDI focuses on the human dimension of development rather than solely on national income. Though it is difficult to describe human well-being using only one composite index, the HDI does provide a basis for comparison between areas of high development, low development, and the extensive continuum in between.

A MODEL OF ECONOMIC DEVELOPMENT

One of the earliest and most influential models of the process of economic development was suggested by the economist Walter Rostow in 1962. Rostow posited that

GROSS DOMESTIC PRODUCT

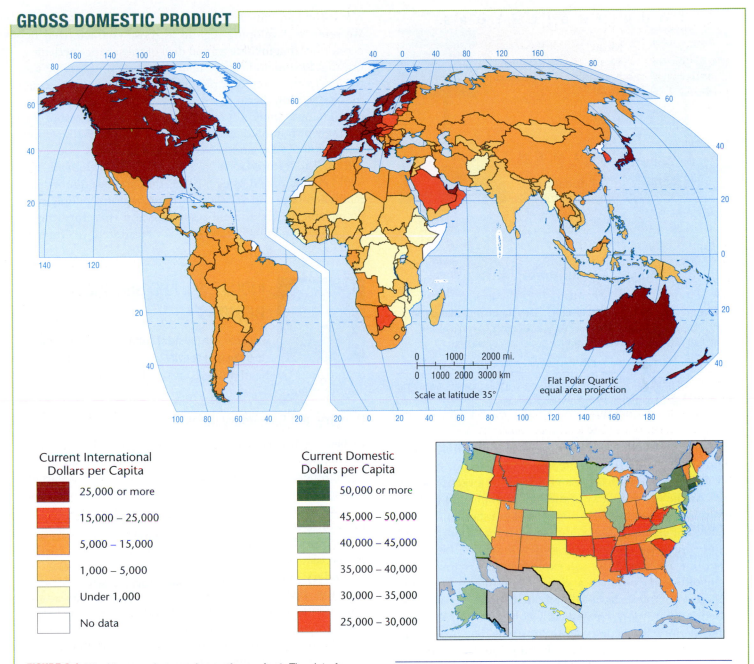

Current International Dollars per Capita

- 25,000 or more
- 15,000 – 25,000
- 5,000 – 15,000
- 1,000 – 5,000
- Under 1,000
- No data

Current Domestic Dollars per Capita

- 50,000 or more
- 45,000 – 50,000
- 40,000 – 45,000
- 35,000 – 40,000
- 30,000 – 35,000
- 25,000 – 30,000

FIGURE 9.1 World map of gross domestic product. The data for this map are based on a set of calculations that estimate the purchasing power parity (PPP) per capita for each country. This estimation therefore takes into consideration the relative cost of living in different countries. (Source: International Monetary Fund Data Mapper.)

THINKING GEOGRAPHICALLY Compare this map with the human development index map in Figure 1.12 on page 14. How are the two maps alike? How do they differ?

economic development was a process that all regions of the world would go through and experience in similar ways. His model assumed that regions would progress in a linear fashion through economic stages. The first stage was what he called a "traditional" economy, one based on agriculture and with limited access to or knowledge of advanced technologies. This first stage was followed by a series of other stages characterized by increasing technological sophistication and the introduction of industrialization. Rostow's final stage is called the "age of high mass consumption," the stage that he said characterized the United States and much of western Europe. In this stage, industrialization has developed to such a point that goods and services are readily available, and most people can

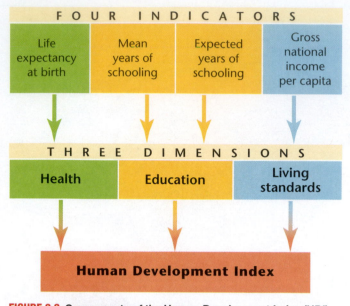

FIGURE 9.2 Components of the Human Development Index (HDI). The HDI allows geographers to measure development using a single index that serves as an indicator of both social and economic development.

THINKING GEOGRAPHICALLY What difficulties do you think might be involved in accurately calculating the HDI for less developed countries?

accumulate wealth to such a degree that they don't need to worry about subsistence needs.

Rostow's model has been criticized for a variety of reasons. While the simplicity of the model is appealing, its assumptions about the world often just don't stand up against reality. For example, the model suggests that countries and regions proceed on this path to development in isolation from one another. As we will explore in this chapter, however, different countries and their economies are interlinked in complex ways, so that if one country's economy is based primarily on services and consumption, it needs other regions and economies to supply its food and manufactured goods and vice versa. In addition, the model assumes that all economies will develop without obstacles from other countries, and we know that this is rarely true, since, for example, some countries may deliberately try to stop economic development in other countries that they think will become their competitors. We also know that different cultures think, feel, and act differently. The proposed goal of Rostow's model—high levels of mass consumption—may not be considered the highest level of development to many people in the world for whom other factors, such as political freedom or decent working conditions, are more important than material goods.

In this chapter we will be using the terms *developing* and *developed* to refer to different regions of the world, but we won't be assuming, as the Rostow model does, that there

is a normal and linear progression from one to the other. We refer to developing regions as those characterized by economies that include a good deal of subsistence activities but that also include some manufacturing and service activities. In these regions, people are often unable to accumulate wealth, because they are often producing just enough, or at times not enough, food and other resources for their own immediate needs. These countries therefore have a lower GDP and are considered relatively poor. We will use the term *developed* to refer to those regions of the world characterized by economies that are based more on manufacturing and services. In these regions, people are better able to accumulate resources, and therefore these regions have a higher GDP and are considered wealthier.

INDUSTRIAL AND SERVICE REGIONS

How can the theme of culture region be applied to industrial activity? Several types of industrial regions can be identified, and each occupies a distinct culture region.

Primary and Secondary Industries

Primary industries refer to industries involved in extracting natural resources from Earth. Fishing, farming, hunting, lumbering, oil extraction, and mining are examples of primary industries (**Figure 9.3**). Although primary industries are located throughout the world, it is in the developing world that these economic activities dominate. **Secondary industries** process the raw materials extracted by primary industries, transforming them into more usable forms. Ore is converted into steel; logs are milled into lumber; and fish are processed and canned. *Manufacturing* is a more common way of referring to secondary industries. Manufacturing activities are found throughout the world, though they tend to cluster in particular areas because of favorable circumstances such as cheap power sources or available labor (**Figure 9.4** on page 254).

primary industry An industry engaged in the extraction of natural resources, such as fishing, farming, hunting, lumbering, and mining.

secondary industry An industry engaged in processing raw materials into finished products; manufacturing.

Services

In many parts of the developed world, where people import the bulk of their manufactured products, economic activities are dominated by **services,** the range of economic activities that provide services to people and in-

services The range of economic activities that provide services to people and industry.

FIGURE 9.3 Oil field at sunset. Primary industries extract natural resources from the Earth. (Bill Ross/Corbis.)

THINKING GEOGRAPHICALLY The tall derricks are an older technology, while the "grasshopper" pump to the right is newer. Why are the derricks spaced so close together? How can you tell that this field is still productive?

dustry. Services include all the different types of work necessary to make goods and resources circulate and get delivered to people. So wide is the range of services that some geographers find it useful to distinguish three different types: transportation/communication services, producer services, and consumer services.

Transportation/communication services include transportation, communication, and utility (such as power companies) services. Highways, railroads, waterways, airlines, pipelines, telephones, radio, television, and the Internet are all transportation/communication services. All facilitate the distribution of goods, services, and information. Modern industries require well-developed transport systems, and every industrial district is served by a network of such facilities.

> **transportation/communication services** The range of economic activities that provide transport, communication, and utilities to businesses.

Major regional differences exist in the relative importance of the various modes of transport. In Russia and Ukraine, for example, highways are not very important to industrial development; instead, railroads—and to a lesser extent waterways—carry much of the transport load. Indeed, Russia still lacks a paved transcontinental highway. In the United States, by contrast, highways reign supreme, while the railroad system has declined. Western European nations rely heavily on a greater balance among rail,

highway, and waterway transport. Meanwhile, electronic transfers of funds and telecommunications between computers continents apart add a new dimension and speed to the exchange of data and ideas.

Producer services are services required by the manufacturers of goods, including insurance, legal services, banking, advertising, wholesaling, retailing, consulting, information generation, and real estate transactions. Such businesses represent one of the major growth

> **producer services** The range of economic activities required by producers of goods, including insurance, legal services, banking, advertising, wholesaling, retailing, information generation, and real-estate transactions.

sectors in developed economies, and a geographical segregation has developed in which manufacturing is increasingly shunted to the peripheries while corporate headquarters and the producer-related service activities remain, for the most part, in the core. Some of these producer service activities, however, are now moving offshore, to such places as the Caribbean and India, to take advantage of educated but low-wage workers. Nevertheless, the main control centers for these large companies are still located within the major cities of the West.

Increasingly important in the producer service industries are the collection, generation, storage, retrieval, and processing of computerized information, including research, publishing, consulting, and forecasting. The

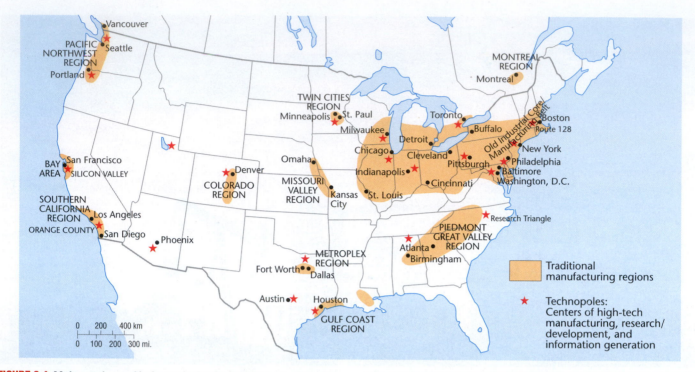

FIGURE 9.4 Major regions of industry in Anglo-America. The largest and most important region is still the American Manufacturing Belt, the traditional industrial core of the United States. Dispersal of manufacturing to other regions occurred after World War II and now involves mainly high-tech and information-based enterprises, or technopoles.

THINKING GEOGRAPHICALLY Why do high-tech industries have a distribution different from that of more traditional manufacturing?

impact of computers has changed and is continuing to change the world dramatically. Many producer service businesses depend on a highly skilled, intelligent, creative, and imaginative labor force. Although information-generating activity in the United States is focused geographically in the old industrial core, the distribution of this activity, if viewed on a more local scale, can be seen to coalesce around major universities and research centers. The presence of Stanford University and the University of California at Berkeley, for example, helped make the San Francisco Bay Area a major center of such industry. Similar technological centers have developed near Harvard and MIT in New England and near the Raleigh–Durham–Chapel Hill Research Triangle of North Carolina (see Figure 9.4). These **high-tech corridors**—or "silicon landscapes," as some have dubbed them—occupy relatively little area. They tend to be located in offices along a limited-access highway. In other words, the information economy is highly focused geographically, contributing to and heightening uneven development spatially. In Europe, for example, the emerging core of producer service industries is even more confined geographically than the earlier concentration of manufacturing.

> **high-tech corridor** An area along a limited-access highway that houses offices and other services associated with high-tech industries.

Consumer services are those provided to the general public, and they include education, government, recreation and tourism, and health care. Many of these activities are also shifting their locations, since the Internet has made it possible to provide these services remotely. For example, some American executives find themselves so busy at work and with their families that they have hired personal assistants to schedule their travel, medical appointments, and so on. Some of these services are now located in India. Similarly, some educational activities can now be done through the Internet. A company called Tutor Vista employs 600 tutors in India to help American students with their homework. The company has over 10,000 subscribers.

> **consumer services** The range of services provided to the general public, including education, government, recreation and tourism, and health care.

One of the most rapidly expanding activities included under consumer services is tourism. As early as 1990, this industry accounted for 5.5 percent of the world's economy, generated $2.5 trillion in income, and employed 112 million workers—more than any other single industrial activity, and amounting to 1 of every 15 workers in the world. Just a decade later, the total income generated had risen to $4.5 trillion, and tourism employed 1 of every 12 workers. This trend toward the increased importance of tourism has continued and spread through most regions

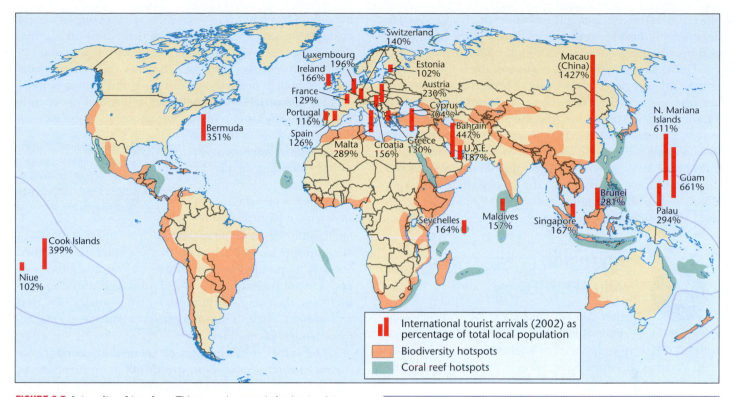

FIGURE 9.5 Intensity of tourism. This map shows only foreign tourists and does not count vacationers who stay within their own country. Notice also how many tourist areas are vulnerable to environmental degradation. (Source: United Nations Environmental Programme.)

THINKING GEOGRAPHICALLY What bias might be built into such a map?

of the world, often in spite of terrorist attacks directed against tourists, as in Bali in 2002 and Mumbai in 2008. By 2012, it was estimated that 235 million people work in the tourism industry, though tourism has suffered in recent years as a result of a prolonged world recession.

Like all other forms of industry, tourism varies greatly in importance from one region or country to another (**Figure 9.5**), with some countries, particularly those in tropical island locations, depending principally on tourism to support their national economies (**Figure 9.6**).

In general, countries that are dominated by primary industries tend to be in the developing category, while countries that are dominated by services are considered to be in the developed category. But, of course, the world and its economies are complex, and there are many exceptions to the general association of different types of economic activities and stages of economic development as outlined earlier. Countries in the Persian Gulf region, like Saudi Arabia, are considered fairly developed, yet their economies are dominated by a primary industry, oil extraction. In contrast, countries like the Russian Federation that were industrialized long ago often are placed in the category of developing regions. So stages of economic activity don't necessarily explain why some countries are rich and some countries are poor; other factors must be considered too,

FIGURE 9.6 Tourism reaches even into remote areas. Small charter buses now deliver climbers to a thatched tourist hut, which lacks running water, at the foot of Mount Wilhelm in highland Papua New Guinea—an area totally unknown to the outside world as late as 1930. Increasing numbers of tourists from Europe, North America, and Japan seek out such places. (Courtesy of Terry G. Jordan-Bychkov.)

THINKING GEOGRAPHICALLY How might these places change as a result of tourism?

like political stability and cultural values. Rich and poor are not always indicators of social development, because money does not always correlate to a higher standard of living.

Understanding why some regions of the world are richer than other regions, and why some groups of people have better access to such things as education and health care, is a difficult affair. We can see that the answers to these questions are related to a region's economic structure—whether its economy is dominated by primary industry, secondary industry, or services—but we know that the economy is not the only factor to consider. In the next section, we take a closer look at the changing geography of industrialization itself, beginning with where it began and how it diffused.

DIFFUSION OF INDUSTRY AND SERVICES

How can we understand the changing locations of industrial activities? The theme of cultural diffusion permits us to begin answering the question. Perhaps the most basic issue is the diffusion of the industrial revolution itself.

Origins of the Industrial Revolution

Until the industrial revolution, society and culture remained overwhelmingly rural and agricultural. Cities certainly existed, as centers of political power, education, and innovation, but the majority of people lived in the countryside working to procure food through agriculture. To be sure, secondary industry already existed in this setting. For as long as *Homo sapiens* has existed, we have fashioned tools, weapons, utensils, clothing, and other objects, but traditionally these items were made by hand, laboriously and slowly. Before about 1700, most such manufacturing was carried out in two rather distinct systems: cottage industry and guild industry.

Cottage industry, by far the more common system, was practiced in farm homes and rural villages, usually as a sideline to agriculture. Objects for family use were made in each household, usually by women. Additionally, most villages had a cobbler, miller, weaver, and smith, all of whom worked part-time at these trades in their homes. Skills passed from parents to children with little formality.

By contrast, the **guild industry** consisted of professional organizations of highly skilled, specialized artisans engaged full-time in their trades and based in towns and cities. Membership in a guild came after a long apprenticeship, during which the apprentice learned the skills of the profession from a master. Although the cottage and guild systems differed in many respects, both depended on hand labor and human power.

The **industrial revolution** began in England in the early 1700s. First, machines replaced human hands in the fashioning of finished products, rendering the word *manufacturing* (meaning "made by hand") technically obsolete. No longer would the weaver sit at a hand loom and painstakingly produce each piece of cloth. Instead, large mechanical looms were invented to do the job faster and cheaper. Second, human power gave way to various other forms of power: water power, the burning of fossil fuels, and later electricity and the energy of the atom fueled the machines. Men and women, once the producers of handmade goods, became tenders of machines.

The initial breakthrough came in the secondary, or manufacturing, sector. More specifically, it occurred in the British cotton textile cottage industry, centered at that time in the district of Lancashire in northwestern England. At first the changes were on a small scale. Mechanical spinners and looms were invented, and flowing water was harnessed to drive the looms. During this stage, manufacturing industries were still largely rural and dispersed, because they were tethered to their individual power sources. Sites where rushing streams could be found, especially those with waterfalls and rapids, were ideal locations. Later in the eighteenth century, the invention of the steam engine provided a better source of power, and a shift away from water-powered machines occurred. The steam engine also allowed textile producers and other manufacturers to move away from their power source—water—and to relocate instead in cities that better suited their labor and transportation needs. Manufacturing enterprises generally began to cluster together in or near established cities.

Traditionally, metal industries had been small-scale, rural enterprises, carried out in small forges situated near ore deposits. Forests provided charcoal for the smelting process. The chemical changes that occurred in the making of steel remained mysterious even to the craftspeople whose job it was, and much ritual, superstition, and ceremony were associated with steel making. Techniques had changed little since the beginning of the Iron Age, 2500 years earlier. The industrial revolution radically altered all of this. In the eighteenth century, a series of inventions by iron makers living in Coalbrookdale, in the English Midlands, allowed the old traditions, techniques, and rituals of steel making to be swept away and replaced by a scientific, large-scale industry. Coke, nearly pure carbon derived from high-grade coal,

industrial revolution A series of inventions and innovations, arising in England in the 1700s, that led to the use of machines and inanimate power in the manufacturing process.

cottage industry A traditional type of manufacturing before the industrial revolution, practiced on a small scale in individual rural households as a part-time occupation and designed to produce handmade goods for local consumption.

guild industry A traditional type of manufacturing before the industrial revolution, involving handmade goods of high quality manufactured by highly skilled artisans who resided in towns and cities.

was substituted for charcoal in the smelting process. Large blast furnaces replaced the forge, and efficient rolling mills took the place of hammer and anvil. Mass production of steel resulted, and that steel was used to make the machines that in turn created more industry.

Primary industries also were revolutionized. Coal mining was the first to feel the effects of the new technology. The adoption of the steam engine required huge amounts of coal to fire the boilers, and the conversion to coke in the smelting process further increased the demand for coal. New mining techniques and tools were invented, and coal mining became a large-scale, mechanized industry. However, coal, which is heavy and bulky, was difficult to transport. As a result, manufacturing industries that relied heavily on coal began flocking to the coalfields to be near the supply. Similar modernization occurred in the mining of iron ore, copper, and other metals needed by rapidly growing industries.

The industrial revolution also affected the service industries, most notably in the development of new forms of transportation. Traditional wooden sailing ships gave way to steel vessels driven by steam engines, and later railroads became more prevalent. The need to move raw materials and finished products from one place to another both cheaply and quickly was the main stimulus that led to these transportation breakthroughs. Without them, the impact of the industrial revolution would have been minimized.

Once in place, the railroads and other innovative modes of transport associated with the industrial revolution fostered additional cultural diffusion (**Figure 9.7**). Ideas spread more rapidly and easily because of this efficient transportation network. In fact, the industrial revolution itself was diffused through these new transportation technologies.

Diffusion of the Industrial Revolution

Great Britain maintained a virtual monopoly on the industrial revolution well into the 1800s. Indeed, the British government actively tried to prevent the diffusion of the various inventions and innovations that made up the industrial revolution. After all, they gave Britain an enormous economic advantage and contributed greatly to the growth and strength of the British Empire. Nevertheless, this technology finally diffused beyond the bounds of the British Isles (**Figure 9.8** on page 258), with continental Europe feeling the impact first. In the last half of the nineteenth century, the industrial revolution took firm root in the coalfields of Belgium, Germany, and other nations of northwestern and central Europe. The diffusion of railroads in Europe provides a good index of the spread of the industrial revolution there. The United States began rapid adoption of this new technology in about 1850, followed half a century later by Japan, the first major nonwestern nation to undergo full industrialization. In the first third of the twentieth century, the diffusion of industry and modern transport spilled over into Russia and Ukraine.

Until fairly recently, most of the world's manufacturing plants were clustered together in pockets within several regions, particularly Anglo-America, Europe, Russia, and Japan. In the United States, secondary industries once clustered mainly in the northeastern part of the country, a region referred to as the American Manufacturing Belt (see Figure 9.4). Across the Atlantic, manufacturing occupied the central core of Europe, which was surrounded by a less industrialized periphery throughout the mid-twentieth century. Japan's industrial complex was located along the shore of the Inland Sea and throughout the southern part of the country.

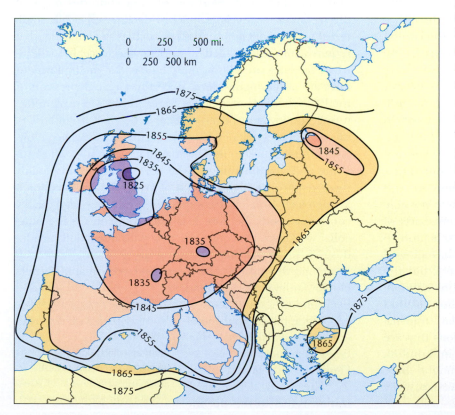

FIGURE 9.7 The diffusion of the railroad in Europe. The industrial revolution and the railroad spread together across much of the continent.

THINKING GEOGRAPHICALLY Can you find evidence of both contagious and hierarchical diffusion?

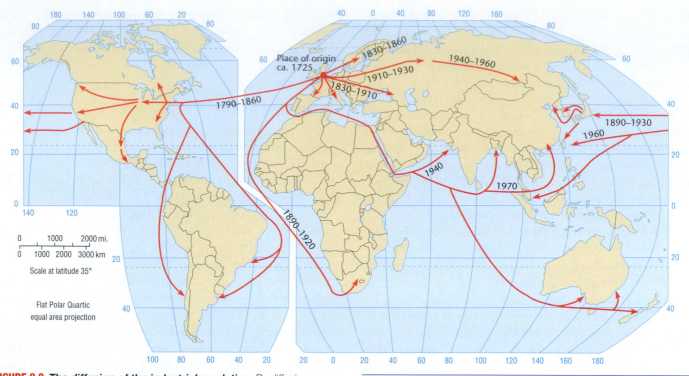

FIGURE 9.8 **The diffusion of the industrial revolution.** By diffusion from Great Britain, the industrial revolution has changed cultures in much of the world.

THINKING GEOGRAPHICALLY   Why might the industrial revolution have originated in so small a country?

The Locational Shifts of Secondary Industry and Deindustrialization

The diffusion of industrialization around the globe continues today, but not in the same manner in which the industrial revolution spread from England to Europe, the United States, and Japan. During the diffusion of industry in the nineteenth and early twentieth centuries, there were very few global transportation or communication networks. Therefore, for the most part, industries were predominantly national in scale. All of their activities were located within national boundaries, from their headquarters to their supply of raw materials to the labor for their plants. However, by the mid to late twentieth century, with the advent of new communication and transportation technologies such as telephones, jet planes, and computers, it became possible to develop industries that could locate different parts of their operation (labor, power, materials, markets, and so on) in places beyond national borders. This has allowed manufacturing companies, for example, to seek out the best locations for their plants without regard to national boundaries, in effect spreading industrialization around the world. Today, **transnational corporations** that are headquartered in London, for example, might have their manufacturing facilities in Vietnam and Mexico, while their materials might be supplied by Indonesia.

transnational corporation   A company that has international production, marketing, and management facilities.

Although the manufacturing dominance of the developed countries persists, a major global geographical shift is under way. In virtually every one of these countries, much of the secondary sector is in marked decline, especially traditional mass-production industries, such as steel making, that require a minimally skilled blue-collar workforce. In such districts, factories are closing and blue-collar unemployment rates are at the highest level since the Great Depression of the 1930s. In the United States, for example, where manufacturing employment began a relative decline around 1950, nine out of every 10 new jobs in recent years have been low-paying service positions. The manufacturing industries surviving and now booming in the core countries are mainly those requiring a highly skilled or artisanal workforce, such as high-tech firms and companies producing high-quality consumer goods. Because it is often difficult for the blue-collar workforce to acquire the new skills needed in such industries, many old manufacturing districts lapse into deep economic depression. Moreover, high-tech manufacturers employ far fewer workers than heavy industries and tend to be geographically concentrated in very small districts, sometimes called **technopoles** (see Figure 9.4 on page 254).

The word **deindustrialization** refers to the decline

technopole   A center of high-tech manufacturing and information-based industry.

deindustrialization   The decline of primary and secondary industry, accompanied by a rise in the service sectors of the industrial economy.

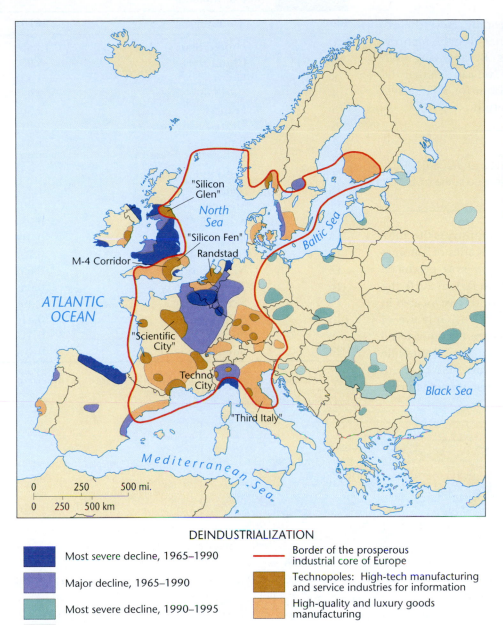

FIGURE 9.9 **Industrial regions and deindustrialization in Europe.** New, prosperous centers of industry specializing in high-quality goods, luxury items, and high-tech manufacture have surpassed older centers of heavy industry—both primary and secondary. The regions in decline were earlier centers of the industrial revolution. (Source: Jordan-Bychkov and Jordan, 2002.)

THINKING GEOGRAPHICALLY Why might the industrial districts in decline not have shared the new prosperity? Why did eastern Europe fall so far behind?

DEINDUSTRIALIZATION

- Most severe decline, 1965–1990
- Major decline, 1965–1990
- Most severe decline, 1990–1995
- Major decline, 1990–1995
- Border of the prosperous industrial core of Europe
- Technopoles: High-tech manufacturing and service industries for information
- High-quality and luxury goods manufacturing

of primary and secondary industry, usually accompanied by a rise in the service sectors. Deindustrialization is responsible for the decline and fall of once-prosperous factory and mining areas, such as the east and west Midlands of England (**Figure 9.9**). Manufacturing industries lost by the core countries relocate to newly industrializing lands that were once the periphery. South Korea, Taiwan, Singapore, Brazil, Mexico, coastal China, the Philippines, and parts of India, among others, have experienced a major expansion of manufacturing. Geographers and others often refer to these countries as *newly industrializing countries (NICs)*. Companies move to these areas for many different reasons, including cheaper labor costs, lower environmen-

tal standards, and the relative proximity of these plants to their expanding markets outside the traditional core. Yet even with this massive diffusion of industry, the geographic pattern of the world's manufacturing sites remains quite uneven. **Table 9.1** on page 260 reflects this unevenness, as long-dominant manufacturing regions in North America, Europe, and East Asia are still represented among countries leading in *gross value added (GVA)*, the value created by production processes. Meanwhile, many parts of Africa, the Middle East, and Central Asia lag well behind. It should be noted that GVA can be calculated in different ways, and it can include varying production activities from country to country. This lack of uniformity makes international comparisons of GVA potentially misleading. Nevertheless, over time, the data can be useful in analyzing the maturity of a country's manufacturing sector. By analyzing GVA along with other economic measures, we can see that the United States, China, and Japan account for a large segment of the world's manufacturing output.

REFLECTING ON **GEOGRAPHY**

Identify and discuss some of the reasons that the "old" manufacturing core still retains its dominance in the global manufacturing system.

A good example of a transnational industry is the automobile industry. In contrast to earlier decades of automobile manufacturing, millions of cars are now produced outside of the United States every year. In fact, in 2010, the top two automobile producers in the world were located in Asia and Europe, respectively (**Figure 9.10**). Only two of the current top 10 auto manufacturers are headquartered in the United States. In comparison to the automobile industry, the major manufacturing plants of which are scattered over several world regions, the manufacturing

TABLE 9.1	Gross Value Added by Country, 2010 (in U.S. Trillions)	
Rank	**Country**	**2010**
1	United States	13.45
2	China	5.93
3	Japan	5.44
4	Germany	2.93
5	France	2.30
6	United Kingdom	2.01
7	Italy	1.84
8	India*	1.28
9	Russian Federation	1.27
10	Spain	1.27
11	Mexico	1.00
12	Australia*	0.93
13	Korea	0.91
14	Indonesia	0.70
15	Netherlands	0.69
16	Turkey	0.65
17	Switzerland	0.49
18	Belgium	0.41
19	Poland	0.41
20	Sweden	0.40

*2009 Data
Note: "Gross value added" is defined as the additional value created by a production process. *Based on data from OECD (2012), OECD.Stat, (database). http://dx.doi.org/10.1787/data-00285-en.*

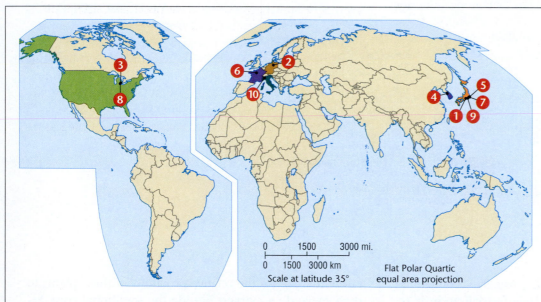

0 1500 3000 mi.
0 1500 3000 km
Scale at latitude 35°

Flat Polar Quartic
equal area projection

Car Company	Head Office	Number of Cars Manufactured in 2010
❶ Toyota Motor Corporation	Aichi, Japan	7,267,535
❷ Volkswagen Group AG	Wolfsburg, Germany	7,120,532
❸ General Motors Company	Detroit, United States	6,266,959
❹ Hyundai Kia Automotive Group	Seoul, South Korea	5,247,339
❺ Honda Motor Company	Tokyo, Japan	3,592,113
❻ PSA Peugeot Citroën S.A.	Paris, France	3,214,810
❼ Nissan	Yokohama, Japan	3,142,126
❽ Ford Motor Company	Dearborn, United States	2,958,507
❾ Suzuki Motor Company	Shizuoka, Japan	2,503,436
❿ Fiat	Turin, Italy	1,781,385

Source: International Organization of Motor Vehicle Manufacturers (OICA)

FIGURE 9.10 The world's top 10 car manufacturing countries. Although it is now a transnational industry, automobile manufacturing remains highly clustered in well-established industrial regions of East Asia, Europe, and the United States. (Source: Organisation Internationale des Constructeurs d'Automobiles [OICA].)

THINKING GEOGRAPHICALLY Why do you think that automobile manufacturing has not become established to any large degree in South America, Africa, or much of Asia?

facilities of the textile and garment industries are disproportionately located in China, which has a large pool of relatively low-wage workers and a government eager to offer incentives. However, not all types of textile production are located in China. While the bulk of mass-produced items are made in China, much high-end designer clothing tends to be produced in different places and under different conditions. This is all part of the process of globalization.

The decline of primary and secondary industries in the older developed core, or deindustrialization, has ushered in an era in these regions widely referred to as the **postindustrial phase,** a phrase used to describe a society characterized by the dominance of the service sectors. Both the United States and Canada are in the postindustrial phase, as are most of Europe and Japan.

> **postindustrial phase** A society characterized by the dominance of the service sectors of economic activity.

INDUSTRIAL-ECONOMIC ECOLOGY

How do economic activities affect the natural environment? Industrialization is a process that transforms natural resources into commodities for human use. The theme of cultural ecology, then, is central to understanding industrialization and economic development.

Renewable Resource Crises

At first glance, it might seem that only finite resources are affected by industrialization, but even renewable resources such as forests and fisheries are endangered. The term **renewable resources** refers to those that can be replenished naturally at a rate sufficient to balance their depletion by human use. That balance is often critically disrupted by the effects of industrialization. So while deforestation is an ongoing process that began at least 3000 years ago, the industrial revolution drastically increased the magnitude of the problem. In just the last half-century, a third of the world's forest cover has been lost. Today, we witness the rapid destruction of one of the last surviving great woodland ecosystems, the tropical rain forest (**Figure 9.11**). The most intensive rain-forest clearing is occurring in the East Indies and Brazil, and commercial lumber interests are largely responsible (**Figure 9.12** on page 262). Although trees are a renewable resource when properly managed, too many countries are in effect mining their forests. Foreign rather than Brazilian interests now hold logging rights to nearly 30 million acres (12 million hectares) of Amazonian rain forest. Even when forests are converted into scientifically managed "tree farms," as is true in most of the developed world, ecosystems often are destroyed. Natural ecosystems have plant and animal diversity that cannot be sustained under the monoculture of commercial forestry.

Similarly, overfishing has brought a crisis to many ocean fisheries, a problem compounded by the pollution of many of the world's seas. The total fish catch of all countries combined rose from 84 million metric tons in 1984 to more than 101 million metric tons by 2009, causing some

> **renewable resource** A resource that can be replenished naturally at a rate sufficient to balance its depletion by human use.

FIGURE 9.11 Destruction of tropical rain forest near Madang in Papua New Guinea by Japanese lumbering interests. The entire forest is leveled to extract a relatively small number of desirable trees. (Courtesy of Terry G. Jordan-Bychkov.)

THINKING GEOGRAPHICALLY Why are overtly destructive practices employed?

Amazon rain forest

Deforested areas by mid-1990s

FIGURE 9.12 The tropical rain forest of the Amazon Basin. In Brazil, the rain forest is under attack by settlers, ranchers, and commercial loggers. Its removal will intensify the impact of greenhouse gases, especially carbon dioxide, because the forest acts to convert those gases into benign forms. About 10,000 square miles (26,000 square kilometers) of Brazil's tropical rain forest are cleared each year. (Source: Worldwatch Institute web site, http://www.worldwatch.org/.)

THINKING GEOGRAPHICALLY Where are the largest deforested areas? Why?

species to decline. Salmon in Pacific coastal North America and cod in the Maritime Provinces of Canada can be said to have reached a marine biological crisis.

Overfishing is a global problem, with some estimates suggesting that if current levels of fishing continue, world fish stocks may be in a total state of collapse by 2048. However, some experts refute these projections, stating that natural dips in fish production always occur and that remaining fish stocks are often underestimated due to difficulties in collecting accurate data. Despite these debates, most experts do agree that fisheries need close oversight to prevent the complete depletion of stocks in areas where fishing is already at or near sustainable levels. Some good news reported in 2009 in the North Sea: for the first time in over a decade, the size of the spawning stock of cod reached a level that scientists agreed was sustainable. Experts credited an array of conservation techniques for the turnaround.

Acid Rain

Secondary and service industries pollute the air, water, and land with chemicals and other toxic substances. The burning of fossil fuels by power plants, factories, and automobiles releases acidic sulfur oxides and nitrogen oxides into the air; these chemicals are then flushed from the atmosphere by precipitation. The resultant rainfall, called **acid rain,** has a much higher acidity than normal rain. Overall, 84 percent of U.S. human-produced energy is generated by burning fossil fuels, making acid rain a prevalent phenomenon.

Acid rain can poison fish, damage plants, and diminish soil fertility. Such problems have been studied intensively in Germany, one of the most completely industrialized nations in the world. German scholars have been

> **acid rain** Rainfall with much higher acidity than normal, caused by sulfur and nitrogen oxides derived from the burning of fossil fuels being flushed from the atmosphere by precipitation, with lethal effects for many plants and animals.

2001

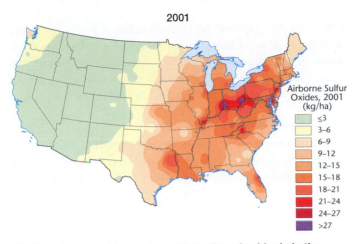

Airborne Sulfur
Oxides, 2001
(kg/ha)

	≤3
	3–6
	6–9
	9–12
	12–15
	15–18
	18–21
	21–24
	24–27
	>27

2006

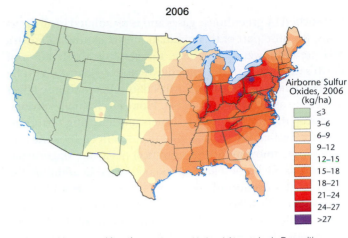

Airborne Sulfur
Oxides, 2006
(kg/ha)

	≤3
	3–6
	6–9
	9–12
	12–15
	15–18
	18–21
	21–24
	24–27
	>27

FIGURE 9.13 **Comparison of the distribution of acid rain in the United States from 2001 to 2006.** Measurements are taken from the annual sulfate deposition, derived from airborne sulfur oxides (SO_2). Deposition levels of 18 pounds per acre (20 kilograms per hectare) are generally regarded as threatening to some aquatic and terrestrial ecosystems. Notice that the intensity of acid rain has decreased in some places and increased in others. (Source: National Atmospheric Deposition Program web site, http://nadp.sws.uiuc.edu/.)

THINKING GEOGRAPHICALLY What factors would you have to consider to explain these changes?

alarmed by the dramatic suddenness with which the effects of acid rain arrived. In 1982, only 8 percent of forests in western Germany showed damage, but by 1990 the proportion had risen to more than half. In North America, the effects of acid rain have accumulated, but not yet with the speed seen in central Europe. By 1980, more than 90 lakes in the seemingly pristine Adirondack Mountains of New York were dead, meaning devoid of fish life, and 50,000 lakes in eastern Canada faced a similar fate.

Since about 1990, the acid-rain problem has become less severe in many parts of the world, especially Europe, but the situation in the Adirondacks did not start improving until very recently. For years, the government of Canada urged U.S. officials to take stringent action to help alleviate acid-rain damage, because much of the problem on the Canadian side of the border derives from American pollution. These pleas have begun to have some effect, as new and cleaner technologies for burning fossil fuels have been adopted by companies within the United States. You can see from **Figure 9.13** that there is a decrease in acid-rain concentrations in some areas, particularly Louisiana and Illinois. Scientists continue to monitor the situation in order to assess future scenarios.

Global Climate Change

Most scientists now agree that we have entered a phase of **global warming** caused by industrial activity and most particularly by the greatly increased amount of carbon dioxide (CO_2) produced by burning fossil fuels. Fossil fuels—coal, petroleum, and natural gas—are burned to produce the energy that powers the

world's factories, and as that burning has increased, so has the amount of carbon dioxide (CO_2) released into the atmosphere. The 10 hottest years on record have all occurred in the period from 1998 to 2010, based on records compiled at more than 14,000 locations.

Climate change has also contributed to the shrinking of the world's glaciers. In 2010, a huge ice mass four times the size of Manhattan (the largest in half a century) broke off from the Petermann Glacier in northern Greenland to drift across the Arctic Ocean and threaten oil platforms and shipping lanes. In 2011, a 19-mile crack formed in the Pine Island Glacier in Antarctica, which will soon cause an even larger ice mass measuring 300 to 350 square miles (up to 900 square kilometers) to break off from the continent. This will be the largest ice mass known to have cleaved from a glacier, and it will form the world's largest iceberg.

These are just two examples of the observed shrinking of the world's glaciers; experts predict further dramatic disappearance of much of the world's ice in coming years, causing widespread climatic effects. For example, scientists expect that rapidly retreating Arctic ice will leave the Arctic Ocean ice free in summers as soon as the middle of this century.

At issue is the so-called **greenhouse effect,** a natural process in which thermal radiation from Earth's surface

global warming The pronounced climatic warming of Earth that has occurred since about 1920 and particularly since the 1970s.

greenhouse effect A process by which thermal radiation from a planet's surface is absorbed by atmospheric greenhouse gases, and is reradiated in all directions. Because part of this reradiation is back toward the planet's surface and the planet's lower atmosphere, the result is an elevation of the planet's average surface temperature above what it would be in the absence of the gases. Earth's natural greenhouse effect makes life as we know it possible. However, human activities, primarily the burning of fossil fuels and clearing of forests, have intensified the natural greenhouse effect, causing global warming.

is absorbed by greenhouse gases and is reradiated in all directions. Because part of this reradiation is back toward Earth's surface and lower atmosphere, the result is an elevation of Earth's average surface temperature above what it would be in the absence of the gases. Earth's natural greenhouse effect makes life as we know it possible. However, human activities, primarily the burning of fossil fuels and clearing of forests, have intensified the natural greenhouse effect. By some estimates, the atmospheric concentration of carbon dioxide has climbed to the highest level in 180,000 years. In addition, the ongoing destruction of the world's rain forests adds huge amounts of carbon dioxide to the atmosphere. Although carbon dioxide is a natural component of Earth's atmosphere, the addition of such huge amounts of the gas is altering the air's chemical composition. Carbon dioxide, only one of the absorbing gases involved in the greenhouse effect, permits solar short-wave heat radiation to reach Earth's surface but acts to block or trap long-wave outgoing radiation, causing a thermal imbalance and global heating.

As the greenhouse effect warms the global climate and melts or partially melts the polar ice caps, sea levels will continue to rise and the world's coastlines may soon be inundated. To address this looming crisis, 38 industrialized countries signed the *Kyoto Protocol* in 2001, agreeing to reduce their emissions of greenhouse gases over a 10-year period. However, the protocol did not impose emissions restrictions on developing countries such as China, Brazil, and India. (These and other developing states have emitted significant amounts of greenhouse gases in recent decades.) Because of these imbalances in emissions restrictions, the United States declined to participate in the Kyoto agreement.

Delegates from over 190 nations met in Doha, Qatar, in late 2012 to ratify an extension of the agreement. However, Canada, Japan, New Zealand, and Russia renounced their obligations under the original agreement. Despite this decreased backing by leading world economic powers, the Doha session ended with the establishment of a second round of environmental commitments for 37 nations. Discussions of a new global pact addressing emissions goals are planned for 2015, and the ensuing agreement will be implemented in 2020, when the Kyoto agreement once again expires.

Overall, environmentalists have been unimpressed by the outcomes of both the Kyoto and the Doha talks. They argue that the talks have not delivered increased cuts in emissions and have not provided feasible ways for the world's poorest countries to finance their environmental commitments. Instead, under the agreement, developed nations are encouraged to help poorer countries cut emissions and adapt to climate change, but many less-developed countries complain that more affluent nations have been reluctant to deliver this aid.

Choosing to pursue its climate-related goals independently, the United States has managed to reduce its emissions of greenhouse gases in recent years, but it has accomplished this goal largely as a result of the decreased industrial activity resulting from the 2008 recession.

Ozone Depletion

Potentially even more serious than the greenhouse effect is the depletion of the upper-atmosphere ozone layer, which acts to shield humans and all other forms of life from the most harmful types of solar radiation. Several chemicals, including the Freon used in refrigeration and air conditioning systems, are almost certainly the main culprits. Refrigeration systems are heavily used in manufacturing plants and are critical for global trade in foods. Because of particular stratospheric conditions, the ozone layer has been most depleted in the Arctic and Antarctic regions. The ozone decrease in the Arctic was first noticed in the early 1990s, with the observation of an ozone hole in that region comparable to the one first detected in the Antarctic during the 1980s. In September 2006, the hole in the ozone layer over Antarctica was the largest ever measured, both in terms of surface area and in the actual loss of mass. In years since, ozone levels in the region have fluctuated but have shown some hopeful signs of a slight recovery.

Other world regions are also affected by the reduction in the ozone layer. In the middle latitudes, where most of the world's population lives, ozone levels have dropped an average of 5 percent per decade since 1979.

Most of the world's industrialized countries disperse large amounts of ozone-depleting chemicals into the air. In a plan to mitigate the resulting damage to the ozone layer, many of these states signed the Montreal Protocol in 1989, which was aimed at reducing ozone-damaging substances. The problem is that there is a significant time lag between when these substances are released into the atmosphere and when they are finally dissipated. Recent measurements show a significant reduction in the level of damaging substances within the lower atmosphere but not yet in the higher atmosphere. Scientific estimates suggest that it will take at least another 50 to 75 years for the ozone layer to recover.

REFLECTING ON **GEOGRAPHY**

If we cannot be certain that global warming and upper-level ozone depletion are caused by industrial activity rather than being natural fluctuations or cycles, should we take action or simply wait and see what happens?

Radioactive Pollution

Like acid rain, radioactive pollution is invisible. It comes from nuclear power plants and waste storage facilities,

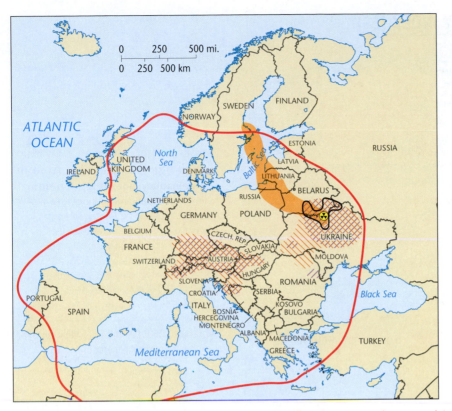

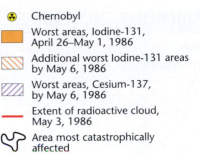

- ☢ Chernobyl
- Worst areas, Iodine-131, April 26–May 1, 1986
- Additional worst Iodine-131 areas by May 6, 1986
- Worst areas, Cesium-137, by May 6, 1986
- Extent of radioactive cloud, May 3, 1986
- Area most catastrophically affected

FIGURE 9.14 The Chernobyl disaster. The most severe radioactive contamination from the Chernobyl catastrophe came from two isotopes, cesium-137 and iodine-131. The map is somewhat speculative because of inadequate data. The future of human habitation in the area most catastrophically affected remains uncertain, but some tracts there have been completely depopulated. (Source: Adapted from Park, 1989: 66, 74, 79, 92.)

THINKING GEOGRAPHICALLY What impact has this catastrophe had on North America's nuclear power industry?

both consequences of the growing need for energy on the part of industrialized societies. The catastrophe at Chernobyl in Ukraine on April 26, 1986, clearly demonstrated the inherent dangers in reliance on nuclear power. When the Chernobyl nuclear reactor core melted down, causing two explosions, a sizable area around Chernobyl in Ukraine and Belarus became heavily contaminated with deadly radiation; everyone within an 18-mile (30 kilometers) radius of the destroyed reactor had to be evacuated (**Figure 9.14**). This area remains uninhabited today, save for a few elderly people who refused to leave. Beyond this zone, sizable swaths across Europe were bombarded with various long-lasting radioactive chemicals that attack the entire human body. Some estimates place the amount of radioactive chemicals released as equivalent to at least 750 Hiroshima atomic bombs. Ultimately, a sizable part of both Ukraine and Belarus may be declared unfit for human habitation, and additional tens of thousands of people could die of exposure to radiation caused by this single catastrophe. Another radioactive contaminant released by the explosion continues to leak into the groundwater table at Chernobyl.

More recently, an accident at the Fukushima Daiichi Nuclear Plant in northeastern Japan caused radioactive pollution to be released after a 2011 earthquake and tsunami. Experts estimate this to be about one-tenth the magnitude of the Chernobyl disaster in terms of its environmental impact. Nevertheless, a 12- to-19-mile (20- to 30-kilometer) radius around the plant was declared an "exclusion zone" following the disaster. This zone may be uninhabitable and unsafe to farm for many decades to come. The ominous term *national sacrifice area* has been used by government officials as a euphemism to describe such districts that have been rendered permanently uninhabitable by radiation pollution. Geographers often refer to such areas as *hazardscapes*. Outside the immediate exclusion zone created by the Fukushima disaster, unsafe levels of radioactive pollution have been detected in the Japanese food supply and in the country's water supplies. Thus far, only trace amounts of radioactive pollution from Fukushima appear to have been carried through the air and water to other countries such as the United States and Canada. However, large amounts of radioactive isotopes were released into the Pacific Ocean during the disaster, and impact on marine life and other ocean resources is not yet known.

ENVIRONMENTAL SUSTAINABILITY INDEX (ESI)

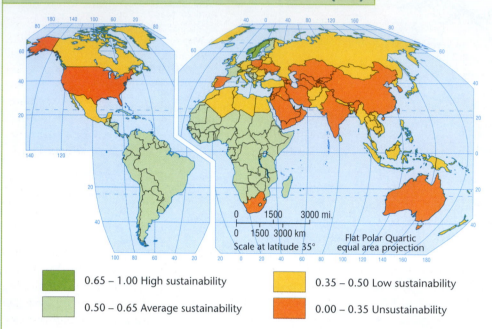

FIGURE 9.15 **Environmental sustainability index (ESI).** The ESI is a multiple-criteria measure of how well countries are doing in their attempts to achieve an ecological balance that might be sustained over centuries. (Source: FEEM Sustainability Index (FEEM SI) – http://www.feemsi.org, a project by Fondazione Eni Enrico Mattei (FEEM) – www.feem.it.)

THINKING GEOGRAPHICALLY What could explain the regional patterns?

- 0.65 – 1.00 High sustainability
- 0.50 – 0.65 Average sustainability
- 0.35 – 0.50 Low sustainability
- 0.00 – 0.35 Unsustainability

0   1500   3000 mi.
0 1500 3000 km
Scale at latitude 35°

Flat Polar Quartic
equal area projection

Environmental Sustainability

The key issue in all these industry-related ecological problems is **sustainability.** A sustainable environment is defined as one that would last for centuries or millennia, providing adequate resources for the population without destruction of the environmental base. Can our present industry-based way of life continue without causing those resources to run out, leading to ecological collapse?

> **sustainability** The survival of a land-use system for centuries or millennia without destruction of the environmental base.

The Environmental Sustainability Index (ESI) has been devised to measure, country by country, the level of progress toward sustainability. The highest possible score on the ESI is 1.00, the lowest 0.00. Eleven major indicators of environmental sustainability, including carbon dioxide emissions, water quality, and biodiversity, are included in each country's score (**Figure 9.15**). In 2011, the highest ranked country was Norway, at 0.8232; the lowest was India at 0.2396. Although far from being a perfect measure, the ESI clearly points to regions where ecosystems are most highly stressed. These regions lie mainly in Asia but also include the United States, Australia, and parts of Europe.

Developing new sources of energy is one of the keys to environmental sustainability. For example, notable progress is being made in the use of wind power to generate electricity. In the United States alone, wind power produced over 94,000 megawatts of electricity by the end of 2010, about 2.3 percent of the total generated. However, the United States still lags far behind such countries as Denmark, Spain, and Germany. Texas, Iowa, Minnesota, and Illinois are the leading wind-power states.

Another development fostering sustainability is **ecotourism,** defined as responsible travel that does not harm ecosystems or the well-being of local people. Ecotourism arose when it was recognized that even seemingly benign industries such as tourism can create ecological problems. Ecotourists tend to visit out-of-the-way places with exotic, healthy ecosystems. They disdain the comforts of large hotels and resorts, preferring more spartan conditions. Revenue from ecotourism is helping to rescue Uganda's mountain gorillas from extinction, especially now that the government has realized that the wildlife serves as a valuable tourist attraction.

> **ecotourism** Responsible travel that does not harm ecosystems or the well-being of local people.

Also notable has been the rise of the **Greens,** political activists who advocate an emphasis on environmental issues. Many countries now have Green political parties. Also active in environmental advocacy are groups such as the Sierra Club, Greenpeace, and the Nature Conservancy.

> **Greens** Activists and organizations, including political parties, whose central concern is addressing environmental issues.

INDUSTRIAL-ECONOMIC CULTURAL **INTERACTION**

Can our understanding of the industrialized world be aided by the theme of cultural interaction? Yes, but in a very different way from cultural ecology. Cultural interaction mainly allows us to understand the location of industries. The causal factors influencing *industrial location* include various economic features, especially labor supply.

Labor Supply

Labor-intensive industries are those industries in which labor costs form a large part of total production costs.

> **labor-intensive industry** An industry whose labor costs constitute a large portion of total production costs.

Examples include industries that depend on skilled workers producing small objects of high value, such as computers, cameras, and watches, and industries that require large numbers of semiskilled workers, such as the textile and garment industries. Manufacturers consider several characteristics of labor in deciding where to locate factories: availability of workers, average wages, necessary skills, and worker productivity. Workers with certain skills tend to live and work in a small number of places, partly as a result of the need for higher education or for person-to-person training in handing down such skills. Consequently, manufacturers often seek locations where these skilled workers live. Geographer Amy Glasmeier has traced the history and contemporary conditions of the watchmaking industry in four countries (Great Britain, the United States, Switzerland, and Japan) and shown how the availability of a pool of skilled labor and the maintenance of that pool through different forms of education and training were critical to the success of these businesses.

In recent decades, with the increasing effects of globalization, two trends have been evident in terms of the relationship between the location of industry and labor supply. On the one hand, the increased mobility of people has in some ways lessened the locational influence of the labor force. Large numbers of workers around the globe have migrated to manufacturing regions. Today, labor migration is at an all-time high, lessening labor's influence on industrial location. On the other hand, particularly for industries that are reliant on large pools of labor and that are transnational in structure, the location of labor has become even more important. These industries are often referred to as *footloose,* in the sense that they shift the location of their facilities in search of cheap labor. Thus we see how globalization is enabling more industries to locate their production facilities close to labor supply.

Some companies are even locating their producer service activities in places where they can take advantage of cheaper labor costs. In the 1990s, many companies began to move what were typically called secretarial jobs—filing, typing, document formatting, and so on—out of their corporate headquarters and into places and spaces that were cheaper, both in terms of rental of office space and especially in terms of labor costs. More recently, some types of businesses, such as accounting firms, Wall Street investment houses, advertising agencies, and insurance companies, have found it cost-effective to **outsource** other white-collar jobs that are generally considered more skilled, such as legal research, financial analysis, and accounting. In general,

> **outsource** The physical separation of some economic activities from the main production facility, usually for the purpose of employing cheaper labor.

outsourcing involves the physical separation of some economic activities from the main production facility. Chennai (formerly known as Madras), India, for example, is a primary site of these more skilled outsourced activities (**Figure 9.16** on page 268). Here, educated and highly skilled workers are trained to do these jobs; they are paid the equivalent of $10,000 to $20,000 a year, compared to the average annual salary of $100,000 for a similar worker in New York City.

Markets

Geographically, a **market** includes the area in which a product may be sold in a volume and at a price profitable to the manufacturer. The size and distribution of markets are generally the most important factors in determining the spatial distribution of industries. With globalization, markets have

> **market** The geographical area in which a product may be sold in a volume and at a price profitable to the manufacturer.

expanded to cover most of the regions in the world, but even so the costs of shipping may prohibit particular kinds of manufactured goods from being sold worldwide. Some manufacturers have to situate their factories among their customers to minimize costs and maximize profits. Such industries include those that manufacture a weight-gaining finished product, such as bottled beverages, or a bulk-gaining finished product, such as metal containers or bottles. In other words, if weight or bulk is added to the raw materials in the manufacturing process, location near the market is economically desirable because of the high transportation costs. Similarly, if the finished product is more perishable or time-sensitive than the raw materials, as with baked goods and local newspapers, a location near the market also is required. In addition, if the product is more fragile than the raw materials that go into its manufacture, as with glass items, the industry will be attracted to locations near its market. In each of these cases—gain in

FIGURE 9.16 Back-office workers in New Delhi, India. This photo was taken inside a call center that supports travel agencies. (Fredrik Renander/Alamy.)

THINKING GEOGRAPHICALLY Although there are still many call centers in the United States, why are these services increasingly being outsourced?

weight or bulk, perishability, or fragility—transportation costs of the finished product are much higher than those of the raw materials. In contrast, items that become easier to transport after manufacturing often are produced in locations close to the site of the raw materials.

As a rule, we can say that the greatest market potential exists where the largest numbers of people live. This is why many manufacturing firms today are looking to the large population centers in such countries as India, China, and Russia for what they consider their emerging markets. The term *emerging markets* refers to places within the global economy that have recently been opened to foreign trade and where populations are just beginning to accumulate capital that they can spend on goods and services. In China, for example, globalization has led to a huge increase in industrialization, and these industries create jobs for people that allow them to purchase consumer goods. China's large population base (over 1.3 billion people) serves as both a great source of labor and an expanding consumer market. Because China has more than 40 cities with populations exceeding a million (**Figure 9.17**) and also has a stable government that encourages foreign investment, transnational companies find the opportunities for low-cost production and expanding consumption there appealing.

The Political Element

So far, we've considered the roles of labor, corporations, and consumers in economic processes, but governments are major players, too. Governmental policies shape economic activities on scales that range from urban areas to regions to nations to the globe. Governments often intervene directly in decisions about industrial location. Such intervention typically results from a desire to encourage for-

eign investment; to create national self-sufficiency by diversifying industries; to bring industrial development and a higher standard of living to poverty-stricken provinces; to establish strategic, militarily important industries that otherwise would not develop; or to halt agglomeration in existing industrial areas. Such governmental influence becomes most pronounced in highly planned economic systems, particularly in certain socialist countries such as China, but it works to some extent in almost every industrial nation.

Other types of government influence come in the form of *tariffs* (taxes on imported goods), import-export quotas, political obstacles to the free movement of labor and capital, and various methods of hindering transportation across borders. Tariffs in effect reduce the size of a market area proportionally to the amount of tariff imposed. A similar effect is produced when the number of border-crossing points is restricted. In some parts of the world, especially Europe, the impact of tariffs and borders on industrial location has been greatly reduced by the establishment of free-trade blocs—groups of nations that have banded together economically and abolished most tariffs. Of these associations, the European Union (EU) is perhaps the most famous. Composed of 27 nations, the EU has succeeded in abolishing tariffs within its area. The North American Free Trade Agreement (NAFTA) among the United States, Canada, and Mexico is a similar arrangement in the Western Hemisphere. At a much larger spatial scale, the World Trade Organization (WTO) administers trade agreements and settles trade disputes, including those over tariffs, throughout much of the world. With nearly 150 member countries, the WTO regulates approximately 97 percent of world trade.

Governments can also exert political influence on economic activity in other ways that distort aspects of their

FIGURE 9.17 Chinese cities with a population over 1 million. As you can see from this map, there are more than 40 cities in China with populations that exceed 1 million. Each of these agglomerations represents a large consumer market and source of labor. (Source: Thomas Brinkhoff: City Population, http://www.citypopulation.de.)

THINKING GEOGRAPHICALLY Why do you think most of these cities are increasing in population?

national economies. For example, many governments establish guaranteed minimum prices for certain domestic commodities and import duties (taxes) on competing products from other countries. This practice is often found in the agricultural sector, as governments set prices for farm commodities at levels higher than the world market would support. The chief goal is to ensure an acceptable standard of living for farmers and to preserve rural heritage. A large-scale example of agricultural price supports is the EU's Common Agricultural Policy (CAP). This program of farm subsidies, price supports, import tariffs, and quotas has cost the EU up to half its annual budget in some years, often exceeding €50 billion per year. Many European residents have protested these expenditures, leading to plans to greatly reduce EU spending on the program by 2013. After that time, farm subsidies and agricultural price supports will likely be used only in emergencies such as natural disasters or outbreaks of plant and animal diseases. These EU budgetary changes are projected to reduce certain distortions of world agricultural markets.

National governments can also influence economic markets through the protection of emerging industries or *infant industries*. Those who support the protection of infant industries argue that new domestic industries often do not have sufficient economies of scale (size) to compete with similar industries in other countries. Governments can attempt to protect these developing industries by insulating them from foreign competition and by protecting

export activity through the lowering or elimination of tariffs. Some governments have even nationalized banks in order to make financing readily available to infant industries needing to borrow investment capital. A well-known case of this type of protectionism occurred in South Korea after the Korean War, when the government acted to shelter many domestic industries (including auto manufacturing, petroleum refining, shipbuilding, and electronics) in order to become competitive on the world stage. These measures led to great successes in many industries, allowing South Korea to take its place among the world leaders in manufacturing today. However, there have been cases in which government protection of infant industries has failed to achieve similar results. For example, in the 1980s, the Brazilian government chose to protect its infant computer industry against outside competitors, but the industry failed to develop and the technological gap between Brazil and the rest of the world at that time actually widened as a result.

Considering the extent of government manipulation of economic markets in agriculture and industry, it would seem that transnational corporations, which scatter their holdings across international borders, would suffer. In reality, however, multinational enterprises are well placed to take advantage of some government policies. Various countries act differently to encourage or discourage foreign investment, creating major spatial discontinuities in opportunities for global corporations. Areas where foreign investment is encouraged

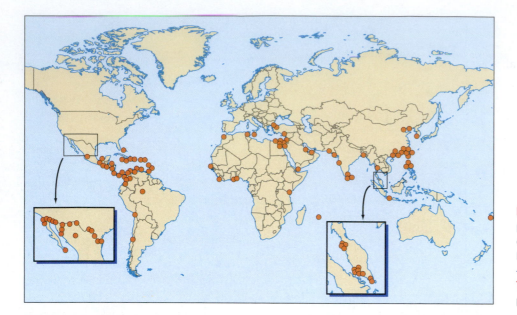

FIGURE 9.18 Global sites of export processing zones. Notice the concentration of these zones along the U.S.–Mexican border. (Source: Adapted from Dicken, 2003.)

THINKING GEOGRAPHICALLY What explains this phenomenon?

are often called **export processing zones** (**EPZs**), although in China they are called special economic zones (SEZs). In general, these zones are designated areas of countries where governments create conditions conducive to export-oriented production,

> **export processing zone (EPZ)** Designated area of a country where the government provides conditions conducive to export-oriented production. Called *special economic zone (SEZ)* in China.

including trade concessions, exemptions from certain types of legislation, provision of physical infrastructure and services, and waivers of restrictions on foreign ownership. According to Peter Dicken, about 90 percent of these zones are located in Latin America, the Caribbean, and Asia (**Figure 9.18**). Those located along the U.S.–Mexican border are populated by U.S.-owned assembly plants called *maquiladoras* (**Figure 9.19**), most of which utilize Mexican low-wage labor, predominantly women, to produce textiles, clothing, and electronics for the export market.

> **maquiladora** U.S.-owned assembly plant located on the Mexican side of the U.S.–Mexican border.

Industrialization, Globalization, and Cultural Change

Industrialization and globalization bring with them enormous changes, and those changes have profound impacts on peoples' lives. As we have already seen in this chapter, the globalization of industry can radically alter regional economies, while the increase in international trade that is a major consequence of globalization allows for the circulation of new and different types of consumer goods and services throughout much of the world. Some very fundamental changes in how people live their everyday lives often accompany these large economic transformations. Increased interregional trade often leads to intercultural contact and the introduction of different ways of living and

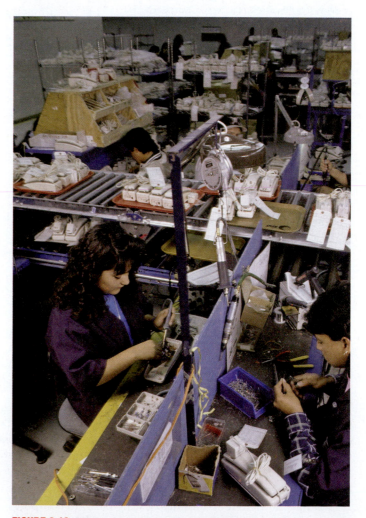

FIGURE 9.19 *Maquiladora* in Nuevo Laredo, Mexico. In this *maquiladora*, Mexican workers are completing telephone repairs for AT&T. (Bob Daemmrich/The Image Works.)

THINKING GEOGRAPHICALLY Why are many of these workers women?

working to formerly isolated regions. The locational shifts of industries into new regions of the world can precipitate large movements of people from rural, farming areas into cities to work in factories. In some places in the world, economic development has restructured families and reshaped gender roles. For example, many transnational companies seek out a female labor force to work in their factories, since they believe women are more reliable and pliant workers, and also because in general they can pay them less than they pay men.

Even though globalization has spread industries and services throughout the world, not all groups of people have benefited equally. Some regions of the world have remained relatively underdeveloped, serving the global economy as sources of raw materials and low-wage labor, while other areas of the world benefit from the cheaper costs of production elsewhere to raise their standards of living. This can be explained by a large range of factors: the particular histories and physical geographies of certain countries, their situation within the geopolitical world (see Chapter 6), international trade agreements, and the types of people who live in different regions and their cultures.

INDUSTRIAL-ECONOMIC LANDSCAPES

In what ways have economic activities altered the cultural landscape? It is difficult to look at any particular landscape and not see some evidence of the production, circulation, and consumption of goods and services. Factory buildings of course provide an obvious example, but perhaps less obvious are the spaces where goods and services circulate (highways, trains, airports, for example) and where they are consumed (for example, malls, shops, offices). Economic landscapes, then, are prominent visible features of our surroundings and form part of daily life.

Each level of industrial activity produces its own distinctive landscape. Primary industries exert perhaps the most drastic impact on the land. The resulting landscapes contain slag heaps, clear-cut commercial forests, strip-mining scars, open-pit mines, and oil derricks (**Figure 9.20**). The destruction of nature is less apparent in other types of primary industrial landscapes. The fishing villages of Portugal or Newfoundland even attract

(a)

(b)

(c)

FIGURE 9.20 **Three primary industrial landscapes.** (a) This bizarrely colored industrial mosaic is on the margins of the Great Salt Lake in Utah, where chemicals and minerals such as metallic magnesium, potassium, and sodium chloride are derived from the water by solar evaporation. (b) This open-pit mine is Bingham Canyon in Utah, the second largest in the world. (c) This mind-boggling artificial alp is the result of potash mining near Kassel in Germany. (Courtesy of Terry G. Jordan-Bychkov.)

THINKING GEOGRAPHICALLY What are the environmental implications of these landscapes?

FIGURE 9.21 **A fishing village in Newfoundland.** Primary industrial landscapes can be pleasing to the eye. (Courtesy of Terry G. Jordan-Bychkov.)

THINKING GEOGRAPHICALLY What problems might such a landscape also suggest?

tourists (**Figure 9.21**). In still other cases, efforts are made to restore the preindustrial landscape. Examples include the establishment of artificial grasslands in old strip-mining areas of the American Midwest and the creation of recreational ponds in old mine pits along interstate highways.

Among the most obvious features of the landscapes of secondary industry, or manufacturing, are factory buildings. Early- to mid-nineteenth-century industrial landscapes are easy to identify, because the technologies that ran these factories were reliant on water power. Hence, the mill buildings that housed the machinery were designed in linear form to take full advantage of the turbines that were connected to waterwheels. What's more, given this locational requirement, these mill complexes were most often built in rural areas. Entire communities, most often with housing for the workers, were thus constructed along with the mills. These mill towns dotted the landscape of England, Scotland, and New England, the site of the first industrial developments in the United States (**Figure 9.22**). Later—as water power was supplanted by steam, coal, and then electric power—factories were located near urban areas, taking advantage of the housing supply already there and the proximity to a large consumer base. These industrial landscapes were often located at the edge of downtowns, lining the railroad routes into the city, and were surrounded by working-class housing. In the second half of the twentieth century—with the development of the interstate highway system and the trucking industry, as well as the switch to high-tech industries such as electronics—industrial landscapes took on a different form. Factories began to move to industrial parks at inter-

state exchanges, and their architecture began to resemble other types of mass-produced architecture, such as big-box stores (**Figure 9.23**).

Given the degree of deindustrialization in certain parts of the old core industrial regions, it is not surprising that many of these factory complexes, particularly those that date from the nineteenth and early twentieth centuries, are derelict or are being retrofitted for housing or commercial uses. Others now house historical museums depicting past industrial technologies and ways of life. Lowell,

FIGURE 9.22 **Mill buildings in central Massachusetts.** Most textile mills in New England were abandoned by the mid-twentieth century, as textile production moved to southern states and then to countries outside the United States. (Courtesy of Mona Domosh.)

THINKING GEOGRAPHICALLY What accounts for the footloose nature of the textile industry?

FIGURE 9.23 **Footwear factory in Picardie, France.** This factory is located right along the highway, providing easy access for its employees and for the trucks that transport its products. (ForrestierYves/Corbis Sygma.)

THINKING GEOGRAPHICALLY Are there any visual clues here that this structure is a factory rather than a big-box store?

Massachusetts, an early (mid-nineteenth-century) planned industrial town that produced textiles, is now the site of a national park. A good percentage of its factories, canals, and housing complexes have been preserved and can be toured by visitors. In Great Britain, many sites of industrial history are now preserved as museums. New Lanark, located outside Glasgow in Scotland, is now a World Heritage Site and provides a particularly interesting example of an industrial landscape, given that it was originally established by Robert Owen as a utopian community. He included in his planning good schools, housing, and even a cooperative food store in order to create a benevolent community of workers (**Figure 9.24**).

FIGURE 9.24 **New Lanark, Scotland.** Robert Owen's planned industrial town is now a World Heritage Site. (Courtesy of Mona Domosh.)

THINKING GEOGRAPHICALLY Compare these buildings to the workers' dormitory in Figure 9.25, and consider the changes that have occurred (or haven't occurred) in the provision of housing for industrial workers.

In other parts of the world industrial landscapes are far from derelict; they are, in fact, being built anew. In Southeast Asia, China, and Mexico, large industrial landscapes are under construction. Some, like the *maquiladoras,* are similar to the early mill towns in that they too are being constructed in nonurban sites without housing or other infrastructure. In the *maquiladoras,* as well as in many industrial regions of China, housing comes either in the form of dormitories or in informal squatter settlements (**Figure 9.25**).

FIGURE 9.25 **Workers' dormitory in China.** Workers are responsible for their own laundry. (Zhang Wei/ImagineChina/AP Images.)

THINKING GEOGRAPHICALLY Compare this image to Figure 9.24.

FIGURE 9.26 Office buildings in suburban Florida. Most of the activities that take place in these offices are related to banking. (Courtesy of Mona Domosh.)

THINKING GEOGRAPHICALLY Why are these buildings set back off the road?

Service industries, too, produce a cultural landscape. Its visual content includes elements as diverse as high-rise bank buildings, hamburger stands, silicon landscapes, gasoline stations, and the concrete and steel webs of highways and railroads. Some highway interchanges can best be described as a modern art form, but perhaps the aesthetic high point of the industrial landscape is found in bridges, which are often graceful and beautiful structures. The massive investment in these transportation systems has even changed the way we view the landscape. As geographer Yi-Fu Tuan commented, "In the early decades of the twentieth century vehicles began to displace walking as the prevalent form of locomotion, and street scenes were perceived increasingly from the interior of automobiles moving staccato-fashion through regularly spaced traffic lights." Los Angeles provides perhaps the best example of the new viewpoints provided by the industrial age. Its freeway system allows individual motorists to observe their surroundings at nonstop speeds. It also allows the driver to look down on the world. The view from the street, on the other hand, is not encouraged. In some areas of Los Angeles, streets actually have no sidewalks at all, so the pedestrian viewpoint is functionally impractical. In addition, the shopping street is no longer scaled to the pedestrian—Los Angeles's Ventura Boulevard extends for 15 miles (24 kilometers).

Producer services related to financial activities, such as legal services, trade, insurance, and banking, traditionally were located in high-rise buildings in urban centers, but with suburbanization they have taken on a nonurban form. Many are now located in five- or six-story buildings along the interstates surrounding cities in what we've called high-tech corridors (**Figure 9.26**). Other producer service industries choose to maintain their downtown location for symbolic reasons. Some consumer services, particularly retailing, have created distinctive and, within the American context at least, socially important landscapes. Shopping malls are now dominant features of the North American suburb and often serve as catalysts to suburban land development, in effect creating entirely new landscapes, all geared toward consumption. Chapter 11 provides more details about these new and emerging landscape elements.

Key Terms

acid rain, 262
consumer services, 254
cottage industry, 256
deindustrialization, 258
economic development, 250
ecotourism, 266
export processing zone (EPZ), 270
global warming, 263
greenhouse effect, 263
Greens, 266
gross domestic product (GDP), 250
guild industry, 256
high-tech corridor, 254
industrial revolution, 256
industrialization, 250

labor-intensive industry, 267
maquiladora, 270
market, 267
outsource, 267
postindustrial phase, 261
primary industry, 252
producer services, 253
renewable resource, 261
secondary industry, 252
services, 252
sustainability, 266
technopole, 258
transnational corporation, 258
transportation/ communication services, 253

Economic Geography on the Internet

You can learn more about economic geography on the Internet at the following web sites:

Atlas of Global Development

http://issuu.com/world.bank.publications/docs/
9780821385838

This online atlas provides extensive information about many aspects of social and economic development. Easy-to-read maps and color graphics highlight the relationships between location and well-being.

Global Policy Forum

http://www.globalpolicy.org

This site provides access to resources about global economic development, poverty, and social justice issues that are relevant for policy makers around the globe.

Inter-American Development Bank

http://www.iadb.org/index.cfm

The Inter-American Development Bank (IDB) provides financing for social and economic development in Latin America and the Caribbean. This site provides information about their various development projects and allows you to download their publications, which include valuable data and sources for data concerning development in the region.

NASA's Ozone Watch

http://ozonewatch.gsfc.nasa.gov/index.html

This site allows observers to view the latest status of the ozone layer over the Antarctic, with a focus on the ozone hole.

United Nations Industrial Development Organization

http://www.unido.org

This specialist agency of the United Nations is devoted to promoting sustainable industrial development in countries with developing and/or transitional economies. The site contains industrial statistics and information on, for example, women in industrial development, and it allows you to access data and maps of the least developed countries in the world.

World Bank

http://www.worldbank.org

In addition to information about the workings of the World Bank, this site contains valuable data on measures of economic development for most countries in the world.

Worldwatch Institute, Washington, D.C.

http://www.worldwatch.org

This nongovernmental watchdog and research institute compiles and analyzes the latest information about such ecological problems as global warning, industrial pollution, and deforestation. It also seeks sustainable alternatives.

Recommended Books on Economic Geography

Dicken, Peter. 2003. *Global Shift: Reshaping the Global Economic Map in the 21st Century.* New York: Guilford Press. An incredibly comprehensive look at the causes and effects of the increasingly global and interlinked industries of the world.

Hall, Colin M., and Stephen J. Page. 1999. *The Geography of Tourism and Recreation: Environment, Place and Space.* New York: Routledge. An excellent comprehensive introduction to the topic, containing case studies from North America, Europe, China, Australia, and the Pacific islands.

Harrington, James W., and Barney Warf. 1995. *Industrial Location: Principles and Practice.* New York: Routledge. A useful basic primer on industrial location written for nonexperts.

Harvey, David. 2006. *Spaces of Global Capitalism: A Theory of Uneven Geographical Development.* New York: Verso. The latest book by a major theorist and critic of the effects of global capitalism throughout the world.

McDowell, Linda. 1997. *Capital Culture: Gender at Work in the City.* 1997. Oxford: Blackwell. An interesting look at the importance of gender identity to the functioning of the financial sector.

Sheppard, Eric, and Trevor Barnes. 2002. *Companion to Economic Geography.* Oxford: Blackwell. A very accessible series of essays that serve to introduce the major concepts and issues in contemporary economic geography.

Journals in Economic Geography

Economic Geography. Published by Clark University. Volume 1 appeared in 1925.

Geoforum. Published by Elsevier Science. Volume 1 appeared in 1969.

Journal of Economic Geography. Published by Oxford University Press. Volume 1 appeared in 2000.

Journal of Transport Geography. Published by Elsevier Science. Volume 1 was published in 1993.

Answers to Thinking Geographically Questions

Figure 9.1: Although there is much similarity, the Southern Cone of South America, for example, ranks higher on the Human Development Index than it does in gross domestic product.

Figure 9.2: Answers will vary, but you may suggest that in less developed countries, data regarding health, education, and other aspects of well-being may not be collected and tabulated as extensively or as accurately as in more developed countries. Therefore, the HDI that is calculated based on this data may not be equally reliable from country to country.

Figure 9.3: In the days when derricks were the common apparatus, each well could tap a relatively small area. The grasshopper well indicates continued use, but fewer wells are needed to tap a broad area.

Figure 9.4: High-tech industries are more footloose, less dependent on moving heavy materials from place to place. They also depend on a highly educated work force. Therefore, they tend to locate in places with universities and with amenities that educated, upper-class people seek, like recreation and cultural attractions. They may also gravitate toward places with pleasant climate.

Figure 9.5: Small countries are more likely to have tourists from outside the country than are large countries.

Figure 9.6: Tourists model different lifestyles to local people, apart from the impact of their activities on the local ecology. They also bring in cash, potentially changing a barter-based economy.

Figure 9.7: Although much of the spread was contagious, the areas around Munich and Zurich adopted railroads earlier than their surroundings, as did St. Petersburg, Russia, and Constantinople (Istanbul).

Figure 9.8: The dense population of Britain had encouraged improvements in agriculture so that it could feed a large nonfarming industrial work force. British people had also explored and colonized other parts of the world, yielding sources of raw materials and food, as well as markets for manufactured goods. Such trade also created a pool of capital to invest in manufacturing.

Figure 9.9: The earliest centers of industrialization had outdated physical plants, and many suffered from contentious labor relations. Eastern Europe underwent 45 years of communist rule with its central planning and subservience to the desires of the Soviet Union.

Figure 9.9: Colonial governments were unwilling to allow colonies to develop manufacturing that would compete with the ruling countries' products. Since independence, these trade relationships have continued, and political corruption has made many former colonies uninviting places to invest.

Figure 9.10: Automobile manufacturing is an industry that demands large supplies of steel and other raw materials, a highly skilled labor force, and cost-effective access to transportation routes. Many developing countries in South America, Africa, and Central Asia would find it challenging to effectively meet these requirements in order to compete on the global market with auto manufacturers in regions where these resources and capital are more readily available.

Figure 9.11: These methods are cheaper than seeking out individual trees.

Figure 9.12: The largest areas are in the eastern part as well as along the Amazon and in the southern part of the forest. These areas are the most accessible.

Figure 9.13: Among the factors would be programs and campaigns to reduce production of the chemicals causing acid rain, such as conversion to different types of fuel. Counteracting that would be growth in population and industrial activity, necessitating additional power generation.

Figure 9.14: Along with the accident at Three Mile Island in Pennsylvania in 1979, it has made the American public extremely fearful of nuclear power. No new nuclear power plants have been constructed in the United States since the two incidents took place.

Figure 9.15: Many of the countries that rank high are already developed and wealthy; others have taken the matter of sustainability seriously. Many countries that rank low are authoritarian.

Figure 9.16: Labor costs are the main reason. In countries like India, wages are much lower, and a large number of English-speaking technical college graduates are well qualified for such jobs.

Figure 9.17: Natural increase in China is low; this is the result of migration. Rural workers are moving to cities in large numbers seeking employment.

Figure 9.18: Countries with a large labor force see this kind of activity as an opportunity to industrialize, and they enact laws and policies to attract such zones.

Figure 9.19: Employers often favor women because they will work for lower wages.

Figure 9.20: The evaporation operation could produce toxic dust as well as ruin the land. The open-pit mine must dump the material it digs out somewhere, and it could flow into populated areas or nearby streams. The alp could experience landslides in heavy rain or snow.

Figure 9.21: Fishing grounds that these people depend on could be overfished. Fishing is also not a very lucrative occupation, compared with urban jobs. The people who live here are also very isolated. Providing services like education and health care may be very expensive.

Figure 9.22: Although the textile industry uses machinery, the tasks are fairly easy to learn, so labor is easy to train and can draw on low-cost sources.

Figure 9.23: There is a high barbed-wire fence around the parking lot with a guardhouse at the entrance.

Figure 9.24: In both cases, the buildings are plain, located near the factory, and provide small spaces as cheaply as possible. This housing is surrounded by trees and other greenery. The buildings seem to be only two stories high and may house families as well as workers.

Figure 9.25: This building has more stories, and probably the units house only single workers. While Figure 9.24 shows chimneys, this building is probably unheated (as are most buildings in southern China).

Figure 9.26: Part of the reason is to allow for large parking lots, as virtually all the workers drive to work in private cars. The setback also allows for some green space and protection of, for example, a small wetland.

A view of Rio de Janeiro, Brazil, looking toward Sugarloaf Mountain. (SIME/eStock Photo.)

Urbanization

10

Imagine the 2 million years that humankind has spent on Earth as a 24-hour day. In this framework, settlements of more than a hundred people came about only in the last half hour. Towns and cities emerged only a few minutes ago. Yet it is during these "minutes" that we see the rise of civilization. *Civitas,* the Latin root word for *civilization,* was first applied to settled areas of the Roman Empire. Later it came to mean a specific town or city. *To civilize* meant literally "to citify."

Urbanization over the past 200 years has strengthened the links among culture, society, and the city. An urban explosion has gone hand in hand with the industrial revolution. According to United Nations assessments, the year 2008 was a momentous one, as it marked the first time that the world's urban population exceeded its rural population. The world's urban population has more than quadrupled since 1950 (733 million in 1950 versus 3.6 billion in 2011) and will reach 6.3 billion by the year 2050. At that time, approximately 67 percent of Earth's population will live in

URBANIZED POPULATION

FIGURE 10.1 **Urbanized population in the world.** This map indicates the percentage of a country's population that lives in urban areas, but this information must be used cautiously due to different international definitions of the word *city*. Notice that, despite recent rapid urbanization in countries of the developing world, many of these countries have less-urbanized populations than countries of the developed world. (Based on the United Nations, "World Urbanization Prospects, the 2011 Revision," Map3, http://esa.un.org/unpd/wup/Maps/maps_urban_2011.htm.)

THINKING GEOGRAPHICALLY Can you identify some of the reasons for this paradox?

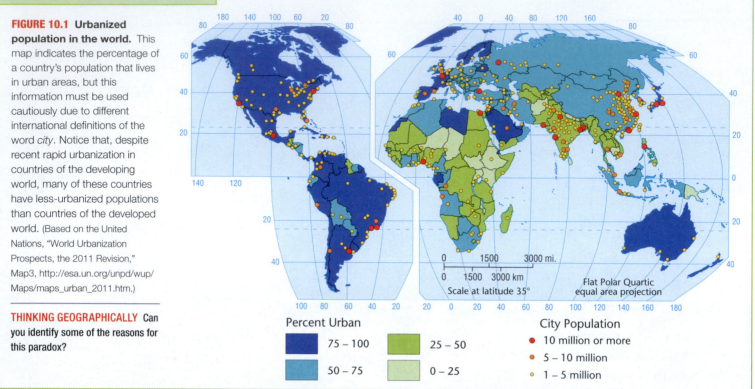

Percent Urban

- 75 – 100
- 50 – 75
- 25 – 50
- 0 – 25

City Population

- 10 million or more
- 5 – 10 million
- 1 – 5 million

cities. The cultural geography of the world will change dramatically as we become a predominantly urban people and the ways of the countryside are increasingly replaced by urban lifestyles.

In this chapter, we consider overall patterns of urbanization, learn how urbanization began and developed, and discuss the differing forms of cities in the developing and developed worlds. In addition, we examine some of the external factors influencing city location. In Chapter 11, we look at the internal aspects of the city as seen through the five themes of cultural geography.

CULTURE **REGIONS**

How are urban areas and urban populations spatially arranged? All of you know from your own travels locally, nationally, or internationally that some regions contain many cities while others contain relatively few and that the size of those cities can vary greatly. How can we begin to understand the location, distribution, and size of cities? We start with a consideration of global patterns of urbanization, examining the general distribution of urban populations around the world. A quick look at **Figure 10.1** reveals differing patterns of **urbanized population**—the percentage of a nation's population living in towns and cities—around the world.

For example, countries in Western Europe, North America, Latin America, and the Caribbean have relatively high levels of urbanization, with more than 75 percent of each country's population living in urban areas. Most of the nations of Africa and Asia, in contrast, are less urbanized, with 25 percent to 75 percent of each country's population residing in urban areas. How do geographers explain these varying regional patterns of urbanization?

Patterns and Processes of Urbanization

According to United Nations estimates, almost all the worldwide population growth in the next 30 years will be concentrated in urban areas, with the cities of the less developed regions responsible for most of that increase. The reasons for this explosion in urban population growth and its uneven distribution around the world vary, as each country's unique history and society present a slightly different narrative of urban and economic development. Making matters more complex is the lack of a standard definition of what constitutes a city. Consequently, the criteria used to calculate a country's urban population differ from nation to nation. Using data based on these varying criteria would result in misleading conclusions. For example, in the United States, according to the U.S. Census Bureau, a city is a densely populated area of 2,500 people or more, and cities of 50,000 or more people constitute *metropolitan statistical areas*. In many European countries, the city designation is granted by government charter and is not based on any par-

urbanized population The proportion of a country's population living in cities.

ticular population criterion. Meanwhile, in India, only urban places of more than 100,000 residents are considered cities. It is important to remember, then, that an international comparison of urban population data can be made only by taking into account the varying definitions of a city.

Nonetheless, several generalizations can be made about the differences in the world's urbanized population. First, there is a close link between urbanized population and the more developed world. Put differently, highly industrialized countries have higher rates of urbanized population than do less developed countries. The second generalization, closely tied to the first, is that developing countries are urbanizing rapidly and that their ratio of urban to rural population is increasing dramatically (**Figure 10.2**).

Urban growth in these countries comes from two sources: the migration of people to the cities (see also the section "Origin and Diffusion of the City") and the higher natural population growth rates of these recent migrants. People move to the cities for a variety of reasons, most of which relate to the effects of uneven economic development in their country. Cities are often the centers of economic growth, whereas opportunities for land ownership and/or farming-based jobs are, in many countries, rare. Because urban employment is unreliable, many migrants continue to have large numbers of children to construct a more extensive family support system. Having a larger family increases the chances of someone's getting work. The demographic transition to smaller families comes later, when a certain degree of security is ensured. Often, this transition occurs as women enter the workforce.

Impacts of Urbanization

Although rural-to-urban migration affects nearly all cities in the developing world, the most visible cases are the extraordinarily large settlements we call **megacities,** those having populations of 10 million or more. **Table 10.1** on page 280 shows the world's 20 most populous cities, over

megacity A particularly large urban center, with a population over 10 million.

half of which are in the developing world. This is a major change from 30 years ago, when the list would have been dominated by Western industrialized cities, a trend that most expect to continue. Projections for future growth, however, must be qualified by two considerations. First, cities of the developing world will continue to explode in size only if economic development expands. If it stagnates because of political or resource problems, city growth will probably slow (although urban migration might increase if rural economies deteriorate). For example, Mexico City's growth is linked to that country's economic growth and more specifically to Mexico's oil industry, which fluctuates according to the world market for oil. Second, because these megacities are plagued by transportation, housing, employment, and ecological problems—such as an inadequate water

(a)
(b)
(c)

FIGURE 10.2 Urbanization. Shown here are scenes from (a) a squatter settlement in Rio de Janeiro, Brazil; (b) the business district in Kolkata (Calcutta), India; and (c) a residential district of Manila, Philippines. As is evident from this scene in Kolkata, the downtowns of cities in the developing world are often more vibrant than those of the developed world. (a: Florian Kopp/imagebro/age fotostock; b: Steve Raymer/National Geographic Society/Corbis; c: © Sean Sprague/The Image Works.)

THINKING GEOGRAPHICALLY Can you think of some factors to explain the vibrant nature of downtown areas in developing countries?

TABLE 10.1   The World's 20 Largest Metropolitan Areas

Rank	Name	Country	Total Population (millions)
1	Tokyo	Japan	34.5
2	Guangzhou	China	25.8
3	Jakarta	Indonesia	25.3
4	Seoul	South Korea	25.3
5	Shanghai	China	25.3
6	Mexico City	Mexico	25.2
7	Delhi	India	23.0
8	New York City	United States	21.5
9	São Paulo	Brazil	21.1
10	Mumbai	India	20.8
11	Manila	Philippines	20.7
12	Karachi	Pakistan	17.4
13	Los Angeles	United States	17.0
14	Osaka	Japan	16.8
15	Beijing	China	16.4
16	Moscow	Russia	16.2
17	Cairo	Egypt	15.7
18	Calcutta	India	15.7
19	Buenos Aires	Argentina	14.3
20	Dacca	Bangladesh	14.0

(Source: http://citypopulation.de/world/Agglomerations.html, April 1, 2012.)

supply, in the case of Mexico City—some countries are trying to control urban migration. The success or failure of these policies will influence city size in the next 10 to 20 years.

Nevertheless, the urban population in the developing world is growing at astounding rates. Even though the developed regions of the world are more urbanized overall than the less developed regions, the sheer scale and rate of growth in absolute numbers reveal a reversal in this pattern. According to geographer David Drakakis-Smith, there are now twice as many urban dwellers in the developing world as there are in developed countries. For example, the population of urbanites in the countries of Europe, North America, Latin America, and the Caribbean (according to the United Nations, 1.2 billion) is smaller than the population of urbanites in Asia (1.5 billion). With this incredible increase in sheer numbers of urban dwellers in the less developed regions of the world comes a large list of problems. Unemployment rates in cities of the developing world are often over 50 percent for newcomers to the city; housing and infrastructure often cannot be built fast enough to keep pace with growth rates; water and sewage systems can rarely handle the influx of new people.

These shortcomings often result in the spread of disease, increased urban flooding, and an increase in preventable

deaths. In other words, the rapid economic growth occurring in many cities of the developing world, although bringing much-needed revenue, can occur at the expense of environmental quality and quality of human life. Additionally, the benefits of this economic growth are often spread unevenly throughout the population, with large numbers of newcomers to cities being forced to reside in slums plagued by a wide variety of health and social hazards. In contrast, in cities of the developed world, environmental problems tend to be related more to overconsumption of resources than to rapid population growth. Urban dwellers in the United States, for example, consume up to 32 times the amount of water and electricity and generate eight times more garbage than do residents of cities in developing countries such as Kenya or India. The massive demand for energy in cities of the developed world contributes significantly to the emission of greenhouse gases. In short: In both the developing world and the developed world, the restructuring of Earth's population and cultures from rural to increasingly urban will provide ongoing challenges.

The target for much urban migration is the **primate city.** This is a settlement that dominates the economic, political, and cultural life of

> **primate city** A city of large size and dominant power within a country.

a country and as a result of rapid growth expands its primacy or dominance. Buenos Aires is an excellent example of a primate city because it far exceeds Rosario, the second-largest city in Argentina, in size and importance. Although many developing countries are dominated by a primate city, often a former center of colonial power, urban primacy is not unique to these countries: think of the way London and Paris dominate their respective countries.

REFLECTING ON GEOGRAPHY

What types of historical, political, and economic factors account for nations that are dominated by a primate city? Why didn't Boston or New York City become the primate city for the United States?

ORIGIN AND **DIFFUSION** OF THE CITY

As we noted in opening this chapter, we are entering an era when more people live in cities than live in rural areas. When cities first began, only a very small portion of the population lived in them. Over time cities began to diffuse throughout the globe, and established cities grew larger. In this section, we analyze two aspects of diffusion in regard to urbanization. First, we take a historical look at the origins of city life and explore how, why, and where cities diffused around the world. Second, we examine the more contemporary processes of rural-to-urban migration.

In seeking explanations for the origin of cities, we ultimately find a relationship among areas of early agricultural development, permanent village settlements, the emergence of new social forms, and urban life. The first cities resulted from a complicated transition that took thousands of years.

As early people, originally hunters and gatherers, became more successful at gathering their resources and domesticating plants and animals, they began to settle, first semipermanently and then permanently. In the Middle East, where the first cities appeared, a network of permanent agricultural villages developed about 10,000 years ago. These farming villages were modest in size, rarely with more than 200 people, and were probably organized on a kinship basis. Jarmo, one of the earliest villages, located in present-day Iraq, had 25 permanent dwellings clustered near grain storage facilities.

Although small farming villages like Jarmo predate cities, it is wrong to assume that a simple quantitative change took place whereby villages slowly grew, first into towns and then into cities. True cities, then and now, differ qualitatively from agricultural villages. All the inhabitants of agricultural villages were involved in some way in food procurement—tending the fields or harvesting and preparing the crops. Cities, however, were more removed, both physically and psychologically, from everyday agricultural activities. Food was supplied to the city, but not all city dwellers were involved in obtaining it. Instead, city dwellers supplied other services, such as technical skills or religious interpretations considered important in a particular society. Cities, unlike agricultural villages, contained a class of people who were not directly involved in agricultural activities.

Two elements were necessary for this dramatic social change: the creation of an **agricultural surplus** and the development of a stratified social system. Surplus food, which is a food supply larger than the everyday needs of the population, is a prerequisite for supporting nonfarmers—people who work at administrative, military, or handicraft tasks. **Social stratification,** the existence of distinct socioeconomic classes, facilitates the collection, storage, and distribution of resources through well-defined channels of authority that can exercise control over goods and people. A society with these two elements—surplus food and a means of storing and distributing it—was set for urbanization.

agricultural surplus The amount of food a society grows that exceeds the demands of its population.

social stratification The existence of distinct socioeconomic classes.

Models for the Rise of Cities

One way to understand the transition from village life to city life is to model the development of urban life assuming that a single factor is the trigger behind the transition. The question that scholars ask is, "What activity could be so important to an agricultural society that its people would be willing to give some of their surplus to support a social class that specializes in that activity?" Next we discuss answers to that question.

Technical Factors The **hydraulic civilization** model, developed by Karl Wittfogel, assumes that the development of large-scale irrigation systems was the prime mover behind urbanization and that a class of technical specialists were the first urban dwellers. Irrigating agricultural crops yielded more food, and this surplus supported the development of a large nonfarming population. A strong, centralized government backed by an urban-based military could expand its power into the surrounding areas. Farmers who resisted the new authority were denied water. Continued reinforcement of the power elite came from the need for organizational coordination to ensure continued operation of the irrigation system. Labor specialization developed. Some people farmed; others worked on the irrigation system. Still others became artisans, creating the implements needed to maintain the system, or administrative workers in the direct employ of the power elite's court.

hydraulic civilization A civilization based on large-scale irrigation.

Although the hydraulic civilization model fits several areas where cities first arose—China, Egypt, and Mesopotamia (present-day Iraq)—it cannot be applied to all urban hearths. In parts of Mesoamerica, for example, an urban civilization blossomed without widespread irrigated agriculture and therefore without a class of technical experts.

Religious Factors Geographer Paul Wheatley suggests that religion led to urbanization. In early agricultural societies, knowledge of such matters as meteorology and climate was considered an element of religion. Such societies depended on their religious leaders to interpret the heavenly bodies before deciding when and how to plant their crops. The propagation of this type of knowledge led to more successful harvests, which in turn allowed for the support of both a larger priestly class and a class of people engaged in ancillary activities. The priestly class exercised the political and social control that held the city together.

In this scenario, early cities were religious spaces. The first urban clusters and fortifications are seen as defenses not against human invaders but against spiritual ones: demons or the souls of the dead. This religious explanation is applicable in some ways to all of the early centers of urbanization, although it seems especially successful in explaining Chinese urbanization.

Political Factors Some scholars suggest that the centralizing force in urbanization was political order. Urban historian

Lewis Mumford described the agent of change in emerging urban centers as the institution of kingship, which involved the centralizing of religious, social, and economic aspects of a civilization around a powerful figure who became known as the king. This figure of authority, who in the pre-urban world was accorded respect for his or her human abilities, ascended to almost superhuman status in early urbanizing societies. By exercising power, the king was able to marshal the labor of others. The resultant social hierarchy enabled the society to diversify its endeavors, with different groups specializing in crafts, farming, trading, or religious activity. The institution of kingship provided essential leadership and organization to this increasingly complex society, which became the city.

Multiple Factors At the onset of urbanization, and even much later in some places, sharp distinctions among economic, religious, and political functions were not always made. The king may also have functioned as priest, healer, astronomer, and scribe, thereby fusing secular and spiritual power. Critics of the kingship theory, therefore, point out that this explanation of urbanization may not be different from the religion-based model. Rather than attempting to isolate one trigger, a wiser course may be to accept the role of multiple factors behind the changes leading to urban life. Technical, religious, and political forces were often interlinked, with a change in one leading to changes in another. Instead of oversimplifying by focusing on one possible development schema, we must appreciate the complexities of the transition period from agricultural village to true city.

Urban Hearth Areas

The first cities appeared in distinct regions, such as Mesopotamia, the Nile River valley, Pakistan's Indus River valley, the Yellow River (or Huang Ho) valley of China, Mesoamerica, and the Andean highlands and coastal areas of Peru. These are called the **urban hearth areas** (**Figure 10.3**).

> **urban hearth area** A region in which the world's first cities evolved.

> **REFLECTING ON GEOGRAPHY**
> Scholars are continually altering the dates for the emergence of urban life, as well as the location of the hearth areas. Why? Can you outline some of the reasons it is so difficult to pinpoint the places and dates for the emergence of urban life?

It is generally agreed that the first cities arose in Mesopotamia, the river valley of the Tigris and Euphrates in what is now Iraq. Mesopotamian cities, small by current standards,

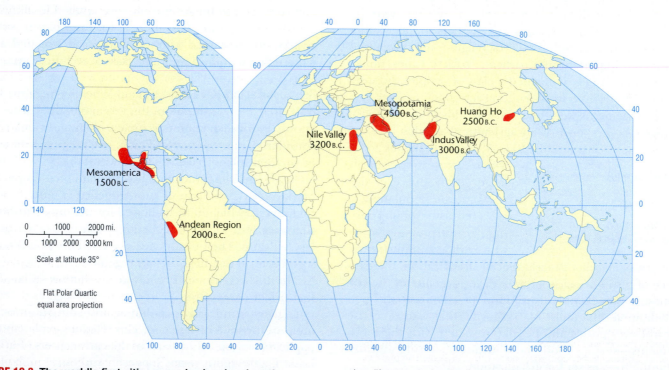

FIGURE 10.3 The world's first cities arose in six urban hearth areas. The dates shown are conservative figures for the rise of urban life in each area. For example, some scholars would suggest that urban life in Mesopotamia existed by 5000 B.C. Ongoing discoveries suggest that urban life appeared earlier in each of the hearth areas and that there are probably other hearth areas—in West Africa, for example.

THINKING GEOGRAPHICALLY Compare this map with Figure 8.16. What similarities and differences do you notice?

covered 0.5 to 2 square miles (1.3 to 5 square kilometers) with populations that rarely exceeded 30,000. Nevertheless, the densities within these cities could easily reach 10,000 people per square mile (4000 per square kilometer), which is comparable to the densities in many contemporary cities.

Scholars have referred to these first cities as **cosmomagical cities,** defined as cities that are spatially arranged according to religious principles. The spatial layouts of cosmomagical cities are similar in three important ways (**Figure 10.4**). First, great

cosmomagical city A type of city that is laid out in accordance with religious principles, characteristic of very early cities, particularly in China.

importance was accorded to the city's symbolic center, which was also thought to be the center of the known world. It was therefore the most sacred spot and was often identified by a vertical structure of monumental scale that represented the point on Earth closest to the heavens. This symbolic center, or **axis mundi,** took the form of the ziggurat in Mesopotamia, the palace or temple in China, and the pyramid in Mesoamerica. Often this elevated structure, which usu-

axis mundi The symbolic center of cosmomagical cities, often demarcated by a large, vertical structure.

ally served a religious purpose, was close to the palace or seat of political power and to the granary. These three structures were often walled off from the rest of the city, forming a symbolic center that both reflected the significance of these societal functions and dominated the city physically and spiritually (**Figure 10.5** on page 284). The Forbidden City in Beijing remains one of the best examples of this guarded, fortresslike city within the city. The second spatial characteristic common to cosmomagical cities is that they were oriented toward the four cardinal directions. By aligning the city in the north-south and east-west directions, the geometric form of the city reflected the order of the universe. This alignment, it was thought, would ensure harmony and order over the known world, which was bounded by the city walls.

In all of these early cities, one sees evidence of a third spatial characteristic: an attempt to shape the form of the city according to the form of the universe. This characteristic may have taken a literal form—a city laid out, for example, in a pattern of a major star constellation. Far more common, however, were cities that symbolically approximated mythical conceptions of the universe. Angkor Thom was an early city in Cambodia that presents one of the best examples of this parallelism. An urban cluster that spread over 6 square miles (15.5 square kilometers), Angkor Thom was a representation in stone of a series of religious beliefs about the nature of the universe.

Nevertheless, regional variations in the basic form certainly existed. For example, the early cities of the Nile were not walled, which suggests that a regional power structure kept individual cities from warring with one another. In the Indus Valley, the great city of Mohenjo-daro was laid out in a grid that consisted of 16 large blocks, and the citadel was located within the block that was central but situated toward the western edge.

The most important variations within the urban hearth areas occurred in Mesoamerican cities (**Figure 10.6** on page 284). Here, cities were less dense and covered large areas. Furthermore, these cities arose without benefit of the technological advances found in the other hearth areas, most notably the wheel, the plow, metallurgy, and draft animals. However, the domestication of maize compensated for these shortcomings. Maize is a grain that in tropical

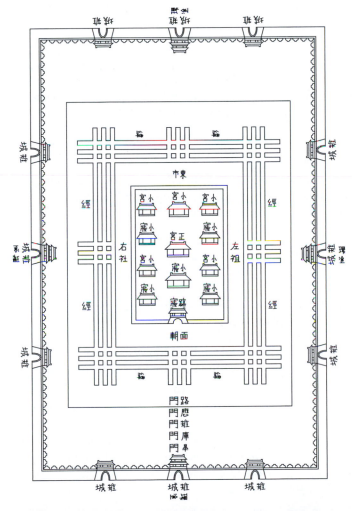

FIGURE 10.4 Plan of the city of Wang-Ch'eng in China. The city as built did not follow this exact design, but the plan itself is of interest because it suggests the symbolic importance of the three spatial characteristics of the Chinese cosmomagical city. The four walls are aligned to the cardinal directions, and the axis mundi is represented by the walled-off center city containing ceremonial buildings. The physical space of the city (microcosmos) replicates the larger world of the heavens (macrocosmos). For example, each of the four walls represents one of the four seasons. (Source: Wheatley, 1971.)

THINKING GEOGRAPHICALLY What do you think the gates to the city represent?

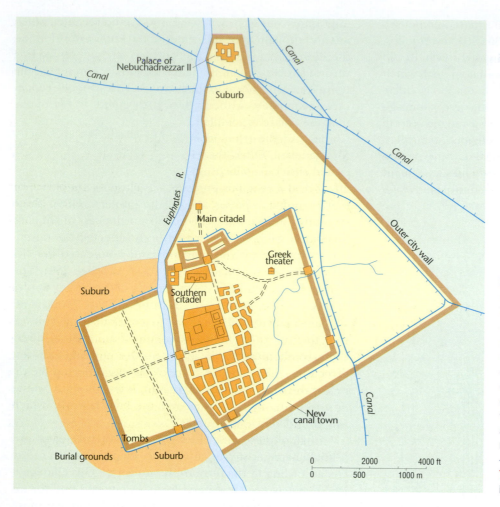

FIGURE 10.5 **Map of Babylon illustrating the urban morphology of early Mesopotamian cities.** The citadel in the inner city was characterized by the ziggurat, main temple, palace, or granary. Beyond the citadel lay the residential areas; we can assume they extended out to the inner walls and occupied both sides of the river. Suburbs grew outside the major gates and were occupied by people not allowed to spend the night in the city, such as traders and noncitizens. (Source: After map in Encyclopaedia Britannica, 1984: 555.)

THINKING GEOGRAPHICALLY Why was the granary in the inner city?

climates yields several crops a year without irrigation; in addition, it can be cultivated without heavy plows or pack animals.

The Diffusion of the City from Hearth Areas

Although urban life originated at several specific places in the world, cities are now found everywhere, including North America, Southeast Asia, Latin America, and Australia. How did city life come to these regions? There are two possible explanations:

1. Cities evolved spontaneously as native peoples created new technologies and social institutions.

2. The preconditions for urban life are too specific for most cultures to have invented without contact with other urban areas; therefore, they must have learned these traits through contact with city dwellers. This scenario emphasizes the diffusion of ideas and techniques necessary for city life.

FIGURE 10.6 **Temple of Warriors, Mayan city of Chichén Itzá.** Monumental and ceremonial architecture often dominated the morphology and landscape of urban hearth areas and reinforced ruling-class power. (Aphra Pia/age fotostock.)

THINKING GEOGRAPHICALLY In what part of the city would you say this monument is located? Can you think of examples from your own daily life of monumental architecture that symbolizes some ruling authority?

Diffusionists argue that the complicated array of ideas and techniques that gave rise to the first cities in Mesopotamia was shared with other people in both the Nile and the Indus River valleys who were on the verge of urban transformation. Indeed, archaeological evidence suggests that these three civilizations had trade ties with one another. Soapstone objects manufactured in Tepe Yahya, 500 miles (800 kilometers) to the east of Mesopotamia, have been uncovered in the ruins of both Mesopotamian and Indus River valley cities, which are separated by thousands of miles. Writings of the Indus civilization have also been found in Mesopotamian urban sites. Although diffusionists use this artifactual evidence to argue that the idea of the city spread from hearth to hearth, an alternative view is that trading took place only after these cities were well established. There is also evidence of contact across the oceans between early urban dwellers of the New World and those of Asia and Africa, although it is unclear whether this means that urbanization diffused to Mesoamerica or simply that some trade routes existed between these peoples.

Nonetheless, there is little doubt that diffusion has been responsible for the dispersal of the city in historical times (**Figure 10.7**), because the city has commonly been used as the vehicle for imperial expansion. Typically, urban life is carried outward in waves of conquest as the borders of an empire expand. Initially, the military controls newly won lands and sets up collection points for local resources, which are then shipped back into the heart of the empire. As the surrounding countryside is increasingly pacified, the new collection points lose some of their military atmosphere and begin to show the social diversity of a city. Artisans, merchants, and bureaucrats increase in number; families appear; the native people are slowly assimilated into the settlement as workers and may eventually control the city. Finally, the process repeats itself as the empire pushes farther outward.

This process, however, did not always proceed without opposition. The imposition of a foreign civilization on native peoples was often met with resistance, both physical and symbolic. Expanding urban centers relied on the surrounding countryside for support. Their food was supplied by farmers living fairly close to the city walls, and tribute was exacted from the agricultural peoples living on the edges of the urban world. The increasing needs of the city required more and more land from which to draw resources. However, the peasants farming that land may not have wanted to change their way of life to accommodate the city. The fierce resistance of many Native American groups to the spread of Western urbanization is testimony to the potential power of folk society to defy urbanization, although the destructive long-term effects of such resistance suggest that the organized military efforts of urban society were difficult to overcome.

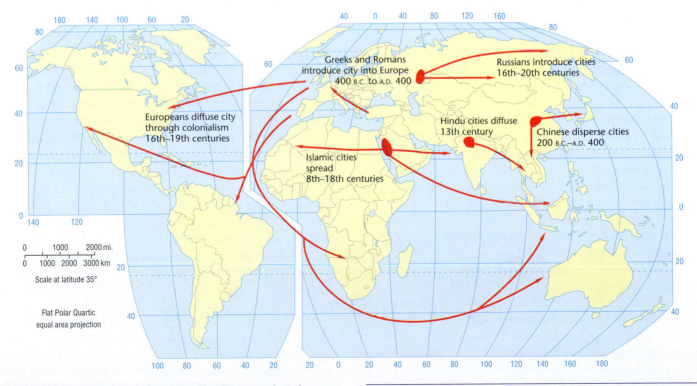

FIGURE 10.7 The diffusion of urban life with the expansion of certain empires.

THINKING GEOGRAPHICALLY What does this map tell us about the importance of urban life to military conquest?

Rural-to-Urban Migration

This analysis of the diffusion of cities throughout the world helps us explain historically the increase in urban population, since it went hand-in-hand with the growing number of cities. In today's world, however, increasing urbanization is caused by two related phenomena: natural population increase (see Chapter 3) and rural-to-urban migrations. While the United Nations estimates that natural increase accounts for the majority of the recent growth in urban population, it was rural-to-urban migrations of the past 20 to 30 years that brought millions of people originally into urban life. In countries throughout Africa and Asia, large numbers of people have left their rural villages and migrated to cities for better economic and social opportunities. Although people flock to the cities in search of jobs, there is often not a perfect correlation between economic growth and urbanization. Cities increase in size not necessarily because there are jobs to lure workers but rather because conditions in the countryside are much worse. In India, for example, rural poverty has exacerbated the rapid population increase in such cities as Mumbai (formerly Bombay) and Kolkata (formerly Calcutta). People leave the countryside in the hope that urban life will offer a slight improvement, and often it does. Given the new global economy, however, many of the jobs that are available in these cities are low-skilled and low-paying manufacturing jobs with harsh working conditions. The result is that rural-to-urban migrants often find themselves either unemployed or with jobs that barely provide them a living.

Sociologist Alejandro Portes argues that the large internal migrations that bring impoverished agricultural people into the city are a phenomenon that can be traced back to colonial times. In colonial Latin America, for example, the city was essentially home to the Spanish elite. When preconquest agricultural patterns were disrupted, peasants came to the city looking to improve their economic situation. These people usually lived on the margins of the city. Moreover, they were completely disenfranchised because only landowners had the right to hold office. The reaction by the elite to this ongoing pattern of movement of large masses of people into the city was a mixture of tolerance and indifference, with no one taking responsibility to integrate the migrants into the city.

China presents a particularly interesting example of rural-to-urban migration. Until the late 1970s, the country was predominantly rural, but state economic policies thereafter encouraged industrialization, and most of those industries were located in urban areas along its southeastern coast. The state lifted its restrictions on internal migration, which enabled its rural population to become more mobile. Today it is estimated that 18 million Chinese migrate to cities each year, a phenomenal rate of urban increase.

These migrations are drastically altering the economy, society, and culture of China, as the country experiences this unprecedented shift from a rural, agricultural nation to one based on cities and industry.

The Globalization of Cities

Many of the globalizing forces we have already discussed in this book—the integration of international economies, the merging of different cultures, and the reshaping of social organization—are centered in urban life. In other words, cities are the places where one can see both the multitude of benefits that come from globalization and the numerous pitfalls that it can bring. In this way, all cities can be considered global cities, places being shaped by the new global forces of diffusion. However, scholars have identified two particular types of cities that are key to understanding globalization.

Global cities are those that have become the control centers of the global economy—the places where major decisions about the world's commercial networks and financial markets are made.

> **global city** A city that is a control center of the global economy.

These cities house a concentration of multinational and transnational corporate headquarters, international financial services, media offices, and related economic and cultural services. According to sociologist Saskia Sassen, there are only three such cities now operating at this level: New York City, London, and Tokyo. These cities have become, in many ways, the headquarters for a global economy and form the top level of a hierarchical global system of cities.

More recently, scholars have begun to broaden their analysis of global cities in order to also understand the next tier of cities within the urban hierarchy—those that contain a large percentage of international producer service firms (law, accountancy, advertising, financial, consultancy). In this new analysis, cities are categorized by the number and type of transnational firms they house—not only headquarters, but regional and national offices as well. This analysis has identified globally and regionally dominant cities and cities that are major participants in the new global economy. Geographers Yefang Huang, Yee Leung, and Jianfa Shen refer to these as *international cities*—places that are significant because they are centers of the new international economy. The result is a very interesting list organized into classes of international cities, dominated by six in class A (London, New York, Hong Kong, Tokyo, Singapore, and Paris), followed by 10 in class B and 44 in class C. **Figure 10.8** maps those cities by class, revealing the dominance of certain regions within the global economy (the United States, Europe, parts of Asia) and revealing another level within the global system of cities.

FIGURE 10.8 **The top 60 international cities in the world of class A, B, and C.** This map indicates the location of cities that contain international producer service firms. Class A cities contain over 88 percent of the 100 international firms that were surveyed, class B cities host over 70 percent of the same firms, and class C cities contain over 44 percent of those firms. (Source: Yefang Huang, Yee Leung, and Jianfa Shen, 2007.)

THINKING GEOGRAPHICALLY Look carefully at the distribution of these cities, and see if you can explain their locations by considering their national and regional contexts.

A **globalizing city** is one that is being shaped by the new global economy and culture. This includes just about every city, present and past, to one degree or another. As geographer Brenda Yeoh indicates, cities have almost always been important hubs of activity beyond the national scale, and therefore it is not surprising that they figure prominently in contemporary discussions of globalization. The degree of globalization in the past 30 years has sharpened and enhanced the ways in which global economies and cultures shape cities. As geographer Kris Olds points out, there are five interrelated dimensions of current globalizing processes that are shaping our cities: the development of an international financial system, the globalization of property markets, the prominence of transnational corporations, the stretching and intensifying of social and cultural networks, and the increased degree of international travel and networking.

Not all cities are affected by these globalizing processes in the same way. Some cities, like London, are the control centers of the world economy, while others, like Jakarta, provide sites of production for the global economy. The differences between these two types of cities emerge out of cultural uniqueness and historical circumstance, particularly the colonial relationships that developed in the eighteenth and nineteenth centuries. With the end of colonialism and the movement toward political and economic independence, developing countries entered a period of rapid, sometimes tumultuous change. Cities have often been the focal point of this change, and as we have seen, millions of people have migrated to cities in search of a better life. Some of these newly globalizing cities are moving into the ranks of international cities, as you can see by looking at Figure 10.8. Other globalizing cities like Gabarone in Botswana are just now beginning to feel the impacts of population growth and global economic integration.

> **globalizing city** A city being shaped by the new global economy and culture.

THE **ECOLOGY** OF URBAN LOCATION

What is the relationship between cities and their physical settings? At first glance, cities seem totally divorced from the natural environment. What possible relationships could the shiny glass office buildings, paved streets, and high-rise apartment buildings that characterize most cities have with forests, fields, and rivers?

In this section we examine several different ways to think about the relationships between urban life and its natural setting.

Site and Situation

There are two components of urban location: site and situation. **Site** refers to the local setting of a city; the **situation** is the regional setting or location. As an example of site and situation, consider San Francisco. The original site of the Mexican settlement that became San Francisco was on a shallow cove on the eastern (inland) shore of a peninsula. The importance of its situation was that it drew on waterborne traffic coming across the bay from other, smaller settlements. Hence the town could act as a transshipment point.

Both site and situation are dynamic, changing over time. For example, in San Francisco during the gold rush period of the 1850s, the small cove was filled to create flatland for warehouses and to facilitate extending wharves into deeper bay waters. The filled-in cove is now occupied by the heart of the central business district (**Figure 10.9**). The geographical situation also changed as patterns of trade and transportation technology evolved. The original transbay situation was quickly replaced during the gold rush by a new role: supplying the mines and settlements of the gold country. Access to the two major rivers leading to the mines, plus continued ties to ocean trade routes, became important components of the city's situation.

San Francisco's situation has changed dramatically in the past two decades, for it is no longer the major port of the bay. The change in technology to containerized cargo was adopted more quickly by Oakland, the rival city on the opposite side of the bay, and San Francisco declined as a port city. Oakland was able to accommodate containerized cargo in part because it filled in huge tracts of shallow bay

site The local setting of a city.
situation The regional setting of a city.

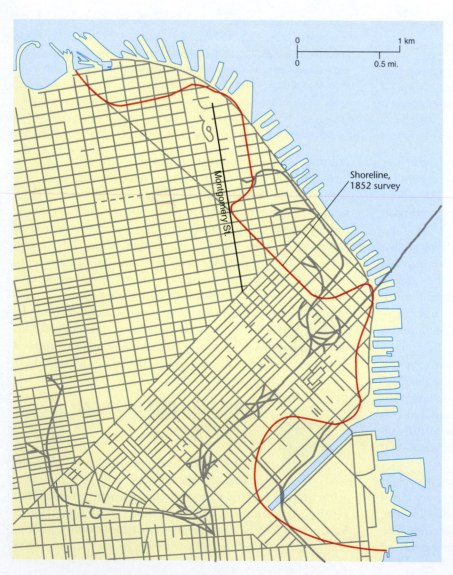

Shoreline, 1852 survey

Montgomery St.

FIGURE 10.9 This map shows how San Francisco's site has been changed by human activity. During the late 1850s, shallow coves were filled, providing easier access to deeper bay waters as well as flat land near the waterfront for warehouses and industry.

THINKING GEOGRAPHICALLY What hazard would such fill land create?

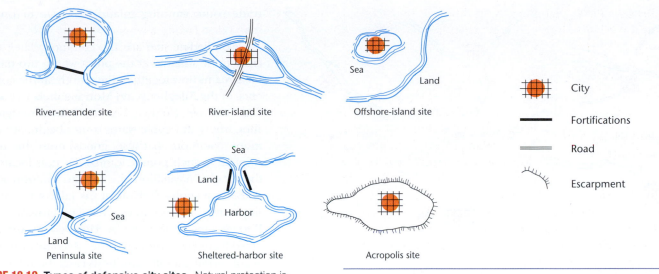

FIGURE 10.10　**Types of defensive city sites.** Natural protection is afforded by physical features.

THINKING GEOGRAPHICALLY　What challenges result for cities established on defensible sites?

lands, creating a massive area for the loading, unloading, and storage of cargo containers.

Certain attributes of the physical environment have been important in the location of cities. Those cities with distinct functions, such as defense or trade, were located because of specific physical characteristics. The locations of many contemporary cities can be partially explained by decisions made in the past that capitalized on the advantages of certain sites. The following classifications detail some of the different location possibilities.

Defensive Sites There are many types of defensive sites for cities (some are diagrammed in **Figure 10.10**). A **defensive site** is a location that can be easily protected from invaders. The river-meander site, with the city located

defensive site　A location from which a city can be easily protected from invaders.

inside a loop where the stream turns back on itself, leaves only a narrow neck of land unprotected by water. Cities such as Bern, Switzerland, and New Orleans are situated inside river meanders. Indeed, the nickname for New Orleans, Crescent City, refers to the curve of the Mississippi River.

Even more advantageous was the river-island site, which, because the stream was split into two parts, often combined a natural moat with an easy river crossing. For example, Montreal is situated on a large island surrounded by the St. Lawrence River and other water channels. The offshore-island site—that is, islands lying off the seashore or in lakes—offered similar defensive advantages (**Figure 10.11**). Mexico City began as an Indian settlement on a lake island. Venice is the classic example of a city built on offshore islands in the sea, as is Hong Kong. Peninsula sites were almost as advantageous as island sites, because they

FIGURE 10.11　**The classic defensive site of Mont St. Michel, France.** A small town clustered around a medieval abbey, which was originally separated from the mainland during high tides, Mont St. Michel now has a causeway that connects the island to shore, allowing armies of tourists to penetrate the town's defenses easily. (Rapho Agence/Science Source.)

THINKING GEOGRAPHICALLY　Why would a defensible site have been important when the abbey was established in the Middle Ages?

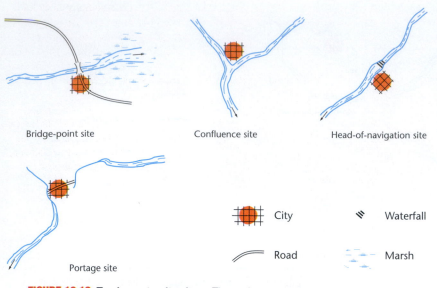

Bridge-point site   Confluence site   Head-of-navigation site

Portage site

⊞ City   ⋙ Waterfall

═ Road   ≈ Marsh

FIGURE 10.12 Trade-route city sites. These sites are strategic positions along transportation arteries.

THINKING GEOGRAPHICALLY Is your city or one near you located on a trade-route site?

offered natural water defenses on all but one side. Boston was founded on a peninsula for this reason, and a wooden palisade wall was built across the neck of the peninsula.

Danger of attack from the sea often prompted sheltered-harbor defensive sites, where a narrow entrance to the harbor could be defended easily. Examples of sheltered-harbor sites include Rio de Janeiro, Tokyo, and San Francisco.

High points were also sought out. These are often referred to as acropolis sites; the word *acropolis* means "high city." Originally the city developed around a fortification on the high ground and then spilled out over the surrounding lowland. Athens is the prototype of acropolis sites, but many other cities are similarly located.

Trade-Route Sites Defense was not always the primary consideration. Sometimes, urban centers were built on **trade-route sites**—that is, at important points along already established trade routes. Here, too, the influence of the physical environment can be detected.

> **trade-route site** A place for a city at a significant point on a transportation route.

Especially common types of trade-route sites (**Figure 10.12**) are bridge-point sites and river-ford sites, places where major land routes could easily cross over rivers. Typically, these were sites where streams were narrow and shallow with firm banks. Occasionally, such cities even bear in their names the evidence of their sites, as in Frankfurt ("ford of the Franks"), Germany, and Oxford, England. The site for London was chosen because it is the lowest point on the Thames River where a bridge—the famous London Bridge—could easily be built to serve a trade route running inland from Dover on the sea.

Confluence sites are also common. They allow cities to be situated at the point where two navigable streams flow together. Pittsburgh, at the confluence of the Allegheny and Monongahela rivers, is a fine example (**Figure 10.13**). Head-of-navigation sites, where navigable water routes begin, are even more common, because goods must be transshipped at such points. Iquitos, Peru, is located at the head-of-navigation site for the Amazon River, and Minneapolis–St. Paul, at the falls of the Mississippi River, also occupies a head-of-navigation site. Portage sites are very similar. Here, goods were portaged from one river to another. (*Portage* means the act of carrying a ship and/or its cargo between navigable waters.) Chicago is near a short portage between the Great Lakes and the Mississippi River drainage basin. In these ways and others, an urban site can be influenced by the physical environment. But so, too, do cities change that physical environment in a multitude of ways.

Natural Disasters

Catastrophic events over the past decade, including tsunamis in Indonesia and Japan, widespread flooding in Europe and Africa, and frequent wildfires in the U.S. Southwest, have become critical reminders of the vulnerability of human settlements to natural disasters. This vulnerability was graphically depicted in photos taken in the aftermath of a major earthquake in the Caribbean island nation of Haiti in January 2010. This disaster killed at least 300,000 people, injured hundreds of thousands, and displaced more than 1 million residents. Densely populated urban areas like the Haitian capital city of Port-au-Prince, with their concentrations of people, are particularly vulnerable to natural disasters because so many peoples' lives and livelihoods are at risk.

In addition, many cities around the world are situated near water, so they are often located in the direct paths of multiple types of disasters, including hurricanes and tsunamis (which are often caused by earthquakes). Making matters even worse, some scientists have recently noted an increase in the occurrence of natural disasters. They attribute this increase partly to global climate change brought about by a range of human activities including deforestation, increasing industrialization, and harmful agricultural practices.

The combination of increasing urbanization and a growing number of natural disasters means that more and more people's lives are being impacted adversely. According to the United Nations Environment Program, since the start of the new millennium, billions of people around the world have been affected by some 2,500 natural disasters (**Figure 10.14**). As with the Haitian earthquake, it is often

FIGURE 10.13 **Pittsburgh's Golden Triangle.** At the confluence of the Allegheny and Monongahela rivers, Pittsburgh is a classic example of how an early trade-route site has evolved into a commercial center. (Comstock Select/Corbis.)

THINKING GEOGRAPHICALLY Before the American Revolution, Pittsburgh was the site of a fort. Why was it a defensible site as well as being a trade-route site?

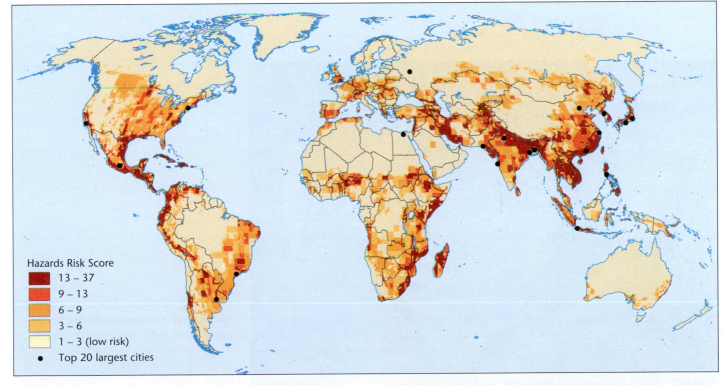

Hazards Risk Score
- 13 – 37
- 9 – 13
- 6 – 9
- 3 – 6
- 1 – 3 (low risk)
- Top 20 largest cities

FIGURE 10.14 **Large cities in relation to current climate-related hazards.** Not all large cities are located in hazardous environments, but you can see from this map that there is significant correlation between urbanization and vulnerability to natural hazards. What makes the situation even more grave is that many of the most vulnerable cities contain large numbers of people living in conditions that add to their vulnerability, such as poorly built housing. (Source: Alex de Sherbenin, Andrew Schiller, and Alex Pulsipher, 2007.)

THINKING GEOGRAPHICALLY Why do people build cities in such vulnerable locations?

poor people who are the most affected, since they tend to live in structures that are not well built and in sections of the city that are particularly vulnerable to environmental forces. In fact, in many cities in the developing world, squatter settlements populated by the poor are often situated on steep hillsides or in poorly drained low-lying areas that are highly prone to flooding, mudslides, and other disasters.

Urbanization and Sustainability

As we saw in Chapter 9, industrialization goes hand in hand with urbanization, so the environmental impacts of industrialization are often found in cities. But urbanization generates its own set of environmental impacts: supplying enough energy, food, and water to large concentrations of people puts an array of stresses on the natural environment. Scholars now refer to these varied impacts of urban areas on the environment as the **urban footprint.** For example, Las Vegas is one of the fastest-growing urban areas in the United States, yet it is located in a desert (**Figure 10.15**). The city's primary water source is the Colorado River, but the costs of delivering that water are extremely

urban footprint The total impact of urban areas on the natural environment.

high. It is expensive to construct the dams and infrastructure necessary to move the water into the city, and the environmental damage to the region is severe. The dams alter the flows of water through the Colorado Valley, harming fish and disrupting aquatic life cycles, and the energy required to divert the water to the city leads to higher sulfur dioxide emissions, which have been shown to contribute to global warming.

But the environmental effects of increased urbanization—the urban footprint—can spread beyond the immediate stresses caused by higher concentrations of population. Urban living often leads to rising levels of consumption, as new urbanites gain access to better jobs and more disposable income. This growing demand for things like better food can impact areas far from urban centers. For example, as the United Nations report *Unleashing the Potential of Urban Growth* suggests, tropical forests in Tobasco, an area 400 miles away from Mexico City, have been turned into cattle-grazing areas in response to urbanites' demands for meat, while, in a second example, a major contributing factor to the deforestation of the Amazon is the increased demand for soybeans from the newly urbanizing regions of China as well as from the urbanites of the United States, Japan, and Europe.

FIGURE 10.15 Las Vegas, Nevada. Las Vegas is sited in the heart of the desert. As such, its growing demands for water are putting stresses on the natural environment, as water has to be diverted from the Colorado River. (Robert Cameron/Getty Images.)

THINKING GEOGRAPHICALLY Why do you think the founders of Las Vegas would choose to locate the city in the middle of a desert?

Urbanization, however, does not necessarily imply environmental degradation. For example, the higher densities of population that cities represent can be seen as a form of sustainable growth. Half the population of the world—the urban half—lives on approximately 3 percent of the landmass. The concentration of people in cities therefore opens up other areas that can be protected and left relatively free from human use. Many countries in the developing and developed world are working on local and regional planning projects that on the one hand maintain urban boundaries and prevent sprawl and on the other hand protect natural environments outside urban areas from environmental degradation. Such urban environmental conservation projects will likely become increasingly important in future years.

CULTURAL **INTERACTION** IN URBAN GEOGRAPHY

How can we understand the location of cities as an integrated system? In recent decades, urban geographers have studied the spatial distribution of towns and cities to determine some of the economic and political factors that influence the pattern of cities. In doing so, they have created a number of models that collectively make up **central-place theory.** These models represent examples of cultural interaction.

> **central-place theory** A set of models designed to explain the spatial distribution of urban service centers.

Central-Place Theory

Most urban centers are engaged mainly in the service industries. The service activities of urban centers include transportation, communication, and utilities—services that facilitate the movement of goods and that provide networks for the exchange of ideas about those goods (see Chapter 9 for a more detailed examination of these different industrial activities). Towns and cities that support such activities are called **central places.**

> **central place** A town or city engaged primarily in the service stages of production; a regional center.
>
> **threshold** In central-place theory, the size of the population required to make provision of goods and services economically feasible.
>
> **range** In central-place theory, the average maximum distance people will travel to purchase a good or service.

In the early 1930s, the German geographer Walter Christaller first formulated central-place theory as a series of models designed to explain the spatial distribution of urban centers. Crucial to his theory is the fact that different goods and services vary both in **threshold,** the size of the population required to make provision of the good or service economically feasible, and in **range,** the average maximum distance people will travel to purchase a good or service. For example, a larger number of people are required to support a hospital, university, or department store than to support a gasoline station, post office, or grocery store. Similarly, consumers are willing to travel a greater distance to consult a heart specialist or purchase an automobile than to buy a loaf of bread or visit a movie theater.

Because the range of central goods and services varies, urban centers are arranged in an orderly hierarchy. Some central places are small and offer a limited variety of services and goods; others are large and offer an abundance. At the top of this hierarchy are regional metropolises—cities such as New York, Beijing, or Mumbai—that offer all services associated with central places and that have very large tributary trade areas, or hinterlands. At the opposite extreme are small market villages and roadside hamlets, which may contain nothing more than a post office, service station, or café. Between these two extremes are central places of various degrees of importance. One regional metropolis may contain thousands of smaller central places in its tributary market area (**Figure 10.16**).

With this hierarchy as a background, Christaller evaluated the individual influence of three forces (the market, transportation, and political borders) in determining the spacing and distribution of urban centers by creating models. His first model measured the influence of the market

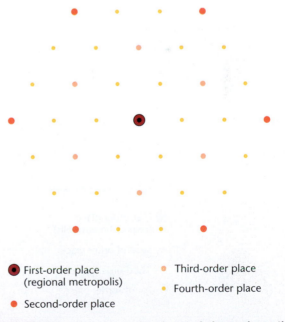

Legend:
- First-order place (regional metropolis)
- Second-order place
- Third-order place
- Fourth-order place

FIGURE 10.16 **Christaller's hierarchy of central places shows the orderly arrangement of towns of different sizes.** This is an idealized presentation of places performing central functions. Each large central place has many smaller central places within its hinterland.

THINKING GEOGRAPHICALLY In what kind of terrain would this arrangement be most likely?

and range of goods on the spacing of cities. To simplify the model, he assumed that the terrain, soils, and other environmental factors were uniform; that transportation was universally available; and that all regions would be supplied with goods and services from the minimum number of central places. If market and range of goods were the only causal forces, the distribution of towns and cities would produce a pattern of nested hexagons, each with a central place at its center (**Figure 10.17a**).

In Christaller's second model, he measured the influence of transportation on the spacing of central places. He no longer assumed that transportation was universally and equally available in the **hinterland.** Instead, he assumed that as many demands for transport as possible would be met with the minimum expenditure for construction and maintenance of transportation facilities. Thus, as many high-ranking central places as possible would be on straight-line routes between the primary central places (**Figure 10.17b**). When transportation is considered to be the influential factor in shaping the spacing of central places, the pattern is rather different from that created by the market

hinterland The area surrounding a city and influenced by it.

factor. This transportation-driven pattern occurs because direct routes between adjacent regional metropolises do not pass through central places of the next-lowest rank. As a result, these second-rank central places are pulled from the corners of the hexagonal market area to the midpoints in order to be on the straight-line routes between adjacent regional metropolises.

Christaller hypothesized that the market factor would be the greater force in rural countries, where goods were seldom shipped throughout a region. In densely settled industrialized countries, however, he believed that the transportation factor would be stronger because there were greater numbers of central places and more demand for long-distance transportation.

Christaller devised a third model to measure the effect of political borders on the distribution of central places. He recognized that political boundaries, especially within independent countries, would tend to follow the hexagonal market-area limits of each central place that was a political center. He also recognized that such borders tend to separate people and retard the movement of goods and services. Such borders necessarily cut through the market areas of many central places below the rank of regional

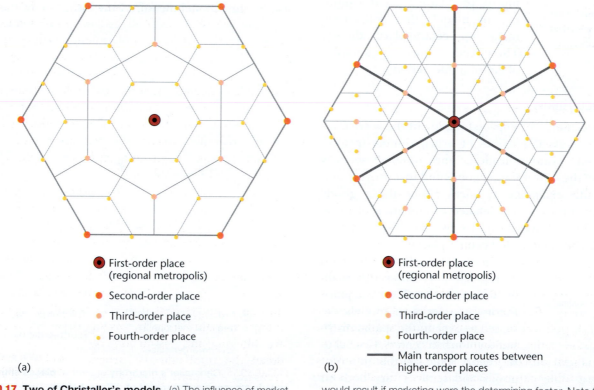

First-order place (regional metropolis)

Second-order place

Third-order place

Fourth-order place

(a)

First-order place (regional metropolis)

Second-order place

Third-order place

Fourth-order place

Main transport routes between higher-order places

(b)

FIGURE 10.17 Two of Christaller's models. (a) The influence of market area on Christaller's arrangement of central places. If market were the only factor controlling the distribution of central places, this diagram would represent the arrangement of towns and cities. (b) The distribution of central places with the influence of transportation taken into account. If the availability of transportation is the determining factor in the location of central places, their distribution will be different from the distribution that

would result if marketing were the determining factor. Note that the second-order central places are pulled away from the apexes of the hexagon to the main transport routes between regional metropolises. (Source: After Christaller, 1966.)

THINKING GEOGRAPHICALLY Why, in this model, would a hexagon be the shape to appear instead of a square, circle, or some other shape?

metropolis. Central places in such border regions lose rank and size because their market areas are politically cut in two. Border towns are thus stunted, and important central places are pushed away from the border, which distorts the hexagonal pattern.

Market area, transportation, and political borders are but three of the many forces that influence the spatial distribution of central places. For example, in all three of these models, it is assumed that the physical environment is uniform and that people are evenly distributed. Of course, neither of these is true, leading to some distortions in the model. Nonetheless, central-place theory has proven to be a very useful tool for understanding the locations of cities and services in relationship to population. For example, with increasing concerns about environmental sustainability, regional planners have begun to use central-place theory to think about how to minimize transportation costs in a particular region by siting key services in central locations. Yet it is important to keep in mind that the model fails to take into account such cultural factors as people's historical attachment to places. For example, if people have shopped in a certain city for several generations, or even several years, and feel attached to that place, it is often the case that they choose not to shop elsewhere, even if another city is located closer to them. In other words, the most economically rational (in terms of saving money by traveling less) path is not always the one most preferred by people.

REFLECTING ON **GEOGRAPHY**

If Christaller's assumptions did not hold, in what ways would central-place theory have to be altered?

URBAN CULTURAL **LANDSCAPES**

As you probably already know from your travels, the look of cities varies from place to place. Some cities have bustling downtowns where most of the residents work, while others are characterized by nodes of activities clustered on the suburban fringe; some cities are built around a set of historical buildings with political or religious significance, while the centers of other cities are dominated by new skyscrapers housing financial offices. Finally, some cities have streets laid out in a gridlike pattern, while in other cities streets are a crisscrossed jumble without apparent order. How can we make sense of this range of urban landscapes?

Globalizing Cities in the Developing World

In cities of the developing world, the combination of large numbers of migrants and widespread unemployment leads to overwhelming pressure for low-rent housing. Governments have rarely been able to meet these needs through housing projects, so one of the most characteristic landscape features of cities in the developing world has been the construction of illegal housing usually made up of temporary shelters: **squatter settlements.** Squatter settlements, or *barriadas* (**Figure 10.18** on page 296), are often referred to as slums—areas of degraded housing. The United Nations defines a *slum household* as a cohabiting group that lacks one or more of the following conditions: an adequate physical structure that protects people from extreme climatic conditions, sufficient living area such that no more than three people share a room, access to a sufficient amount of water, access to sanitation, and secure tenure or protection from forced eviction. Current estimates suggest that more than 1 billion people in the world live in slums. In greater Cairo alone, for example, it is estimated that 5.5 million people live in slum conditions in squatter settlements. In Manila, it is estimated that almost half of the city's 19 million inhabitants live in slum conditions in squatter settlements.

> **squatter settlement** or ***barriada*** An illegal housing settlement, usually made up of temporary shelters, that surrounds a large city and may eventually become a permanent part of the city.

Squatter settlements usually begin as collections of crude shacks constructed from scrap materials; gradually, they become increasingly elaborate and permanent. Paths and walkways link houses, vegetable gardens spring up, and often water and electricity are bootlegged into the area so that a common tap or outlet serves a number of houses. At later stages, such economic pursuits as handicrafts and small-scale artisan activities take place in the squatter settlements. In many instances, these supposedly temporary settlements become permanent parts of the city and function as many neighborhoods do—that is, as social, economic, and cultural centers.

Squatter settlements are located not only in downtown areas but also close to places where people work. In many cities, this means places that just a few years ago were considered rural. Because of growing populations and pressures on space, many globalizing cities in the developing world are expanding outward. Some of these cities, such as Mexico City, Hanoi, and Manila, are expanding so rapidly into the countryside that the city is swallowing up smaller villages and rural areas. In this way, these new urban forms are blurring the distinction between the rural and the urban, creating what some scholars have called **extended metropolitan regions (EMRs).** Multinational corporations often rely on the labor of the new urban migrants, so the city grows rapidly at the margins as factories are built in export processing zones (EPZs; see

> **extended metropolitan region (EMR)** A new type of urban region, complex in both landscape form and function, created by the rapid spatial expansion of a city in the developing world.

(a)

(b)

FIGURE 10.18 Squatter settlements in (a) Mexico City and (b) Kuala Lumpur. Migration to cities has been so rapid that often illegal squatter settlements have been the only solution to housing problems.
(a: Environmental Images/Universal Images Group/age fotostock; b: AP Photo/ C. F. Tham.)

THINKING GEOGRAPHICALLY What upgrading is evident in the Mexico City settlement?

Chapter 9). At the same time, the downtowns of these cities often experience not dispersal but concentration: highrises accommodate the regional offices of corporations and the services associated with them (media, advertising, personnel management, and so forth), and upper-class housing developments accommodate this new class of white-collar workers.

These new urban regions are complex in both landscape form and function because they developed rapidly, without any planning, and have in many cases incorporated preexisting places. Scholars who write about Mexico City, for example, find it impossible to speak of the city as one place; rather, they suggest that it can now only be represented as a pastiche of different places, each with its own center and outlying neighborhoods.

In addition, landscape forms associated with the global consumer, entertainment, and tourist economy are emerging in the new urban regions: American-style shopping districts (with American stores) and new airports, hotels, restaurants, and entertainment facilities. Many of these landscape developments are funded by international investments because entrepreneurs and governments in many parts of the world look to the new globalizing cities of the developing world as good places to make financial and real estate investments. Much of the extended metropolitan region of Hanoi, for example, was built with funds from a range of countries: an export processing zone and a golf and entertainment facility were partly funded by Malaysia; South Korea invested in a hotel, another golf course, and a business center; Taiwan was involved in a tourism project and an industrial park. Hanoi's downtown is being developed with more hotels and office construction funded by Japan and Singapore. Finland provided the infra-

structure for the water supply to one of these new developments.

Latin American Urban Landscapes

To describe the landscapes of Latin American cities, urban geographers have developed a model, which is shown in Figure 1.17 (page 21), so you might want to refer to it as you read this section. In contrast to many contemporary cities in the United States, the central business districts of Latin American cities are vibrant, dynamic, and increasingly specialized. The dominance of the central district is explained partly by widespread reliance on public transit and partly by the existence of a large and relatively affluent population close to it. Outside the central district, the dominant component is a commercial spine surrounded by an elite residential sector. Because these two zones are interrelated, they are referred to as the *spine/sector*. This combination is an extension of the central business district down a major boulevard, along which the city's important amenities, such as parks, theaters, restaurants, and even golf courses, are located. Strict zoning and land controls ensure continuation of these activities and protect the elite from incursions by low-income squatters.

Somewhat less prestigious is the inner-city zone of maturity, a collection of homes occupied by people unable to afford housing in the spine/sector. The *zone of maturity* is an area of upward mobility. The *zone of accretion* is a diverse collection of housing types, sizes, and quality, which can be thought of as a transition between the zone of maturity and the next zone. It is an area of ongoing construction and change, emblematic of the explosive population growth that characterizes the Latin American city. Although some neighborhoods within this zone have city-provided utilities, other blocks must rely on trucks to deliver essentials such as water and butane.

The most recent migrants to the Latin American city are found in the zone of peripheral squatter settlements. This fringe of poor people and inadequate housing contrasts dramatically with the affluent and comfortable suburbs that ring North American cities. Streets are unpaved, open trenches carry waste, residents haul water from distant locations, and electricity is often pirated by attaching illegal wires to the closest utility pole. Although this zone's quality of life seems marginal, many residents transform these squatter settlements over time into permanent neighborhoods with minimal amenities.

Landscapes of the Apartheid and Postapartheid City

Racism and the residential segregation that often results from it can have profound effects on landscapes. In South Africa, the state-sanctioned policy of segregating "races,"

known as **apartheid,** significantly altered urban patterns. Although racial segregation was not the only force shaping the apartheid city, it certainly was a dominant one. The intended effects of this policy on urban form are delineated in **Figure 10.19**. To understand this illustration fully, we need to outline some of the important components of the apartheid state.

> **apartheid** In South Africa, a policy of racial segregation and discrimination against non-European groups.

The policies of economic and political discrimination against non-European groups in South Africa were formalized and sharpened under National Party rule after 1948. To segregate the "races," the government passed two major pieces of legislation in 1950. The first was the Population Registration Act, which mandated the classification of the population into discrete racial groups. The three major groups were white, black, and colored, each of which was subdivided into smaller categories. The second piece of major legislation was called the Group Areas Act. Its goal was, in the words of geographer A. J. Christopher, "to effect the total urban segregation of the various population groups defined under the Population Registration Act." Cities were thus divided into sections that were to be inhabited only by members of one population group. Although the effects of these acts on the form of South African cities did not appear overnight, they were massive

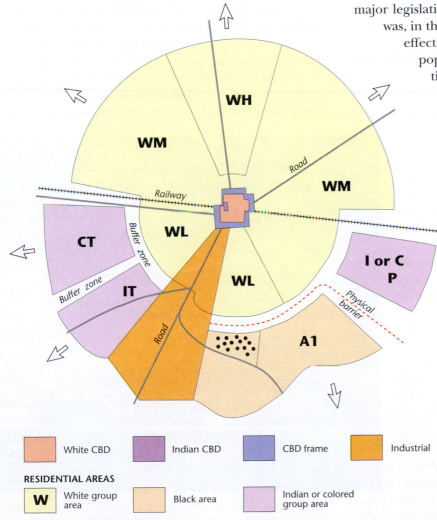

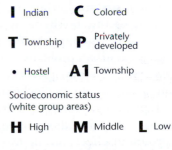

I Indian		**C** Colored	
T Township		**P** Privately developed	
• Hostel		**A1** Township	

Socioeconomic status
(white group areas)

H High   **M** Middle   **L** Low

Domestic servants' quarters not shown

☐ White CBD   ☐ Indian CBD   ☐ CBD frame   ☐ Industrial

RESIDENTIAL AREAS

W White group area   ☐ Black area   ☐ Indian or colored group area

FIGURE 10.19 The model apartheid city. Townships were areas set aside by the South African government for members of nonwhite groups to live in, often located close to industrial areas of the city. Hostels were built to house black men from the rural areas who were needed to work in particular industries.

(Source: From Christopher, 1994: 107. Copyright © 1994 by A. J. Christopher. Reprinted by permission of Routledge.)

THINKING GEOGRAPHICALLY Why are nonwhite groups located close to industry?

nonetheless. Members of nonwhite groups were by far the ones most adversely affected. Almost without exception, the downtowns of cities were restricted to whites, whereas those areas set aside for nonwhites were peripheral and restricted, often lacking any urban services, such as transportation and shopping. Large numbers of nonwhite families were displaced with little or no compensation (estimates suggest that only 2 percent of the displaced families were white). Buffer zones were established between residential areas to curtail contact between groups, further hindering access to the central city for those groups pushed into the periphery.

After the change of government in 1994, apartheid came to an end as an official policy, and the laws that enforced racial segregation in South Africa were abolished. Results from the 1996 census revealed a marked decline in degrees of racial segregation within South Africa's cities, a trend that characterizes the postapartheid city. However, historical patterns of residential segregation and racial prejudice are very difficult to change. A. J. Christopher found that despite a general overall decline in levels of segregation, particularly among the colored population, the majority of the white population in South Africa remained isolated from their nonwhite neighbors, while the black population remained relatively isolated due to the lack of mobility caused by poverty.

Landscapes of the Socialist and Postsocialist City

The effects of centralized state policies on urban land-use patterns are also evident in the cities of the former Soviet Union and of other former socialist countries in eastern Europe. In the Soviet Union, for example, the Bolshevik Revolution of 1917 brought with it attempts by the state to confront and solve the urban problems brought by industrialization. Socialist principles called for the nationalization of all resources, including land. Further, central planning replaced market forces as the means for allocating those resources. These ideas had profound effects on the form of Soviet cities. The principles that underlie capitalist cities—that land is held privately and that the economic market dictates urban land use—were no longer the prime influences on urban form. Instead, Soviet policies attempted to create a more equitable arrangement of land uses in the city. The results of those policies included (1) a relative absence of residential segregation according to socioeconomic status, (2) equitable housing facilities for most citizens, (3) relatively equal accessibility to sites for the distribution of consumer items, (4) cultural amenities (theater, opera, and so on) located and priced to be accessible to as many people as possible, and (5) adequate and accessible public

transportation. Although these results created better living and working conditions for many people, the situation was far from ideal. By the 1970s and 1980s, many Soviet citizens realized that their standards of living were well below those in the West and that the central planning system was not successful.

National policies of economic restructuring introduced in the late 1980s, referred to as *perestroika*, led to mandates to privatize resources and the land market. In the postsocialist city, market forces are once again dominant in shaping urban land uses, and the pace and scale of urban change are unprecedented. One of the most significant of those changes is the privatization of the housing market. In Moscow, for example, an increasing segment of the housing stock is in private hands, as opposed to being government owned. However, this privatization does not necessarily mean a better housing market for the city's residents. In the past two decades, much new construction in Moscow has taken the form of high-priced residential and commercial properties. At the same time, rental prices in the city rose beyond affordable limits for many families. This combination of factors led to a collapse of the real estate market in Moscow, so today many newly constructed luxury buildings stand empty of tenants, while many other buildings were never completed at all. Mean-

FIGURE 10.20 **Moskva River, Moscow.** Pictured here is one of the "Seven Sisters" skyscrapers commissioned and built under Josef Stalin between 1947 and 1955. (© Franz-Marc Frei/Corbis.)

THINKING GEOGRAPHICALLY What evidence of globalization does this photograph show?

while, many of Moscow's residents continue to live in the communal apartments assigned to them under the Soviet system (**Figure 10.20**). In countries formerly part of or dominated by the Soviet Union, processes akin to gentrification (see Chapter 11) are taking place in the center of cities such as Prague, Czech Republic; Krakow, Poland; and Budapest, Hungary, displacing residents to the peripheral portions of the city while housing others in Western-style elegance.

Postsocialist cities are also taking on the look of Western cities. The downtowns are increasingly dominated by retailing outlets of familiar Western companies, such as Nike and McDonald's. Tall office buildings that house financial service businesses are replacing industrial buildings. In other postsocialist cities, the new landscapes of finance capitalism are located outside the city center, where land is cheaper. In Prague, for example, most of the city center is of great historical significance and is therefore protected from new building construction. Large financial institutions (for example, international banks) have found locations farther out from the city to be ideal for the construction of modern office buildings that suit their needs. These new buildings are, in many cases, built to standards that make them more energy efficient and environmentally friendly than older historic buildings in the city center. These new financial centers bring with them economic growth, and some are now resembling the suburban sprawl of North American cities (**Figure 10.21**).

FIGURE 10.21 A recently built "green" bank building in the Czech Republic. Because most of downtown Prague is protected from new development by its historical significance, the many financial institutions that want to locate in the Czech Republic have constructed modern buildings to house their offices on the edge of the city or in the suburbs. (© Profimedia.CZ a.s./Alamy.)

THINKING GEOGRAPHICALLY Compare this situation to the location of bank office buildings in North American cities.

Key Terms

agricultural surplus, 281	hydraulic civilization, 281
apartheid, 297	megacity, 279
axis mundi, 283	primate city, 280
barriada, 295	range, 293
central place, 293	site, 288
central-place theory, 293	situation, 288
cosmomagical city, 283	social stratification, 281
defensive site, 289	squatter settlement, 295
extended metropolitan region (EMR), 295	threshold, 293
	trade-route site, 290
global city, 286	urban footprint, 292
globalizing city, 287	urban hearth area, 282
hinterland, 294	urbanized population, 278

The City on the Internet

You can learn more about the city on the Internet at the following web sites:

Globalization and World Cities

http://www.lboro.ac.uk/gawc/
This fantastic web site details the work of the globalization and world cities research group and network, including data sets that document relationships among 100 world cities, and recent publications of the group.

United Nations Statistics Division

http://www.un.org/Depts/unsd/
The source for world population statistics, this site contains a comprehensive section on urbanization as a social indicator.

U.S. Department of Housing and Urban Development

http://www.hud.gov
Here you can find information about housing issues and urban economic development and about how to get involved personally in your own community.

Recommended Books on Urban Geography

Brenner, Neil, and Roger Keil (eds.). 2006. *The Global Cities Reader.* New York: Routledge. A rich collection of essays that explore the rise of the concept of the global city and examine a range of case studies of global cities throughout the world.

Colten, Craig E. 2006. *An Unnatural Metropolis: Wresting New Orleans from Nature.* Baton Rouge: Louisiana State University Press. An award-winning study of how backers of New Orleans attempted to conquer nature, marginalizing the poor and powerless in the process and leading to one of the worst "natural" disasters in American history.

Davis, Mike. 2007. *Planet of Slums.* New York: Verso. A devastating overview of people's everyday struggles to maintain decent lives in the world's urban slums.

King, Anthony D. 2004. *Spaces of Global Culture: Architecture, Urbanism, Identity.* New York: Routledge. A fascinating examination of transnational urban forms.

Krueger, Rod, and David Gibbs (eds.). 2010. *The Sustainable Development Paradox: Urban Political Economy in the United States and Europe.* New York: Guilford. An incisive overview of the politics of sustainability within urban contexts, explored through a series of case studies.

Legates, Richard T., and Frederic Stout (eds.). 1999. *The City Reader,* 2nd ed. New York: Routledge. An extensive edited collection of readings covering the evolution of cities and the contemporary forces that are restructuring them.

Olds, Kris. 2001. *Globalization and Urban Change: Capital, Culture, and Pacific Rim Mega-Projects.* New York: Oxford University Press. An in-depth examination of how globalization actually operates in terms of large-scale urban developments in Vancouver and Shanghai.

Sassen, Saskia (ed.). 2003. *Global Networks, Linked Cities.* New York: Routledge. A collection of essays examining the emerging networks of global commerce and communication that are reshaping the world's cities.

Taylor, Peter. 2003. *World City Network: A Global Urban Analysis.* New York: Routledge. An empirically rich accounting of the myriad commercial and financial connections between cities that form a global network.

Journals in Urban Geography

Environment and Urbanization. Published by Sage, Beverly Hills. Volume 1 appeared in 1989.

International Journal of Urban and Regional Research. Published by Edward Arnold, London. Volume 1 appeared in 1987.

Urban Geography. Published by Bellwether Press, Lanham, MD. Volume 1 appeared in 1976.

Urban Studies. Published by Routledge, New York. Volume 1 appeared in 1964.

Answers to Thinking Geographically Questions

Figure 10.1: Although less developed countries are seeing massive movements of people to cities, they started with very low urbanization rates. In many such countries, farming is still very labor intensive, requiring that many farmers remain on the land in order to meet even minimal food needs.

Figure 10.2: More people continue to live close to the central business district in developing countries, and there are fewer shopping malls and big-box stores. If there are such commercial outlets, they tend to be more centrally located than is the case in many North American cities.

Figure 10.3: Comparison of Figures 10.3 and 8.16 shows that the Nile Valley was a place of early urbanization but not of early plant domestication. Plant cultivation diffused to the Nile Valley from Mesopotamia. Similarly, there was a large center of early food production in north-central South America but no center of early urbanization there.

Figure 10.4: With each wall containing three gates, making 12 in all, they could represent months.

Figure 10.5: Protecting the stored food supply was one of the most important roles of a city.

Figure 10.6: The building is in the middle of the city. Examples would include state and national capitols and courthouses in the United States. The residences of leaders are often large and imposing, such as the White House, royal palaces, and many governors' mansions.

Figure 10.7: Conquerors often establish urban centers first as military camps and then as centers of authority and power.

Figure 10.8: Class A cities are the world's leading cities in the eastern United States and Western Europe, as well as Tokyo, Hong Kong, and Singapore. Other cities are in developed countries, with Class C cities in emerging countries of East Asia and Latin America—where multinational corporations are building plants.

Figure 10.9: With San Francisco's location on a major geological fault line, the use of fill makes it vulnerable to earthquake damage.

Figure 10.10: As they grow, cities may face limited room for expansion. Connections with the hinterland, which may be across water, are also difficult and expensive to provide.

Figure 10.11: There was a great deal of warfare among small powers in Europe at that time, and the abbey needed to be able to defend itself.

Figure 10.12: Many cities in the East were established at bridging points or heads of navigation. In the West and Midwest, towns were often located where railroads crossed rivers.

Figure 10.13: The point has water on two sides, with only one landward side to be defended.

Figure 10.14: Most of these cities were sited in places that were favorable for trade, especially for seaports. That makes them vulnerable to coastal storms and flooding and in some cases to earthquakes.

Figure 10.15: Las Vegas started as a stopover point on pioneer trails to the west and became a popular railroad town in the early twentieth century. It was a staging point for local mines but declined when the railroads were built. The completion of Hoover Dam and the resulting Lake Mead increased tourism, and gambling was legalized in 1931. Casino developers were likely attracted by the flat, extremely inexpensive land; harsh desert conditions would cause people to remain indoors, where they would be more likely to gamble. Las Vegas also grew largely because of the influx of scientists and staff involved with the Manhattan Project, the atomic bomb research project.

Figure 10.16: It is most likely in flat terrain, where topographical features offer little or no interference.

Figure 10.17: Since the model is based on distance from a central point, the ideal shape would be a circle. However, circles don't tessellate; there would be gaps or overlaps. The closest shape to a circle that does tessellate is a hexagon.

Figure 10.18: The houses are built of cinder blocks and have regular windows and doors. There is a paved sidewalk in front. Two television antennae are also visible.

Figure 10.19: Nonwhites are the low-paid labor force for the industries. They are also relegated to the land that is likely to be subject to fumes and noise from the industrial activities.

Figure 10.20: This modern commercial and residential complex has been built in an architectural style that would be equally at home in other global cities such as New York, Paris, or Sydney.

Figure 10.21: In North American cities, high-rise bank office buildings would likely be built in the center of the city regardless of the historic sites displaced. In some cities (for example, Boston and Philadelphia), they are built in a different part of the center city from historic sites. In any case, they are similarly located in being close to public transportation such as a Metro or subway.

A street in Chinatown, Chicago. (Photolibrary/Getty Images.)

11

Inside the City

Finding and understanding patterns in a city is a difficult matter. As you walk or drive through a city, its intricacy may dazzle you and its form can seem chaotic. It is often hard to imagine why city buildings are where they are, why people cluster where they do. Why does one block have high-income housing and another slum tenements? Why are ethnic neighborhoods next to the central business district? Why does the highway run through one neighborhood and around another? Just when you think you are beginning to understand some patterns in your city, you note that those patterns are swiftly changing. The house you grew up in is now part of the business district. The central city that you roamed as a child looks abandoned. A suburban shopping center thrives on what was once farmland.

Chapter 10 focused on cities as points in geographical space. In this chapter, we try to orient ourselves within cities to gain some perspective on their spatial patterns. In other words, the two chapters differ in scale. Chapter 10 presented cities from afar, as small dots diffusing across space and interacting with one another and with their environment. In this chapter, we use the five themes of cultural geography to study the city as if we were walking its streets.

URBAN CULTURE **REGIONS**

How are areas within a city spatially arranged? Much of the fascination with urban life comes from its diversity, from the excitement created by different groups of people and different types of activities packaged in a fairly small area. Yet within this diversity, it is possible to discern regional patterns, for cities are composed of a series of districts, each of which is defined by a particular set of land uses.

Downtowns

In the center of the typical city is the **central business district (CBD),** a dense cluster of offices and shops. The CBD is formed around the point within the city that is most accessible. As such, businesses and services in the CBD experience the most action

> **central business district (CBD)** The central portion of a city, characterized by high-density land uses.

as measured by the volume of people, money, and ideas moving through this space. Competition for this space often leads to the construction of skyscrapers, creating a skyline that characterizes and symbolizes a city's CBD and the activities that are often located there: financial services, corporate headquarters, and related services such as advertising and public relations firms. The tallest buildings in the world are located in Asian cities such as Shanghai, Kuala Lumpur, and Taipei—cities that are vying to be major financial centers (**Figure 11.1**). Just beyond the skyscrapers are often four- and five-story buildings that make up the city's main shopping district, traditionally centered on several department stores. Also within the CBD are concentrations of smaller retail establishments, transportation hubs such as railroad stations, and often civic centers such as a city hall and main library. Surrounding the CBD is a transitional zone, because it is situated between the core commercial area and the outlying residential areas. It is a district of mixed land uses, characterized by older residential buildings, warehouses, small factories, and apartment buildings.

Residential Areas and Neighborhoods

Beyond the transitional zone are various types of residential communities, or regions. Geographers have studied these culture regions in depth, trying to discern patterns of the distribution of diverse peoples. Some have focused on the idea of a **social culture region:** a residential area characterized by socioeconomic traits, such as income, education, age, and family structure (**Figures 11.2** and **11.3**). Other researchers, who use the notion of **ethnic culture region,** highlight traits of ethnicity, such as language and migration history. Obviously the two concepts overlap, because social regions can exist within ethnic regions and vice versa. In addition, some researchers treat both social and

> **social culture region** An area in a city where many of the residents share social traits such as income, education, age, and family structure.
>
> **ethnic culture region** An area occupied by people of similar ethnic background who share traits of ethnicity, such as language and migration history.

FIGURE 11.1 Skyline of Shanghai, China. The skyline is a clear marker of the global importance of Shanghai as a new financial center. (Photographer's Choice/Getty.)

THINKING GEOGRAPHICALLY Compare these very recent skyscrapers to those of an American or Canadian city.

FIGURE 11.3 Middle-income neighborhood in Sacramento, California. Social areas within the city can be delimited by certain traits taken from the census, such as income, education, age, and family structure. (Cameron Davidson/Corbis.)

THINKING GEOGRAPHICALLY How would the social characteristics of this neighborhood differ from those in Figure 11.2?

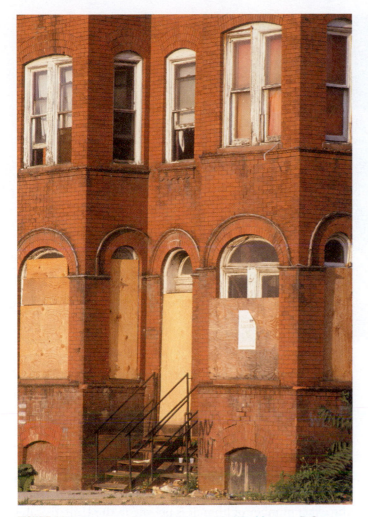

FIGURE 11.2 An inner-city neighborhood in Washington, D.C. One of the most pressing problems facing the United States is reversing the continued decay of inner cities. (Visions of America/SuperStock.)

THINKING GEOGRAPHICALLY What factors have led to this decay of the inner city?

ethnic culture regions as functions of the political and economic forces underlying and reinforcing residential segregation and discrimination. (More information on ethnic areas is found in Chapter 5.)

One way to define social culture regions is to isolate one social trait, such as income, and plot its distribution within the city. The U.S. census is a common source of such information because the districts used to count population, called **census tracts,** are small enough to allow the subtle texture of social regions to show. These maps of social traits form the basis of many urban land-use models, which we discuss later in this chapter.

census tract A small district used by the U.S. Census Bureau to survey the population.

Ethnicity often distinguishes one urban residential area from another. This is not surprising, given the history of immigration to North America and the propensity of immigrants to move to cities. During the middle to late nineteenth century, when waves of immigrants left eastern and southern Europe, North American port cities were a common destination, and employment as laborers in factories was the norm. With limited affordable housing available, most immigrants settled close to one another, forming ethnic communities or pockets within the mosaic of the city. These ethnic urban regions—with names such as Little Italy and Chinatown—often allowed the immigrants to maintain their native languages, holidays, foodways, and religions. Although the names of these urban enclaves still exist today, most of them have been transformed by new and different waves of immigrants arriving in North America in the past 20 years (**Figure 11.4** on page 304). New York's Little Italy, for example, is now home to people from East and Southeast Asia, with Italian restaurants vying for space with noodle houses.

REFLECTING ON GEOGRAPHY

As we look at these urban regions, various questions concern geographers: Why do people of similar social traits cluster together? What subtle patterns might be found within these regions?

Social culture regions are not merely statistical definitions. They are also areas of shared values and attitudes, of

FIGURE 11.4 Woodside, Queens, New York City. Once home to Irish immigrants, Woodside is now the destination of immigrants from a wide range of countries. (James Estrin/NYT Pictures.)

THINKING GEOGRAPHICALLY In what ways are these new ethnic neighborhoods different from those of the Irish-Americans? In what ways do you think they are similar?

interaction and communication. **Neighborhood** is a concept often used to describe small social culture regions where people with shared values and concerns interact daily. For example, if we consider only census figures, we might find that parents between 30 and 45 years of age with two or three children and earning between $65,000 and $90,000 a year cover a fairly wide area in any given city. Yet, from our own observations, we know intuitively that this broad social area is probably composed of numerous neighborhoods where people associate a sense of community with a specific locale.

neighborhood A small social area within a city where residents share values and concerns and interact with one another on a daily basis.

A conventional sociological explanation for neighborhoods is that people of similar values cluster together to reduce social conflict. Where a social consensus exists about such mundane issues as home maintenance, child rearing, acceptable behavior, and public order, there is little daily worry about these matters. People who deviate from this consensus will face social coercion that could force them to seek residence elsewhere, thus preserving the values of the neighborhood. However, many neighborhoods have more heterogeneity than this traditional definition would allow. Although a neighborhood might be ethnically and socially diverse, its residents may think of themselves as a community that shares similar political concerns, holds neighborhood meetings to address these problems, and achieves recognition at city hall as a legitimate group with political standing.

Homelessness

Neighborhoods are usually composed of people who have access to a permanent or semipermanent place of residence. In the cities of the United States, however, homelessness, and with it the loss of neighborhood, is increasingly common. It is nearly impossible to determine the exact number of homeless people in the United States because definitions of **homelessness** vary. For example, does living in a friend's house for more than a month constitute a homeless condition? How permanent does a shelter have to be before it is considered a home? To some people, home connotes a suburban middle-class house; to others, it simply refers to a room in a city-owned shelter.

homelessness A temporary or permanent condition of not having a legal home address.

REFLECTING ON GEOGRAPHY
Why is home such a difficult concept to define? How does it differ from the concept of a house?

Recent studies suggest that there are up to 3 million homeless people in the United States, concentrated in the downtown areas of large cities, often in what we call the transitional zone. The causes of homelessness are varied and complex. Many homeless people suffer from some type of disorder or handicap that contributes to their inability to maintain a job and obtain adequate

FIGURE 11.5 **The distribution of services for the homeless in the Skid Row district of Los Angeles.** The population of the area is difficult to estimate, ranging from 6000 to 30,000. There are approximately 2000 shelter beds in the area, half of which are available to women. Single-room-occupancy hotels provide about 6700 units of longer-term housing. More than 50 social service programs are run through agencies, missions, and shelters. Love Camp and Justiceville are the sites of informal street encampments of homeless people. (Source: After Rowe and Wolch, 1990.)

THINKING GEOGRAPHICALLY What kinds of services do you think homeless people obtain from informal encampments like Love Camp and Justiceville?

housing. Deprived of the social networks that a permanent neighborhood provides, the homeless are left to fend for themselves. Most cities have tried to provide temporary shelter, but many homeless people prefer to rely on their own social ties for support to maintain some sense of personal pride and privacy. In a study of the Los Angeles skid row district, Stacy Rowe and Jennifer Wolch explored how homeless women formed new types of social networks and established a sense of community to cope with the day-to-day needs of physical security and food (**Figure 11.5**).

CULTURAL **DIFFUSION** IN THE CITY

How can we understand the spatial movement of people and activities in the city? The patterns of activities we see in the city are the result of thousands of individual decisions about location: Where should we locate our store—in the central city or the suburbs? Where should we live—downtown or outside the city? And so on. The result of such decisions might be expansion at the city's edge or the relocation of activities from one part of the city to another. The cultural geographer looks at such decisions in terms of expansion and relocation diffusion (see Chapter 1).

To understand the role of diffusion, let us divide the city into two major areas—the inner city and the outer city. Those diffusion forces that result in residences, stores, and factories locating in the inner or central city are **centralizing forces.** Those that result in activities locating outside the central city are called **decentralizing forces,** or suburbanizing forces. The pattern of homes, neighborhoods, offices, shops, and factories in the city results from the constant interplay of these two forces.

> **centralizing force** Diffusion force that encourages people or businesses to locate in the central city.
>
> **decentralizing force** A diffusion force that encourages people or businesses to locate outside the central city. Also called suburbanizing force.

Centralization

Centralization has two primary advantages: economic and social.

Economic Advantages An important economic advantage of central-city location has always been accessibility. For example, imagine that a department store seeks a new location. If its potential market area is viewed as a full circle, then naturally the best location is in the center. There, customers from all parts of the city can gain access with equal ease. Before the automobile, a central-city location was particularly important because public transportation—such as the streetcar—was usually focused there. A central location is also important to those who must deliver their goods to customers. Bakeries and dairies were usually located as close to the center of the city as possible to maximize the efficiency of their delivery routes.

Location near regional transportation facilities is another aspect of accessibility and is thus an economic advantage. Many a North American city developed with the railroad at its center. Hence, any activity that needed access to the railroad had to locate in the central city. In many urban areas, giant wholesale and retail manufacturing districts grew up around railroad districts, producing "freight-yard and terminal cities" that supplied the produce of the nation. Although many of these areas have been abandoned by their original occupants, a walk by the railroad tracks today will give the most casual pedestrian a view of the modern ruins of the railroad city.

Another major economic advantage of the inner city is **agglomeration,** or clustering, which results in mutual benefits for businesses. Agglomeration is a snowballing geographical process by which secondary and service industrial activities become clustered in cities and compact industrial regions in order to share infrastructure and markets. For example, retail stores locate near one another to take advantage of the pedestrian traffic each generates. Because a large department store generates a good deal of foot traffic, any nearby store will benefit.

> **agglomeration** A snowballing geographical process by which secondary and service industrial activities become clustered in cities and compact industrial regions in order to share infrastructure and markets.

Historically, offices clustered in the central city because of their need for communication. Remember, the telephone was not invented until 1875. Before that, messengers hand-carried the work of banks, insurance firms, lawyers, and many other services. Clustering was essential for rapid communication. Even today, office buildings tend to cluster because face-to-face communication is still important for businesspeople. In addition, central offices take advantage of the complicated support system that grows up in a central city and aids everyday efficiency. Service providers such as printers, bars, restaurants, travel agents, and office suppliers are within easy reach.

Social Advantages Three social factors have traditionally reinforced central-city location: historical momentum, prestige, and the need to locate near work. The strength of historical momentum should not be underestimated. Many activities remain in the central city simply because they began there long ago. For example, the financial district in San Francisco is located mainly on Montgomery Street. This street originally lay along the waterfront, and San Francisco's first financial institutions were established there in the mid-nineteenth century because it was the center of commercial action. In later years, however, landfill extended the shoreline (see Figure 10.9, p. 288). Today, the financial district is several blocks from the bay; consequently, the district that began at the wharf head remained at its original location, even though other activity moved with the changing shoreline.

The prestige associated with the downtown area is also a strong centralizing force. Some activities still necessitate a central-city address. Think how important it is for some advertising firms to be on New York's Madison Avenue or for a stockbroker to be on Wall Street. This factor

extends to many activities in cities of all sizes. The "downtown lawyer" and the "uptown banker" are examples.

Residences have often been located in the central city because of the prestige associated with it. For example, in Chicago, the Magnificent Mile along the north end of Michigan Avenue has long attracted affluent residents desiring a prestigious address. Here, luxury high-rise apartments with stunning lake views are interspersed with high-end retail stores and restaurants, elite law firms and advertising agencies, and major television and media firms (**Figure 11.6**). Many cities around the world feature similarly prestigious districts in downtown areas. The Avenue des Champs-Élysées in Paris, Severn Road in Hong Kong, and Ostozhenka Street in Moscow all offer residents the opportunity to live in some of the most expensive and exclusive downtown properties in the world.

FIGURE 11.6 Chicago's Magnificent Mile along Michigan Avenue. With proximity to Lake Michigan as well as to elite shopping and dining, this street has become synonymous with luxury living. (Henryk Sadura/age fotostock.)

THINKING GEOGRAPHICALLY Why do you think that prestigious addresses in many downtown areas have retained their elite status despite the increasing movement of affluent people to suburban areas around the world?

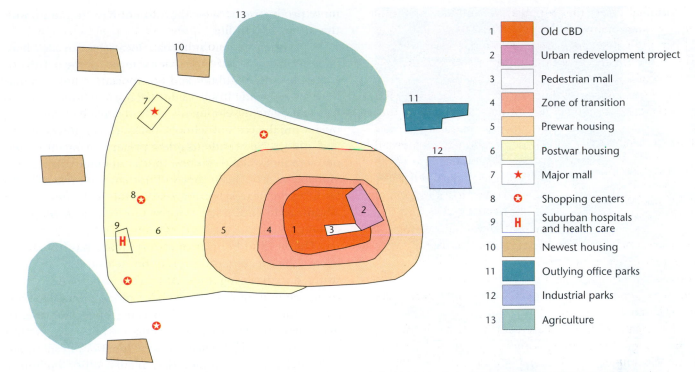

1		Old CBD
2		Urban redevelopment project
3		Pedestrian mall
4		Zone of transition
5		Prewar housing
6		Postwar housing
7	★	Major mall
8	✪	Shopping centers
9	H	Suburban hospitals and health care
10		Newest housing
11		Outlying office parks
12		Industrial parks
13		Agriculture

FIGURE 11.7 A hypothetical decentralized city. While the old central business district (CBD) struggles (vacant stores and upper floors), newer activities locate either in the urban redevelopment project (offices, convention center, hotel) or in outlying office parks, malls, or shopping centers. However, some new specialty shops might be found around the new downtown pedestrian mall. New industry locates in suburban industrial parks that, along with outlying office areas, form major destinations for daily lateral commuting.

THINKING GEOGRAPHICALLY In what ways does your city follow this hypothetical pattern? In what ways does it differ?

Probably the strongest social force for centralization has been the desire to live near one's place of employment. Until the development of the electric trolley in the 1880s, most urban dwellers had little alternative but to walk to work, and as most employment was in the central city, people had no choice but to live nearby. Upper-income people had their carriages and cabs, but others had nothing. Even after the introduction of electric streetcar lines in the 1880s, which made possible the exodus of some middle-class residents, many people continued to walk to work, particularly those who could not afford the new housing being constructed in what Sam Bass Warner has called "streetcar suburbs."

Suburbanization and Decentralization

The past 50 years have witnessed massive changes in the form and function of most Western cities (**Figure 11.7**). In the United States in particular, the suburbanization of residences and the decentralization of workplaces have emptied many downtowns of economic vitality. How and why has this happened? Geographer Neil Smith argues that the processes of suburbanization and decline of the inner city are fundamentally linked: capital investment in the suburbs is often made possible by disinvestment, or the removal of money, from the central city. In the post–World War II United States, investors found greater returns on their money in the new suburbs than they did in the inner city, and therefore much of the economic boom of this period took place in the suburbs at the expense of the city. Smith refers to these processes as **uneven development** (**Figure 11.8** on page 308). This type of explanation gives us a broad picture of the economic reasons that many cities are now decentralized. We will now look more closely at decentralization, examining the specific socioeconomic and public policy causes for the decentralization of our cities and for the problems that have resulted.

> **uneven development** The tendency for industry to develop in a core-periphery pattern, enriching the industrialized countries of the core and impoverishing the less industrialized periphery. This term is also used to describe urban patterns in which suburban areas are enriched while the inner city is impoverished.

Socioeconomic Factors Changes in accessibility have been a major reason for decentralization. The department store that was originally located in the central city may now find that its customers have moved to the suburbs and no longer shop downtown. As a result, the department store may move to a suburban shopping mall. The same process affects many other industries as well. The activities that were

FIGURE 11.8 **Abandoned row houses in North Philadelphia.** Often the economic neglect of these areas is directly linked to economic investment in the suburbs. (Associated Press.)

THINKING GEOGRAPHICALLY Under what circumstances were these houses once homes to the middle or upper classes?

located downtown because of its proximity to the railroad may now find trucking more cost effective. They relocate closer to a freeway system that skirts the downtown area. Finally, many offices now locate near airports so that their executives and salespeople can fly in and out more easily.

Although agglomeration once served as a centralizing force, its former benefits have now become liabilities in many downtown areas. These disadvantages include rising rents as a result of the high demand for space; congestion in the support system, which may cause delays in the supply chain or long lines at lunchtime; and traffic congestion, which makes delivery to market time-consuming and costly. Some downtown areas are so congested that traffic moves more slowly today than it did at the turn of the twentieth century. Often dissatisfied with the inconveniences of central-city living, employees may demand higher wages as compensation. This adds to the cost of doing business in the central city, and many firms choose to leave instead. For example, many firms have left New York City for the suburbs. They claim that it costs less to locate there and that their employees are happier and

more productive because they do not have to put up with the turmoil of city life.

Clustering in new suburban locations can also have benefits. In industrial parks, for example, all the occupants share the costs of utilities and transportation links. Similar benefits can come from residential agglomeration. Suburban real estate developments take advantage of clustering by sharing the costs of schools, parks, road improvements, and utilities. New residents much prefer moving into a new development when they know that a full range of services is available nearby. Then they will not have to drive miles to find, say, the nearest hardware store. It is to the developer's advantage to encourage construction of nearby shopping centers.

The need to be near one's workplace has historically been a great centralizing force, but it can also be a very strong decentralizing force. At first the suburbs were "bedroom communities" from which people commuted to their jobs in the downtown area. This is no longer the case. In many metropolitan areas, most jobs are not in the central city but in outlying districts. Now people work in suburban industrial parks, manufacturing plants, office buildings, and shopping centers. Thus, a typical journey to work involves **lateral commuting:** travel from one suburb to another. As a result, most people who live away from the city center actually live closer to their workplaces.

> **lateral commuting** Traveling from one suburb to another between home and work.

The prestige of the downtown area might once have lured people and businesses into the central city. But once it begins to decay, once shops close and offices are empty, a certain stigma develops that may drive away residents and commercial activities. Investors will not sink money into a downtown area that they think has no chance of recovery, and shoppers will not venture downtown when streets are filled with vacant stores, transients, pawnshops, and secondhand stores. One of the persistent problems faced by cities is how to reverse this image of the downtown area so that people once again consider it the focus of the city.

REFLECTING ON GEOGRAPHY

Identify some of the efforts that your city has undertaken to create a better image of itself. Have these efforts been successful? Does this reimagining of the city actually help residents of the inner city?

Public Policy Many public policy decisions, particularly at the national level, have contributed greatly to the decentralization and abandonment of our cities. Both the Federal Road Act of 1916 and the Interstate Highway Act of 1956 directed government spending on transportation to the advantage of automobiles and trucks. Urban expressways, in combination with the emerging trucking industry, led to massive decentralization of industry and housing. In addition, the ability to deduct mortgage interest from income for tax purposes favors individual homeownership, which has tended to support a move to the suburbs.

The federal government in the United States has also intervened more directly in the housing market. In *Crabgrass Frontier,* Kenneth Jackson outlines the implications of two federal housing policies for the spatial patterning of our metropolitan areas. The first was the establishment of the Federal Housing Administration (FHA) in 1934 and its supplement, known as the GI Bill, enacted in 1944. These federal acts, which insured long-term mortgage loans for home construction, were meant to provide employment in the building trades and to help house soldiers returning from World War II. Although the FHA legislation contained no explicit antiurban bias, most of the houses it insured were located in new residential developments in the suburbs, thereby neglecting the inner city.

Jackson identifies three reasons this happened. First, by setting particular terms for its insurance, the FHA favored the development of single-family over multifamily projects. Second, FHA-insured loans for repairs were of short duration and were generally small. Most families, therefore, were better off buying a new house that was probably in the suburbs than updating an older home in the city.

Jackson regards the third factor as the most important. To receive an FHA-insured loan, the applicant and the neighborhood of the property had to be assessed by an "unbiased professional." This requirement was intended to guarantee that the property value of the house would be greater than the debt. This policy, however, encouraged bias against any neighborhood that was considered a potential risk in terms of property values. The FHA explicitly warned against neighborhoods with a racial mix, assuming that such a social climate would bring property values down, and it encouraged the inclusion of **restrictive covenants** in property deeds. These covenants restricted the use of the land in some way and often prohibited certain "undesirable" groups from buying property. The agency also prepared extensive maps of metropolitan areas depicting the locations of African-American families and predicting the spread of that population. These maps often served as the basis for **redlining,** a practice whereby banks and mortgage companies demarcated areas (often by drawing a red line around them on maps) considered to be at high risk for defaulting on housing loans.

restrictive covenant A statement written into a property deed that limits the use of the land in some way; often used to prohibit certain groups of people from buying property.

redlining A practice by banks and mortgage companies of demarcating areas considered to be at high risk for defaulting on housing loans.

These policies had two primary effects. First, they encouraged construction of single-family homes in suburban areas while discouraging central-city locations. Second, they intensified the segregation of residential areas and actively promoted homogeneity in the new suburbs.

The second federal housing policy that had a major impact on the patterning of metropolitan areas, the United States Housing Act, was intended to provide public housing for those who could not afford private housing. Originally implemented in 1937, the legislation did encourage the construction of many low-income housing units. Yet most of those units were built in the inner city, thereby contributing to the view of the suburbs as the refuge of the white middle class. This growing pattern of racial and economic segregation arose in part because public housing decisions were left up to local municipalities. Many municipalities did not need federal dollars and therefore did not want public housing. In addition, the legislation required that for every unit of public housing erected, one inferior housing unit had to be eliminated. Thus, only areas with inadequate housing units could receive federal dollars, again ensuring that public housing projects would be constructed in the older downtown areas, not the newer suburbs. As Jackson claims, "The result, if not the intent, of the public housing program of the United States was to segregate the races, to concentrate the disadvantaged in inner cities, and to reinforce the image of suburbia as a place of refuge from the problems of race, crime, and poverty."

The Costs of Decentralization Decentralization has taken its toll on North American cities. Many of the problems they now face are the direct result of the rapid decentralization that has taken place in the last 50 years. Those people who cannot afford to live in the suburbs are forced to live in inadequate and rundown housing in the inner city, areas that currently do not provide good jobs. Vacant storefronts, empty offices, and deserted factories testify to the movement of commercial functions from central cities to suburbs. Retail stores in North American central cities have steadily lost sales to suburban shopping centers. Even offices are finding advantages to suburban location when they can capitalize on lower costs and easier access to new transportation networks.

Decentralization has also cost society millions of dollars in problems brought to the suburbs. Where rapid suburbanization has occurred, sprawl has usually followed. A common pattern is leapfrog or **checkerboard development:** housing tracts jump over parcels of farmland, resulting in a mixture of open lands with built-up areas. This pattern occurs because developers buy cheaper land farther away from built-up areas,

checkerboard development A mixture of farmlands and housing tracts. Also called leapfrog development.

FIGURE 11.9 Suburban sprawl, Charlotte, North Carolina.
Suburbanization gives us a familiar landscape of look-alike houses and yards; automobile-efficient street and transportation patterns; and, in the background, remnants of agriculture, awaiting the day that they are converted into housing tracts. [Masterfile (Royalty-Free Div.).]

THINKING GEOGRAPHICALLY Why are such street patterns sometimes called "loop and lollipop"? What problems does such a pattern cause for provision of services?

thereby cutting their costs. Furthermore, home buyers are often willing to pay premium prices for homes in subdivisions surrounded by farmland (**Figure 11.9**). The developer's gain is the area's loss, for it is more expensive to provide city services—such as police, fire protection, sewers, and electrical lines—to those areas that lie beyond open parcels that are not built up. Obviously, the most cost-efficient form of development is the addition of new housing directly adjacent to built-up areas so that the costs of providing new services are minimal.

Sprawl also extracts high costs because of the increased use of cars. Public transportation is extremely costly and inefficient when it must serve a low-density checkerboard development pattern—so costly that many cities and transit firms cannot extend lines into these areas. This means that the automobile is the only form of transportation. More energy is consumed for fuel, more air pollution is created by exhaust, and more time is devoted to commuting and everyday activities in a sprawling urban area than in a centralized city.

Moreover, we should not overlook the costs of losing valuable agricultural land to urban development. Farmers cultivating the remaining checkerboard parcels have a hard time earning a living. They are usually taxed at extremely high rates because their land has high potential for development, and few can make a profit when taxes eat up all their resources. Often the only recourse is to sell

out to subdividers. So the cycle of leapfrog development goes on.

Many cities are now taking strong measures to curb this kind of sprawling growth. San José, California, for example, one of the fastest-growing cities of the 1960s, is now focusing new development on empty parcels of the checkerboard pattern. This is called **in-filling**. Other cities are tying the number of building permits grant-

> **in-filling** New building on empty parcels of land within a checkerboard pattern of development.

ed each year to the availability of urban services. If schools are already crowded, water supplies inadequate, and sewage plants overburdened, the number of new dwelling units approved for an area will reflect this lower carrying capacity.

Gentrification

Beginning in the 1970s, urban scholars observed what seemed to be a trend opposite to suburbanization. This trend, called **gentrification,** is the movement of middle-class people into deteriorated areas of city centers. Gentrification often begins in an inner-city residential district, with gentrifiers moving into an area that had been run

> **gentrification** The displacement of lower-income residents by higher-income residents as buildings in deteriorated areas of city centers are restored.

down and is therefore more affordable than suburban housing (**Figure 11.10**). The infusion of new capital into the housing market usually results in higher property values, and this, in turn, often displaces residents who cannot afford the higher prices. Displacement opens up more housing for gentrification, and the gentrified district continues its spatial expansion.

Commercial gentrification usually follows residential, as new patterns of consumption are introduced into the inner city by the middle-class gentrifiers. Urban shopping malls and pedestrian shopping corridors bring the conveniences of the suburbs into the city, and bars and restaurants catering to this new urban middle class provide entertainment and nightlife for the gentrifiers.

The speed with which gentrification has proceeded in many of our downtowns and the scale of landscape changes that accompany it are causing dramatic shifts in the urban mosaic. What factors have led to this reshaping of our cities?

Economic Factors Some urban scholars look to broad economic trends in the United States to explain gentrification. We already know from our discussion of suburbanization that throughout the post–World War II era most investments

in metropolitan land were made in the suburbs; as a result, land in the inner city was devalued. By the 1970s, many home buyers and commercial investors found land in the city much more affordable, and a better economic investment, than in the higher-priced suburbs. This situation brought capital into areas that had been undervalued and accelerated the gentrifying process.

In addition, most Western countries have been experiencing **deindustrialization,** a process whereby the economy is shifting from one based on primary and secondary industry to one based on the service sector. This shift has led to the abandonment of older industrial districts in the inner city, including waterfront areas. Many of these areas are prime targets of gentrifiers, who convert the waterfront from a noisy, commercial port area into an aesthetic asset. In Buenos Aires, for example, one of the new gentrified neighborhoods is Puerto Madero, an area that was once home to docking facilities and a wholesale market (**Figure 11.11**). The shift to an economy based on the service sector also means that the new productive areas of the city will be dedicated to white-collar activities. These activities often take place in relatively clean and quiet office buildings, contributing to a view of the city as a more livable environment.

> **deindustrialization** The decline of primary and secondary industry, accompanied by a rise in the service sectors of the industrial economy.

Social Factors Other scholars look to changes in social structure to explain gentrification. The maturing of the baby-boom generation has led to significant modifications of our traditional family structure and lifestyle. With a majority of women in the paid labor force and many young couples choosing not to have children or to delay that decision, a suburban residential location looks less appealing. A gentrified location in the inner city attracts this new class because it is close to managerial or professional jobs downtown, is usually easier to maintain, and is considered more interesting than the bland suburban areas of childhood. Living in a newly gentrified area is also a way to display social status. Many suburbs have become less exclusive, while older neighborhoods in the inner city frequently exploit their historical associations as a status symbol.

FIGURE 11.10 Gentrification in San Francisco, California. Gentrification often occurs in older neighborhoods with historic buildings. These Victorian houses have been carefully restored to reflect their new owners' interest in preserving the past, although this might have come at a cost of displacing the previous tenants. (© Rudy Sulgan/Corbis.)

THINKING GEOGRAPHICALLY Why does gentrification often take place in a neighborhood close to downtown?

FIGURE 11.11 Puerto Madero, Buenos Aires. These expansive dock facilities in Buenos Aires have recently been converted by the city and private entrepreneurs into a center of nightlife for the city, complete with clubs, restaurants, and shopping. New condominiums and hotels are now under construction. (Courtesy of Mona Domosh.)

THINKING GEOGRAPHICALLY Is there anything that marks this gentrified district as being located in Buenos Aires, instead of, say, New York City?

Political Factors Many metropolitan governments in the United States, faced with the abandonment of the central city by the middle class and therefore the erosion of their tax base, have enacted policies to encourage commercial and residential development in downtown areas. Some policies provide tax breaks for companies willing to locate downtown; others furnish local and state funding to redevelop central-city residential and commercial buildings.

At a more comprehensive level, some larger metropolitan areas have devised long-term planning agendas that target certain neighborhoods for revitalization. Often this is accomplished by first condemning the targeted area, thereby transferring control of the land to an urban-development authority or other planning agency. Such areas are often older residential neighborhoods that were originally built to house people who worked in nearby factories, which are usually torn down or transformed into lofts or office space. The redevelopment authority might locate a new civic or arts center in the neighborhood. Public-sector initiatives often lead to private investment, thereby increasing property values. These higher property values in turn lead to further investment and the eventual transformation of the neighborhood into a middle- to upper-class gentrified district.

Sexuality and Gentrification Gentrified residential districts are often correlated with the presence of a significant gay and lesbian population. It is fairly easy to understand this correlation. First, the typical suburban life tends not to appeal to people whose lifestyle is often regarded as different and whose community needs are often different from those of people living in the suburbs. Second, gentrified inner-city neighborhoods provide access to the diversity of city life and amenities that often include gay cultural institutions. In fact, the association of urban neighborhoods with gays and lesbians has a long history. For example, urban historian George Chauncey has documented gay culture in New York City between 1890 and World War II, showing that a gay world occupied and shaped distinctive spaces in the city, such as neighborhood enclaves, gay commercial areas, and public parks and streets.

Yet, unlike this earlier period, when gay cultures were often forced to remain hidden, the gentrification of the postwar period has provided gay and lesbian populations with the opportunity to reshape entire neighborhoods actively and openly. Urban scholar Manuel Castells argues that in cities such as San Francisco, the presence of gay men in institutions directly linked to gentrification, such as the real estate industry, significantly influenced that city's gentrification processes in the 1970s.

Geographers Mickey Lauria and Lawrence Knopp emphasize the community-building aspect of gay men's involvement in gentrification, recognizing that gays have seized an opportunity to combat oppression by creating neighborhoods over which they have maximum control and that meet long-neglected needs. Similarly, geographer Gill Valentine argues that the limited numbers and types of lesbian spaces in cities also serve as community-building centers for lesbian social networks.

According to geographer Tamar Rothenberg, the gentrified neighborhood of Park Slope in Brooklyn is home to the heaviest concentration of lesbians in the United States. Its extensive social networks are marking the neighborhood as both a center of lesbian identity and a visible lesbian social space.

The Costs of Gentrification Gentrification often results in the displacement of lower-income people, who are forced to leave their homes because of rising property values. This displacement can have serious consequences for the city's social fabric. Because many of the displaced people come from disadvantaged groups, gentrification frequently contributes to racial and ethnic tensions. Displaced people are often forced into neighborhoods more peripheral to the city, a trend that only adds to their disadvantages. In addition, gentrified neighborhoods usually stand in stark contrast to surrounding areas where investment has not taken place, thus creating a very visible reminder of the uneven distribution of wealth within cities.

The success of a gentrification project is usually measured by its appeal to an upper-middle-class clientele. This suggests that gentrified neighborhoods are completely homogeneous in their use of land. Residential areas are consciously planned to be separate from commercial districts and are themselves sorted by cost and tenure type (homeownership versus rental). Thus, gentrification often draws on the suburban notion of residential homogeneity and eliminates what many people consider to be a great asset of urban life—its diversity and heterogeneity.

Immigration and New Ethnic Neighborhoods

According to the United States Department of Homeland Security, approximately 3.2 million people immigrated to the United States as "legal permanent residents" between 2009 and 2011: 39.5 percent came from Latin America and the Caribbean, 39.7 percent from Asia, 9.9 percent from Europe and Canada, and 10.9 percent from other regions. The vast majority of these immigrants found job opportunities and cultural connections that drew them to major metropolitan regions in six states—California, New York, Florida, Texas, New Jersey, and Illinois (**Figure 11.12**). This spatial concentration of America's new immigrants has created diverse communities with distinctive landscapes, both within the downtown areas of these cities and in the suburban regions. Miami's Little Havana, for example, is easily recognized by

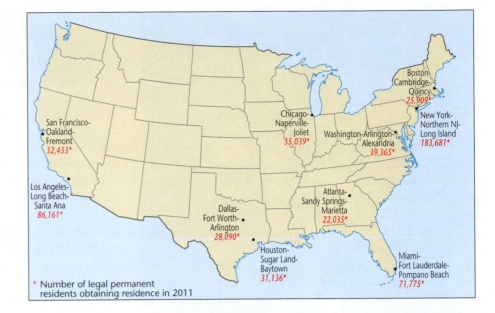

FIGURE 11.12 Map indicating the location of 10 metropolitan areas in the United States with the largest number of persons obtaining legal permanent resident status, 2011. (Source: U.S. Department of Homeland Security, 2011.)

THINKING GEOGRAPHICALLY Why is immigration focused on these particular cities?

the commercial signs in Spanish, Spanish street names, and the colors and styles of buildings. Parts of what were once run-down neighborhoods of the city have been remade into vibrant commercial and residential communities.

These new ethnic landscapes are not limited to the central city. Portions of America's suburbs have also become diverse. The decentralization of the downtown has created new economic centers in suburban regions, and immigrants are drawn to these centers. According to the 2010 census, nationwide approximately 51 percent of new immigrants to the United States are now living in suburbs of large metropolitan areas. Geographer Wei Li refers to these new immigrant communities located in the suburbs as **ethnoburbs.** Instead of the dense residential

ethnoburb A suburban ethnic neighborhood, sometimes home to a relatively affluent immigration population.

and commercial districts that characterize downtown ethnic enclaves, the new suburban ethnicity is proclaimed within shopping centers, suburban cemeteries, and dispersed houses of worship. According to geographer Joseph Wood, their presence in the landscape is often not visible to observers precisely because the landscape is suburban. For example, most of the 50,000 Vietnamese-Americans who migrated to the Washington, D.C., area from the 1970s through the 1990s settled in suburbs in northern Virginia. The focal point of the Vietnamese community there is the Eden Center, a typical L-shaped shopping center that has been transformed into a Vietnamese-American economic and social center (**Figure 11.13**). This pattern of shopping plazas serving as ethnic community markers for a dispersed immigrant population is not peculiar to northern Virginia; it is commonplace in the metropolitan areas of most major North American cities.

FIGURE 11.13 Eden Shopping Center in Fairfax, **Virginia.** This shopping plaza serves as a social and cultural center for the Vietnamese community of northern Virginia. (Courtesy of Joseph Wood.)

THINKING GEOGRAPHICALLY Compare this center to a traditional urban ethnic neighborhood.

FIGURE 11.14 A Muslim family in the neighborhood of Belleville in Paris. Many cities have ethnic enclaves where immigrants from other countries live, creating a different scale of community within the larger scale of the city. (Peter Turnley/Corbis.)

THINKING GEOGRAPHICALLY How are these ethnic enclaves related to globalization?

Urban ethnic enclaves are not limited to U.S. and Canadian cities. Many cities around the world are experiencing the effects of immigration as globalization makes it easier for people from one country to seek employment in thriving metropolitan areas of a different country. There are North African communities in Paris (**Figure 11.14**), Turkish neighborhoods in Berlin, and Kurdish neighborhoods in Istanbul, and each of these is influencing the culture, economy, and landscape of these cities in profound ways.

THE CULTURAL **ECOLOGY** OF THE CITY

How can we understand the relationships between the urban mosaic and the physical environment? The physical environment affects cities, just as urbanization profoundly alters natural environmental processes (**Figure 11.15**). The theme of cultural ecology helps us to organize information about these city-nature relationships.

Urban Weather and Climate

Cities alter virtually all aspects of local weather and climate. Temperatures are higher in cities; rainfall is increased; the incidence of fog and cloudiness is greater; and levels of atmospheric pollution are much higher.

The causes of these changes are no mystery. Because cities cover large areas of land with streets, buildings, parking lots, and rooftops, about 50 percent of the urban area is a hard surface. Rainfall is quickly carried into gutters and sewers, so that little standing water is available for evaporation. Because evaporation removes heat from the air, when moisture is reduced, evaporation is lessened and air temperatures are higher.

Moreover, cities generate enormous amounts of heat. This heat comes not just from the heating systems of buildings but also from automobiles, industry, and even human bodies. One study showed that on a winter day in Manhattan, the amount of heat produced in the city is two and a half times the amount that reaches the ground from the sun. This results in a large mass of warmer air sitting over the city, called the urban **heat island** (**Figure 11.16**). The heat island causes yearly temperature averages in cities to be 3.5°F (2°C) higher than in the countryside;

heat island An area of warmer temperatures at the center of a city, caused by the urban concentration of heat-retaining concrete, brick, and asphalt.

during the winter, when there is more city-produced heat, the average difference can easily reach 7°F to 10°F (4°C to 5.6°C).

Urbanization also affects precipitation (rainfall and snowfall). Because of higher temperatures in the urban area, snowfall will be about 5 percent less than in the surrounding countryside. However, rainfall can be 5 percent to 10 percent higher. The increased rainfall results from two factors: the large number of dust particles in urban air and the higher city temperatures. Dust particles are a necessary precondition for condensation, offering a nucleus to which moisture can adhere. An abundance of dust particles, then, facilitates condensation. That is why fog and clouds are usually more frequent around cities.

FIGURE 11.15 **Suburban homes built on landfills, Treasure Island, Florida.** When land values are high and pressure for housing intense, terrain rarely stands in the way of the developer. In fact, particular physical site characteristics can actually increase land values. (© 2013 Alex S. MacLean/Landslides.)

THINKING GEOGRAPHICALLY What site characteristics evident in this photo tell you that Treasure Island is a very expensive place to live?

dust dome A layer of pollution over a city that is thickest at the center of the city.

The term **dust domes** is used to refer to layers of pollution over a city. The dust dome is thickest at the center of the city (**Figure 11.17** on page 316).

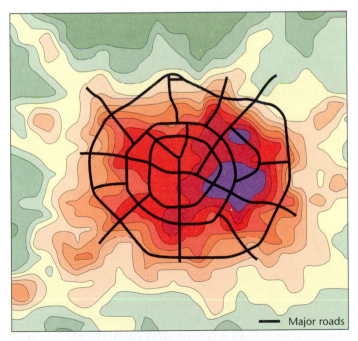

FIGURE 11.16 **Diagram of the urban heat island in Chengdu, China.** The deep colors (purples and red) indicate higher temperatures, while the shades of yellow and green indicate lower temperatures. Notice the marked contrast in temperature between the built-up part of the city and the surrounding rural areas. (Source: Shangming and Bo, 2001.)

— Major roads

THINKING GEOGRAPHICALLY What might cause the center of the heat island to be located slightly east of the center of the city?

Urban Hydrology

Not only is the city a great consumer of water, but it also alters runoff patterns in a way that increases the frequency and magnitude of flooding. Within the city, residential areas are usually the greatest consumers of water. Water consumption can vary, but generally each person in the United States uses about 60 gallons (264 liters) per day in a residence. Of course, residential demand varies. It is greater in drier climates as well as in middle- and high-income neighborhoods. Higher-income groups usually have a larger number of water-using appliances, such as washing machines, dishwashers, and swimming pools.

Urbanization can increase both the frequency and the magnitude of flooding because cities create large impervious areas where water cannot soak into the earth. Instead, precipitation is converted into immediate runoff. It is forced into gutters, sewers, and stream channels that have been straightened and stripped of vegetation, which results in more frequent high-water levels than are found in a comparable area of rural land. Furthermore, the time between rainfall and peak runoff is reduced in cities; there is more lag in the countryside, where water runs across soil and vegetation into stream channels and then into rivers. So, because of hard surfaces and artificial collection channels, runoff in cities is concentrated and immediate.

Urban Vegetation

Until a decade ago, it was commonly thought that cities were made up mostly of artificial materials: asphalt, concrete, glass, and steel. Studies, however, show that about two-thirds of a typical North American city is composed

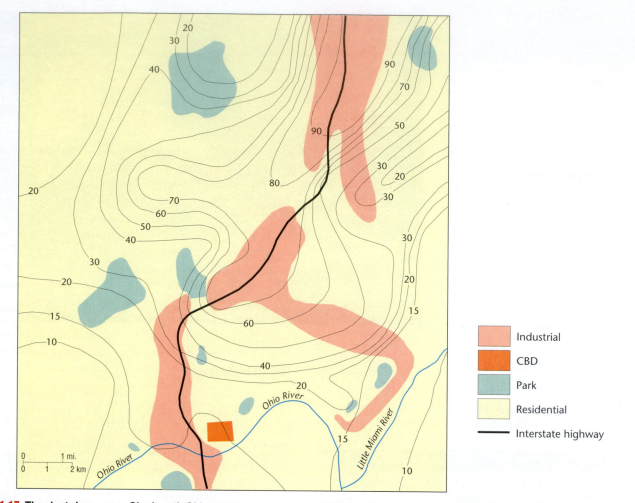

FIGURE 11.17 The dust dome over Cincinnati, Ohio. Numbers show the concentration of particulate matter in the air at an elevation of 3000 feet (914 meters). The higher the value, the greater the amount of particulate matter. (Source: After Bach and Hagedorn, 1971.)

THINKING GEOGRAPHICALLY Does land use (industrial, central business district, and so on) have an effect on concentration of particulate matter in the air?

of trees and herbaceous plants (mostly weeds in vacant lots and cultivated grasses in lawns). This urban vegetation is usually a mix of natural and introduced species and is a critical component of the urban ecosystem because it affects the city's topography, hydrology, and meteorology.

More specifically, urban vegetation influences the quantity and quality of surface water and groundwater; reduces wind velocity and turbulence and temperature extremes; affects the pattern of snow accumulation and melting; absorbs thousands of tons of airborne particulates and atmospheric gases; and offers a habitat for mammals, birds, reptiles, and insects, all of which play some useful role in the urban ecosystem. Furthermore, urban vegetation influences the propagation of sound waves by muffling much of the city's noise; affects the distribution of natural and artificial light; and, finally, is an extremely important

component in the development of soil profiles—which, in turn, control hillside stability.

Our urban settlements are still closely tied to the physical environment. Cities change these natural processes in profound ways, and we must understand these disturbances in order to make better decisions about adjustments and control.

CULTURAL **INTERACTION** AND MODELS OF THE CITY

Are there generalizable spatial patterns or models that describe types of cities? Geographers are interested in understanding both the uniqueness of each city's urban pattern and the similarities that a particular city may have with others. To provide insights

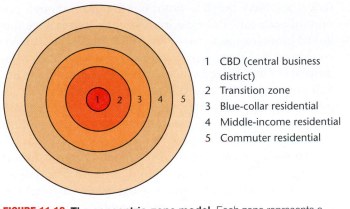

FIGURE 11.18 **The concentric-zone model.** Each zone represents a different type of land use in the city.

1 CBD (central business district)
2 Transition zone
3 Blue-collar residential
4 Middle-income residential
5 Commuter residential

THINKING GEOGRAPHICALLY Can you identify examples of each zone in your community?

about the diverse human mosaic, geographers have tried to describe some spatial models of land use within cities, each of which helps us understand a particular type of city. Let us remember, though, that these models were constructed to examine single cities and do not necessarily apply to the metropolitan coalescences so common in the world today.

Concentric-Zone Model

The **concentric-zone model** was developed in 1925 by Ernest W. Burgess, a sociologist at the University of Chicago.

concentric-zone model A social model developed by Ernest Burgess that depicts a city as five areas bounded by concentric rings.

Figure 11.18 shows the concentric-zone model with its five zones. Although not shown in the figure, there is a distinct pattern of income

levels from zone 1, the CBD, out to the commuter residential zone. This pattern shows that even at the beginning of the automobile age, American cities expressed a clear separation of social groups. The extension of trolley lines into the surrounding countryside had a lot to do with this pattern.

Zone 2, a transitional area between the CBD and residential zone 3, is characterized by a mixed pattern of industrial and residential land use. Rooming houses, small apartments, and tenements attract the lowest-income segment of the urban population. Often this zone includes slums and skid rows. In the past, many ethnic ghettos took root here as well. Landowners, while waiting for the CBD to reach their land, erected shoddy tenements to house a massive influx of foreign workers. An aura of uncertainty was characteristic of life in zone 2, because commercial activities rapidly displaced residents as the CBD expanded. Today, this area is often characterized by physical deterioration (**Figure 11.19**).

Zone 3, the "workingmen's quarters," is a solid blue-collar arc, located close to the factories of zones 1 and 2.

Yet zone 3 is more stable than the zone of transition around the CBD. It is often characterized by ethnic neighborhoods: blocks of immigrants who broke free from the ghettos in zone 2 and moved outward into flats or single-family dwellings. Burgess suggested that this working-class area, like the CBD, was spreading outward because of pressure from the zone of transition and because blue-collar workers demanded better housing.

Zone 4 is a middle-class area of better housing. From here, established city dwellers—many of whom moved out of the central city with the construction of the first streetcar network—commute to work in the CBD.

Zone 5, the commuters' zone, consists of higher-income families clustered in suburbs, either on the farthest extension of the trolley or on commuter railroad lines. This zone of spacious lots and large houses is the growing edge of the city. From here, the rich press outward to avoid the increasing congestion and social heterogeneity brought to their area by an expansion of zone 4.

Burgess's concentric-zone theory represented the American city in a new stage of development. Before the 1870s, an American metropolis, such as New York, was a city of mixed neighborhoods where merchants' stores and sweatshop factories were intermingled with mansions and hovels. Rich and poor, immigrant and native-born rubbed shoulders in the same neighborhoods. However, in Chicago, Burgess's hometown, something else occurred. In 1871, the Great Chicago Fire burned down the core of the city, leveling almost one-third of its buildings. As the city was rebuilt, it was influenced by late-nineteenth-century market forces: real estate speculation in the suburbs, inner-city industrial development, new streetcar systems, and the need for

FIGURE 11.19 **An abandoned building in Detroit, Michigan.** The transitional zone in the city contains vacant and deteriorated buildings. (Timothy Fadek/Corbis.)

THINKING GEOGRAPHICALLY Why has this area become a likely target for gentrification?

low-cost working-class housing. The result was more clearly demarcated social patterns than existed in other large cities. Chicago became a segregated city with a concentric pattern working its way out from the downtown in what one scholar called "rings of rising affluence." It was this rebuilt city that Burgess used as the basis for his concentric zone model.

However, as you can see from **Figure 11.20**, the actual residential map of Chicago does not exactly match the simplicity of Burgess's concentric zones. For instance, it is evident that the wealthy continue to monopolize certain high-value sites within the other rings, especially Chicago's Gold Coast along Lake Michigan on the Near North Side. According to the concentric-zone theory, this area should have been part of the zone of transition. Burgess accounted for some of these exceptions by noting how the rich tended to monopolize hills, lakes, and shorelines, whether they were close to or far from the CBD. Critics of Burgess's

model also were quick to point out that even though portions of each zone did exist in most cities, rarely were they linked in such a way as to totally surround the city. Burgess countered that there were distinct barriers, such as old industrial centers, that prevented the completion of the arc. Still other critics felt that Burgess, as a sociologist, overemphasized residential patterns and did not give proper credit to other land uses—such as industry, manufacturing, and warehouses—in describing urban patterns.

Though the concentric-zone model retains its historical legacy as a key contribution to theories of urban development, contemporary urban geographers highlight its failure to accurately describe many modern cities, particularly those outside of the United States. Different historical contexts in other countries caused cities to develop in quite different ways from U.S. cities. For example, in European cities, historically, the highest-rent zones are close to city centers, while lower-rent areas are often farther away. Even in the United States, due to changes in transportation and information technology, there are often no longer clear zones of economic activity and residence. For example, telecommuters may conveniently live in many different areas of a city, and many businesses are now meeting their locational needs in the suburbs rather than in the CBD. So as businesses, retail stores, and restaurants decentralize to accommodate increasingly diverse residential patterns, the concentric-zone model becomes less applicable.

Sector Model

Homer Hoyt, an economist who studied housing data for 142 American cities, presented his **sector model** of urban land use in 1939. He maintained that high-rent residential districts (*rent* meaning capital outlay for the occupancy of space, including purchase, lease, or rent in the popular sense) were instrumental in

> **sector model** An economic model, developed by Homer Hoyt, that depicts a city as a series of sectors or wedges shaped like the slices of a pie.

shaping the land-use structure of the city. Because these areas were reinforced by transportation routes, the pattern of their development was one of sectors or wedges shaped like the slices of a pie (**Figure 11.21**), not concentric zones.

Hoyt suggested that the high-rent sector would expand according to four factors. First, a high-rent sector moves from its point of origin near the CBD along established routes of travel toward another nucleus of high-rent buildings. That is, a high-rent area directly next to the CBD will naturally head in the direction of a high-rent suburb, eventually linking the two in a wedge-shaped sector. Second, this sector will progress toward high ground or along waterfronts when these areas are not used for industry. The rich have always preferred such environments for their residences. Third, a high-rent sector will move along the route of fastest transportation. Fourth, it will move toward

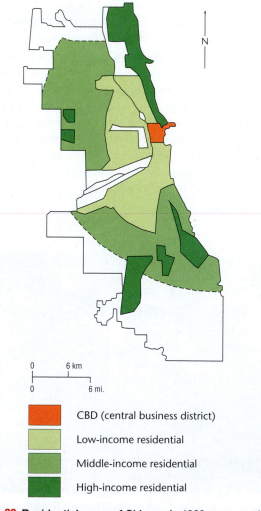

0   6 km

0   6 mi.

■ CBD (central business district)

□ Low-income residential

■ Middle-income residential

■ High-income residential

FIGURE 11.20 Residential areas of Chicago in 1920 were used as the basis for many studies and models of the city.

THINKING GEOGRAPHICALLY Compare this pattern with the concentric-zone and sector models.

transportation arteries, which are generally freeways, we see that the areas surrounding them are often low-rent districts. According to Hoyt's theory, they should be high-rent districts. Freeways are rather recent additions to the city, coming only after World War II, which were imposed on an existing urban pattern. To minimize the economic and political costs of construction, they were often built through low-rent areas, where the costs of land purchase for the rights-of-way were less and where political opposition was kept to a minimum because most people living in these low-rent areas had little political clout. This is why so many freeways rip through ethnic ghettos and low-income areas. Economically speaking, this is the least expensive route.

As with the concentric-zone model, emerging patterns of telecommuting and the growth of suburban "commuter villages" have reduced the applicability of the sector model to modern urban development. In fact, rather than following the sector-type pattern that Hoyt predicted, "leapfrog" urban development has occurred in many areas, resulting in districts of economic activity and residential status that are quite irregular in pattern. Furthermore, following World War II, many cities around the world began to adopt urban planning policies that restricted certain land uses to particular areas of a city. Therefore, patterns of land use in contemporary cities have often strayed significantly from a sector pattern in order to accommodate the preservation of green areas and historical districts, for example.

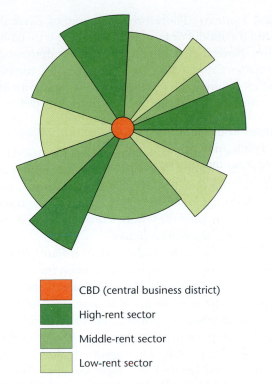

CBD (central business district)

High-rent sector

Middle-rent sector

Low-rent sector

FIGURE 11.21   The sector model. In this model, zones are pie-shaped wedges radiating along main transportation routes.

THINKING GEOGRAPHICALLY Does this model reflect reality more than the concentric-zone model? How?

open space. A high-income community rarely moves into an occupied lower-income neighborhood. Instead, the wealthy prefer to build new structures on vacant land where they can control the social environment.

As high-rent sectors develop, the areas between them are filled in. Middle-rent areas move directly next to them, drawing on their prestige. Low-rent areas fill in the remaining areas. Thus, moving away from major routes of travel, rents go from high to low.

There are distinct patterns in today's cities that echo Hoyt's model. He had the advantage over Burgess in that he wrote later in the automobile age and could see the tremendous impact that major thoroughfares were having on cities. However, when we look at today's major

Multiple-Nuclei Model

Both Burgess and Hoyt assumed that a strong central city affected patterns throughout the urban area. However, as cities increasingly decentralized, districts developed that were not directly linked to the CBD. In 1945, two geographers, Chauncey Harris and Edward Ullman, suggested a new model: the **multiple-nuclei model.** They maintained that a city developed with equal intensity around various points, or multiple nuclei (**Figure 11.22**). In their eyes, the CBD was not the only focus of activity. Equal

multiple-nuclei model A model, developed by Chauncey Harris and Edward Ullman, that depicts a city growing from several separate focal points.

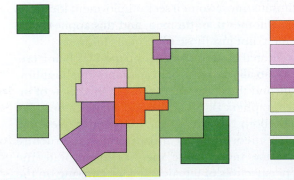

CBD (central business district)

Light industry and warehouses

Heavy industry

Low-rent residential

Middle-rent residential

High-rent residential

FIGURE 11.22   The multiple-nuclei model. This model was devised to show that the CBD is not the sole force in creating land-use patterns within the city. Rather, land-use districts may evolve for specific reasons at specific points elsewhere in the city—hence the name *multiple nuclei.*

THINKING GEOGRAPHICALLY Compare this model to Figure 11.21. How and why are they different?

weight had to be given to an old community on the city outskirts around which new suburban developments clustered; to an industrial district that grew from an original waterfront location; or to a low-income area that developed because of some social stigma attached to the site.

Harris and Ullman rooted their model in four geographical principles. First, certain activities require highly specialized facilities, such as accessible transportation for a factory or large areas of open land for a housing tract. Second, certain activities cluster because they profit from mutual association. Car dealers, for example, are commonly located near one another because automobiles are very expensive and so people will engage in comparative shopping—moving from one dealer to another until their decisions are made. Third, certain activities repel each other and will not be found in the same area. Examples would be high-rent residences and industrial areas, or slums and expensive retail stores. Fourth, certain activities could not make a profit if they paid the high rent of the most desirable locations and would therefore seek lower-rent areas. For example, furniture stores may like to locate where pedestrian traffic is greatest to lure the most people into their showrooms. However, they need large amounts of space for showrooms and storage. Thus, they cannot afford the high rents that the most accessible locations demand. They compromise by finding an area of lower rent that is still relatively accessible.

The multiple-nuclei model, more than the other models, seems to take into account the varied factors of decentralization in the structure of the North American city. Many geographers criticize the concentric-zone and sector models as being rather simplistic, for they emphasize a single factor (residential differentiation in the concentric-zone model and rent in the sector model) to explain the pattern of the city. The multiple-nuclei model encompasses a larger spectrum of economic and social factors. Harris and Ullman could probably accommodate the variety of forces working on the city because they did not confine themselves to seeking simply a social or economic explanation.

The multiple-nuclei model has three additional strengths. First, it takes into account the fragmentation of urban areas so common in many developed countries today. Second, it considers the process of suburbanization. Third, the multiple-nuclei model can take into account specialized economic functions such as heavy versus light manufacturing. The multiple-nuclei model's ability to incorporate these contemporary realities increases the applicability of Harris and Ullman's work to many modern cities. However, though the multiple-nuclei model integrates many different elements of culture and economics, caution must always be used when applying it to real-world situations. Cities within the United States and around the world have developed differently in many historical and cultural contexts. Therefore, the study of cities through *any* land-use model or models must take this variability into account. Nevertheless, due to its acknowledgment of modern transportation, technology, and lifestyles, the multiple-nuclei model may offer greater insight into contemporary urban studies than earlier models of the city did.

Critiques of the Models

Most of the criticisms of the models just discussed focus on their simplification of reality or their inability to account for all the complexities of actual urban forms. More recently, feminist geographers have noticed some flaws in the models and in how they were constructed that call into question their descriptive power.

All three models assume that urban patterns are shaped by an economic trade-off between the desire to live in a suburban neighborhood appropriate to one's economic status and the need to live relatively close to the central city for employment opportunities. These models assume that only one person in the family is a wage worker—the male head of the family. They ignore dual-income families and households headed by single women, who contend with a larger array of factors in making locational decisions, including distances to child-care and school facilities and other services important for other members of a family. For many of these households, the traditional urban models that assume a spatial separation of workplace and home are no longer appropriate.

For example, a study of the activity patterns of working parents shows that women living in a city have access to a wider array of employment opportunities and are better able to combine domestic and wage labor than are women who live in the suburbs. Many of these middle-class women will choose to live in a gentrified inner-city location, hoping that this type of area will offer the amenities of the suburbs (good schools and safety) while also accommodating their work schedules. Other research has shown that some businesses will locate their offices in the suburbs because they rely on the labor of highly educated, middle-class women who are spatially constrained by their domestic work. As geographers Susan Hanson and Geraldine Pratt found in their study of employment practices and gender in Worcester, Massachusetts, most women seek employment locations closer to their homes than do men, and this applies to almost all women, not just those with small children.

The traditional models are also criticized for being created by men who all shared certain assumptions about how cities operate and thus presented a very limited view of urban life. Geographers David Sibley and Emily Gilbert, for instance, have both brought to our attention the development of other theories about urban form and structure during the same time. These theories incorporate the alternative perspectives of female scholars. Drawing on the urban reform work done by Jane Addams at Hull House in Chicago, scholars in the first decades of the twentieth century examined the causes of and possible solutions to urban problems. For example, Edith Abbott, Sophonisba Breckinridge, and Helen Rankin Jeter, faculty at the School of Social Service Administration at the University of Chicago, worked with their mostly female students to produce a number of studies about "race," ethnicity, class, and housing in Chicago. These studies differed in several ways from those of such theorists as Burgess and Hoyt. For instance, they emphasized the role of landlords in shaping the housing market and included an awareness of how racism is related to the allocation of housing and a sensitivity to the different urban experiences of ethnic groups.

Much of what these researchers at the School of Social Service uncovered in the 1930s is applicable to urban areas today. For example, a study by urban historian Raymond Mohl chronicles the making of black ghettos in Miami between 1940 and 1960. His research reveals the role of public policy decisions, landlordism, and discrimination in that process—forces identified by Abbott and others that continue to operate today.

URBAN **LANDSCAPES**

What do the urban patterns we have been discussing look like? How can we recognize different types of cities from their three-dimensional forms, and how are these forms changing? Cities, like all places humans inhabit, demonstrate an intriguing array of cultural landscapes, the reading of which gives varied insights into the complicated interactions between people and their surroundings. In this section, we offer some thoughts about how to view and read urban landscapes. We begin by discussing some geographical reference points—one might say helpful hints—for investigating

> **cityscape** An urban landscape.

and reading **cityscapes.** You will find that you have a great deal of intuitive knowledge about cityscapes.

Ways of Reading Cityscapes

What do cities look like? Understanding urban landscapes requires an appreciation of large-scale urbanizing processes and local urban environments both past and present. The patterns we see today in the city, such as building forms, architecture, street plans, and land use, are a composite of past and present cultures. They reflect the needs, ideas, technology, and institutions of human occupancy. Two concepts underlie our examination of urban landscapes. The first is **urban morphology,** or

> **urban morphology** The form and structure of cities, including street patterns, building sizes and shapes, architecture, and density.

the physical form of the city, which consists of street patterns, building sizes and shapes, architecture, and density. The second concept is **functional zonation,** which refers to the pattern of land uses within a city or, put another way, the existence of areas with differing functions, such as residential, commercial, and governmental. Functional zonation also includes social patterns—whether, for example, an area is occupied by the power elite or by people of low status, by Jews or by Christians, by the wealthy or by the poor. Both concepts are central to understanding the cultural landscape of cities, because both make statements about how cultures occupy and shape space.

> **functional zonation** The pattern of land uses within a city; the existence of areas with differing functions, such as residential, commercial, and governmental.

Cultural geographers look to cityscapes for many different kinds of information. Here we discuss four interconnected themes that are commonly used as organizational frameworks for landscape research (**Figure 11.23**).

FIGURE 11.23 Boston's central city. There are various ways of looking at cityscapes: as indicators of change, as palimpsests, as expressions of visual biases, and as manifestations of symbolic traditions. This photo offers evidence of all approaches. (Steve Dunwell/The Image Bank.)

THINKING GEOGRAPHICALLY Which clues would you select to illustrate each cityscape theme?

Landscape Dynamics Think of some familiar features of the cityscape: downtown activities creeping into residential areas, deteriorated farmland on a city's outskirts, older buildings demolished for the new. These are all signs of specific processes that create urban change; the landscape faithfully reflects these dynamics.

When these visual clues are systematically mapped and analyzed, they offer evidence for the currents of change expressed in our cities. Of equal interest is where change is *not* occurring—those parts of the city that for various reasons remain relatively static. An unchanging landscape also conveys an important message. Perhaps that part of the city is stagnant because it is removed from the forces that produce change in other parts. Or perhaps there is a conscious attempt by local residents to inhibit change—to preserve open space by resisting suburban development, for example, or to preserve a historic landmark. Documenting landscape changes over time gives valuable insight into the paths of settlement development.

The City as Palimpsest Because cityscapes change, they offer a rich field for uncovering remnants of the past. A **palimpsest** is an old parchment used repeatedly for written messages. Before a new missive was written, the old was erased, yet rarely were all the previous characters and words obliterated—so remnants of earlier messages were still visible. This record of old and new is called a *palimpsest*, a word geographers use fondly to describe the visual mixture of past and present in cultural landscapes.

> **palimpsest** A term used to describe cultural landscapes with various layers and historical messages. Geographers use this term to reinforce the notion of the landscape as a text that can be read; a landscape palimpsest has elements of both modern and past periods.

Cities are full of palimpsestic offerings, scattered across the contemporary landscape. How often have you noticed a Victorian farmhouse surrounded by new tract homes, or a historic street pattern obscured or highlighted by a recent urban redevelopment project, or a brick factory shadowed by new high-rise office buildings? All of these give clues to past settlement patterns, and all are mute testimony to the processes of change in the city.

Our interest in this historical accumulation is more than romantic nostalgia. A systematic collection of these urban remnants provides us with glimpses of the past that might otherwise be hidden. All societies pick and choose, consciously or not, what they wish to preserve for future generations, and, in this process, a filtering takes place that often excludes and distorts information, but the landscape does not lie.

Symbolic Cityscapes Landscapes contain much more than literal messages. They are also loaded with figurative or metaphorical meaning and can elicit emotions and memories. To some people, skyscrapers are more than high-rise office buildings: they are symbols of progress, economic vitality, downtown renewal, or corporate identities. Similarly, historical landscapes—those parts of the city where the past has been preserved—help people to define themselves in time; establish social continuity with the past; and codify a largely forgotten, yet sometimes idealized, past.

D. W. Meinig, a geographer who has given much thought to urban landscapes, maintains that there are three highly symbolic townscapes in the United States: the New England village, with its white church, common, and tree-lined neighborhoods; Main Street of Middle America, a string street of a small midwestern town, with storefronts, bandstand, and park (**Figure 11.24**); and what Meinig calls

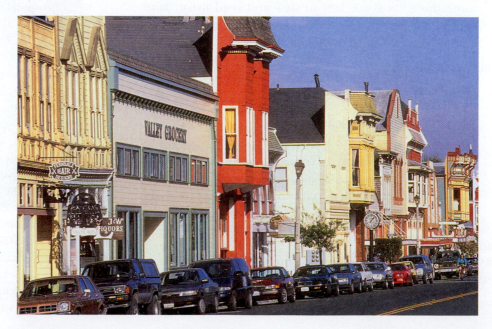

FIGURE 11.24 Main Street, Ferndale, California. The symbolism of Main Street, USA, is a powerful force that still plays a role in shaping communities today. (ChromoSohm/Sohm/Photo Researchers, Inc.)

THINKING GEOGRAPHICALLY In what ways is Main Street used symbolically in art, literature, film, and television? What messages and emotions are conveyed by this landscape?

California Suburbia, suburbs of quarter-acre lots, effusive garden landscaping, swimming pools, and ranch-style houses. As Meinig explains: "Each is based upon an actual landscape of a particular region. Each is an image derived from our national experience...simplified...and widely advertised so as to become a commonly understood symbol. Each has...influenced the shaping of the American scene over broader areas."

More politically and problematically, the cultural landscape is an important vehicle for constructing and maintaining, subtly and implicitly, certain social and ethnic distinctions. For example, geographers James and Nancy Duncan have found that because conspicuous consumption is a major way of conveying social identity in our culture, elite landscapes are created through large-lot zoning, imitation country estates, and the preservation of undeveloped land. They see the residential landscapes in upper-income areas as controlled and managed to reinforce class and status categories. Their study of elite suburbs near Vancouver and New York sensitizes us to how the cultural landscape can be thought of as a repository of symbols used by our society to differentiate itself and protect vested interests.

Perception of the City During the last 25 years, geographers have been concerned with measuring people's perceptions of the urban landscape. They assume that if we really know what people see and react to in the city, we can ask architects and urban planners to design and create a more humane urban environment.

Kevin Lynch, an urban designer, pioneered a method for recording people's images of the city. On the basis of interviews conducted in Boston, Jersey City (New Jersey), and Los Angeles, Lynch suggested five important elements in mental maps (images) of cities:

1. *Pathways* are the routes of frequent travel, such as streets, freeways, and transit corridors. We experience the city from the pathways, and they become the threads that hold our maps together.

2. *Edges* are boundaries between areas or the outer limits of our image. Mountains, rivers, shorelines, and even major streets and freeways are commonly used as edges. They tend to define the extremes of our urban vision. Then we fill in the details.

3. *Nodes* are strategic junction points, such as breaks in transportation, traffic circles, or any place where important pathways come together.

4. *Districts* are small areas with a common identity, such as ethnic areas and functional zones (for instance, the CBD or a row of car dealers).

5. *Landmarks* are reference points that stand out because of shape, height, color, or historical importance. The city hall in Los Angeles, the Washington Monument, and the golden arches of a McDonald's are all landmarks.

Using these concepts, Lynch saw that some parts of the cities were more **legible,** or easier to decipher, than others. Lynch discovered that in general legibility increases when the urban landscape offers clear pathways, nodes, districts, edges, and landmarks. Further, some cities are more legible than others. For example, Lynch found that Jersey City is not very legible. Wedged between New York City and Newark, Jersey City is fragmented by railroads and highways. Residents' mental maps of Jersey City have large blank areas in them. When questioned, they can think of few local landmarks. Instead, they tend to point to the New York City skyline just across the Hudson River.

> **legible** (A city that is) easy to decipher, with clear pathways, edges, nodes, districts, and landmarks.

Megalopolis and Edge Cities

In the nineteenth century, U.S. and Canadian cities grew at unprecedented rates because of the concentration of people and commerce. The inner city became increasingly dominated by commerce and the working class. In the twentieth century, particularly after World War II, new forms of transportation and communication led to the decentralization of many urban functions. As a result, one metropolitan area blends into another, until supercities stretch for hundreds of miles. The geographer Jean Gottmann coined the term **megalopolis** to describe these supercities.

> **megalopolis** A large urban region formed as several urban areas spread and merge, such as Boswash, the region including Boston, New York, Philadelphia, and Washington, D.C.

This term is now used worldwide in reference to giant metropolitan regions such as Boston–New York–Philadelphia–Washington, D.C. in the United States and Tokyo-Yokohama in Japan. These urban regions are characterized by high population densities extending over hundreds of square miles or kilometers; concentrations of numerous older cities; transportation links formed by freeways, railroads, air routes, and rapid transit; and an extremely high proportion of the nation's wealth, commerce, and political power.

The past 25 years have witnessed an explosion in metropolitan growth in areas that had once been peripheral to the central city. Many of the so-called bedroom communities of the post–World War II era have been transformed into urban centers, with their own retail, financial, and entertainment districts (**Figure 11.25** on page 324). Author Joel Garreau refers to these new centers of urban activity, which surround nineteenth-century downtowns,

FIGURE 11.25 Edge city. These new centers of economic activity are located on the edges of traditional downtowns. (Damir Frkovic/Masterfile.)

THINKING GEOGRAPHICALLY Why have such developments grown up along circumferential highways and interstates?

as **edge cities,** although many other terms have been used in the past, including *suburban downtowns, galactic cities,* and *urban villages.* As Garreau mentions, most Americans now live, work, play, worship, and study in this type of settlement. What differentiates an edge city from the suburbs is that it is a place of work, of productive economic activity, and therefore is the destination of many commuters. In fact, the conventional work commute from the suburbs to the inner city has been replaced by commuting patterns that completely encircle the inner city. People live in one part of an edge city and commute to their workplace in another part of that city.

This commuting pattern, now common in many North American urban areas, is explained by the **peripheral model.** First conceived by Chauncey Harris (co-creator of the multiple-nuclei

edge city A new urban cluster of economic activity that surrounds a nineteenth-century downtown.

peripheral model A geographic model created by Chauncey Harris that explains how suburban edge cities are linked by ring roads or beltways that allow for commuter traffic to conveniently avoid passing through major urban centers.

model), the peripheral model describes how large suburban residential and business areas are functionally tied together by beltways or ring roads. These transportation corridors allow residents to commute within or between edge cities without ever having to pass through the major city center nearby. Though convenient for suburban commuters and advantageous for businesses located in edge cities, this pattern of suburban transportation flow brings with it some disadvantages as well. The cost of constructing new roads through and between edge cities is significant, and green spaces and agricultural lands are often lost in the process. In European suburbs, these issues have been addressed by planning for circular "greenbelts" of open land between roadways and for the preservation of agricultural lands.

The New Urban and Suburban Landscapes

Within the past 30 to 35 years, our cityscapes have undergone massive transformations. The impact of suburbanization and decentralization has led to edge cities, while in older downtown areas, we have seen that gentrification, redevelopment, and immigration have also created novel urban forms. What we see in today's cityscapes, then, is a new urban landscape, composed of distinctive elements that we now will discuss.

Shopping Malls Many people consider the image of the shopping mall, surrounded by mass-produced suburbs, to be one of the most distinctive landscape symbols of modern urbanity. Yet, oddly, most malls are not designed to be seen from the outside. As a matter of fact, without appropriate signs, a passerby could proceed past a mall without noticing any visual display. Unlike the retail districts of nineteenth- and early-twentieth-century cities, where grand architectural displays along major boulevards were the norm, shopping malls are enclosed, private worlds that are meant to be seen from the inside. Often located near an off-ramp of a major freeway or beltway of a metropolitan area and close to the middle- and upper-class residential neighborhoods, shopping malls can be distinguished more by their extensive parking lots than by their architectural design.

For all that, shopping malls do have a characteristic form. The early malls of the 1960s tended to have a simple, linear form, with department stores at both ends that functioned as anchors and 20 to 30 smaller shops connecting the two ends. In the 1970s and 1980s, much larger malls were built, and their form became more complex.

Malls today are often several stories high and may contain five or six anchor stores and up to 400 smaller shops. In addition, many malls now serve more than a retail function—they often contain food courts and restaurants, professional offices, movie complexes, hotels, chapels, and amusement arcades and centers. A shopping center in Kuala Lumpur, known simply as The Mall, contains Malaysia's largest indoor amusement park, a replica of the historic city of Malacca, a cineplex, and a food court, in addition to hundreds of stores. Many scholars consider these megamalls to be the new centers of urban life because they seem to be the major sites for social interaction. For example, after workplace, home, and school, the shopping mall is where most Americans spend their time.

However, unlike the open-air marketplaces of an earlier era, shopping malls are private, not public, spaces. The use of the shopping mall as a place for social interaction, therefore, is always of secondary importance to its private, commercial function. If a certain group of people were considered a nuisance to shoppers, the mall owners could prevent them from what Jeffrey Hopkins calls "'mallingering'—the act of lingering about a mall for economic and noneconomic social purposes."

Office Parks With the connection of metropolitan areas by major interstate highways and with the development of new communication technologies, office buildings no longer have to be located in the central city. Cheaper rent in suburban locations, combined with the convenience of easy-access parking and the privacy of a separate location, has led to the construction of **office parks** throughout suburban America.

Many of these office parks are occupied by the regional or national headquarters of large corporations or by local sales and professional offices. To take advantage of economies of scale, many of these

office park A cluster of office buildings, usually located along an interstate highway, often forming the nucleus of an edge city.

offices will cluster together and rent or buy space from a land development company. These clusters of office buildings are usually located along an interstate highway and often form the nuclei of edge cities.

The use of the word *park* to identify this new landscape element points to the consciously antiurban imagery of these complexes. Many of these developments are surrounded by a well-landscaped outdoor space, often incorporating artificial lakes and waterfalls (**Figure 11.26**). Jogging paths, fitness trails, and picnic tables all cater to the new lifestyle of professional and managerial employees.

Many office parks are located along what have been called **high-tech corridors**: areas along a limited-access highway that contain offices and other services associated with high-tech industries. As our discussion of edge cities suggests, this new type of commercial landscape is gradually replacing our downtowns as the workplace for most Americans.

high-tech corridor An area along a limited-access highway that contains offices and other services associated with high-tech industries.

Master-Planned Communities Many newer residential developments on the suburban fringe are planned and built as complete neighborhoods by private development companies. These **master-planned communities** include not only architecturally compatible housing units but also recreational facilities (such as tennis courts, fitness centers, bike paths, and swimming pools), schools, and security measures (gated or guarded entrances).

master-planned community A large-scale residential development that includes, in addition to architecturally compatible housing units, planned recreational facilities, schools, and security measures.

FIGURE 11.26 A suburban office building. Compare this suburban office building in Yonkers, New York, with your image of downtown skyscrapers. Notice the green space around the building and its horizontal rather than vertical appearance. (Michael Melford/The Image Bank.)

THINKING GEOGRAPHICALLY Why do you rarely find very tall office buildings in the suburbs?

In Weston, a master-planned community that covers approximately 10,000 acres (4000 hectares) in southern Florida, land use is completely regulated not only within the gated residential complexes but also along the road system that connects Weston to the interstate highway. Shrubbery is planted strategically to shield residents from views of the roadway, and the road signs are uniform in style and encased in stylish gray weathered-wood frames. This massive community—Weston is now home to 50,000 people—contains various complexes catering to particular lifestyles, ranging from smaller patio homes to equestrian estates. In the mid-1990s, homes in one such complex—Tequesta Point—cost $250,000 to $300,000 and came with gated entranceways, split-level floor plans, and Roman bathtubs. Those in Bermuda Springs, on the other hand, which originally cost $115,000 to $120,000, were significantly smaller and did not offer the same interior features. Typical of master-planned communities, the name of the development itself was chosen to convey a hometown feeling. Developers carried out an extensive marketing survey before they settled on the name Weston.

Festival Settings   In many cities, gentrification efforts focus on a multiuse redevelopment scheme that is built around a particular setting, often one with a historical association. Waterfronts are commonly chosen as focal points for these large-scale projects, which Paul Knox has referred to as **festival settings.** These complexes integrate retailing, office, and entertainment facilities and incorporate trendy shops, restaurants, bars and nightclubs, and hotels. Knox suggests that these developments are "distinctive as new landscape elements merely because of their scale and their consequent ability to stage—or merely to be—the spectacular." Such festival settings as Faneuil Hall in Boston, Bayside in Miami, and Riverwalk in San Antonio serve as sites for concerts, ethnic festivals, and street performances; they also serve as focal points for the more informal human interactions that we usually associate with urban life (**Figure 11.27**). In this sense, festival settings do perform a vital function in the attempt to revitalize our downtowns. Yet, like many other gentrification efforts, these massive displays of wealth and consumption often stand in direct contrast to neighboring areas of the inner city that have received little, if any, monetary or other social benefit from these projects.

"Militarized" Space and Gated Communities   Considered together, these new elements in the urban landscape suggest some trends that many scholars find disturbing. Urbanist Mike Davis has called one such trend the "militarization" of urban space, meaning that increasingly space is used to set up defenses against people the city considers undesirable. This includes landscape developments that range from the lack of street furniture to guard against the homeless living on the streets to gated and guarded residential communities (**Figure 11.28**) to the complete segregation of classes and "races" within the city. Particularly in downtown redevelopment schemes, the goal of

> **festival setting**   A multiuse redevelopment project built around a particular setting, often a waterfront or a location with a historical association.

FIGURE 11.27 Quincy Market, an early-nineteenth-century wholesale market near Faneuil Hall in Boston. The market is re-created as a festival setting, complete with upscale shops, restaurants, and bars. (Ulrike Welsch/Photo Researchers.)

THINKING GEOGRAPHICALLY   In what ways do you think this cultural landscape differs from the original Quincy Market in the early nineteenth century?

FIGURE 11.28 A gated community in China.
Gated communities in China are common as signs of
status and for security. (Macduff Everton/Corbis.)

THINKING GEOGRAPHICALLY Why does this community
look from the outside so similar to what you would see in a
North American city?

city planners and others is usually to provide safe and homogeneous environments, segregated from the diversity
of cultures and lifestyles that often characterizes the
central area of most cities. As Davis says, "Cities of all sizes
are rushing to apply and profit from a formula that links
together clustered development, social homogeneity, and
a perception of security." Although this "militarization" is
not completely new, it has taken on epic proportions as
whole sections of such cities as Los Angeles, Atlanta, Dallas, Houston, and Miami have become "militarized" spaces
(**Figure 11.29**).

Decline of Public Space Related to the increase in "militarized" space is the decline of public spaces in most of our
cities. For example, the change in shopping patterns from
the downtown retailing area to the suburbs indicates a
change of emphasis from the public space of city streets to
the privately controlled and operated shopping malls. Similarly, many city governments, often joined by private developers, have built enclosed walkways either above or below
the city streets. These walkways serve partly to provide a
climate-controlled means of passing from one building to
another and partly to provide pedestrians with a "safe" environment that avoids possible confrontations on the
street. Again, the public space of the street is being replaced by controlled spaces that do not provide the same
access to all members of the urban community. It remains
to be seen whether this trend will continue or whether new
public spaces will accommodate all groups of people within
the urban landscape.

FIGURE 11.29 **The Metro Dade Cultural Center in downtown
Miami.** Inside this literal fortress are Dade County's public library and
Center for the Fine Arts. The building was designed to be completely
removed from the street, with the few entrances heavily monitored by
cameras. (Courtesy of Mona Domosh.)

THINKING GEOGRAPHICALLY Does this seem like a welcoming space?
Should it be?

Key Terms

The Urban Mosaic on the Internet

You can learn more about cities and suburbs on the Internet at the following web sites:

The Centre for Urban History at the University of Leicester

http://www.le.ac.uk/ur/index.html

This site is a wonderful portal for researching the history of urban form online.

Levittown: Documents of an Ideal American Suburb

http://www.uic.edu/~pbhales/Levittown

Here you can find a history of Levittown, New York, the first mass-produced suburb, with interesting historical and contemporary photographs.

The Skyscraper Museum

http://www.skyscraper.org/home.htm

The Skyscraper Museum in New York City runs this site and provides extremely useful information and virtual tours of tall buildings around the world.

The State of the Nation's Cities: A Comprehensive Database on American Cities and Suburbs

http://policy.rutgers.edu/cupr/sonc/sonc.htm

This site is a comprehensive database on 77 American cities and suburbs that you can download onto your computer and display in graphic form.

Recommended Books on the Urban Mosaic

Blunt, Alison, and Robyn Dowling. 2006. *Home.* New York: Routledge. An insightful and concise examination of the concept and everyday experiences of home.

Crutcher, Michael E., Jr. 2010. *Tremé: Race and Place in a New Orleans Neighborhood.* Athens: University of Georgia Press. A case study of a neighborhood in New Orleans where issues of race and gentrification come to the fore.

Ford, Larry. 2003. *America's New Downtowns: Revitalization or Reinvention?* Baltimore: Johns Hopkins Press. An interesting assessment of 16 contemporary American downtowns.

Harvey, David. 2003. *Paris, Capital of Modernity.* New York: Routledge, 2003. A revealing look at Paris during its mid-nineteenth-century political turmoil.

Hayden, Dolores. 2003. *Building Suburbia: Green Fields and Urban Growth, 1820–2000.* New York: Pantheon. A critical historical treatment of American suburban development.

Lees, Loretta, Tom Slater, and Elvin Wylie. 2007. *Gentrification.* New York: Routledge. An overview of the major theoretical and conceptual frameworks for understanding gentrification.

Marshall, Richard. 2002. *Global Urban Projects in the Asia Pacific Rim.* New York: Routledge. A fascinating and well-illustrated look at how global cities in the Asia Pacific Rim are creating and embodying a new urban vision.

Pattilo, Mary. 2007. *Black on the Block: The Politics of Race and Class in the City.* Chicago: University of Chicago Press. An in-depth examination of the many conflicts faced by residents in an African-American Chicago community undergoing gentrification.

Answers to Thinking Geographically Questions

Figure 11.1: The architecture is even more flamboyant, with the tower (Pearl of the Pacific) and globe-shaped buildings. The buildings are all very sleek and modern, like new skyscrapers in North American cities.

Figure 11.2: Money was invested in suburbs rather than inner cities, including subsidies for mortgages so that people could buy homes, construction of freeways and expressways, and other infrastructure. It was therefore not invested in the inner city, which was allowed to decay.

Figure 11.3: Families are likely to be wealthier and more highly educated, and most probably have children, although relatively few of them.

Figure 11.4: Woodside today would have considerably more diversity of ethnic groups. Some of the immigrants are probably professionals. The mixed population may still consider themselves a neighborhood. Early Irish neighborhoods were more homogeneous, establishing their own churches and other organizations that defined the community.

Figure 11.5: A great deal of information is probably exchanged about sources of help with various needs, including food, shelter, and medical care. Friendship ties are also formed. Unfortunately, with the high rate of substance abuse among the homeless, such encampments are probably also sources of drugs or information about where drugs may be obtained.

Figure 11.6: Although many suburban areas have become quite exclusive and are home to affluent residents, historically prestigious downtown districts retain their advantageous locations near the most elite shopping, dining, and cultural activities that cities have to offer. Subsequently, many affluent residents choose to remain in these downtown areas rather than relocate to more private and secluded suburban locations.

Figure 11.7: Answers will vary.

Figure 11.8: In times when city residents depended on foot travel or public transportation, such housing was efficient use of land and provided easy access to transit lines, such as street cars.

Figure 11.9: It is based on dead-end cul-de-sacs surrounded by looping streets that usually do not front on major arterial roads. While this pattern may reduce through traffic, it makes provision of services, especially emergency police and fire response, difficult. Some communities are now prohibiting cul-de-sacs for that reason.

Figure 11.10: Historic neighborhoods are more likely to be located near downtown; neighborhoods farther out would be newer. People interested in such preservation are likely to have a relatively high income (and perhaps no children) and want to live near downtown jobs and amenities.

Figure 11.11: Apart from a few architectural details (the style of windows, perhaps), it looks similar to projects in North American cities. Living in such converted buildings ("lofts") is very chic in some circles.

Figure 11.12: All these cities are centers of high-tech industry. One of the largest groups of immigrants to the United States is workers in high-tech industries. Other jobs would also open up in such cities, based on the incomes of the high-tech workers.

Figure 11.13: Like traditional neighborhoods, this shopping center probably has stores that sell goods of interest to Vietnamese, such as foods and Vietnamese-language videos. It may also have a community center and places of worship. However, it is farther from the immigrants' residences, which may be widely scattered.

Figure 11.14: These ethnic enclaves originate because people move long distances across national boundaries. They also maintain ties with their former homes, often sending back money (remittances).

Figure 11.15: Apart from the information given that this is built on fill, the streets are long, the yards and houses large, and each house backs onto open land, which would command a premium.

Figure 11.16: Apart from a variation in land use (e.g., a factory district), westerly winds may blow the warmer air eastward.

Figure 11.17: There is an effect, but it is slightly shifted from the land use, probably by winds from the southwest.

Figure 11.18: Answers will vary, but most towns have some kind of CBD and residential areas with a range of income levels.

Figure 11.19: The buildings are of an elegant architectural style and are probably basically solidly built. They are near cultural amenities and the lakefront, attractive locations for young professionals.

Figure 11.20: All are focused on a CBD (although because of the lake, Chicago includes only half a circle instead of a complete circle). Higher-income residential areas tend to be on the outskirts as in the concentric-zone model, but as with the sector model, there is a wedge of high-income housing along the lakefront north and south of the CBD. Low-income areas surround the CBD in reality and in both models.

Figure 11.21: It reflects both high- and low-income sectors that follow transportation routes. Along railroads would be industrial zones that would have working-class residential areas nearby.

Figure 11.22: The sector model revolves around a single CBD, with sectors showing arterial routes to and from that center. The multiple-nuclei model also shows industrial and warehousing districts and outlying suburban development.

Figure 11.23: The evidence of change is in the skyscrapers of varying dates; visual biases might focus on how the parks (Boston Common and the Public Garden) take up most of the picture. Manifestations of symbolic traditions are apparent in the statehouse (upper left corner of the park) and the historic area of the North End (barely visible in the upper left corner of the picture).

Figure 11.24: Among the uses in literature are Sinclair Lewis's novel *Main Street* and numerous others that portray small-town life. Many paintings depict Main Streets, and 1950s situation comedies on television, such as the *Andy Griffith Show* and *Leave It to Beaver*, showed Main Street. It conveys nostalgia for small-town life with a close-knit community and families.

Figure 11.25: Not only do they share such facilities as tables, chairs, and waste disposal systems, but the agglomeration benefits all of them for marketing purposes.

Figure 11.26: Ultimately, zoning laws prohibit such structures, but that is based on the desire to keep buildings low and unobtrusive. Businesses also like to have as much as possible of their activity on one level.

Figure 11.27: There would have been no colored banners, no globe lights and other fancy light fixtures, and no tents. Signs would have been small.

Figure 11.28: Things Western, and especially American, are seen as reflecting wealth and high status in China.

Figure 11.29: Although it may have been designed to protect the workers and visitors against criminal attack, it certainly does not present a very welcoming appearance.

GLOSSARY

absolute population density A country's population divided by its total land area. Also known as arithmetic population density. [Chapter 3, p. 72]

absorbing barrier A barrier that completely halts diffusion of innovations and blocks the spread of cultural elements. [Chapter 1, p. 12]

acculturation The adoption by an ethnic group of enough of the ways of the host society to be able to function economically and socially. [Chapter 5, p. 123]

acid rain Rainfall with much higher acidity than normal, caused by sulfur and nitrogen oxides derived from the burning of fossil fuels being flushed from the atmosphere by precipitation, with lethal effects for many plants and animals. [Chapter 9, p. 262]

adaptive strategy The unique way in which each culture uses its particular physical environment; those aspects of culture that serve to provide the necessities of life—food, clothing, shelter, and defense. [Chapter 1, p. 17]

agglomeration A snowballing geographical process by which secondary and service industrial activities become clustered in cities and compact industrial regions in order to share infrastructure and markets. [Chapter 11, p. 306]

agribusiness Highly mechanized large-scale farming, usually under corporate ownership. [Chapter 8, p. 222]

agricultural landscape The cultural landscape of agricultural areas. [Chapter 8, p. 242]

agricultural region A geographic region defined by a distinctive combination of physical environmental conditions; crop type; settlement patterns; and labor, cultivation, and harvesting practices. [Chapter 8, p. 214]

agricultural surplus The amount of food grown by a society that exceeds the demands of its population. [Chapter 10, p. 281]

agriculture The tilling of crops and the raising of domesticated animals to produce food, feed, drink, and fiber. [Chapter 8, p. 213]

amenity landscape Landscape that is prized for its natural and cultural aesthetic qualities by the tourism and real estate industries and their customers. [Chapter 2, p. 52]

Anatolian hypothesis A theory of language diffusion holding that the movement of Indo-European languages from the area in contemporary Turkey known as Anatolia followed the spread of plant domestication technologies. [Chapter 4, p. 101]

animist An adherent of animism, the idea that souls or spirits exist not only in humans but also in animals, plants, rocks, natural phenomena such as thunder, geographic features such as mountains or rivers, or other entities of the natural environment. [Chapter 7, p. 189]

apartheid In South Africa, a policy of racial segregation and discrimination against non-European groups. [Chapter 10, p. 297]

aquaculture The cultivation and harvesting of aquatic organisms under controlled conditions. [Chapter 8, p. 223]

architectural style The exterior and interior designs and layouts of the cultural and physical landscape. [Chapter 1, p. 26]

assimilation The complete blending of an ethnic group into the host society resulting in the loss of many or all distinctive ethnic traits. [Chapter 5, p. 123]

axis mundi The symbolic center of cosmomagical cities, often demarcated by a large, vertical structure. [Chapter 10, p. 283]

barriada See *squatter settlement.* [Chapter 10, p. 295]

bilingualism The ability to speak two languages fluently. [Chapter 4, p. 96]

biofuel Broadly, any form of energy derived from biological matter, increasingly used in reference to replacements for fossil fuels in internal combustion engines, industrial processes, and the heating and cooling of buildings. [Chapter 8, p. 236]

border zone An area where different regions meet and sometimes overlap. [Chapter 1, p. 6]

buffer state An independent but small and weak country lying between two powerful countries. [Chapter 6, p. 152]

cadastral pattern The shapes formed by property borders; the pattern of land ownership. [Chapter 8, p. 242]

carrying capacity The maximum number of people who can be supported in a given area. [Chapter 3, p. 60]

census tract A small district used by the U.S. Census Bureau to survey the population. [Chapter 11, p. 303]

central business district (CBD) The central portion of a city, characterized by high-density land uses. [Chapter 11, p. 302]

central place A town or city engaged primarily in the service stages of production; a regional center. [Chapter 10, p. 293]

central-place theory A set of models designed to explain the spatial distribution of urban service centers. [Chapter 10, p. 293]

centralizing force A diffusion force that encourages people or businesses to locate in the central city. [Chapter 11, p. 305]

centrifugal force Any factor that disrupts the internal order of a country. [Chapter 6, p. 155]

centripetal force Any factor that supports the internal order of a country. [Chapter 6, p. 155]

chain migration The tendency of people to migrate along channels, over a period of time, from specific source areas to specific destinations. [Chapter 5, p. 133]

channelization A migration process in which a specific source location becomes linked to a particular destination, so that neighbors in the old place become neighbors in the new place. [Chapter 5, p. 135]

checkerboard development A mixture of farmlands and housing tracts. Also called leapfrog development. [Chapter 11, p. 309]

cityscape An urban landscape. [Chapter 11, p. 321]

cleavage model A political-geographic model suggesting that persistent regional patterns in voting behavior, sometimes leading to separatism, can usually be explained in terms of tensions pitting urban against rural, core against periphery, capitalists against workers, and power group against minority culture. [Chapter 6, p. 171]

colonialism The building and maintaining of colonies in one territory by people based elsewhere; the forceful appropriation of a territory by a distant state, often involving the displacement of indigenous populations to make way for colonial settlers. [Chapter 6, p. 152]

commemorative toponym A place name that honors a famous or important person. [Chapter 4, p. 113]

commercial agriculture The growing of crops, including nonfood crops, for sale rather than strictly for consumption by one's own family. [Chapter 8, p. 217]

concentric-zone model A social model, developed by Ernest Burgess, that depicts a city as five areas bounded by concentric rings. [Chapter 11, p. 317]

consumer nationalism Local consumers' favoring of nationally produced goods over imported goods as part of a nationalist political agenda. [Chapter 2, p. 47]

consumer services The range of services provided to the general public, including education, government, recreation and tourism, and health care. [Chapter 9, p. 254]

contact conversion The spread of religious beliefs by personal contact. [Chapter 7, p. 191]

contagious diffusion A type of expansion diffusion in which cultural innovation spreads by person-to-person contact, moving wavelike through an area and population without regard to social status. [Chapter 1, p. 11]

convergence hypothesis A hypothesis holding that cultural differences among places are being reduced by improved transportation and communications systems and networks, leading to a homogenization of popular culture. [Chapter 2, p. 47]

cool chain The refrigeration and transport technology that allows for the distribution of perishables. [Chapter 8, p. 238]

core area The territorial nucleus from which a country grows in area and over time, often containing the national capital and the main center of commerce, culture, and industry. [Chapter 6, p. 161]

core-periphery A concept based on the tendency of both formal and functional culture regions to consist of a core in which defining traits are purest or functions are headquartered, and a periphery that is tributary and displays fewer of the defining traits. [Chapter 1, p. 6]

cosmomagical city A type of city that is laid out in accordance with religious principles, characteristic of very early cities, particularly in China. [Chapter 10, p. 283]

cottage industry A traditional type of manufacturing before the industrial revolution, practiced on a small scale in individual rural households as a part-time occupation and designed to produce handmade goods for local consumption. [Chapter 9, p. 256]

creole A language derived from a pidgin language that has acquired a fuller vocabulary and become the native language of its speakers. [Chapter 4, p. 96]

crude birthrate The annual number of births per thousand population. [Chapter 3, p. 61]

crude death rate The annual deaths per thousand persons in the population. [Chapter 3, p. 63]

cultural adaptation The adaptation of humans and cultures to the challenges posed by the physical environment. [Chapter 1, p. 17]

cultural diffusion The spread of elements of culture from the point of origin over an area. [Chapter 1, p. 10]

cultural ecology Broadly defined, the study of the relationships between the physical environment and culture; narrowly (and more commonly) defined, the study of culture as an adaptive system that facilitates human adaptation to nature and environmental change. [Chapter 1, p. 16]

cultural geography The study of spatial variations among cultural groups and the spatial functioning of society. [Chapter 1, p. 3]

cultural interaction The relationship of various elements within a culture. [Chapter 1, p. 20]

cultural landscape The artificial landscape; the visible human imprint on the land. [Chapter 1, p. 22]

cultural maladaptation Poor or inadequate adaptation that occurs when a group pursues an adaptive strategy that, in the short run, fails to provide the necessities of life or, in the long run, destroys the environment that nourishes it. [Chapter 5, p. 137]

cultural preadaptation A set of adaptive traits and skills possessed in advance of migration by a group, giving it survival ability and competitive advantage in occupying the new environment. [Chapter 5, p. 136]

cultural simplification The process by which immigrant ethnic groups lose certain aspects of their traditional culture in the process of settling elsewhere, creating a new culture that is less complex than the old. [Chapter 5, p. 135]

culture A total way of life held in common by a group of people, including such learned features as speech, ideology, behavior, livelihood, technology, and government; or the local, customary way of doing things—a way of life; an ever-changing process in which a group is actively engaged; a dynamic mix of symbols, beliefs, speech, and practices. [Chapter 1, p. 2]

culture hearth A focused geographic area where important innovations are born and from which they spread. [Chapter 7, p. 190]

culture region A geographical unit based on characteristics and functions of culture. [Chapter 1, p. 4]

cybergeography A branch of geography that studies the Internet as a virtual place. Cybergeographers examine locations in cyberspace as sites of human interaction with structures that can be mapped. [Chapter 1, p. 12]

decentralizing force A diffusion force that encourages people or businesses to locate outside the central city. Also called suburbanizing force. [Chapter 11, p. 305]

defensive site A location from which a city can be easily protected from invaders. [Chapter 10, p. 289]

deindustrialization The decline of primary and secondary industry, accompanied by a rise in the service sectors of the industrial economy. [Chapters 9, 11, pp. 258, 311]

demographic transition The movement from high birth and death rates to low birth and death rates. [Chapter 3, p. 67]

dependency ratio The number of dependent individuals (those under age 15 or over age 64) in a population divided by the number of productive individuals, expressed as a percentage. [Chapter 3, p. 73]

depopulation A decrease in population that sometimes occurs as the result of sudden catastrophic events, such as natural disasters, disease epidemics, and warfare. [Chapter 3, p. 88]

descriptive toponym A place name that describes a physical feature or environmental characteristic of a place. [Chapter 4, p. 113]

desertification A process whereby human actions unintentionally turn productive lands into deserts through agricultural and pastoral misuse, destroying vegetation and soil to the point where they cannot regenerate. [Chapter 8, p. 234]

dialect Distinctive local or regional variant of a language that remains mutually intelligible to speakers of other dialects of that language. [Chapter 4, p. 95]

dispersed A settlement form in which people live relatively distant from each other. [Chapter 1, p. 25]

domesticated animal An animal that depends on people for food and shelter and that differs from wild species in physical appearance and behavior as a result of controlled breeding and frequent contact with humans. [Chapter 8, p. 228]

domesticated plant A plant deliberately planted, protected, cared for, and used by humans; it is genetically distinct from its wild ancestors as a result of selective breeding. [Chapter 8, p. 227]

double-cropping Harvesting twice a year from the same parcel of land. [Chapter 8, p. 217]

dust dome A layer of pollution over a city that is thickest at the center of the city. [Chapter 11, p. 315]

economic development The process by which an agricultural society moves toward industrialization and (usually) higher patterns of income. [Chapter 9, p. 250]

ecosystem A territorially bounded system consisting of interacting organic and inorganic components. [Chapter 1, p. 16]

ecotheology The study of the interaction of religious belief and habitat modification. [Chapter 7, p. 195]

ecotourism Responsible travel that does not harm ecosystems or the well-being of local people. [Chapter 9, p. 266]

edge city A new urban cluster of economic activity that surrounds a nineteenth-century downtown. [Chapter 11, p. 323]

electoral geography The study of the interactions among space, place, and region and the conduct and results of elections. [Chapter 6, p. 158]

enclave A piece of territory surrounded by, but not part of, a country. [Chapter 6, p. 152]

environmental determinism The belief that cultures are directly or indirectly shaped by the physical environment. [Chapter 1, p. 17]

environmental perception The belief that culture depends more on what people perceive the environment to be than on the actual character of the environment; perception, in turn, is colored by the teachings of culture. [Chapter 1, p. 19]

environmental racism The targeting of areas where ethnic or racial minorities live with respect to environmental contamination or failure to enforce environmental regulations. [Chapter 5, p. 138]

ethnic cleansing The removal of unwanted ethnic minority populations from a nation-state through mass killing, deportation, or imprisonment. [Chapter 5, p. 133]

ethnic culture region An area occupied by people of similar ethnic background who share traits of ethnicity, such as language and migration history. [Chapter 11, p. 302]

ethnic flag A readily visible marker of ethnicity on the landscape. [Chapter 5, p. 142]

ethnic geography The study of the spatial aspects of ethnicity. [Chapter 5, p. 123]

ethnic group A group of people who share a common ancestry and cultural tradition, often living as a minority group in a larger society. [Chapter 5, p. 122]

ethnic homeland A sizable area inhabited by an ethnic minority that exhibits a strong sense of attachment to the region and often exercises some measure of political and social control over it. [Chapter 5, p. 124]

ethnic island A small ethnic area in the rural countryside; sometimes called a folk island. [Chapter 5, p. 125]

ethnic neighborhood A voluntary community where people of like origin reside by choice. [Chapter 5, p. 127]

ethnic religion A religion identified with a particular ethnic or tribal group; it does not seek converts. [Chapter 7, p. 180]

ethnic substrate Regional cultural distinctiveness that remains following the assimilation of an ethnic homeland. [Chapter 5, p. 126]

ethnicity See *ethnic group*. [Chapter 5, p. 121]

ethnoburb A suburban ethnic neighborhood, sometimes home to a relatively affluent immigration population. [Chapters 5, 11, pp. 129, 313]

ethnographic boundary A political boundary that follows some cultural border, such as a linguistic or religious border. [Chapter 6, p. 153]

ethnomedicine An interdisciplinary area of study that focuses on the natural ingredients and traditional practices used in folk cultures to treat illness and disease, as well as on cultural differences in perceptions of health and disease. [Chapter 2, p. 44]

European Union (EU) A supranational organization composed of 27 European nations. [Chapter 6, p. 155]

exclave A piece of national territory separated from the main body of a country by the territory of another country. [Chapter 6, p. 152]

expansion diffusion The spread of innovations within an area in a snowballing process, so that the total number of knowers or users becomes greater and the area of occurrence grows. [Chapter 1, p. 10]

export processing zone (EPZ) Designated area of a country where the government provides conditions conducive to export-oriented production. Called *special economic zone (SEZ)* in China. [Chapter 9, p. 270]

extended metropolitan region (EMR) A new type of urban region, complex in both landscape form and function, created by the rapid spatial expansion of a city in the developing world. [Chapter 10, p. 295]

extensive agriculture A less capital-intensive form of agriculture that is highly influenced by local environmental characteristics, and often produces low crop yields per unit of land being farmed. [Chapter 8, p. 215]

farm village A clustered, tightly bunched rural settlement inhabited by people who are engaged in farming. [Chapter 3, p. 84]

farmstead The center of farm operations, containing the house, barn, sheds, livestock pens, and garden. [Chapter 3, p. 84]

federal state An independent country that gives considerable powers and even autonomy to its constituent parts. [Chapter 6, p. 154]

feedlot A factorylike farm devoted to livestock feeding or dairying; typically all feed is imported and no crops are grown on the farm. [Chapter 8, p. 220]

festival setting A multiuse redevelopment project built around a particular setting, often a waterfront or a location with a historical association. [Chapter 11, p. 326]

folk Traditional, rural; the opposite of "popular." [Chapter 2, p. 30]

folk architecture Structures built by members of a folk society or culture in a traditional manner and style, without the assistance of professional architects or blueprints, using locally available raw materials. [Chapter 2, p. 48]

folk culture Small, cohesive, stable, isolated, nearly self-sufficient group that is homogeneous in custom and ethnicity; characterized by a strong family or clan structure, order maintained through sanctions based in the religion or family, little division of labor other than that between the sexes, frequent and strong interpersonal relationships, and material cultures consisting mainly of handmade goods. [Chapters 2, 8, pp. 31, 218]

folk fortress A stronghold area with natural defensive qualities, useful in the defense of a country against invaders. [Chapter 6, p. 166]

folk geography The study of the spatial patterns and ecology of traditional groups. [Chapter 2, p. 31]

foodway Customary behavior associated with food preparation and consumption. [Chapter 5, p. 145]

forced migrant A migrant whose movements are coerced by threats to life and livelihood or by threats arising from natural or human-made causes. [Chapter 3, p. 74]

formal culture region A cultural region inhabited by people who have one or more cultural traits in common. [Chapter 1, p. 4]

formal demographic region A demographic region based on the single trait of population density. [Chapter 3, p. 60]

functional culture region A cultural area that functions as a unit politically, socially, or economically. [Chapter 1, p. 7]

functional zonation The pattern of land uses within a city; the existence of areas with differing functions, such as residential, commercial, and governmental. [Chapter 11, p. 321]

fundamentalism A movement to return to the founding principles of a religion, which can include literal interpretation of sacred texts, or the attempt to follow the ways of a religious founder as closely as possible. [Chapter 7, p. 181]

gateway city A city that acts as a port of entry, early settlement area, and distribution center for immigrants to a country because of its relative proximity to the homeland. [Chapter 5, p. 130]

gender role What it means to be a man, and what it means to be a woman, in different cultural and historical contexts. [Chapter 3, p. 70]

generic toponym The descriptive part of many place names, often repeated throughout a culture area. [Chapter 4, p. 114]

genetically modified (GM) crop New organism whose genetic characteristics have been programmed through gene splicing between plant and/or animal species. [Chapter 8, p. 240]

gentrification The displacement of lower-income residents by higher-income residents as buildings in deteriorated areas of city centers are restored. [Chapter 11, p. 310]

geodemography Population geography; the study of the spatial and ecological aspects of population, including distribution, density per unit of land area, fertility, gender, health, age, mortality, and migration. [Chapter 3, p. 57]

geography The study of spatial patterns and of differences and similarities from one place to another in environment and culture. [Chapter 1, p. 2]

geometric boundary A political border drawn in a regular, geometric manner, often a straight line, without regard for environmental or cultural patterns. [Chapter 6, p. 153]

geopolitics The influence of geography and the habitat on political entities. [Chapter 6, p. 166]

gerrymandering The drawing of electoral district boundaries in an awkward pattern to enhance the voting impact of one constituency at the expense of another. [Chapter 6, p. 159]

ghetto Traditionally, an area within a city where an ethnic group lives, either by choice or by force. Today in the United States, the term typically indicates an impoverished African-American urban neighborhood. [Chapter 5, p. 127]

global city A city that is a control center of the global economy. [Chapter 10, p. 286]

global warming The pronounced climatic warming of Earth that has occurred since about 1920 and particularly since the 1970s. [Chapter 9, p. 263]

globalization The binding together of all the lands and peoples of the world into an integrated system driven by capitalistic free markets, in which cultural diffusion is rapid, independent states are weakened, and cultural homogenization is encouraged. [Chapter 1, p. 13]

globalizing city A city being shaped by the new global economy and culture. [Chapter 10, p. 287]

green revolution The fairly recent introduction of high-yield hybrid crops and chemical fertilizers and pesticides into traditional agricultural systems, most notably paddy rice farming, with attendant increases in production and ecological damage. [Chapter 8, p. 229]

greenhouse effect A process by which thermal radiation from a planet's surface is absorbed by atmospheric greenhouse gases, and is reradiated in all directions. Because part of this reradiation is back towards the planet's surface and the planet's lower atmosphere, the result is an elevation of the planet's average surface temperature above what it would be in the absence of the gases. Earth's natural greenhouse effect makes life as we know it possible. However, human activities, primarily the burning of fossil fuels and clearing of forests, have intensified the natural greenhouse effect, causing global warming. [Chapter 9, p. 263]

Greens Activists and organizations, including political parties, whose central concern is addressing environmental issues. [Chapter 9, p. 266]

gross domestic product (GDP) A measurement of the total value of all goods and services produced within a particular region or country within a set time period. [Chapter 9, p. 250]

guild industry A traditional type of manufacturing before the industrial revolution, involving handmade goods of high quality manufactured by highly skilled artisans who resided in towns and cities. [Chapter 9, p. 256]

hamlet A small rural settlement, smaller than a farm village. [Chapter 8, p. 242]

heartland The interior of a sizable landmass, removed from maritime connections; in particular, the interior of the Eurasian continent. [Chapter 6, p. 166]

heartland theory A 1904 proposal by Mackinder that the key to world conquest lay in control of the interior of Eurasia. [Chapter 6, p. 166]

heat island An area of warmer temperatures at the center of a city, caused by the urban concentration of heat-retaining concrete, brick, and asphalt. [Chapter 11, p. 314]

hierarchical diffusion A type of expansion diffusion in which innovations spread from one important person to another or from one urban center to another, temporarily bypassing other persons or rural areas. [Chapter 1, p. 10]

high-tech corridor An area along a limited-access highway that contains offices and other services associated with high-tech industries. [Chapters 9, 11, pp. 254, 325]

hinterland The area surrounding a city and influenced by it. [Chapter 10, p. 294]

homelessness A temporary or permanent condition of not having a legal home address. [Chapter 11, p. 304]

hunting and gathering The killing of wild game and the harvesting of wild plants to provide food in traditional cultures. [Chapter 8, p. 226]

hydraulic civilization A civilization based on large-scale irrigation. [Chapter 10, p. 281]

in-filling New building on empty parcels of land within a checkerboard pattern of development. [Chapter 11, p. 310]

independent invention A cultural innovation that is developed in two or more locations by individuals or groups working independently. [Chapter 1, p. 10]

industrial revolution A series of inventions and innovations, arising in England in the 1700s, that led to the use of machines and inanimate power in the manufacturing process. [Chapter 9, p. 256]

industrialization The transformation of raw materials into commodities or the process by which a society moves from subsistence agriculture toward mass production based on machinery and industry. [Chapter 9, p. 250]

infant mortality rate The number of infants per 1000 live births who die before reaching one year of age. [Chapter 3, p. 71]

intensive agriculture A form of agriculture using mechanization, labor, and other capital to produce large crop yields relative to the amount of land being farmed. [Chapter 8, p. 214]

intercropping (intertillage) The practice of growing two or more different types of crops in the same field at the same time. [Chapter 8, p. 216]

intergovernmental organization An international organization that does not require member states to relinquish sovereignty in order to benefit from cooperation between states in the association. [Chapter 6, p. 156]

international organization A general term used to describe varying types of organizations that attempt to promote cooperation and coordination among their member states. [Chapter 6, p. 155]

involuntary migration Also called forced migration, refers to the forced displacement of a population, whether by government policy (such as a resettlement program), warfare or other violence, ethnic cleansing, disease, natural disaster, or enslavement. [Chapter 5, p. 133]

isogloss The border of usage of an individual word or pronunciation. [Chapter 4, p. 99]

Kurgan hypothesis A theory of language diffusion holding that the spread of Indo-European languages originated with animal domestication; originated in the Central Asian steppes; and was later, more violent, and swifter than proponents of the Anatolian hypothesis maintain. [Chapter 4, p. 101]

labor-intensive industry An industry whose labor costs represent a large portion of total production costs. [Chapter 9, p. 267]

land-division pattern The spatial pattern of various types of land use. [Chapter 1, p. 25]

language A mutually agreed-on system of symbolic communication that has a spoken and usually a written expression. [Chapter 4, p. 93]

language family A group of related languages derived from a common ancestor. [Chapter 4, p. 97]

language hotspot Place on Earth that is home to a unique, misunderstood, or endangered language. [Chapter 4, p. 110]

lateral commuting Traveling from one suburb to another in going from home to work. [Chapter 11, p. 308]

legible city A city that is easy to decipher, with clear pathways, edges, nodes, districts, and landmarks. [Chapter 11, p. 323]

leisure landscape A landscape that is planned and designed primarily for entertainment purposes, such as ski and beach resorts. [Chapter 2, p. 52]

lingua franca A language of communication and commerce used widely where it is not a mother tongue. [Chapter 4, p. 96]

linguistic refuge area An area protected by isolation or inhospitable environmental conditions in which language or dialect has survived. [Chapter 4, p. 106]

livestock feeding A commercial type of agriculture that produces cattle and hogs for meat. [Chapter 8, p. 220]

local consumption cultures A set of distinct consumption practices and preferences in food, clothing, music, and so forth, formed in a specific place and/or historical moment. [Chapter 2, p. 47]

Malthusian Those who hold the views of Thomas Malthus, who believed that overpopulation is the root cause of poverty, illness, and warfare. [Chapter 3, p. 65]

manufactured toponym A place name that is "made up," often by early founders of a settlement or an influential member of a community. [Chapter 4, p. 113]

maquiladora A U.S.-owned assembly plant located on the Mexican side of the U.S.–Mexico border, [Chapter 9, p. 270]

mariculture A branch of aquaculture specific to the cultivation of marine organisms, often involving the transformation of coastal environments and the production of distinctive new landscapes. [Chapter 8, p. 224]

market The geographical area in which a product may be sold in a volume and at a price profitable to the manufacturer. [Chapter 9, p. 267]

market gardening Farming devoted to specialized nontropical fruit, vegetable, or vine crops for sale rather than consumption. Also known as *truck farming*. [Chapter 8, p. 220]

mass culture A form of culture that is produced, distributed, and marketed through mass media, art, and other forms of communication. [Chapter 2, p. 32]

master-planned community A large-scale residential development that includes, in addition to architecturally compatible housing units, planned recreational facilities, schools, and security measures. [Chapter 11, p. 325]

material culture All physical, tangible objects made and used by members of a cultural group. Examples are clothing, buildings, tools and utensils, musical instruments, furniture, and artwork; the visible aspect of culture. [Chapter 2, p. 30]

mechanistic view of nature The view that humans are separate from nature and hold dominion over it and that the habitat is an integrated mechanism governed by external forces that the human mind can understand and manipulate. [Chapter 1, p. 20]

medical geography The branch of geography that deals with health and disease-related topics. [Chapter 3, p. 76]

megacity A particularly large urban center, one with a population over 10 million. [Chapter 10, p. 279]

megalopolis A large urban region formed as several urban areas spread and merge, such as Boswash, the region including Boston, New York, and Washington, D.C. [Chapter 11, p. 323]

migration The large-scale movement of people between different regions of the world. [Chapter 1, p. 13]

model An abstraction, an imaginary situation, proposed by geographers to simulate laboratory conditions so that they may isolate certain causal forces for detailed study. [Chapter 1, p. 21]

monoculture The raising of only one crop on a huge tract of land in agribusiness. [Chapter 8, p. 240]

monotheism The worship of only one god. [Chapter 7, p. 180]

multiple-nuclei model A model, developed by Chauncey Harris and Edward Ullman, that depicts a city growing from several separate focal points. [Chapter 11, p. 319]

nation-state An independent country dominated by a relatively homogeneous culture group. [Chapter 6, p. 168]

nationalism The sense of belonging to and self-identification with a national culture. [Chapter 6, p. 155]

natural boundary A political border that follows some feature of the natural environment, such as a river or mountain ridge. [Chapter 6, p. 153]

natural decrease The birthrate minus the death rate. [Chapter 3, p. 72]

natural hazard An inherent danger present in a given habitat, such as flooding, hurricanes, volcanic eruptions, or earthquakes; often perceived differently by different peoples. [Chapter 1, p. 19]

neighborhood A small social area within a city where residents share values and concerns and interact with one another on a daily basis. [Chapter 11, p. 304]

neighborhood effect Microscale diffusion in which acceptance of an innovation is most rapid in small clusters around an initial adopter. [Chapter 1, p. 13]

node A central point in a functional culture region where functions are coordinated and directed. [Chapter 1, p. 7]

nomadic livestock herder A member of a group that continually moves with its livestock in search of forage for its animals. [Chapter 8, p. 223]

nonmaterial culture The wide variety of tales, songs, lore, beliefs, superstitions, and customs that pass from generation to generation as part of an oral or written tradition. [Chapter 2, p. 30]

North American Free Trade Association (NAFTA) An intergovernmental organization composed of Canada, Mexico, and the United States, designed to serve as a trading bloc. [Chapter 6, p. 157]

nucleation A settlement form characterized by density. [Chapter 1, p. 25]

office park A cluster of office buildings, usually located along an interstate highway, often forming the nucleus of an edge city. [Chapter 11, p. 325]

organic agriculture A form of farming that relies on manuring, mulching, and biological pest control and rejects the use of synthetic fertilizers, insecticides, herbicides, and genetically modified crops. [Chapter 8, p. 235]

organic view of nature The view that humans are part of, not separate from, nature and that the habitat possesses a soul and is filled with nature-spirits. [Chapter 1, p. 20]

orthodox religion A strand within most major religions that emphasizes purity of faith and is not open to blending with elements of other belief systems. [Chapter 7, p. 180]

outsource The physical separation of some economic activities from the main production facility, usually for the purpose of employing cheaper labor. [Chapter 9, p. 267]

paddy rice farming The cultivation of rice on a paddy, or small flooded field enclosed by mud dikes, practiced in the humid areas of the Far East. [Chapter 8, p. 217]

palimpsest A term used to describe cultural landscapes with various layers and historical "messages." Geographers use this term to reinforce the notion of the landscape as a text that can be read; a landscape palimpsest has elements of both modern and past periods. [Chapter 11, p. 322]

peasant A farmer belonging to a folk culture and practicing a traditional system of agriculture. [Chapter 8, p. 218]

peripheral model A geographic model, created by Chauncey Harris, that explains how suburban edge cities are linked by ring roads or beltways that allow for commuter traffic to conveniently avoid passing through major urban centers. [Chapter 11, p. 324]

permeable barrier A barrier that permits some aspects of an innovation to diffuse through it but weakens and retards continued spread. An innovation can be modified in passing through a permeable barrier. [Chapter 1, p. 12]

personal space The amount of space that individuals feel "belongs" to them as they move about their daily business. [Chapter 3, p. 81]

physical environment All aspects of the natural physical surroundings, such as climate, terrain, soils, vegetation, and wildlife. [Chapter 1, p. 3]

physiological population density A country's population divided by its total arable land. [Chapter 3, p. 72]

pidgin A composite language consisting of a small vocabulary borrowed from the linguistic groups involved in trade and commerce. [Chapter 4, p. 96]

pilgrimage A journey to a place of religious importance. [Chapter 7, p. 199]

place A term used to connote the subjective, humanistic, culturally oriented type of geography that seeks to understand the unique character of individual regions and places, rejecting the principles of science as flawed and unknowingly biased. [Chapter 1, p. 22]

placelessness A spatial standardization that diminishes regional variety; may result from the spread of popular culture, which can diminish or destroy the uniqueness of place through cultural standardization on a local, national, or worldwide scale. [Chapter 2, p. 33]

plantation A large landholding devoted to specialized production of a single tropical cash crop. [Chapter 8, p. 219]

plantation agriculture A system of monoculture for producing export crops requiring relatively large amounts of land and capital; originally dependent on slave labor. [Chapter 8, p. 219]

political geography The geographic study of politics and political matters. [Chapter 6, p. 149]

polyglot A mixture of different languages. [Chapter 4, p. 98]

polytheism The worship of many gods. [Chapter 7, p. 180]

popular culture A dynamic culture based in large, heterogeneous societies permitting considerable individualism, innovation, and change; features include a money-based economy, division of labor into professions, secular institutions of control, and weak interpersonal ties; used to describe a common set of cultural ideas, values, and practices that are consumed by a population. [Chapter 2, p. 31]

population density A measurement of population per unit area (for example, per square mile). [Chapter 3, p. 59]

population doubling time The time it takes a population to double in size from any given numerical point. Determined by dividing the growth rate for a population into the number 72. [Chapter 3, p. 73]

population explosion The rapid, accelerating increase in world population since about 1650 and especially since 1900. [Chapter 3, p. 65]

population geography Geodemography; the study of the spatial and ecological aspects of population, including distribution, density per unit of land area, fertility, gender, health, age, mortality, and migration. [Chapter 3, p. 57]

population growth rate The increase in a country's population during a period of time (usually one year), expressed as a percentage of the population at the start of that period. [Chapter 3, p. 73]

population pyramid A graph used to show the age and sex composition of a population. [Chapter 3, p. 70]

possibilism A school of thought based on the belief that humans, rather than the physical environment, are the primary active force; that any environment offers a number of possible ways for a culture to develop; and that the choices among these possibilities are guided by cultural heritage. [Chapter 1, p. 17]

postindustrial phase A society characterized by the dominance of the service sectors of economic activity. [Chapter 9, p. 261]

primary industry An industry engaged in the extraction of natural resources, such as fishing, farming, hunting, lumbering, and mining. [Chapter 9, p. 252]

primate city A city of large size and dominant power within a country. [Chapter 10, p. 280]

producer services The range of economic activities required by producers of goods, including insurance, legal services, banking, advertising, wholesaling, retailing, information generation, and real-estate transactions. [Chapter 9, p. 253]

proselytic religion A religion that actively seeks new members and aims to convert all humankind. [Chapter 7, p. 180]

push-and-pull factors Unfavorable, repelling conditions and favorable, attractive conditions that interact to affect migration and other elements of diffusion. [Chapter 3, p. 76]

race A classification system that is sometimes understood as arising from genetically significant differences among human populations, or visible differences in human physiognomy, or as a social construction that varies across time and space. [Chapter 5, p. 121]

racism The belief that certain individuals are inferior because they are born into a particular ethnic, racial, or cultural group. Racism often leads to prejudice and discrimination, and it reinforces relationships of unequal power between groups. [Chapter 5, p. 122]

ranching The commercial raising of herd livestock on a large landholding. [Chapter 8, p. 223]

range In central-place theory, the average maximum distance people will travel to purchase a good or service. [Chapter 10, p. 293]

redlining A practice by banks and mortgage companies of demarcating areas considered to be high risk for defaulting on housing loans. [Chapter 11, p. 309]

refugee A person fleeing from persecution in his or her country of nationality. The persecution can be religious, political, racial, or ethnic. A refugee is a forced migrant. [Chapter 3, p. 74]

regional trading bloc Entity formed by an agreement made among geographically proximate countries to reduce trade barriers and to better compete with other regional markets. [Chapter 6, p. 156]

relic boundary A former political border that no longer functions as a boundary. [Chapter 6, p. 154]

religion A relatively structured set of beliefs and practices through which people achieve mental and physical harmony with the universe and attempt to accommodate or influence the forces of nature, life, and death. [Chapter 7, p. 179]

relocation diffusion The spread of an innovation or other element of culture that occurs with the bodily relocation (migration) of the individual or group responsible for the innovation. [Chapter 1, p. 10]

renewable resource A resource that can be replenished naturally at a rate sufficient to balance its depletion by human use. [Chapter 9, p. 261]

restrictive covenant A statement written into a property deed that limits the use of the land in some way; often used to prohibit certain groups of people from buying property. [Chapter 11, p. 309]

return migration A type of ethnic diffusion that involves the voluntary movement of a group of migrants back to its ancestral or native country or homeland. [Chapter 5, p. 135]

rimland The maritime fringe of a country or continent; in particular, the western, southern, and eastern edges of the Eurasian continent. [Chapter 6, p. 166]

sacred space An area recognized by one or more religious groups as worthy of devotion, loyalty, esteem, or fear to the extent that it is sought out, avoided, inaccessible to the nonbeliever, and/or removed from economic use. [Chapter 7, p. 208]

satellite state A small, weak country dominated by one powerful neighbor to the extent that some or much of its independence is lost. [Chapter 6, p. 152]

secondary industry An industry engaged in processing raw materials into finished products; manufacturing. [Chapter 9, p. 252]

sector model An economic model, developed by Homer Hoyt, that depicts a city as a series of sectors or wedges shaped like the slices of a pie. [Chapter 11, p. 318]

secular Characterized by little or no religious belief or practice. [Chapter 7, p. 202]

sedentary cultivation Farming in fixed and permanent fields. [Chapter 8, p. 223]

services The range of economic activities that provide services to people and industry. [Chapter 9, p. 252]

settlement form The spatial arrangement of buildings, roads, towns, and other features that people construct while inhabiting an area. [Chapter 1, p. 25]

sex ratio The numerical ratio of males to females in a population. [Chapter 3, p. 70]

shantytown A deprived area on the outskirts of a town consisting of a large number of crude dwellings. [Chapter 3, p. 88]

sharia A legal framework, derived from the religious interpretation of Islamic texts, that governs almost all aspects of public life, and some aspects of private life, for Muslims living under this system of jurisprudence. [Chapter 7, p. 200]

site The local setting of a city. [Chapter 10, p. 288]

situation The regional setting of a city. [Chapter 10, p. 288]

slang Words and phrases that are not part of a standard, recognized vocabulary for a given language but that are nonetheless used and understood by some of its speakers. [Chapter 4, p. 100]

social culture region An area in a city where many of the residents share social traits such as income, education, age, and family structure. [Chapter 11, p. 302]

social stratification The existence of distinct socioeconomic classes. [Chapter 10, p. 281]

sovereignty The right of individual states to control political and economic affairs within their territorial boundaries without external interference. [Chapter 6, p. 151]

space A term used to connote the objective, quantitative, theoretical, model-based, economics-oriented type of geography that seeks to understand spatial systems and networks through application of the principles of social science. [Chapter 1, p. 21]

special economic zone (SEZ) See *export processing zone (EPZ)*.

squatter settlement An illegal housing settlement, usually made up of temporary shelters, that surrounds a large city and may eventually become a permanent part of the city. [Chapter 10, p. 295]

state An independent political unit with a centralized authority that makes claims to sole jurisdiction over a bounded territory, within which a central authority controls and enforces a single system of political and legal institutions; often used synonymously with "country." [Chapter 6, p. 150]

stimulus diffusion A type of expansion diffusion in which a specific trait fails to spread but the underlying idea or concept is accepted. [Chapter 1, p. 11]

subculture A group of people with norms, values, and material practices that differentiate them from the dominant culture to which they belong. [Chapter 2, p. 30]

subsistence agriculture Farming to supply the minimum food and materials necessary to survive. [Chapter 8, p. 216]

suitcase farm In U.S. commercial grain agriculture, a farm on which no one lives; planting and harvesting is done by hired migratory crews. [Chapter 8, p. 221]

supranationalism Occurs when states willingly relinquish some degree of sovereignty in order to gain the benefits of belonging to a larger political-economic entity. [Chapter 6, p. 155]

survey pattern A pattern of original land survey in an area. [Chapter 8, p. 242]

sustainability The survival of a land-use system for centuries or millennia without destruction of the environmental base. [Chapters 8, 9, pp. 232, 266]

swidden cultivation A type of agriculture characterized by land rotation in which temporary clearings are used for several years and then abandoned to be replaced by new clearings; also known as *slash-and-burn agriculture*. [Chapter 8, p. 215]

symbolic landscape A landscape that expresses the values, beliefs, and meanings of a particular culture. [Chapter 1, p. 24]

syncretic religion A religion or strand within a religion that combines elements of two or more belief systems. [Chapter 7, p. 180]

technopole A center of high-tech manufacturing and information-based industry. [Chapter 9, p. 258]

teleology The belief that Earth was created especially for human beings, who are separate from and superior to the natural world. [Chapter 7, p. 196]

territoriality A learned cultural response, rooted in European history, that produced the external bounding and internal territorial organization characteristic of modern states. [Chapter 6, p. 152]

threshold In central-place theory, the size of the population required to make provision of goods and services economically feasible. [Chapter 10, p. 293]

time-distance decay The decrease in acceptance of a cultural innovation with increasing time and distance from its origin. [Chapter 1, p. 12]

toponym A place name, usually consisting of two parts, the generic and the specific. [Chapter 4, p. 112]

topophilia Love of place; used to describe people who exhibit a strong sense of place. [Chapter 1, p. 22]

total fertility rate (TFR) The number of children the average woman will bear during her reproductive lifetime (15–44 years old). A TFR of less than 2.1, if maintained, will cause a natural decline of population. [Chapter 3, p. 61]

trade-route site A place for a city at a significant point on a transportation route. [Chapter 10, p. 290]

transnational corporation A company that has international production, marketing, and management facilities. [Chapter 9, p. 258]

transnationalism A phenomenon in which immigrants maintain social and/or economic ties to their place of origin. Often, these relationships include visits "home" and the circular exchange of money and goods. [Chapter 1, p. 13]

transportation/communication services The range of economic activities that provide transport, communication, and utilities to businesses. [Chapter 9, p. 253]

uneven development The tendency for industry to develop in a core-periphery pattern, enriching the industrialized countries of the core and impoverishing the less-industrialized periphery. This term is also used to describe urban patterns in which suburban areas are enriched while the inner city is impoverished. [Chapters 1, 11, pp. 15, 307]

unitary state An independent state that concentrates power in the central government and grants little authority to the provinces. [Chapter 6, p. 154]

universalizing religion Also called proselytic religion, a universalizing religion expands through active conversion of new members and aims to encompass all of humankind. [Chapter 7, p. 180]

urban agriculture The raising of food, including fruit, vegetables, meat, and milk, inside cities, especially common in developing countries. [Chapter 8, p. 224]

urban footprint The total impact of urban areas on the natural environment. [Chapter 10, p. 292]

urban hearth area A region in which the world's first cities evolved. [Chapter 10, p. 282]

urban morphology The form and structure of cities, including street patterns, building sizes and shapes, architecture, and density. [Chapter 11, p. 321]

urbanized population The proportion of a country's population living in cities. [Chapter 10, p. 278]

vernacular culture region A cultural region perceived to exist by its inhabitants; based in the collective spatial perception of the population at large; bearing a generally accepted name or nickname, such as Dixie. [Chapters 1, 2, pp. 8, 37]

voluntary migrant A person leaving his or her place of residence in order to improve his or her quality of life, often for the purpose of employment. [Chapter 3, p. 74]

zero population growth A condition of population stability; zero population growth is achieved when an average of only 2.1 children per couple survive to adulthood, so that eventually the number of deaths equals the number of births. [Chapter 3, p. 62]

SOURCES

Chapter 1

Anderson, Kay, and Fay Gale. 1999. *Cultural Geography*. Melbourne: Pearson Education.

Carson, Rachel. 1962. *Silent Spring*. Boston: Houghton Mifflin.

Department of the Interior, U.S. Geological Survey (USGS). 2011 (January 3). "Deaths from Earthquakes in 2011." Available online at *USGS Earthquakes Hazards Program*, http://earthquake.usgs.gov/earthquakes/eqarchives/year/2011/2011_deaths.php.

Gould, Peter. 1993. *The Slow Plague: A Geography of the AIDS Pandemic*. Cambridge: Blackwell.

Gordon, Eric (ed.). 2008. *Space and Culture* 11(3).

Griffin, Ernst, and Larry Ford. 1980. "A Model of Latin American City Structure." *Geographical Review* 70: 397–422.

Gumilev, Leo. *Ethnogenesis and the Biosphere*. Moscow: Progress Publishers, 1990.

Hägerstrand, Torsten. 1967. *Innovation Diffusion as a Spatial Process*. Allan Pred (trans.). Chicago: University of Chicago Press.

Holloway, Sarah, and Gill Valentine. 2001. "Placing Cyberspace: Processes of Americanization in British Children's Use of the Internet." *Area* 33(2): 153–160.

Jones, Steve. 2002. "Music That Moves: Popular Music, Distribution and Network Technologies." *Cultural Studies* 16(2): 213–232.

Lewis, Pierce F. 1979. "Axioms for Reading the Landscape: Some Guides to the American Scene." In D. W. Meinig and J. B. Jackson (eds.), *The Interpretation of Ordinary Landscapes*. New York: Oxford University Press, 12–32.

Lewis, Peirce. 1983. "Learning from Looking: Geographic and Other Writing About the American Cultural Landscape." *American Quarterly* 35: 242–261.

Norwine, Jim, and Thomas D. Anderson. 1980. *Geography as Human Ecology?* Lanham, MD: University Press of America.

Parker, Geoffrey. 2000. "Ratzel, the French School and the Birth of Alternative Geopolitics." *Political Geography* 19: 957–969.

Ratzel, Friedrich. 1882–1891. *Antropogeographie*, (2 vol). Stuttgart: Engelhorn.

Ratzel, Friedrich. 1896. "Die Gesetze des raumlichen Wachtums der Staaten: Ein Beitrag zur wissenschaft-lichen politischen Geographie." *Petermanns Geographische Mitteilungen* 42: 97–107.

Relph, Edward. 1981. *Rational Landscapes and Humanistic Geography*. New York: Barnes & Noble.

Staudt, Amanda, Nancy Huddleston, and Sandi Rudenstein. 2006. *Understanding and Responding to Climate Change, a Report Prepared by the National Research Council Based on National Academies' Reports*. Washington, D.C.: National Academy of Science.

Tuan, Yi-Fu. 1974. *Topophilia: A Study of Environmental Perception, Attitudes, and Values*. Englewood Cliffs, NJ: Prentice-Hall.

Chapter 2

Abler, Ronald F. 1973. "Monoculture or Miniculture? The Impact of Communication Media on Culture in Space." In David A. Lanegran and Risa Palm (eds.), *An Invitation to Geography*. New York: McGraw-Hill, 186–195.

Alcorn, J. 1994. "Noble Savage or Noble State? Northern Myths and Southern Realities in Biodiversity Conservation." *Ethnoecologica* 2(3): 7–19.

Bebbington, A. 1996. "Movements, Modernizations, and Markets: Indigenous Organizations and Agrarian Strategies in Ecuador." In R. Peet and M. Watts (eds.), *Liberation Ecologies: Environment, Development, Social Movements*. London: Routledge, 86–109.

Bebbington, A. 2000. "Reencountering Development: Livelihood Transitions and Place Transformations in the Andes." *Annals of the Association of American Geographers* 90(3): 495–520.

Brownell, Joseph W. 1960. "The Cultural Midwest." *Journal of Geography* 59: 81–85.

Clevinger, Woodrow R. 1938. "The Appalachian Mountaineers in the Upper Cowlitz Basin." *Pacific Northwest Quarterly* 29: 115–134.

Clevinger, Woodrow R. 1942. "Southern Appalachian Highlanders in Western Washington." *Pacific Northwest Quarterly* 33: 3–25.

Curtis, James R. 1982. "McDonald's Abroad: Outposts of American Culture." *Journal of Geography* 81: 14–20.

de Blij, Harm, and Peter Muller. 2004. *Geography: Realms, Regions and Concepts*, 11th ed. New York: Wiley.

Edmonson, Brad, and Linda Jacobsen. 1993. "Elvis Presley Memorabilia." *American Demographics* 15(8): 64.

Evans, E. Estyn. 1957. *Irish Folk Ways*. London: Routledge & Kegan Paul.

Francaviglia, Richard V. 1973. "Diffusion and Popular Culture: Comments on the Spatial Aspects of Rock Music." In David A. Lanegran and Risa Palm (eds.), *An Invitation to Geography*. New York: McGraw-Hill, 87–96.

Frederickson, Kristine. 1984. *American Rodeo from Buffalo Bill to Big Business*. College Station: Texas A&M University Press.

Gade, Daniel W. 1982. "The French Riviera as Elitist Space." *Journal of Cultural Geography* 3: 19–28.

Glassie, Henry. 1968. *Pattern in the Material Folk Culture of the Eastern United States*. Philadelphia: University of Pennsylvania Press.

Godfrey, Brian J. 1993. "Regional Depiction in Contemporary Film." *Geographical Review* 83: 428–440.

Goss, Jon. 1999. "Once-upon-a-Time in the Commodity World: An Unofficial Guide to the Mall of America." *Annals of the Association of American Geographers* 89: 45–75.

Graff, Thomas O., and Dub Ashton. 1994. "Spatial Diffusion of Wal-Mart: Contagious and Reverse Hierarchical Elements." *Professional Geographer* 46: 19–29.

Hecock, Richard D. 1987. "Changes in the Amenity Landscape: The Case of Some Northern Minnesota Townships." *North American Culture* 3(1): 53–66.

Holder, Arnold. 1993. "Dublin's Expanding Golfscape." *Irish Geography* 26: 151–157.

Hopkins, Jeffrey. 1990. "West Edmonton Mall: Landscape of Myths and Elsewhereness." *Canadian Geographer* 34: 2–17.

IDRC. 2004. Web site of the International Development Research Centre, available online at http://web.idrc.ca/en/ev-1248-201-1-DO_TOPIC.html.

International Cancún Declaration of Indigenous Peoples. 2003. Prepared during the fifth WTO Ministerial Conference, Cancún, Mexico. Available online at http://www.ifg.org/programs/indig/CancunDec.html.

International Labor Organization. 1989. Indigenous and Tribal Peoples Convention. Available online at http://members.tripod.com/PPLP/ILOC169.html.

Jackson, Peter. 2004. "Local Consumption Cultures in a Globalizing World." *Transactions of the Institute of British Geographers* NS 29: 165–178.

Jakle, John A., and Richard L. Mattson. 1981. "The Evolution of a Commercial Strip." *Journal of Cultural Geography* 1(2): 12–25.

Jett, Stephen C. 1991. "Further Information on the Geography of the Blowgun and Its Implications for Early Transoceanic Contacts." *Annals of the Association of American Geographers* 81: 89–102.

Kay, Jeanne, et al. 1981. "Evaluating Environmental Impacts of Off-Road Vehicles." *Journal of Geography* 80: 10–18.

Kimber, Clarissa T. 1973. "Plants in the Folk Medicine of the Texas-Mexico Borderlands." *Proceedings, Association of American Geographers* 5: 130–133.

Kimmel, James P. 1997. "Santa Fe: Thoughts on Contrivance, Commoditization, and Authenticity." *Southwestern Geographer* 1: 44–61.

Kniffen, Fred B. 1951. "The American Agricultural Fair." *Annals of the Association of American Geographers* 42: 45–51.

Kniffen, Fred B. 1965. "Folk-Housing: Key to Diffusion." *Annals of the Association of American Geographers* 55: 549–577.

Kunstler, James H. 1993. *The Geography of Nowhere: The Rise and Decline of America's Man-Made Landscape.* New York: Simon & Schuster.

Lowenthal, David. 1968. "The American Scene." *Geographical Review* 58: 61–88.

McDowell, L. 1994. "The Transformation of Cultural Geography. In D. Gregory, R. Martin, and G. Smith (eds.), *Human Geography: Society, Space, and Social Science.* Minneapolis: University of Minnesota Press, 146–173.

Miller, E. Joan Wilson. 1968. "The Ozark Culture Region as Revealed by Traditional Materials." *Annals of the Association of American Geographers* 58: 51–77.

Mings, Robert C., and Kevin E. McHugh. 1989. "The RV Resort Landscape." *Journal of Cultural Geography* 10: 35–49.

Nietschmann, B. 1979. "Ecological Change, Inflation, and Migration in the Far West Caribbean." *Geographical Review* 69: 1–24.

Pillsbury, Richard. 1990a. "Striking to Success." *Sport Place* 4(1): 16, 35.

Pillsbury, Richard. 1990b. "Ride'm Cowboy." *Sport Place* 4: 26–32.

Pillsbury, Richard. 1998. *No Foreign Food: The American Diet in Time and Place.* Boulder, CO: Westview Press.

Price, Edward T. 1960. "Root Digging in the Appalachians: The Geography of Botanical Drugs." *Geographical Review* 50: 1–20.

Raitz, Karl B. 1975. "Gentleman Farms in Kentucky's Inner Bluegrass." *Southeastern Geographer* 15: 33–46.

Raitz, Karl B. 1987. "Place, Space and Environment in America's Leisure Landscapes." *Journal of Cultural Geography* 8(1): 49–62.

Relph, Edward. 1976. *Place and Placelessness.* London: Pion.

Richards, P. 1975. "'Alternative' Strategies for the African Environment. 'Folk Ecology' as a Basis for Community Oriented Agricultural Development." In P. Richards (ed.), *African Environment, Problems and Perspectives.* London: International African Institute, 102–117.

Roark, Michael. 1985. "Fast Foods: American Food Regions." *North American Culture* 2(1): 24–36.

Robertson, David S. 1996. "Oil Derricks and Corinthian Columns: The Industrial Transformation of the Oklahoma State Capitol Grounds." *Journal of Cultural Geography* 16: 17–44.

Rooney, John F., Jr. 1969. "Up from the Mines and Out from the Prairies: Some Geographical Implications of Football in the United States." *Geographical Review* 59: 471–492.

Rooney, John F., Jr. 1988. "Where They Come From." *Sport Place* 2(3): 17.

Rooney, John F., Jr., and Paul L. Butt. 1978. "Beer, Bourbon, and Boone's Farm: A Geographical Examination of Alcoholic Drink in the United States." *Journal of Popular Culture* 11: 832–856.

Rooney, John F., Jr., and Richard Pillsbury. 1992. *Atlas of American Sport.* New York: Macmillan.

Sack, Robert D. 1992. *Place, Modernity, and the Consumer's World: A Rational Framework for Geographical Analysis.* Baltimore: Johns Hopkins University Press.

Scott, Damon. 2008. "The Gay Geography of San Francisco." PhD dissertation. University of Texas at Austin.

Shortridge, Barbara G. 1987. *Atlas of American Women.* New York: Macmillan.

Shortridge, James R. 1989. *The Middle West: Its Meaning in American Culture.* Lawrence: University Press of Kansas.

Shuhua, Chang. 1977. "The Gentle Yamis of Orchid Island." *National Geographic* 151(1): 98–109.

Tuan, Yi-Fu. *Topophilia.* Englewood Cliffs, NJ: Prentice-Hall, 1974.

Weiss, Michael J. 1988. *The Clustering of America.* New York: Harper & Row.

Wilhelm, Eugene J., Jr. 1968. "Field Work in Folklife: Meeting Ground of Geography and Folklore." *Keystone Folklore Quarterly* 13: 241–247.

Williams, R. 1976. *Keywords: A Vocabulary of Culture and Society.* New York: Oxford University Press.

Zelinsky, Wilbur. 1974. "Cultural Variation in Personal Name Patterns in the Eastern United States." *Annals of the Association of American Geographers* 60: 743–769.

Zelinsky, Wilbur. 1980a. "North America's Vernacular Regions." *Annals of the Association of American Geographers* 70: 1–16.

Zelinsky, Wilbur. 1980b. "Selfward Bound? Personal Preference Patterns and the Changing Map of American Society." *Economic Geography* 50: 144–179.

Zonn, Leo (ed.). 1990. *Place Images in Media: Portrayal, Experience, and Meaning.* Savage, MD: Rowman & Littlefield.

Chapter 3

BBC News. July 8, 2011. "South Sudan Counts Down to Independence." http://www.bbc.co.uk/news/world-africa-14077511.

Central Intelligence Agency. 2012. *World Factbook* 2012.

de Macgregor, María T. de Gutiérrez. 1984. "Population Geography in Mexico." In John J. Clarke (ed.), *Geography and Population.* Oxford: Pergamon.

D'Emilio, Frances. 2004. "Italy's Seniors Finding Comfort with Strangers," *Miami Herald,* 30 October: 1A, 2A.

Girish, Uma. 2005. "For India's Daughters, a Dark Birth Day," *Christian Science Monitor,* 9 February. Available online at http://www.csmonitor.com/2005/0209/p11s01-wosc.html.

Gould, Peter. 1993. *The Slow Plague: A Geography of the AIDS Pandemic.* Oxford: Blackwell.

Hooper, Edward. 1999. *The River: A Journey to the Source of HIV and AIDS.* Boston: Little, Brown.

International Organization for Migration web site, www.iom.int.

Malthus, Thomas R. 1989 [1798]. *An Essay on the Principle of Population.* Patricia James (ed.). Cambridge: Cambridge University Press.

Montero, David. 2002. "Charting the World's Oil," Frontline/World, *Colombia—The Pipeline War.* Available online at http://www.pbs.org/frontlineworld/stories/colombia/oila.html.

NBC News. 2004. "China Grapples with Legacy of its 'missing girls.'" http://www.msnbc.msn.com/id/5953508/ns/world_news/t/china-grapples-legacy-its-missing-girls/.

Paul, Bimal K. 1994. "AIDS in Asia." *Geographical Review* 84: 367–379.

Shannon, Gary W., Gerald F. Pyle, and Rashid L. Bashshur. 1991. *The Geography of AIDS: Origins and Course of an Epidemic.* New York: Guilford.

Statistics Canada. "Births and Total Fertility Rate, by Province and Territory." Available online at www.statcan.gc.ca.

Tyner, James A. 1996. "Filipina Migrant Entertainers." *Gender, Place and Culture* 3: 77–93.

United Nations Population Information Network web site, www.un.org/popin.

U.S. Bureau of the Census. 2011. Housing and Household Economic Statistics, Working Paper No. 2011-18. Available online at http://www.census.gov/hhes/www/housing/housing_patterns/pdf/Housing%20by%20Year%20Built.pdf.

U.S. Centers for Disease Control and Prevention. National Vital Statistics Reports, "Births: Preliminary Data for 2009." Available online at www.cdc.gov.

U.S. Department of Energy. 2012. Energy Information Administration Web site, www.eia.gov.

World Bank. "Statistics in Africa." Available online at http://www.worldbank.org/afr/stats.

World Population Balance web site, available online at http://www.worldpopulationbalance.org/3_times_sustainable.

World Population Data Sheet. 2011. Washington, D.C: Population Reference Bureau.

Chapter 4

Anderson, Greg, and David Harrison. 2007. "Language Hotspots." Model developed at the Living Tongues Institute for Endangered Languages, available online at http://www.nationalgeographic.com/mission/enduringvoices/.

Basso, Keith H. 1996. *Wisdom Sits in Places: Landscape and Language Among the Western Apache.* Albuquerque: University of New Mexico Press.

Bennett, Charles J. 1980. "The Morphology of Language Boundaries: Indo-Aryan and Dravidian in Peninsular India." In David E. Sopher (ed.), *An Exploration of India: Geographical Perspectives on Society and Culture.* Ithaca, NY: Cornell University Press, 234–251.

Bomhard, Allan R., and John C. Kerns. 1994. *The Nostratic Macrofamily.* Berlin: Mouton de Gruyter.

Carver, Craig M. 1986. *American Regional Dialects: Word Geography.* Ann Arbor: University of Michigan Press.

Collin, Richard Oliver. 2005. "Revolutionary Scripts: The Politics of Writing." Paper presented at the *Vernacular 2005 Conference on Language and Society,* Puebla, Mexico. Available online at http://www.omniglot.com/language/articles/revolutionary_scripts.doc.

Cutler, Charles. 1994. *O Brave New Words! Native American Loanwords in Current English.* Norman: University of Oklahoma Press.

Dingemanse, Mark. Wikipedia. Available online at http://en.wikipedia.org/wiki/Image:Bantu_expansion.png.

Encyclopaedia Britannica. 2000. *2000 Britannica Book of the Year.* Chicago: Encyclopaedia Britannica.

Ethnologue. "Nearly Extinct Languages." Available online at http://www.ethnologue.com/nearly_extinct.asp.

Ford, Clark. Available online at http://www.public.iastate.edu/~cfford/342worldhistoryearly.html.

Herman, R. D. K. 1999. "The Aloha State: Place Names and the Anti-Conquest of Hawaii." *Annals of the Association of American Geographers* 89: 76–102.

Hill, Robert T. 1986. "Descriptive Topographic Terms of Spanish America." *National Geographic* 7: 292–297.

Houston, James M. 1967. *The Western Mediterranean World.* New York: Praeger.

Jett, Stephen C. 1997. "Place-Naming, Environment, and Perception Among the Canyon de Chelly Navajo of Arizona." *Professional Geographer* 49: 481–493.

Krantz, Grover S. 1988. *Geographical Development of European Languages.* New York: Peter Lang.

Kurath, Hans. 1949. *Word Geography of the Eastern United States.* Ann Arbor: University of Michigan Press.

Latrimer Clarke Corporation Pty Ltd. Available online at http://www.altapedia.com.

Levison, Michael, R., Gerard Ward, and John W. Webb. 1973. *The Settlement of Polynesia: A Computer Simulation.* Minneapolis: University of Minnesota Press.

Lewis, M. Paul (ed.), 2009. *Ethnologue: Languages of the World,* sixteenth edition. Dallas: SIL International. Online version: http://www.ethnologue.com/.

Renfrew, Colin. 1989. "The Origins of Indo-European Languages." *Scientific American* 261(4): 106–114.

Sappenfield, Mark. 2006. "Tear Up the Maps: India's Cities Shed Colonial Names." *Christian Science Monitor,* September 7, online edition.

Stavans, Ilan. 2003. *Spanglish: The Making of a New American Language.* New York: Rayo.

U.S. Census, 1990; 2000; 2010.

U.S. English, Inc. Available online at http://www.us-english.org/inc/official/states.asp.

Wixman, Ronald. 1980. *Language Aspects of Ethnic Patterns and Processes in the North Caucasus.* Research Paper No. 191. University of Chicago, Department of Geography.

World Almanac Books. 2001. *World Almanac and Book of Facts 2001.* New York: World Almanac Books.

Yoon, Hong-key. 1986. "Maori and Pakeha Place Names for Cultural Features in New Zealand," in *Maori Mind, Maori Land: Essays on the Cultural Geography of the Maori People from an Outsider's Perspective.* Bern, Switzerland: Peter Lang, 98–122.

Chapter 5

Allen, James, and Eugene Turner. 1988. *We the People: An Atlas of America's Ethnic Diversity.* New York: Macmillan.

Arreola, Daniel D. 1984. "Mexican-American Exterior Murals." *Geographical Review* 74: 409–424.

Arreola, Daniel D. 2002. *Tejano South Texas: A Mexican American Cultural Province.* Austin: University of Texas Press.

Associated Press, 2006 (March 26). "500,000 Rally Immigration Rights in Los Angeles." Available online at http://www.msnbc.msn.com/id/11442705/ns/politics/t/immigration-issue-draws-thousands-streets/#.UMjqj-Qr2So.

BBC News. 2007 (February 7). "UN Warns of Iraq Refugee Disaster." Available online at http://news.bbc.co.uk/2/hi/middle_east/6339835.stm.

Bjorklund, Elaine M. 1964. "Ideology and Culture Exemplified in Southwestern Michigan." *Annals of the Association of American Geographers* 54: 227–241.

Bullard, Robert D. (ed.). 1993. *Confronting Environmental Racism: Voices from the Grassroots.* Boston: South End Press.

Carlson, Alvar W. 1990. *The Spanish American Homeland: Four Centuries in New Mexico's Rio Arriba.* Baltimore: Johns Hopkins University Press.

Carter, Timothy J., et al. 1980. "The Peoples of China." Map supplement, *National Geographic* (July): 158.

Catania, Sara. 2006 (October 16). "From Fish Sauce to Salsa—New Orleans Vietnamese Adapt to Influx of Latinos." Available online at *New American Media*, http://www.bestofneworleans.com/gambit/from-fish-sauce-to-salsa/Content?oid=1246596.

Chacko, Elizabeth. 2003. "Ethiopian Ethos and the Making of Places in the Washington Metropolitan Area." *Journal of Cultural Geography*, Spring/Summer, 2003. 20(2): 21–42.

Cobo, Leila. 2007 (September 30). "Mexican Music Enjoys Strong Sales, Timeless Appeals." Available online at http://www.reuters.com/article/2007/10/01/music-mexican-dc-idUSN3045871320071001.

Dawson, C. A. 1936. *Group Settlement: Ethnic Communities in Western Canada.* Toronto: Macmillan.

Eisenhower Research Project. 2011. "The Costs of War since 2001: Iraq, Afghanistan, and Pakistan," Executive Summary, June, 2011. Watson Institute for International Studies, Brown University.

Frey, William H. 2001. "Micro Melting Pots." *American Demographics* (June): 20–23. Abstract available at http://www.psc.isr.umich.edu/pubs/abs/1459.

Harries, Keith D. 1971. "Ethnic Variation in Los Angeles Business Patterns." *Annals of the Association of American Geographers* 61: 736–743.

Harris, R. Colebrook. 1977. "The Simplification of Europe Overseas." *Annals of the Association of American Geographers* 67: 469–483.

Hill, G. W. 1942. "The People of Wisconsin According to Ethnic Stocks, 1940." *Wisconsin's Changing Population.* Bulletin, Serial No. 2642. Madison: University of Wisconsin.

Johnson, James H., Jr., and Curtis C. Roseman. 1990. "Recent Black Outmigration from Los Angeles: The Role of Household Dynamics and Kinship Systems." *Annals of the Association of American Geographers* 80: 205–222.

Jordan, Terry G., with John L. Bean, Jr., and William M. Holmes. 1984. *Texas: A Geography.* Boulder, CO: Westview Press.

Kent, Noel Jacob. 2000. "The New Racism on Campus: What's Going On?" *The NEA Higher Education Journal* (Fall): 83–94.

Massey, Doreen. 1994. *Space, Place, and Gender.* Minneapolis: University of Minnesota Press.

Matwijiw, Peter. 1979. "Ethnicity and Urban Residence: Winnipeg, 1941–71." *Canadian Geographer* 23: 45–61.

Meigs, Peveril, III. 1941. "An Ethno-Telephonic Survey of French Louisiana." *Annals of the Association of American Geographers* 31: 243–250.

Meinig, Donald W. 1965. "The Mormon Culture Region." *Annals of the Association of American Geographers* 55: 191–220.

Mitchell, Rick. 1995. "Selena: In Life, She Was the Queen of *Tejano* Music. In Death, the 23-Year-Old Singer Is Becoming a Legend." *Houston Chronicle,* May 21. Available online at http://www.chron.com/CDA/archives/archive.mpl/1995_1275097/selena-in-life-she-was-the-queen-of-tejano-music-i.html.

Motzafi-Haller, Pnina. 1997. "Writing Birthright: On Native Anthropologists and the Politics of Representation," in Deborah E. Reed-Danahay (ed.), *Auto/Ethnography: Rewriting the Self and the Social.* Oxford: Berg, 195–222.

National Geographic. "The Genographic Project." Available online at http://www3.nationalgeographic.com/genographic/.

Nostrand, Richard L., and Lawrence E. Estaville, Jr. (eds.). 2001. *Homelands: A Geography of Culture and Place Across America.* Baltimore: Johns Hopkins University Press.

Oberle, Alex P., and Daniel D. Arreola. 2008. "Resurgent Mexican Phoenix." *Geographical Review* 98(2): 171–196.

O'Loughlin, John, Frank Witmer, Thomas Dickinson, Nancy Thorwardson, and Edward Holland. 2007. "Preface to Special Issue and Caucasus Map Supplement." *Eurasian Geography and Economics,* 48(1): 127–134.

Pillsbury, Richard. 1998. *No Foreign Food: American Diet in Time and Place.* Boulder, CO: Westview Press.

Rechlin, Alice. 1976. *Spatial Behavior of the Old Order Amish of Nappanee, Indiana.* Geographical Publication No. 18. Ann Arbor: University of Michigan.

Scott, James C. 1998. *Seeing Like a State: How Certain Schemes to Improve the Human Condition Have Failed.* New Haven, CT: Yale University Press.

Shannon, Gary W., and Gerald F. Pyle. 1992. *Disease and Medical Care in the United States: A Medical Atlas of the Twentieth Century.* New York: Macmillan.

Shortridge, Barbara G., and James R. Shortridge (eds.). 1998. *The Taste of American Place: A Reader on Regional and Ethnic Foods.* Lanham, MD: Rowman & Littlefield.

Sivanandan, A. 2001 (August 17). "Poverty Is the New Black." Available online at http://www.guardian.co.uk/politics/2001/aug/17/globalisation.race.

Toxic Release Inventory (U.S. EPA). 1996. Available online at http://www.inmotionmagazine.com/auto/mapofla.html.

U.S. Bureau of the Census, 2010. United States Census 2010: It's In Our Hands. Available online at http://2010.census.gov/2010census/.

U.S. Bureau of the Census. 2000. MLA Language Map Data Center. Available online at http://www.mla.org/map_data.

Winsberg, Morton D. 1986. "Ethnic Segregation and Concentration in Chicago Suburbs." *Urban Geography* 7: 142–143.

Zelinsky, Wilbur. 1985. "The Roving Palate: North America's Ethnic Restaurant Cuisines." *Geoforum* 16: 66.

Chapter 6

Agnew, John. 1998. *Geopolitics: Re-Visioning World Politics*. London: Routledge.

Agnew, John. 2002. *Making Political Geography*. London: Arnold.

Blaikie, Piers, and Harold Brookfield. 1987. *Land Degradation and Society*. London: Methuen.

Blouet, Brian W. 1987. *Halford Mackinder: A Biography*. College Station: Texas A&M University Press.

Dalby, Simon. 2002. "Environmental Governance," in R. Johnston, P. Taylor, and M. Watts (eds.), *Geographies of Global Change: Remapping the World*. London: Routledge, 427–440.

Elazar, Daniel J. 1994. *The American Mosaic: The Impact of Space, Time, and Culture on American Politics*. Boulder, CO: Westview Press.

Gastner, M., C. Shalizi, and M. Newman. 2004. "Maps and Cartograms of the 2004 U.S. Presidential Election Results," Ann Arbor: University of Michigan Department of Physics and Center for the Study of Complex Systems.

Jones, Martin, and Rhys Jones. 2004. "Nation States, Ideological Power and Globalization: Can Geographers Catch the Boat?" *Geoforum* 35: 409–424.

Jordan, Bella Bychkova, and Terry G. Jordan-Bychkov. 2001. *Siberian Village: Land and Life in the Sakha Republic*. Minneapolis: University of Minnesota Press.

Jordan-Bychkov, Terry G., and Bella Bychkova Jordan. 2002. *The European Culture Area: A Systematic Geography*, 4th ed. Lanham, MD: Rowman & Littlefield, Chapter 7.

Lean, Geoffrey, et al. 1990. *Atlas of the Environment*. Upper Saddle River, NJ: Prentice-Hall.

Lipset, Seymour Martin, and Stein Rokkan (eds.). 1967. *Party Systems and Voter Alignments: Cross-National Perspectives*. New York: Free Press.

Mackinder, Halford J. 1904. "The Geographical Pivot of History." *Geographical Journal* 23: 421–437.

Mitchell, Katharyne. 1998. "Fast Capital, Race, and the Monster House." In R. George (ed.), *Burning Down the House: Recycling Domesticity*. Boulder, CO: Westview Press, 187–212.

Mitchell, Katharyne. 2002. "Cultural Geographies of Transnationality." In K. Anderson, M. Domosh, S. Pile, and N. Thrift (eds.), *Handbook of Cultural Geography*. London: Sage, 74–87.

Morrill, Richard L. 1981. *Political Redistricting and Geographic Theory*. Washington, D.C.: Association of American Geographers, Resource Publications.

Neumann, Roderick P. 1995. "Local Challenges to Global Agendas: Conservation, Economic Liberalization, and the Pastoralists' Rights Movement in Tanzania." *Antipode* 27(4): 363–382.

Neumann, Roderick P. 2002. "The Postwar Conservation Boom in British Colonial Africa." *Environmental History* 7(1): 22–47.

Neumann, Roderick P. 2004. "Nature-State-Territory: Toward a Critical Theorization of Conservation Enclosures," In R. Peet and M. Watts (eds.), *Liberation Ecologies*, 2nd ed. London: Routledge, 195–217.

Oldale, John. 1990. "Government-Sanctioned Murder." *Geographical Magazine* 62: 20–21.

O'Reilly, Kathleen, and Gerald R. Webster. 1998. "A Sociodemographic and Partisan Analysis of Voting in Three Anti-Gay Rights Referenda in Oregon." *Professional Geographer* 50: 498–515.

Ó Tuathail, Gearóid. 1996. *Critical Geopolitics*. Minneapolis: University of Minnesota Press.

Paulin, C., and John K. Wright. 1932. *Atlas of the Historical Geography of the United States*. New York: American Geographical Society and the Carnegie Institute.

Reitsma, Hendrik J. 1971. "Crop and Livestock Production in the Vicinity of the United States–Canada Border." *Professional Geographer* 23: 216–223.

Reitsma, Hendrik J. 1988. "Agricultural Changes in the American–Canadian Border Zone, 1954–1978." *Political Geography Quarterly* 7: 23–38.

Rumley, Dennis, and Julian V. Minghi (eds.). 1991. *The Geography of Border Landscapes*. London: Routledge.

Sack, Robert D. 1986. *Human Territoriality: Its Theory and History*. Studies in Historical Geography, No. 7. Cambridge: Cambridge University Press.

Smith, Dan. 1992. "The Sixth Boomerang: Conflict and War." In Susan George (ed.), *The Debt Boomerang*. London: Pluto Press.

Smith, Dan, et al. 1997. *The State of War and Peace Atlas*, 3rd ed. New York: Penguin.

Smith, Graham. 1999. "Russia, Geopolitical Shifts and the New Eurasianism." *Transactions of the Institute of British Geographers* 24: 481–500.

Spykman, Nicholas J. 1944. *The Geography of the Peace*. New York: Harcourt Brace.

Toal, Gerard, and John Agnew. 2002. "Introduction: Political Geographies, Geopolitics and Culture," in K. Anderson, M. Domosh, S. Pile, and N. Thrift (eds.), *Handbook of Cultural Geography*. London: Sage, 455–461.

Vanderbei, R. J. 2004. "Election 2004 Results." Available online at www.princeton.edu/~rvdb/JAVA/election2004/.

Chapter 7

Bradley, Martin B., et al. 1992. *Churches and Church Membership in the United States*. Atlanta: Glenmary Research Center.

Curran, Claude W. 1991. "Mt. Shasta, California, and the I Am Religion," in *Abstracts, The Association of American Geographers 1991 Annual Meeting, April 13–17, Miami, Florida*. Washington, D.C.: Association of American Geographers, 42.

Doughty, Robin W. 1981. "Environmental Theology: Trends and Prospects in Christian Thought." *Progress in Human Geography* 5: 234–248.

Eliade, Mircea. 1987 [1957]. *The Sacred and the Profane: The Nature of Religion*. Willard R. Trask (trans.). San Diego: Harcourt, Inc.

Foley, Jonathan A., et al. "Solutions for a Cultivated Planet." *Nature* 478, 337–342 (20 II).

Foote, Kenneth E. 1997. *Shadowed Ground: America's Landscapes of Violence and Tragedy*. Austin: University of Texas Press.

Government of Canada. 1974. *The National Atlas of Canada*, 4th ed. Ottawa: Government of Canada, Surveys and Mapping Branch.

Griffith, James S. 1992. *Beliefs and Holy Places: A Spiritual Geography of the Pimería Alta*. Tucson: University of Arizona Press.

Hobbs, Joseph J. 1995. *Mount Sinai*. Austin: University of Texas Press.

Huntsinger, Lynn, and María Fernández-Giménez. 2000. "Spiritual Pilgrims at Mount Shasta, California." *Geographical Review* 90: 536–558.

Jordan-Bychkov, Terry G., and Bella Bychkova Jordan. 2002. *The European Culture Area: A Systematic Geography,* 4th ed. Lanham, MD: Rowman & Littlefield.

Klaer, Wendelin. 1996. "Survey: Lebanon." *The Economist.* February 24, l996: 333.

Lodrick, Deryck O. 1981. *Sacred Cows, Sacred Places: Origins and Survivals of Animal Homes in India.* Berkeley: University of California Press.

Raban, Jonathan. 1996. *Bad Land: An American Romance.* New York: Pantheon Books.

Simoons, Frederick J. 1994. *Eat Not This Flesh: Food Avoidances in the Old World,* 2nd ed. Madison: University of Wisconsin Press.

Singh, Rana P. B. 1994. "Water Symbolism and Sacred Landscape in Hinduism." *Erdkunde* 48: 210–227.

Stump, Roger W. (ed.). 1986. "The Geography of Religion." Special issue, *Journal of Cultural Geography* 7: 1–140.

2010 Britannica Book of the Year. Chicago: Encyclopedia Britannica.

World Almanac and Book of Facts 2012. New York: World Almanac Books.

Chapter 8

Andrews, Jean. 1993. "Diffusion of Mesoamerican Food Complex to Southeastern Europe." *Geographical Review* 83: 194–204.

Binns, T. 1990. "Is Desertification a Myth?" *Geography* 75: 106–113.

Bourne, Joel, Jr. 2007. "Biofuels: Boon or Boondoggle." *National Geographic Magazine* 212(4): 38–59.

Boyd, William, and Michael Watts. 1997. "Agroindustrial Just-In-Time: The Chicken Industry and Postwar American Capitalism." In M. Watts and D. Goodman (eds.), *Globalizing Food: Agrarian Questions and Global Restructuring.* London: Routledge, 192–225.

Carney, Judith. 2001. *Black Rice: The African Origins of Rice Cultivation in the Americas.* Cambridge, MA: Harvard University Press.

Chakravarti, A. K. 1973. "Green Revolution in India." *Annals of the Association of American Geographers* 63: 319–330.

Chuan-jun, Wu (ed.). 1979. "China Land Utilization." Map. Beijing: Institute of Geography of the Academica Sinica.

Cowan, C. Wesley, and Patty J. Watson (eds.). 1992. *The Origins of Agriculture: An International Perspective.* Washington, D.C.: Smithsonian Institution Press.

Cross, John A. 1994. "Agroclimatic Hazards and Fishing in Wisconsin." *Geographical Review* 84: 277–289.

Darby, H. Clifford. 1956. "The Clearing of the Woodland in Europe." In William L. Thomas, Jr. (ed.), *Man's Role in Changing the Face of the Earth.* Chicago: University of Chicago Press, 183–216.

Demangeon, Albert. 1946. *La France.* Paris: Armand Colin.

Diamond, Jared. 2002. "Evolution, Consequences and Future of Plant and Animal Domestication." *Nature* 418: 700–707.

Dillehay, T., J. Rossen, T. Andres, and D. Williams. 2007. "Preceramic Adoption of Peanut, Squash, and Cotton in Northern Peru." *Science* 316(5833): 1890–1893.

Ewald, Ursula. 1977. "The von Thünen Principle and Agricultural Zonation in Colonial Mexico." *Journal of Historical Geography* 3: 123–133.

Freidberg, Susanne. 2001. "Gardening on the Edge: The Conditions of Unsustainability on an African Urban Periphery." *Annals of the Association of American Geographers* 91(2): 349–369.

Griffin, Ernst. 1973. "Testing the von Thünen Theory in Uruguay." *Geographical Review* 63: 500–516.

Grigg, David B. 1969. "The Agricultural Regions of the World: Review and Reflections." *Economic Geography* 45: 95–132.

Griliches, Zvi. 1960. "Hybrid Corn and the Economics of Innovation." *Science* 132(July 26): 275–280.

Guthman, Julie. 2004. *Agrarian Dreams: The Paradox of Organic Farming in California.* Berkeley: University of California Press.

Hewes, Lewlie. 1973. *The Suitcase Farming Frontier: A Study in the Historical Geography of the Central Great Plains.* Lincoln: University of Nebraska Press.

Hidore, John J. 1963. "Relationship Between Cash Grain Farming and Landforms." *Economic Geography* 39: 84–89.

Horvath, Ronald J. 1969. "von Thünen's Isolated State and the Area Around Addis Ababa, Ethiopia." *Annals of the Association of American Geographers* 59: 308–323.

ISAAA. 2007. International Service for the Acquisition of Agribiotech Applications web site. http://www.isaaa.org.

Johannessen, Carl L. 1966. "The Domestication Processes in Trees Reproduced by Seed: The Pejibaye Palm in Costa Rica." *Geographical Review* 56: 363–376.

Lofgren, Clifford S., and Robert K. Vander Meer, eds. 1986. *Fire Ants and Leaf-Cutting Ants.* Boulder, CO: Westview Press, 38–39, 44.

Mathews, K., J. Bernstein, and J. Buzby. 2003. "International Trade of Meat and Poultry Products and Food Safety Issues," in J. Buzby (ed.), *International Trade and Food Safety: Economic Theory and Case Studies.* Washington, D.C.: U.S. Department of Agriculture Economic Research Service.

Millstone, Erik, and Tim Lang. 2003. *The Penguin Atlas of Food.* New York: Penguin Books.

Murphey, Rhoads. 1951. "The Decline of North Africa Since the Roman Occupation: Climatic or Human?" *Annals of the Association of American Geographers* 41: 116–131.

Norberg-Hodge, Helena, Todd Merrifield, and Steven Gorelick. 2002. *Bringing the Food Economy Home: Local Alternatives to Global Agribusiness.* London: Zed.

Popper, Deborah E., and Frank Popper. 1987. "The Great Plains: From Dust to Dust." *Planning* 53(12): 12–18.

Saarinen, Thomas F. 1966. *Perception of Drought Hazard on the Great Plains.* University of Chicago, Department of Geography, Research Paper No. 106. Chicago: University of Chicago Press.

Sauer, Carl O. 1952. *Agricultural Origins and Dispersals.* New York: American Geographical Society.

Sauer, Jonathan D. 1993. *Historical Geography of Crop Plants.* Boca Raton, FL: CRC Press.

Thomas, David S. G. 1993. "Sandstorm in a Teacup? Understanding Desertification." *Geographical Journal* 159(3): 318–331.

Thomas, David S. G., and Nicholas J. Middleton. 1994. *Desertification: Exploding the Myth.* New York: John Wiley.

Thrower, Norman J. W. 1966. *Original Survey and Land Subdivision.* Chicago: Rand McNally.

U.S. Department of Agriculture. 2011. Economic Research Service web site www.ers.usda.gov.

Vogeler, Ingolf. 1981. *The Myth of the Family Farm: Agribusiness Dominance of United States Agriculture.* Boulder, CO: Westview Press.

von Thünen, Johann Heinrich. 1966. *von Thünen's Isolated State: An English Edition of Der Isolierte Staat.* Carla M. Wartenberg (trans.). Elmsford, NY: Pergamon Press.

Westcott, Paul. 2007. *Ethanol Expansion in the United States: How Will the Agricultural Sector Adjust?* Washington, D.C.: U.S. Department of Agriculture Economic Research Service.

Whittlesey, Derwent S. 1936. "Major Agricultural Regions of the Earth." *Annals of the Association of American Geographers* 26: 199–240.

Wilken, Gene C. 1987. *Good Farmers: Traditional Agricultural and Resource Management in Mexico and Central America.* Berkeley: University of California Press.

Zimmerer, Karl. 1996. *Changing Fortunes: Biodiversity and Peasant Livelihood in the Peruvian Andes.* Berkeley: University of California Press.

Chapter 9

Airriess, Christopher A. 2001. "Regional Production, Information-Communication Technology, and the Developmental State: The Rise of Singapore as a Global Container Hub." *Geoforum* 32: 235–254.

Braudel, Fernand. 1972. *The Mediterranean and the Mediterranean World in the Age of Phillip II.* New York: Harper & Row.

Branch, T., Jensen, O., Ricard, D., Ye, Y., and Hilborn, R. 2011. "Contrasting Global Trends in Marine Fishery Status Obtained from Catches and from Stock Assessments." *Conservation Biology* 25: 777–786.

Brown, Laurie, Martha Ronk, and Charles E. Little. 2000. *Recent Terrains: Terraforming the American West.* Baltimore: Johns Hopkins University Press.

Dean, Cornelia. "Study Sees 'Global Collapse' of Fish Species." *New York Times,* November 3, 2006.

Dicken, Peter. 2003. *Global Shift: Reshaping the Global Economic Map in the 21st Century.* New York: Guilford Press.

Food and Agriculture Organization of the United Nations (FAO). Fisheries and Aquaculture Department. 2010. *The State of World Fisheries and Aquaculture.* Rome, Italy.

Francaviglia, Richard V. 1991. *Hard Places: Reading the Landscape of America's Historic Mining Districts.* Iowa City: University of Iowa Press.

Glasmeier, Amy. 2000. *Manufacturing Time: Global Competition in the Watch Industry, 1795–2000.* New York: Guilford Press.

Holden, Andrew. 2000. *Environment and Tourism.* New York: Routledge.

Hudson, John C., and Edward B. Espenshade, Jr. (eds.). 2000. *Goode's World Atlas,* 20th ed. Chicago: Rand McNally.

Jakle, John A., and Keith A. Sculle. 1994. *The Gas Station in America.* Baltimore: Johns Hopkins University Press.

Jordan-Bychkov, Terry G., and Bella Bychkova Jordan. 2002. *The European Culture Area,* 4th ed. Lanham, MD: Rowman & Littlefield.

Lovett, Richard. 2012. "Giant Crack in Antarctica About to Spawn New York Sized Iceberg," *National Geographic News,* 2 February. Available online at http://news.nationalgeographic.com/news/2012/120202-crack-antarctica-iceberg-science-glacier.

Met Office of the Hadley Research Center and Climatic Research Unit web site. http://www.metoffice.gov.uk/climate-change/resources/hadley.

NASA Goddard Space Flight Center web site. http://ozonewatch.gsfc.nasa.gov.

National Atmospheric Deposition Program web site. http://nadp.sws.uiuc.edu.

O'Hare, Greg. 2000. "Reviewing the Uncertainties in Climate Change Science." *Area* 32: 357–368.

Organization for Economic Co-operation and Development (OECD) web site. http://www.oecd.org.

Park, Chris C. 1989. *Chernobyl: The Long Shadow.* London: Routledge, 66–92.

Power, Thomas M. 1996. *Lost Landscapes and Failed Economies: The Search for the Value of Place.* Washington, D.C.: Island Press.

Rostow, W. W. 1952. *The Process of Economic Growth.* New York: Norton and Co.

Tuan, Yi-Fu. 1989. "Cultural Pluralism and Technology." *Geographical Review* 79: 269–279.

United Nations. 2006. *Statistical Yearbook.* New York: United Nations.

United States Energy Information Administration Web site. http://www.eia.gov.

Warrick, Richard, and Graham Farmer. 1990. "The Greenhouse Effect, Climatic Change and Rising Sea Level: Implications for Development." *Transactions of the Institute of British Geographers* 15: 5–20.

Weber, Alfred. 1929. *Theory of the Location of Industries.* Carl J. Friedrich (trans. and ed.). Chicago: University of Chicago Press.

"Wind Power: Maybe This Time." (2001). *The Economist* (March 10): 30–31.

Yale Center for Environmental Law and Policy and the Center for International Earth Science Information Network. 2005. *2005 Environmental Sustainability Index.* New Haven, CT: Yale University.

Chapter 10

Agnew, John A., and James S. Duncan (eds.). 1989. *The Power of Place: Bringing Together Geographical and Sociological Imaginations.* Boston: Unwin Hyman.

Bater, James H. 1980. *The Soviet City: Ideal and Reality.* Beverly Hills: Sage Publications.

Bater, James H. 1996. *Russia and the Post-Soviet Scene.* New York: John Wiley.

Christaller, Walter. 1966. *The Central Places of Southern Germany.* C. W. Baskin (trans.). Englewood Cliffs, NJ: Prentice-Hall.

Christopher, A. J. 1994. *The Atlas of Apartheid.* London: Routledge.

Christopher, A. J. 2001. "Urban Segregation in Post-Apartheid South Africa." *Urban Studies* 38: 449–466.

Cosgrove, Denis. 1984. *Social Formation and Symbolic Landscape.* London: Croom Helm.

Daniell, Jennifer, and Raymond Struyk. 1997. "The Evolving Housing Market in Moscow: Indicators of Housing Reform." *Urban Studies* 34: 235–254.

De Sherbenin, Alex, Andrew Schiller, and Alex Pulsipher. 2007. "The Vulnerability of Global Cities to Climate Hazards." *Environment and Urbanization* 19: 39–64.

Diamond, Jared. "What's Your Consumption Factor?" *New York Times,* January 2, 2008. Online: http://www.nytimes.com/2008/01/02/opinion/02diamond.html?pagewanted=all

Drakakis-Smith, David. 1987. *The Third World City.* London: Methuen.

Drakakis-Smith, David. 2000. *Third World Cities,* 2nd ed. London: Routledge.

Encyclopaedia Britannica. 1984. "Babylon." *New Encyclopaedia Britannica*, vol. 2. Chicago: Encyclopaedia Britannica.

Fowler, Melvin. 1975. "A Pre-Columbian Urban Center on the Mississippi." *Scientific American* (August): 93–102.

Griffin, Ernst, and Larry Ford. 1983. "Cities of Latin America." In Stanley Brunn and Jack Williams (eds.), *Cities of the World: World Regional Urban Development*. New York: Harper & Row, 199–240.

Huang, Yefang, Yee Leung, and Jianfa Shen. 2007. "Cities and Globalization: An International Perspective." *Urban Geography* 28: 209–231.

Mumford, Lewis. 1961. *The City in History*. New York: Harcourt Brace Jovanovich.

Population Division of the United Nations Secretariat. 2007. *World Urbanization Prospects: The 2007 Revision Population Database*.

Portes, Alejandro. 1977. "Urban Latin America: The Political Condition from Above and Below." In Janet Abu-Lughod and Richard Hag, Jr. (eds.), *Third World Urbanization*. New York: Methuen, 59–70.

Pounds, Norman J. G. 1969. "The Urbanization of the Classical World" *Annals of the Association of American Geographers* 59: 135–157.

Robinson, Jennifer. 1997. "The Geopolitics of South African Cities: States, Citizens, Territory." *Political Geography* 16: 365–386.

Samuels, Marwyn S., and Carmencita Samuels. 1989. "Beijing and the Power of Place in Modern China," In John A. Agnew and James S. Duncan (eds.), *The Power of Place*. Boston: Unwin Hyman, 202–227.

Sassen, Saskia. 1991. *The Global City: New York, London, Tokyo*. Princeton, NJ.: Princeton University Press.

Simon, D. 1984. "Third-World Colonial Cities in Context: Conceptual and Theoretical Approaches with Particular Interest to Africa." *Progress in Human Geography* 8: 493–514.

United Nations Department of Economic and Social Affairs, Population Division. 2012. *World Urbanization Prospects, The 2011 Revision: Highlights*. Online: http://esa.un.org/unpd/wup/pdf/WUP2011_Highlights.pdf.

United Nations Environment Program (UNEP). 2010. *Disasters and Conflicts Fact Sheet*. Online: http://www.unep.org/pdf/brochures/DisastersAndConflicts.pdf

United Nations Human Settlements Programme (UN-HABITAT). 2010. *State of the World's Cities 2010/11: Cities for All: Bridging the Urban Divide*.

United Nations Population Fund. 2007. *State of the World Population 2007: Unleashing the Potential of Urban Growth*. Online: http://www.unfpa.org/public/publications/pid/408

Wheatley, Paul. 1971. *The Pivot of the Four Quarters*. Chicago: Aldine Publishing.

Wittfogel, Karl. 1957. *Oriental Despotism: A Comparative Study of Total Power*. New Haven, CT: Yale University Press.

Yeoh, Brenda S. A. 1999. "Global/Globalizing Cities." *Progress in Human Geography* 23: 607–616.

Chapter 11

Abbott, Edith, and Mary Zahrobsky. 1936. "The Tenement Areas and the People of the Tenements." in E. Abbott (ed.), *The Tenements of Chicago*. Chicago: University of Chicago Press, 72–169.

Agnew, John, John Mercer, and David Sopher (eds.). 1984. *The City in Cultural Context*. Boston: Allen & Unwin.

Anderson, Kay. 1987. "The Idea of Chinatown: The Power of Place and Institutional Practice in the Making of a Racial Category." *Annals of the Association of American Geographers* 77: 580–598.

Bach, Wilfred, and Thomas Hagedorn. 1971. "Atmospheric Pollution: Its Spatial Distribution over an Urban Area." *Proceedings of the Association of American Geographers* 3: 22.

Bell, David, and Gill Valentine (eds.). 1995. *Mapping Desire: Geographies of Sexualities*. London: Routledge.

Bell, Morag. 1995. "A Woman's Place in 'a White Man's Country.' Rights, Duties and Citizenship for the 'New' South Africa, c. 1902." *Ecumene* 2: 129–148.

Benevolo, Leonardo. 1980. *The History of the City*. Cambridge, MA.: MIT Press.

Brown, Elsa Barkley, and Gregg D. Kimball. 1995. "Mapping the Terrain of Black Richmond." *Journal of Urban History* 21: 296–346.

Burgess, Ernest W. 1925. "The Growth of the City: An Introduction to a Research Project," in Robert E. Park, Ernest W. Burgess, and Roderick D. McKenzie (eds.), *The City*. Chicago: University of Chicago Press, 47–62.

Castells, Manuel. 1983. *The City and Grassroots: A Cross-Cultural Theory of Urban Social Movements*. Berkeley: University of California Press.

Chauncey, George. 1994. *Gay New York: Gender, Urban Culture, and the Making of the Gay Male World, 1890–1940*. New York: Basic Books.

Clay, Grady. 1973. *Close-Up: How to Read the American City*. New York: Praeger.

Davis, Mike. 1992. "Fortress Los Angeles: The Militarization of Urban Space." In Michael Sorkin (ed.), *Variations on a Theme Park: The New American City and the End of Public Space*. New York: Hill & Wang, 154–180.

Domosh, Mona. 1995. "The Feminized Retail Landscape: Gender Ideology and Consumer Culture in Nineteenth-Century New York City." In Neil Wrigley and Michelle Lowe (eds.), *Retailing, Consumption and Capital*. Essex, U.K.: Longman, 257–270.

Doxiades, Constantine A. 1972. *Architectural Space in Ancient Greece*. Cambridge, MA.: MIT Press.

Drakakis-Smith, David. 2000. *Third World Cities*, 2nd ed. London: Routledge.

Duncan, James, and Nancy Duncan. 1984. "A Cultural Analysis of Urban Residential Landscapes in North America: The Case of the Anglophile Elite," in John Agnew, John Mercer, and David Sopher (eds.), *The City in Cultural Context*. Boston: Allen and Unwin, 255–276.

Duncan, James, and Nancy Duncan. 2003. *Landscapes of Privilege: The Politics of the Aesthetic in an American Suburb*. New York: Routledge.

Garreau, Joel. 1991. *Edge City: Life on the New Frontier*. New York: Doubleday.

Gilbert, Emily. 1994. "Naturalist Metaphors in the Literatures of Chicago, 1893–1925." *Journal of Historical Geography* 20: 283–304.

Gordon, Margaret T., et al. 1981. "Crime, Women, and the Quality of Urban Life." In Catherine R. Stimpson, Elsa Dixler, Martha J. Nelson, and Kathryn B. Yatrakis (eds.), *Women and the American City*. Chicago: University of Chicago Press, 141–157.

Gottmann, Jean. 1961. *Megalopolis*. Cambridge, MA.: MIT Press.

Hanson, Susan. 1992. "Geography and Feminism: Worlds in Collision?" *Annals of the Association of American Geographers* 82: 569–586.

Hanson, Susan, and Geraldine Pratt. 1995. *Gender, Work, and Space.* New York: Routledge.

Harris, Chauncey D., and Edward L. Ullman. 1945. "The Nature of Cities." *Annals of the Association of American Academy of Political and Social Science* 242: 7–17.

Hartshorn, Truman A., and Peter O. Muller. 1989. "Suburban Downtowns and the Transformation of Metropolitan Atlanta's Business Landscape." *Urban Geography* 10: 375–395.

Hayes, John. 1969. *London: A Pictorial History.* New York: Arco Publishing.

Hopkins, Jeffrey S. P. 1991. "West Edmonton Mall as a Centre for Social Interaction." *The Canadian Geographer* 35: 268–279.

Hoyt, Homer (ed.). 1939. *Structure and Growth of Residential Neighborhoods in American Cities.* Washington, D.C.: Federal Housing Administration.

Jackson, Kenneth T. 1985. *Crabgrass Frontier.* New York: Oxford University Press.

Kenny, Judith T. 1995. "Climate, Race and Imperial Authority: The Symbolic Landscape of the British Hill Station in India." *Annals of the Association of American Geographers* 85: 694–714.

Knopp, Lawrence. 1995. "Sexuality and Urban Space: A Framework for Analysis." In David Bell and Gill Valentine (eds.), *Mapping Desire.* London: Routledge, 149–161.

Knox, Paul L. 1991. "The Restless Urban Landscape: Economic and Sociocultural Change and the Transformation of Metropolitan Washington, D.C." *Annals of the Association of American Geographers* 81: 181–209.

Lauria, Mickey, and Lawrence Knopp. 1985. "Toward an Analysis of the Role of Gay Communities in the Urban Renaissance." *Urban Geography* 6: 152–169.

Li, Wei. 1998. "Anatomy of a New Ethnic Settlement: The Chinese Ethnoburb in Los Angeles." *Urban Studies* 35: 479–501.

Lynch, Kevin. 1960. *The Image of the City.* Cambridge, MA: MIT Press.

McDowell, Linda. 1983. "Towards an Understanding of the Gender Division of Urban Space." *Environment and Planning D: Society and Space* 1: 59–72.

Meinig, D. W. (ed.). 1979. *The Interpretation of Ordinary Landscapes: Geographical Essays.* New York: Oxford University Press.

Mills, Sara. 1996. "Gender and Colonial Space." *Gender, Place and Culture* 3: 125–148.

Mohl, Raymond A. 1995. "Making the Second Ghetto in Metropolitan Miami, 1940–1960." *Journal of Urban History* 21: 395–427.

Mumford, Lewis. 1961. *The City in History.* New York: Harcourt Brace Jovanovich.

Pirenne, Henri. 1996 [1927]. *Medieval Cities.* Garden City, NY: Doubleday (Anchor Books).

Portes, Alejandro, and Ruben G. Rumbaut. 1996. *Immigrant America: A Portrait,* 2nd ed. Berkeley: University of California Press.

Pounds, Norman J. G. 1969. "The Urbanization of the Classical World." *Annals of the Association of American Geographers* 59: 135–157.

Pow, Choon-Piew, and Lily Kong. 2007. "Marketing the Chinese Dream House: Gated Communities and Representations of the Good Life in (Post-)Socialist Shanghai." *Urban Geography* 28: 129–169.

Pratt, Geraldine. 1990. "Feminist Analyses of the Restructuring of Urban Life." *Urban Geography* 11: 594–605.

Rothenberg, Tamar. 1995. "'And She Told Two Friends': Lesbians Creating Urban Social Space." In David Bell and Gill Valentine (eds.), *Mapping Desire.* London: Routledge, 165–181.

Rowe, Stacy, and Jennifer Wolch. 1990. "Social Networks in Time and Space: Homeless Women in Skid Row, Los Angeles." *Annals of the Association of American Geographers* 80: 184–204.

Secor, Anna. 2004. "'There Is an Istanbul That Belongs to Me': Citizenship, Space and Identity in the City." *Annals of the Association of American Geographers* 94: 352–368.

Shangming, Dan, and Dan Bo. 2001. "Analysis of the effects of urban heat island by satellite remote sensing." Paper presented at the 22nd Asian Conference on Remote Sensing, November 2001, Singapore, 1–6.

Sibley, David. 1995. "Gender, Science, Politics and Geographies of the City." *Gender, Place and Culture: A Journal of Feminist Geography* 2: 37–49.

Sjoberg, Gideon. 1960. *The Preindustrial City.* New York: Free Press.

Smith, Neil. 1984. *Uneven Development.* New York: Blackwell.

Smith, Neil. 1996. *The New Urban Frontier: Gentrification and the Revanchist City.* New York: Routledge.

Sorkin, Michael (ed.). 1992. *Variations on a Theme Park: The New American City and the End of Public Space.* New York: Hill & Wang.

Spain, Daphne. 1992. *Gendered Spaces.* Chapel Hill: University of North Carolina Press.

Summerson, John. 1946. *Georgian London.* New York: Charles Scribner's Sons.

U.S. Department of Homeland Security. 2009. *Yearbook of Immigration.* Washington, D.C.: U.S. Government Printing Office.

Valentine, Gill. 1993. "Desperately Seeking Susan: A Geography of Lesbian Friendships." *Area* 25: 109–116.

Vance, James E., Jr. 1971. "Land Assignment in the Pre-Capitalist, Capitalist and Post-Capitalist City." *Economic Geography* 47: 101–120.

Vance, James E., Jr. 1990. *The Continuing City: Urban Morphology in Western Civilization.* Baltimore: Johns Hopkins University Press.

Ward, David. 1971. *Cities and Immigrants.* New York: Oxford University Press.

Warner, Sam Bass. 1962. *Streetcar Suburbs: The Process of Growth in Boston, 1870–1900.* Cambridge, MA: Harvard University Press.

Warr, Mark. 1985. "Fear of Rape Among Urban Women." *Social Problems* 32: 238–250.

Watson, Sophie, with Helen Austerberry. 1986. *Housing and Homelessness: A Feminist Perspective.* London: Routledge & Kegan Paul.

Women and Geography Study Group of the Institute of British Geographers. 1997. *Feminist Geographies: Explorations in Diversity and Difference.* London: Prentice Hall.

Wood, Joseph. 1997. "Vietnamese-American Place Making in Northern Virginia." *The Geographical Review* 87: 58–72.

INDEX

Note: Page numbers followed by f indicate figures; those followed by m indicate maps; those followed by t indicate tables.